Earth Science: Current Studies

Earth Science: Current Studies

Editor: Jasper O'Brien

R CALLISTO REFERENCE

www.callistoreference.com

Callisto Reference,
118-35 Queens Blvd., Suite 400,
Forest Hills, NY 11375, USA

Visit us on the World Wide Web at:
www.callistoreference.com

ISBN: 978-1-64116-199-2 (Hardback)

Cataloging-in-Publication Data

Earth science : current studies / edited by Jasper O'Brien.
   p. cm.
Includes bibliographical references and index.
ISBN 978-1-64116-199-2
1. Earth sciences. 2. Geology. 3. Geography. I. O'Brien, Jasper.
QE26.3 .E27 2019
550--dc23

Table of Contents

Preface

Earth science is the science that studies the physical constitution of the Earth and its atmosphere. This includes the analysis of the physical characteristics of the Earth, such as earthquakes, floods, fossils, etc. The lithosphere, atmosphere, hydrosphere and the biosphere are the fundamental spheres of the Earth. The studies of these are approached in a multidisciplinary manner involving the integration of the tools of geography, physics, chronology, chemistry, etc. The study of the Earth's atmosphere, its magnetic field and the Earth's interior are some of the areas of interest in this domain. This book presents the complex subject of Earth science in the most comprehensible and easy to understand language. Some of the diverse topics covered in this book address the varied branches that fall under this category. A number of latest researches have been included to keep the readers up-to-date with the global concepts in this area of study.

Various studies have approached the subject by analyzing it with a single perspective, but the present book provides diverse methodologies and techniques to address this field. This book contains theories and applications needed for understanding the subject from different perspectives. The aim is to keep the readers informed about the progresses in the field; therefore, the contributions were carefully examined to compile novel researches by specialists from across the globe.

Indeed, the job of the editor is the most crucial and challenging in compiling all chapters into a single book. In the end, I would extend my sincere thanks to the chapter authors for their profound work. I am also thankful for the support provided by my family and colleagues during the compilation of this book.

Editor

Syn-thrusting, near-surface flexural-slipping and stress deflection along folded sedimentary layers of the Sant Corneli-Bóixols anticline (Pyrenees, Spain)

Stefano Tavani[1], Pablo Granado[2,3], Pau Arbués[2,3], Amerigo Corradetti[1], and J. Anton Muñoz[2,3]

[1]DiSTAR, Università degli Studi di Napoli "Federico II", Largo S. Marcellino 10, 80138 Naples, Italy
[2]Institut de Recerca Geomodels, Universitat de Barcelona, Martí i Franquès s/n, 08028 Barcelona, Spain
[3]Departament de Dinàmica de la Terra i de l'Oceà, Universitat de Barcelona, Martí i Franquès s/n, 08028 Barcelona, Spain

Correspondence to: Stefano Tavani (stefano.tavani@unina.it)

Abstract. In the Spanish Pyrenees, the Sant Corneli-Bóixols thrust-related anticline displays an outstandingly preserved growth strata sequence. These strata lie on top of a major unconformity exposed at the anticline's forelimb that divides and decouples a lower pre-folding unit from an upper syn-folding one. The former consists of steeply dipping to overturned strata with widespread bedding-parallel slip indicative of folding by flexural slip, whereas the syn-folding strata above define a 200 m amplitude fold. In the inner and outer sectors of the forelimb, both pre- and syn-folding strata are near vertical to overturned and the unconformity angle ranges from 10 to 30°. In the central portion of the forelimb, syn-folding layers are gently dipping, whereas the angular unconformity is about 90° and the unconformity surface displays strong S–C shear structures, which provide a top-to-foreland slip sense. This sheared unconformity is offset by steeply dipping faults, which are at low angles to the underlying layers of the pre-folding unit. Strong shearing along the unconformity surface also occurred in the inner sector of the forelimb, with S–C structures providing an opposite, top-to-hinterland slip sense. Cross-cutting relationships and slip senses along the pre-folding bedding surfaces and the unconformity indicate that regardless of its orientation, layering in the pre- and syn-folding sequences of the Sant Corneli-Bóixols anticline were continuously slipped. This slipping promoted an intense stress deflection, with the maximum component of the stress tensor remaining at low angles to bedding during most of the folding process.

1 Introduction

Templates used to describe the state of stress of growing regional-scale thrust-related anticlines (e.g. Hancock, 1985; Lisle, 1994; Fischer and Wilkerson, 2000; Belayneh and Cosgrove, 2004; Tavani et al., 2015) typically integrate punctual strain data (e.g. Engelder and Geiser, 1980; Laubach, 1989; Lacombe, 2012; Balsamo et al., 2016) and indirect information provided by the large-scale geometry of the structure, such as curvature or strata thinning or thickening (e.g. Price and Cosgrove, 1990). Widespread documentation of bedding-parallel slip and preservation of layer thickness provides key information for modelling the distribution of stress in actively growing anticlines. These observations indicate flexural-slip folding in the multilayered portions of reservoir-scale thrust-related folds (e.g. Donath and Parker, 1964; Ramsay, 1967; Tanner, 1989; Suppe, 1983; Fowler, 1996; Erslev and Mayborn, 1997). The assumption and/or observation of flexural slipping has important consequences for the stress distribution:

- Slipping along numerous bedding surfaces with a wide range of bedding dip is only possible where the bedding surfaces have a low friction and a low cohesion.

- Conversely, the reactivation of these closely spaced low-friction surfaces should prevent the layer-parallel shear stress to exceed a certain value, which in turn constrains the direction of the maximum component of the stress field to be at low angle to bedding, i.e. at low angle to the slipping surface (Wiltschko et al., 1985; Ohlmacher and Aydin, 1997; Tavani et al., 2015).

The process of layer-parallel slipping is, however, discontinuous in time and space, making it unclear whether the regional stress only reorients locally or if the layer-parallel slipping is sufficiently dense, both in time and space, to promote the reorientation of the stress in wide, actively folding areas. Field data are in agreement with the second hypothesis, i.e. that slipping along low-friction bedding surfaces promotes deflection of the principal directions of the remote stress field, so that the direction of the maximum compressive stress maintains at low angle to the bedding; thus, the maximum stress in active fold and thrust belts may not always be strictly horizontal. In fact, syn-folding layer-parallel shortening structures are reported in the folded pre-growth strata of many thrust-related anticlines (e.g. Tavani et al., 2006, 2012). However, palaeo-stress and/or palaeo-strain indicators cannot easily and unequivocally constrain the range of bedding dip values at which such stress deflection mechanisms can operate (e.g. Callot et al., 2010; Beaudoin et al., 2012). In fact, almost all published datasets come from pre-growth sequences of thrust-related folds. In these cases, determining the bedding dip when a given set of deformation structures are formed (i.e. fractures or slickenlines along a bedding surface) remains difficult, and any assumptions made carry significant uncertainty.

Conversely, observations made in syn-growth layers of thrust-related anticlines serve to drastically reduce uncertainties related to the timing of deformation (e.g. Shackleton et al., 2005, 2011). In fact, the study of growth strata sequences are by far the most commonly used approach for understanding the kinematics of fault-related folds in contractional settings (e.g. Suppe et al., 1992, 1997; Burbank et al., 1996; Ford et al., 1997; Vergés et al., 2002). In contrast with the abundance of detailed geometrical studies (Suppe, 1983; Medwedeff, 1989; Mitra, 1990; Suppe and Medwedeff, 1990; Zapata and Allmendinger, 1996; Poblet et al., 1997; Suppe et al., 1997), only a few contributions dealing with the dynamics of folding inferred from syn-kinematic layers have been published (e.g. Ford et al., 1997; Nicol and Nathan, 2001; Shackleton et al., 2011), mostly because of the lack of well-preserved and accessible exposures. In this work we have focused on the macro- and meso-structures developed within a growth strata wedge and a related major syn-kinematic unconformity exposed at the forelimb of the Sant Corneli-Bóixols anticline. Bedding-parallel slip occurs along pre- and syn-kinematic strata, which are oriented obliquely to each other, together with mesoscale faults cutting across strata and the unconformity. Thus, this area provides an excellent, almost unique field example to observe, describe and analyse how oblique anisotropies, i.e. layers and unconformity, respond to progressive shortening and related folding in a contractional setting. In addition, the studied area allowed us to determine the threshold dip value at which flexural slip is of sufficient magnitude to deflect the maximum principal stress direction from the regional stress field.

2 Geological setting

The Pyrenean Belt is a doubly vergent orogenic wedge (Fig. 1a) formed during the Late Cretaceous to Miocene subduction of the Iberian lithosphere beneath the Eurasian plate (e.g. Choukroune et al., 1990; Muñoz, 1992; Teixell, 1998). It largely deformed and inverted the Mesozoic extensional basins developed between Iberia and Eurasia during the Mesozoic separation of these two plates (Muñoz, 2002). The Early Cretaceous Organyà basin developed on the thinned continental crust of the Iberian plate to the south of the exhumed mantle domain of the Pyrenean rift (e.g. Tugend et al., 2014). Upon convergence and shortening, the Organyà basin was positively inverted and incorporated into the hanging wall of the Bóixols thrust starting in Late Cretaceous times (e.g. Mencos et al., 2015 and references therein). The positive inversion of the inherited extensional structures occurred under oblique, NNW–SSE oriented convergence (Tavani et al., 2011) and was responsible for the development of the E–W-striking Sant Corneli-Bóixols anticline (Fig. 1a). The location and geometry of this anticline is controlled by the orientation of the Early Cretaceous extensional border fault system of the Organyà basin (e.g. Bond and McClay, 1995; García-Senz, 2002; Mencos et al., 2015).

Several detailed studies of the stratigraphy of the Organyà basin have been carried out in the last 50 years (e.g. Rosell, 1963; Garrido, 1973; Simó, 1986; Berástegui et al., 1990; García-Senz; 2002; Mencos, 2011). The pre-rift Mesozoic stratigraphy is represented by clays and evaporites belonging to the Triassic Keuper facies, followed by Jurassic shallow marine carbonates and deeper water marls. The Early Cretaceous syn-rift megasequence consists of platform carbonates that thicken towards the north and change laterally (i.e. toward the north) into basinal marls (e.g. García-Senz, 2002). In the hanging wall of the Bóixols thrust, the maximum thickness of the syn-rift megasequence is about 4500 m. It thins southwards around the hinge zone of the Sant Corneli-Bóixols anticline across the extensional fault system at the southern margin of the Organyà basin (Lanaja et al., 1987; Berástegui et al., 1990; Arbués et al., 1996; García-Senz, 2002; Muñoz et al., 2010; Mencos et al., 2015). The Upper Cenomanian to lower Santonian post-rift megasequence consists of carbonates with lesser clastics and can be up to 700 m thick (García-Senz, 2002; Mencos, 2011). The syn-orogenic strata are exposed in the leading syncline (i.e. Tremp-Sallent syncline) of the Sant Corneli-Bóixols anticline (Fig. 1a–b) and include more than 1000 m of Upper Santonian to Palaeocene deep water and continental strata that thin abruptly to a few tens of metres northwards, i.e. towards the Sant Corneli-Bóixols anticline (e.g. Arbués et al., 1996; Roma et al., 2011).

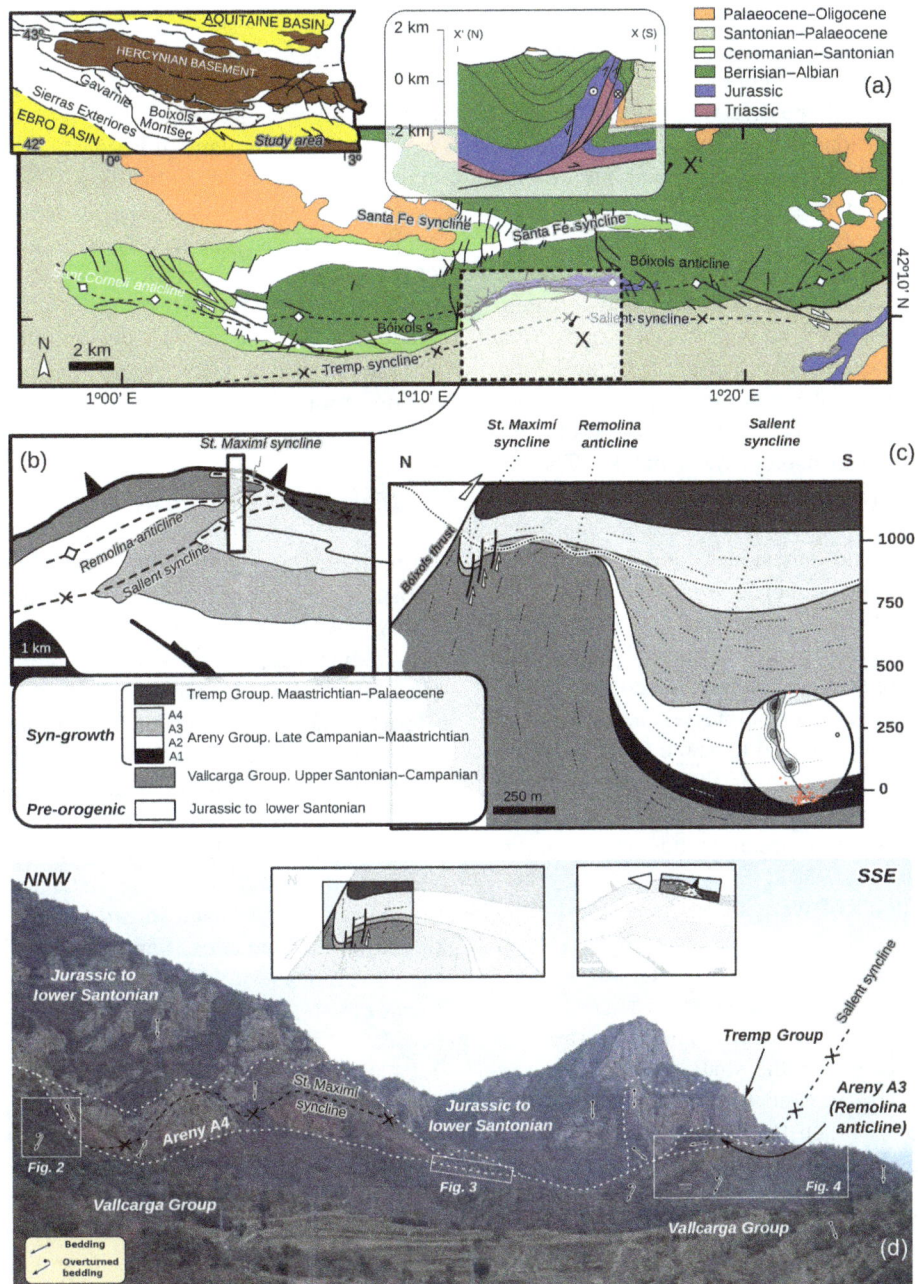

Figure 1. (a) Geological maps of the eastern Pyrenees, with detail and cross section (modified from García-Senz, 2002) of the Sant Corneli-Bóixols anticline. Geological map **(b)** and schematic cross section **(c)** of the study area, with cumulative contouring of poles to bedding and best-fit beta axis of syn-folding beds and poles of pre-folding beds (red circles). **(d)** Panoramic view of the study area, with insets showing the location of Figs. 2 to 4 and the photographed area in cross section and map view respectively.

In the studied area, the syn-orogenic succession can be subdivided into two units (Fig. 1b–c): (1) the lower, Upper Santonian to Campanian Vallcarga Group (Nagtegaal, 1972) is constituted by a multilayered marine sequence of thin to medium bedded limestones and mudstones. Around the Sant Corneli-Bóixols anticline, the Vallcarga Group was deposited during folding, as evidenced by growth geometries at the eastern tip of the Sant Corneli-Bóixols anticline (Men-

cos et al., 2015). However, no clear indications of deposition during the early stages of folding are visible in the studied area and hence it is here geometrically considered as pre-folding but within a regional syn-orogenic scenario. (2) The Late Campanian to Maastrichtian Areny Group (Arbués et al., 1996) unconformably overlies the Vallcarga Group. Its thickness exceeds 1000 m in the Tremp-Sallent syncline depocentre (Fig. 1a–b) but thins abruptly to a few tens of metres towards the Sant Corneli-Bóixols anticline; thus, it is considered to have been deposited during folding (i.e. syn-folding). The Areny Group records sedimentation in deep water and neritic conditions coeval with the inversion and related folding of the Organyà extensional basin. The Areny Group has been divided into four depositional sequences (A1 to A4 from older to younger; Arbués et al., 1996; Roma et al., 2011) that, overall, display a marked regressive character. The A1 depositional sequence includes deep-water olistostromes, deep-water mudstones, outer-shelf mudstones and calcarenites, and fan-deltaic conglomerates and sandstones. The A2 depositional sequence includes fan-deltaic conglomerates and sandstones, deep-water mudstones, and outer-shelf mudstones–calcarenites and also includes re-sedimented, rudist-bearing limestones in boulder-sized blocks. The A3 depositional sequence includes deep-water slump-deformed marls, outer-shelf mudstones–calcarenites, inner-shelf and near-shore calcarenites and sandstones, and in situ, rudist build ups. Depositional sequence A4 is constituted by fan-deltaic conglomerates and sandstones, lagoonal to alluvial mudstones and sandstones, and alluvial fanglomerates. (3) The syn-folding Maastrichtian to Palaeocene Tremp Group (Cuevas, 1992) includes continental facies associations, which are commonly referred to as the Garumnian facies (Cuevas, 1992). These consist of alluvial and colluvial conglomerates and breccias, passing southwards to alluvial plain and fluvial reddish sandstones and mudstones (e.g. Arbués et al., 1996; Roma et al., 2011).

From a structural point of view, the major structures in the studied area are the Bóixols thrust and related splays, the E–W trending Sant Corneli-Bóixols anticline and the associated Santa Fe syncline to the north and the Tremp-Sallent syncline to the south (Fig. 1a). The main ramp of the Bóixols thrust crops out in the study area, whereas to the west and to the east it remains blind along most of the frontal limb of the Sant Corneli-Bóixols anticline. In its exposed sector, the Bóixols thrust has Triassic to Upper Cretaceous pre-growth rocks in its hanging wall and syn-orogenic and syn-folding strata in its footwall (Fig. 1c). A thin sheet of overturned post-rift limestones, mainly the upper Cenomanian ones, defines two thrusts. The lower thrust remains blind beneath the vertical beds of the Garumnian succession on the northern limb of the Tremp-Sallent syncline. Conversely, the upper thrust truncates the Garumnian beds. This upper one is the Bóixols thrust, and according to magnetostratigraphic and thermochronological studies it would have been reactivated during Palaeogene times (Beamud et al., 2011). The pre-folding beds of the Vallcarga Group in the footwall of the Bóixols thrust are folded into a syncline with overturned strata immediately below the thrust fault. These strata progressively acquire sub-horizontal attitudes in the Tremp-Sallent syncline to the south (Fig. 1c). Conversely, the unconformably overlying Areny and Tremp groups display a series of folded structures, namely the Sant Maximí syncline and the Tremp-Sallent syncline, with the Remolina anticline in

between. The two synclines disappear toward the east, where the two synclines join (Fig. 1b). The three folds display a significant eastward plunge (Roma et al., 2011) of 24°, with a N72° plunge direction, as derived by the direction normal to the best-fit plane of bedding data of the Areny Group (Fig. 1c). These poles to bedding are well clustered along a great circle, thus defining the axis of a cylindrical fold. This suggests that the plunge was acquired after the deposition of the syn-folding Areny Group. In addition, it is to be noted that poles to pre-folding bedding are clustered along the same great circle, indicating that the folding axis was parallel to the intersection between the pre-unconformity beds and the unconformity (Ramsay, 1967).

3   Macro- and meso-structures

In the following, we describe the structural assemblages occurring along and around the major unconformity dividing the Areny Group from the Vallcarga Group, i.e. the upper syn-folding from the lower pre-folding strata respectively (Fig. 1d). The macro- and meso-structures are described from north to south in three subsections, corresponding to the three limbs of the S-shaped fold that define the Sant Maximí syncline, the Remolina anticline and the Tremp-Sallent syncline (Fig. 1). We will present and discuss stereoplots of bedding attitude, fault orientations and kinematic indicators from faults. In these stereoplots the plane normal to the structural plunge is also displayed in order to ease the interpretation of the fault kinematics. For each stereoplot, we also show the two graphs resulting from the removal of plunge first and then of the residual bedding dip (Ramsay, 1967).

3.1   Northern limb

On the inner (northern) limb of the Sant Maximí syncline, the marls and limestones of the Vallcarga Group are overturned, while the unconformably overlying strata of the Areny Group are steeply south dipping to near vertical (Fig. 2). In the northern limb of the syncline, the strata of the Areny Group include siltstones, sandstones and conglomerates belonging to the A4 depositional sequence; the A1 to A3 depositional sequences are missing (Fig. 2a–b), either because they were never deposited there or because they were eroded before the deposition of the A4 sequence. Overturned bedding surfaces of the Vallcarga Group display evidence of slipping. Movements along bedding are mostly toward the NW, with normal sense of slip in the present overturned bedding orientation (Fig. 2b). Stereoplots show that slickenlines along bedding surfaces of the Vallcarga Group are mostly perpendicular to the local fold axis. The unconformity between the Vallcarga and the Areny groups has been reactivated as a thrust and displays an intense S–C fabric that affects a few metres of the Vallcarga Group (Fig. 2c). Similar to other S–C tectonites developed in carbonates and at shallow depth (e.g. Tesei et

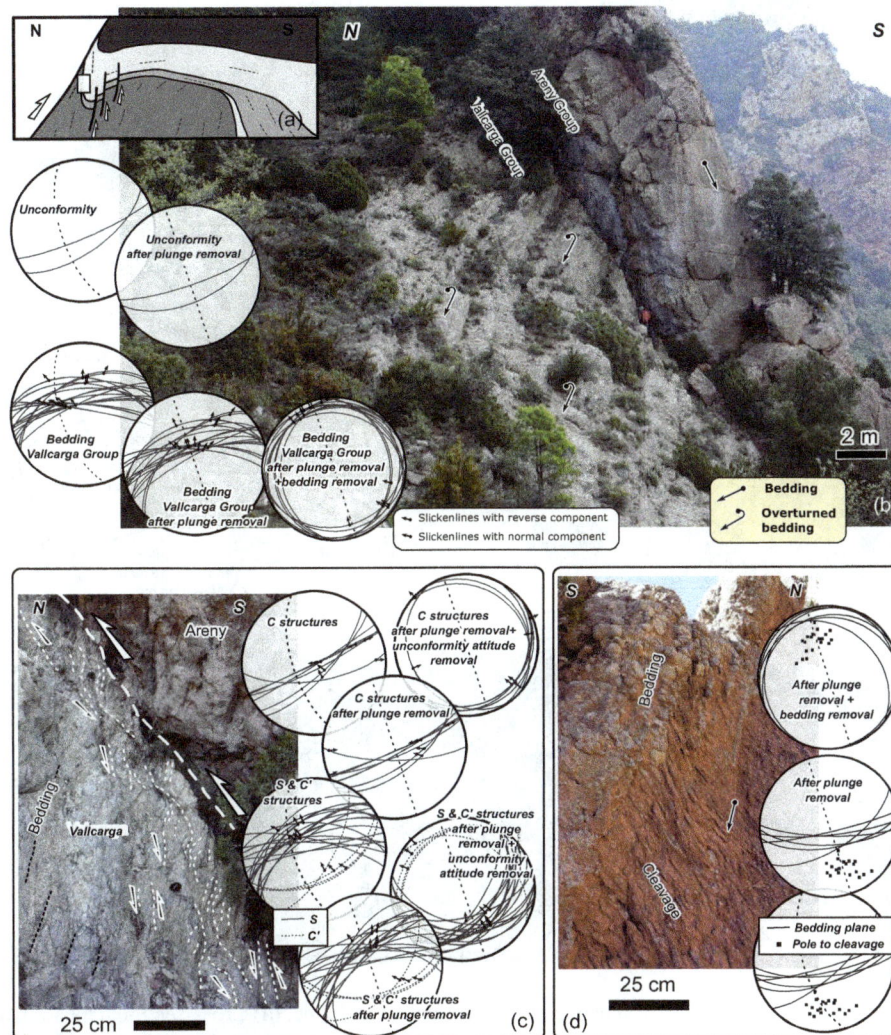

Figure 2. Structures exposed at the northern limb of the San Maximí syncline. **(a)** Cross-sectional location of the site. **(b)** South-dipping conglomerates of the Areny Group unconformably overlying the overturned north-dipping strata of the Vallcarga Group, with stereoplots of the unconformity and bedding surfaces in the Vallcarga Group. **(c)** Detail of the unconformity, showing S–C–C' fabric, with corresponding stereoplots. **(d)** South-dipping alternating conglomerates and siltstones of the Areny Group, with pervasive cleavage at high angle to bedding.

al., 2013; Vitale et al., 2014), the S–C structures found in the first 2–3 m of the Vallcarga Group immediately below the unconformity formed as a consequence of pressure solution of marly limestones and marls, thus indicating a ductile to brittle-ductile behaviour of the Vallcarga Group associated with this major fault. Slip directions provided by C, S and C' structures are top to NW and, comparable to the bedding-parallel slip surfaces, the average slip direction lies along the plunge-normal plane (Fig. 2c). The Areny Group conglomerates immediately above the unconformity are not affected by such S–C fabric, whereas siltstones occurring a few metres above the unconformity are affected by a penetrative cleavage (Fig. 2d). This cleavage is at a high angle to bedding, as seen in the field and as evidenced by the fact that poles to cleavage in the stereoplot occur close to the bedding

plane great circles (Fig. 2d). However, cleavage is not strictly bedding-perpendicular. Once bedding dip is restored to the horizontal, the poles to cleavage still lie along the plunge normal plane and cleavage becomes SE dipping. These relationships are indicative of a minor top-to-NW slip component during cleavage development. It is worth remarking that despite the importance for stress direction reconstruction, the cleavage described above is a localised feature, which affects only the silty beds of the uppermost portion of the Areny Group in some outcrops.

3.2 Central limb

Immediately to the south of the axial surface of the Sant Maximí syncline, strata of the Areny Group are shallow dip-

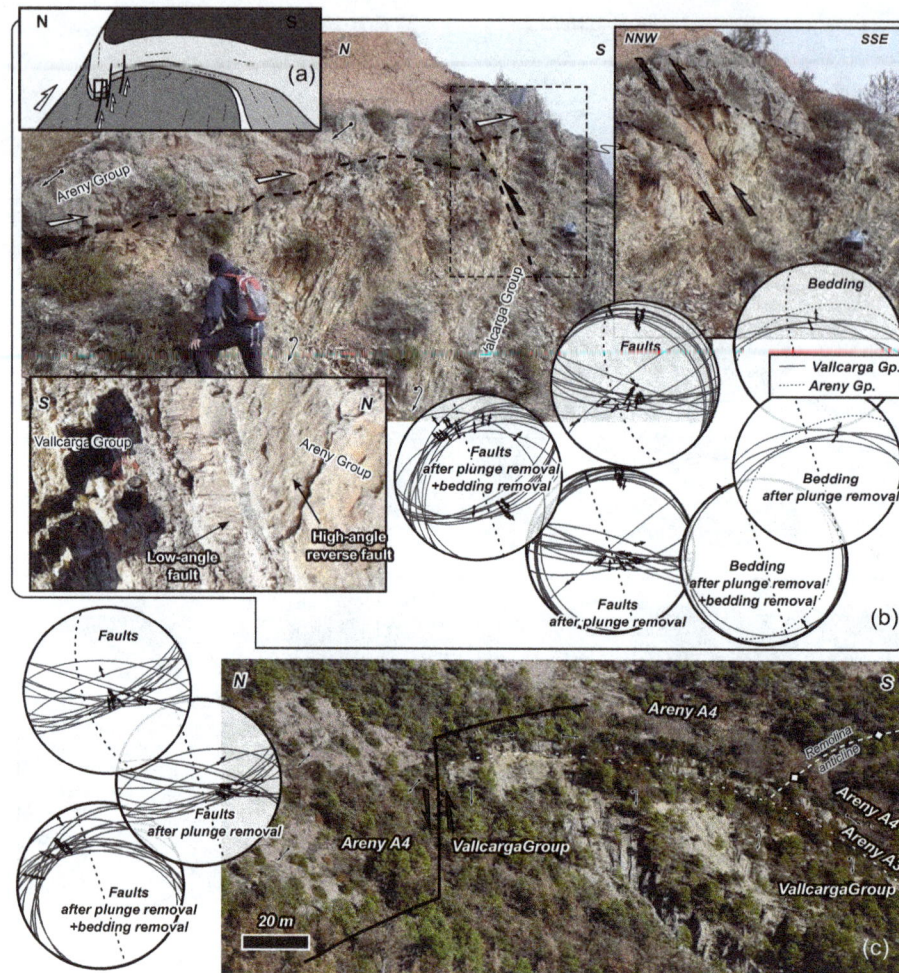

Figure 3. Transitional area between the San Maximí syncline and the Remolina anticline. **(a)** Cross-sectional location of the site. **(b)** Shallow-dipping unconformity between the Areny and the Vallcarga groups reactivated as a low-angle fault and displaced by a high-angle fault. Details of the low- and high-dipping faults are shown, together with stereoplots of faults and bedding surfaces of the Vallcarga Group. **(c)** Panoramic view and stereoplot of a near-vertical fault system uplifting the Remolina anticline, which has folded strata of the Areny Group unconformably on top of near-vertical strata of the Vallcarga Group.

ping to sub-horizontal (Fig. 1c–d). Locally, however, strata of the A4 depositional sequence of the Areny Group are steeply north dipping and the Sant Maximí syncline forms a tight structure with a north-dipping to near-vertical axial surface. In the hinge zone, the unconformable strata of the Areny Group are separated from the underlying overturned strata of the Vallcarga Group by a sub-horizontal slip zone, which corresponds to the sheared syn-folding unconformity (Fig. 3). In addition, the sheared unconformity is offset by a series of high-angle faults that uplift the southern block (Fig. 3b). Striae along bedding surfaces of the Vallcarga Group indicate top-to-N movements with normal kinematics (Fig. 3b). Striae within the sub-horizontal slip zone indicate top-to-S movement, whereas the high-angle faults that offset it have slickenlines lying along the plunge-normal plane and show a top-to-N movement. A few strike-slip slickenlines also oc-

cur along high-angle faults and are indicative of left-lateral movements. In other places, the system of high-angle reverse faults uplifting the southern limb of the Sant Maximí syncline displays a top-to-NW movement, including some right-lateral kinematic indicators (Fig. 3c). Looking at the system of high-angle faults in natural cross sections at a larger scale of observation (Fig. 3c), it is evident that these faults are approximately (i.e. angular difference is less than 10°) parallel to the bedding of the Vallcarga Group and that the amount of uplift of the southern block is a few tens of metres.

The uplifted southern block is well exposed at the Sallent hill (Fig. 4). There, the sub-horizontal rudist-bearing units of the A3 depositional sequence sit unconformable on top of the overturned north-dipping strata of the Vallcarga Group (Fig. 4a–b). Slickenlines are consistently found on bedding surfaces of these strata. As shown in the stereoplots

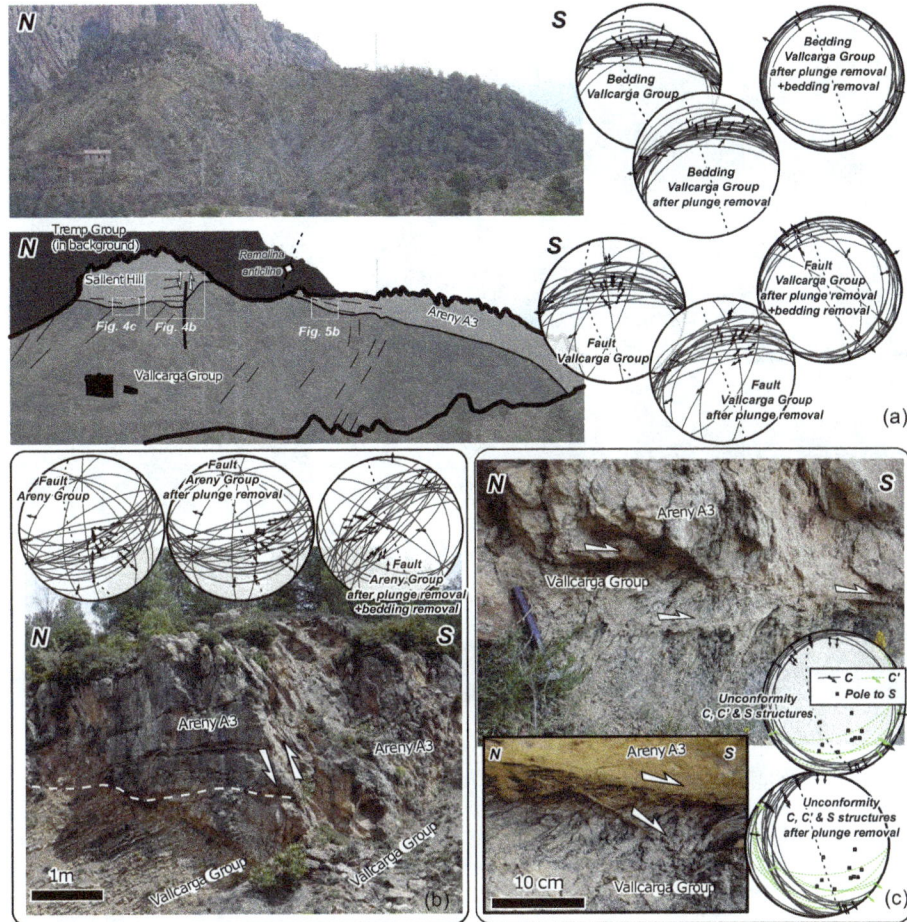

Figure 4. Macro- and meso-structures of the Remolina anticline. **(a)** Panoramic view and line-drawing of the Remolina anticline (with insets showing the location of **(b)** and **(c)**, and Fig. 5b), with stereoplots of faults and bedding measured in the Vallcarga Group. **(b)** Detail of a south-dipping reverse fault with sub-horizontal carbonates of the Areny Group on top of overturned strata of the Vallcarga Group in its footwall. Stereoplots show fault data in the Areny carbonates. **(c)** Details of the unconformity, with S–C–C' illustrated and plotted.

of Fig. 4a, most of slickenlines display top-to-N normal kinematics, whereas a few are characterised by strike-slip (both left and right lateral) or reverse kinematics. The slickenlines displaying strike-slip and reverse kinematics postdate the top-to-N normal ones. Faults oblique to bedding have also been found in the Vallcarga Group in this area (Fig. 4a). The faults are at low angle with the bedding surface and display normal and, subordinately, reverse kinematics. After bedding dip removal, both bedding-parallel slip surfaces and bedding-oblique faults show a top-to-NW slip sense. As mentioned above, strata of the rudist-bearing A3 sequence are shallow dipping (Fig. 4a–b) and the unconformity between the Areny and the Vallcarga groups is also sub-horizontal. The unconformity is affected by a pervasive shear fabric (Fig. 4c), with S, C and C' structures providing a top-to-SSE shear sense. In addition, the sheared unconformity is cross-cut by SSE-dipping and NNW-verging reverse faults (Fig. 4b).

3.3 Southern limb

On the southern limb of the Remolina anticline, strata of the Areny Group are overturned to steeply south dipping (Fig. 5a). These strata are still unconformable on top of the overturned strata of the Vallcarga Group, but the unconformity angle between the two groups becomes significantly reduced to about 20°. The unconformity preserves its stratigraphic origin and, as opposed to the northern and central limbs of the Sant Maximí syncline, no appreciable evidence of shear occurs (Fig. 5b). Instead, striae along the bedding surfaces of the Areny Group are observed (Fig. 5c). These striae indicate normal top-to-NNW and reverse top-to-NNE movements along north-dipping overturned and steeply south-dipping strata respectively (Fig. 5c). In both cases, striae lie along the bedding surfaces at the intersection between bedding and the plunge-normal plane (Fig. 5c). Faults at a low angle to bedding have the same behaviour as the bedding: south- and north-dipping faults are reverse and normal

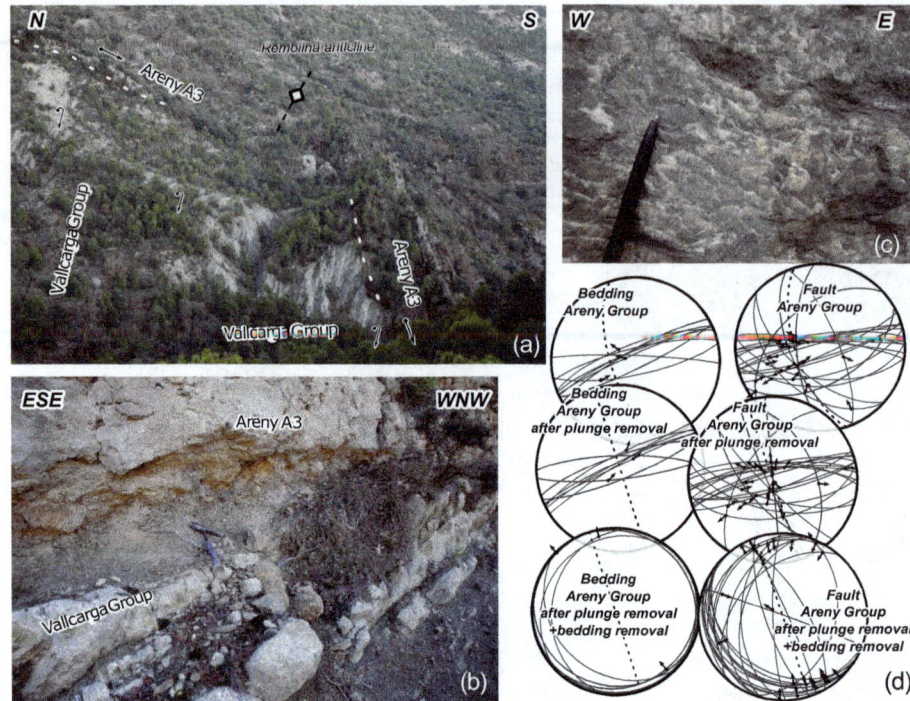

Figure 5. (a) Panoramic view of the hinge zone of the Remolina anticline, visible in the Areny strata that are on top of constantly dipping strata of the Vallcarga Group. **(b)** Detail of the unconformity between the Areny and Vallcarga groups at the southern limb of the anticline, where no evidence of shear occurs. **(c)** Detail of slickenlines along a near-vertical bedding surface of the Areny strata, providing a top-to-north slip sense for the upper bed. **(d)** Stereoplots of bedding surfaces and faults collected in the Areny Group strata of the southern limb of the Remolina anticline.

respectively with slickenlines lying at the intersection between the fault and the plunge-normal plane (Fig. 5c). Once bedding dip is restored to the horizontal, a top-to-NNW slip sense is exhibited by both faults and bedding-parallel slip surfaces.

3.4   Structural summary

The deformation structures observed along and around the unconformity separating the upper syn-folding strata of the Areny Group from the underlying multilayered limestones and marls of the pre-folding Vallcarga Group can be summarised as follows:

In the Vallcarga Group, many of the E–W-striking bedding surfaces of near-vertical to overturned strata have been reactivated as slip surfaces. Most of these bed-parallel slip surfaces exhibit dip-slip kinematics, with only a few beds showing strike-slip movements. After removing the plunge of the structure and then restoring the local bedding to the horizontal, most of the slickenlines measured along the bedding surfaces provide a slip sense ranging from top-to-NW to top-to-N, with an average top-to-NNW movement. Faults are roughly E–W to WSW–ENE striking and show very low cut-off angles to bedding. After removing the fold plunge and the bedding dip, these faults provide the same top-to-

NNW slip sense as the bedding-parallel slickenlines. Some faults, which are presently steeply dipping to near vertical, have cut-off angles ranging from 20 to 40°, and after removing plunge and bedding dip, they show normal kinematics. Still, the slip sense provided by them after bedding dip removal is top to NNW. This fault pattern and the illustrated kinematics of bedding surfaces are observed all across the studied thrust-related fold profile, i.e. in the northern, central and southern limbs.

The syn-folding unconformity is characterised by an intense S–C fabric (showing also some C' structures) in the northern limb and in the sub-horizontal central limb. In both cases, slip direction is roughly NNW–SSE, although some strike-slip movements are occasionally observed. However, the slip sense is opposite in the two limbs, being top to NNW in the northern limb (i.e. where the sheared unconformity strikes about E–W and has a near-vertical attitude) and top to SSE in the central limb (i.e. where the sheared unconformity is offset by the steeply dipping to near-vertical faults surging from the underlying Vallcarga Group). Further to the south, in the southern limb of the Remolina anticline, the unconformity shows little evidence of deformation.

Strata of the Areny Group exposed at the northern limb of the San Maximí syncline are affected by an intense cleavage at high angle to bedding. The cleavage-bedding angle is not exactly 90°, however, indicating the occurrence of a top-to-NNW bedding-parallel slip component. Slickenlines are observed along the bedding surfaces of the near-vertical to overturned strata of the Areny Group exposed at the southern limb of the Remolina anticline. In this area, some faults at very low angle to bedding occur. For both faults and bedding surfaces, the slip sense measured after removing the plunge and the dip of the bedding is roughly top to NNW.

4 Chronology of deformation stages

The syn-folding strata of the Areny Group exposed at the forelimb of the Sant Corneli-Bóixols anticline unconformably overlie the north-dipping overturned strata of the pre-folding Vallcarga Group. The unconformity between both groups is clearly a syn-folding feature. Its unconformity angle varies across the studied area, and in addition, its surface shows unequivocal evidence of strong shearing, with an average NNW–SSE-oriented shear direction. Such a shear direction is parallel to the slip directions measured along both faults and bedding surfaces of the Vallcarga and Areny groups. However, the NNW–SSE direction is not perpendicular to the average strike of the hosting anticline, although the local fold axis in the study area is WSW–ENE striking (Fig. 1a). This structural relationship reaffirms that the E–W-striking Sant Corneli-Bóixols anticline has developed under an oblique convergence setting where the shortening direction was NNW–SSE (Tavani et al., 2011). In this sense, the observed meso-structures are interpreted as developed during the growth of the Sant Corneli-Bóixols anticline, and they cannot be attributed to a subsequent tectonic event. The fact that these structures occur in syn-folding strata also rules out a pre-folding origin. In agreement with this, the observed opposite shear senses along the unconformity surface to the north and to the south of the San Maximí syncline axial surface have a syn-folding origin. As illustrated in the next section, these opposite senses of slip can be used to unravel the kinematic evolution of the unconformity, the unconformity angle itself and therefore that of the Sant Corneli-Bóixols anticline.

5 Modelling the folding of the angular unconformity

The existence of the Sant Maximí syncline and the Remolina anticline in syn-folding strata, with their axial planes being parallel to bedding in the pre-folding units, is a challenging geological feature. Neither shortening perpendicular to bedding of the pre-folding units nor inhomogeneous flexural slip can be invoked to explain their existence. In the first case, pervasive strain, indicating a remarkable amount of bedding-perpendicular shortening, should be found in the strata of the pre-folding Vallcarga Group, which is not observed. In the second case, localised layer-parallel slip should be observed, but is also not observed. There are no remarkable lithologic

changes in the pre-folding Vallcarga Group that could justify an inhomogeneous (i.e. discrete) flexural slip. In fact, significant bed-parallel slip is found distributed through the Vallcarga Group multilayer. In agreement with this, the existence of the Sant Maximí syncline and the Remolina anticline can be better explained in the framework of flexural folding of an angular unconformable sequence, which included different initial unconformity angles across the structure. Guidelines about flexural folding of angular unconformable sequences were firstly given by Alonso (1989). These include the progressive variation of the unconformity angle during tilting of pre-unconformity layers, synchronous with progressive slipping along the unconformity surface (Alonso, 1989). Figure 6a illustrates the folding of an unconformable sequence using a kink-band template with synclinal geometry. The position of six points undergoing folding is illustrated. The points P_0 and P_1 are fixed and inactive, i.e. they do not move during folding and the rock does not pass through them. The point P_1 is the origin of our reference system. The point P_2 is located at the intersection between the axial surface and the unconformity and, as the axial surface moves during folding, this point is an active point that migrates through the rock. The remaining points are mobile but inactive material points, which are attached to the rock. In detail, the points P_3 and P_5 are attached to the base of the post-unconformity unit, while the point P_4 is immediately below the unconformity, and it is attached to the layer corresponding to the stratigraphic elevation of the point P_0. The simple kink-band construction is used to quantify how the unconformity angle and the amount of slip along the unconformity are modified during folding, where D_0 is the initial dip of layers, U_0 the initial unconformity angle, H_0 the stratigraphic elevation of the unconformity and L_0 the distance from the origin of an arbitrarily placed pin line (note that the L_0 parameter will disappear from the final equations used here). The X and Y coordinates of the six key points and the length of the segments joining them can be expressed as a function of L_0, H_0, U_0, D_0 and D, as provided in Fig. 6b. In particular, the length of the segment joining points P_5 and P_4 provides the amount of slip (ΔS, considered positive when it is top to hinterland), while points P_3 and P_4 allow calculation of the unconformity angle (U).

As folding takes place, the unconformity angle increases and the sense of slip along the unconformity is initially top to hinterland, i.e. in the same sense as the flexural slip along the pre-unconformity layers. When the pre-unconformity strata become overturned, the unconformity angle continues to increase and the flexural slip in the pre-unconformity layers continues to be top to hinterland; instead, the sense of slip along the unconformity flips and becomes top to foreland. Close to the leading syncline, i.e. along the P_2P_3 segment, the sense of the incremental slip is top to hinterland, whereas along the P_3P_4 segment it becomes to top to foreland, despite the cumulative slip may continue to be top to hinterland.

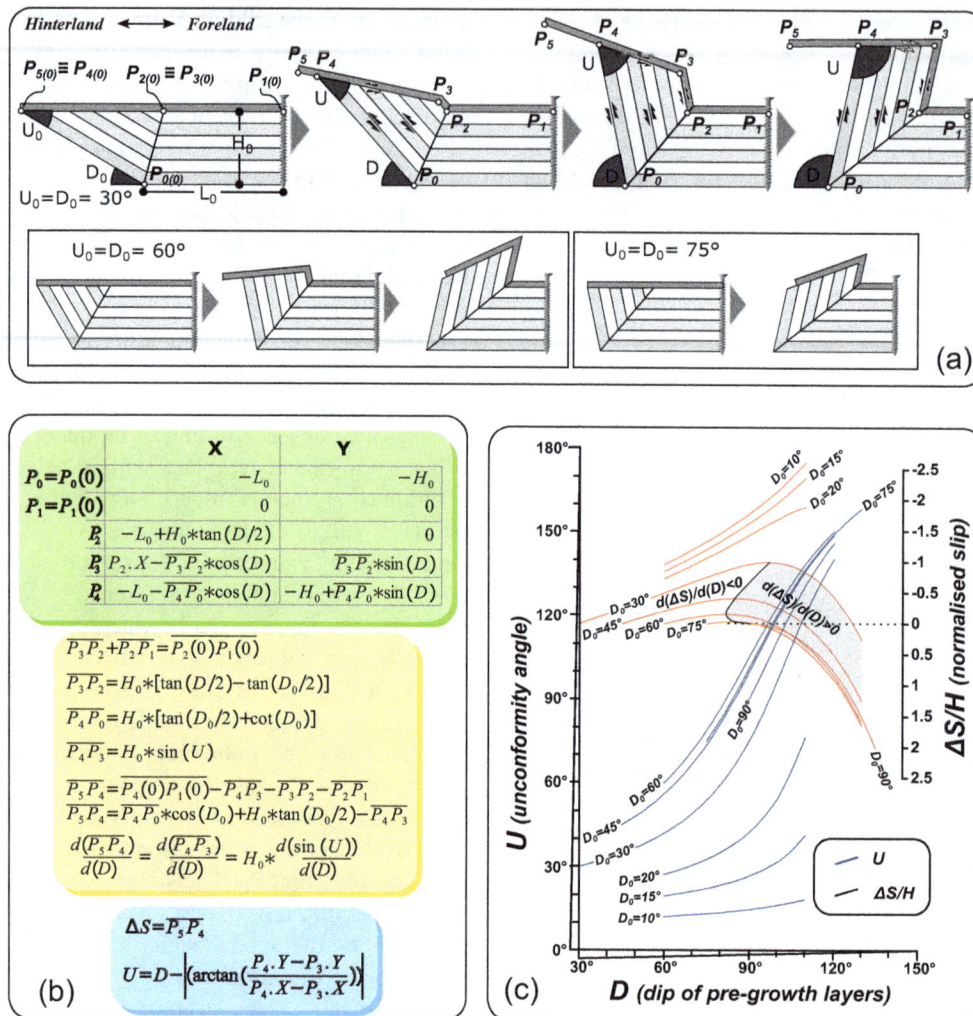

Figure 6. (a) Evolving angular relationships between unconformable sequences during flexural folding in the inner limb of a syncline, with incremental slip senses along pre-unconformity layers and along the unconformity indicated for different initial unconformity angles. The position of six points undergoing folding is illustrated, as well as the dip of pre-unconformity layers (D) and of the unconformity angle (U), which is the angle between the unconformity and the underlying layers. **(b)** X and Y coordinates of the six points of **(a)**, with length of segments, and derived amount of slip along the unconformity (ΔS) and unconformity angle (U). **(c)** Graphical solution of equations in **(b)**. Blue lines relate the unconformity angle (U) to the dip of pre-growth strata (D) for different initial unconformity angles (U_0). Red lines relates the normalised slip along the unconformity in the inner portion (i.e. $\Delta S = P_4P_5$ segment divided H) to D for different initial unconformity angles (U_0). Notice that the y axis for red lines is on the right and that positive and negative values are flipped. The lines indicate the cumulative slip along the unconformity, while the grey area bordered by the black line indicates the area where the incremental slip is negative.

The relationship between U, U_0 and D derived in Fig. 6b is graphed in Fig. 6c. The relationships between U and D (blue lines in the figure) for different U_0 values indicate that

1. it is possible to develop overturned pre-folding strata and a nearly sub-horizontal unconformity, as observed in the central limb of our study area, (i.e. the unconformity angle is roughly equal to the dip of the pre-unconformity layer) for a wide range of initial unconformity angles (from 30 to 90°).

2. in order to obtain an unconformity angle of less than 20° where the pre-unconformity strata are near vertical, as observed in the northern and southern limbs of the study area, the initial unconformity angle cannot exceed 10–15°.

The predicted amount and sense of slip (normalised to H_0) at different tilting stages (i.e. for different D) and for different initial unconformity angles (i.e. U_0) are plotted in red in Fig. 6c, where positive and negative values indicate top-to-hinterland and top-to-foreland slip respectively. For small initial unconformity angles, the unconformity angle and the cumulative top-to-hinterland slip along the unconformity increase during folding. This occurs until the dip of pre-growth strata attains a near-vertical attitude. From this point, further folding would imply overturning of strata and the decrease of the cumulative top-to-hinterland slip, which eventually becomes negative (i.e. top-to-foreland sense), while the unconformity angle U exceeds 90°. Where the initial unconformity angle is close to 90°, progressive folding would imply a short period of top-to-foreland slipping along the unconformity, followed by top-to-hinterland slipping when strata become nearly vertical to overturned. At this point it is important to remark that the progressive and incremental slip senses do not coincide. The D value at which the cumulative slip passes from top-to-hinterland to top-to-foreland largely depends on the initial unconformity angle U_0. Conversely, for an initial unconformity angle U_0 between 45 and 90°, the incremental slip changes its sign for values of D ranging from 80 to 90°, almost regardless of U_0 value (the regions where the incremental slip has opposite directions are in white and grey in Fig. 6c).

The absence of any kinematic indicator of top-to-hinterland slip in the central limb of the study area indicates that the initial unconformity angle had to be high at that position, and if any, the initial stage of top-to-hinterland slip was negligible. This can be achieved when the initial unconformity angle is at least 70–75°. This represents a key argument for unravelling the deformation sequence as structures postdating the top-to-foreland slipping have to be interpreted as developed synchronously with layer tilting, and in particular, as developed at least after layers have become near vertical. The top-to-hinterland slip sense and the small unconformity angle observed in the northern limb, instead, point out a small initial unconformity angle.

It is worth noting that the position of the axial surface is determined by the position of P_0, and thus by the value of L_0 and H_0. These two parameters do not influence the value of the unconformity angle. Instead, the amount of slip is directly proportional to the value of H_0. However, we are interested in the sign of the slip, which is independent of H_0. In agreement, provided results are unrelated to the position of the axial surface, and thus of P_0, which can be arbitrarily set everywhere below the unconformity.

Two cautionary notes must be added for these conclusions. (1) They are based on a purely geometric model, in which both bed thickness and line length are preserved during folding. However, and despite the occurrence of local penetrative strain at some places (Fig. 2d), based on the field observations reported here and in Tavani et al. (2011), pressure solution cleavage is an extremely localised phenomenon in this anticline, and deformation structures pointing out folding-related bed thickness variations do not occur. In agreement, and despite the intrinsic simplification of any geometric model, information provided by the model of Fig. 6 can be applied to our case study. (2) The model described above al-

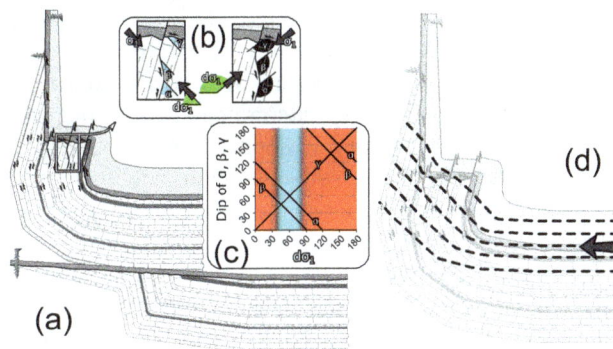

Figure 7. (a) Scheme showing the present-day geometry of the frontal limb of the Sant Corneli-Bóixols anticline (upper part), the geometry of the unconformity, and the pre- and syn-folding strata before folding (lower part). (b) Details showing the structural assemblages observed at the Remolina anticline, with two alternative configurations for the maximum stress orientation. The maximum stress forms the following clockwise angles: $d\sigma_1$ with the horizontal, γ with the unconformity, α with the bedding-parallel steps of flexural-slip faults in the Vallcarga Group, and β with the oblique to bedding strands of the flexural-slip fault in the Vallcarga Group. Red and cyan colours indicate angles not compatible and compatible with the observed slip pattern respectively. (c) Relationships between $d\sigma_1$ and α, β and γ, with the red area indicating the orientation of the maximum stress not compatible with the slip pattern observed at the Remolina anticline. (d) Inferred maximum stress trajectories during the late stages of folding.

lows the reconstruction of the initial, pre-unconformity configuration of the Vallcarga Group units. However, like in all the retro-deformed models produced by cross-section balancing, we have no direct access to the "real" initial configuration. We cannot thus discuss the discrepancy between the model and the reality. We can merely observe that the assumptions of the model, i.e. line-length and bed thickness preservation and flexural slipping, are largely consistent with field observations.

6 Discussion

6.1 Relative timing between slipping and layer tilting

Strata of the Vallcarga Group exposed at the forelimb of the Sant Corneli-Bóixols anticline display a rather constant attitude. However, according to the model described in Fig. 6, slip senses and angles of the unconformity on top of these strata indicate that the pre-unconformity layers (i.e. the Vallcarga Group) were not homoclinally dipping when the Areny Group was unconformably deposited on top them. The scheme of Fig. 7a illustrates the present-day simplified geometry of the studied structure, together with the balanced (i.e. line length is preserved; Dahlstrom, 1969; Brandes and Tanner, 2014) reconstruction at a time immediately after the unconformity development. The reconstructed dip of pre-unconformity layers in the three limbs is obtained according to what is illustrated in Sect. 5. As previously mentioned, the top-to-foreland shearing along the unconformity of the central limb has to be interpreted as occurring when strata of the Vallcarga Group attained a near-vertical to overturned attitude. Conversely, the absence of any evidence of top-to-hinterland shearing along the unconformity surface at the central limb suggests that such an unconformity had to be developed when the layers of the Vallcarga Group were steeply dipping. In agreement with this, the north-dipping faults offsetting the sheared unconformity have to be regarded as syn-folding structures developed when strata of the Vallcarga Group were overturned. These 10 to 30 m spaced faults mostly consist of bedding-parallel segments (Figs. 4a and 5d), with some strands showing cut-off angles between 20 and 40° (see stereoplots of Fig. 3b–c), and are interpreted as flexural-slip faults, like those offsetting the topographic surface of growth folds (e.g. Burbank and Anderson, 2011; Gutiérrez et al., 2014; Li et al., 2015). These faults are therefore late-stage flexural-slip features and, as detailed in the next subsection, cannot be compatible with a sub-horizontal maximum stress.

6.2 Maximum stress orientation

The studied natural stratigraphic units includes centimetre to metre thick strata of limestones, marls, sandstones and conglomerates exposed across a 500 m wide area (Fig. 1). The large number of strata involved in the deformation, coupled with their high compositional variability, prevent the collection of a representative dataset of friction and cohesion of both layers and interlayers. It thus makes it impossible to carry out a quantitative dynamic (i.e. stress) reconstruction. However, many stress configurations can be easily discarded due to their kinematic inconsistency with the observed shearing pattern. In particular, we consider that the maximum palaeo-stress lies on the plane oriented perpendicular to the fault and/or flexural-slip plane and containing the slip direction, and it forms an obtuse angle with the slip sense (e.g. Etchecopar et al., 1981). The following observations can thus restrict the range of possible solution and the sources of stress during folding: (1) if we consider faults and strata in their present orientation and after plunge and bedding dip removal, the top-to-foreland layer-parallel slipping and the south-verging reverse faulting are rare features in the northern limb of the San Maximí syncline, which is located at a distance of less than 100 m from the Bóixols thrust. The scarcity of these structures, and the occurrence of flexural-slip surfaces forming a low angle with the thrust and having an opposite slip sense (i.e. normal kinematics), indicates the limited role of faulting-related stress, sourced from the process zone (e.g. Cowie and Scholz, 1992) of the upward propagating Bóixols thrust, in controlling the pattern of syn-folding slipping. This is contrary to what has instead been documented in other thrust-related anticlines (e.g. Bel-

lahsen et al., 2006). (2) The top-to-hinterland (i.e. top to crest of the Sant Corneli-Bóixols anticline) layer-parallel slipping observed along the bedding surfaces of the Vallcarga Group for all three limbs has to be regarded as syn-folding. With the exception of a few bedding-oblique strands of flexural-slip faults, no significant evidence of strata thinning and/or thickening has been observed in the Vallcarga Group. This points out that folding has been almost entirely produced by layer-parallel slipping with bed-thickness preservation until late stage flexural-slip faulting took place (e.g. Donath and Parker, 1964). (3) Deformation structures such as the nearly layer-parallel shortening-related cleavage measured in the silty levels of the Areny Group along the axial zone of the San Maximí syncline indicate a maximum stress oriented at a low angle to bedding (Fig. 2d). (4) The fourth key observation concerns the steeply dipping faults, with high cut-off angles, cutting and displacing the unconformity. These faults include steps with cut-off angles of about 30° and steps parallel to the overturned bedding surfaces. Under the assumption that late-stage flexural-slip faulting caused the arrest of shearing along the unconformity, a range of possible maximum stress orientation during the transition from top-to-foreland shearing along the unconformity to the late-stage flexural-slip faulting can be defined for the central limb, as shown in Fig. 7b. The angle between the maximum stress and the bedding-parallel steps of flexural-slip faults in the Vallcarga Group is α and the angle between the maximum stress and the flexural-slip fault strands oblique to bedding in the Vallcarga Group is β, whereas the angle between the maximum stress and the unconformity is γ. These three angles must be comprised of a value between 0 and 90° to produce the observed slip pattern and for it to be kinematically compatible. When using the average dip of the unconformity (i.e. 0°) and the dip of the Vallcarga Group strata in the central limb (i.e. 60° overturned), a maximum stress dip (labelled $d\sigma_1$ in Fig. 7b) ranging from 30 to 90° is obtained.

At this stage, one may argue that the maximum stress was inclined only during the latest stage of folding, when pre-folding strata were overturned, while the maximum stress was sub-horizontal during most of the folding process. Such a scenario, in which the reorientation of the maximum stress is a discontinuous process, contrasts with the fact that in order to produce slipping along bedding surfaces, the maximum stress should have been south dipping, not only when strata were 60° overturned. In fact, dip-slip slipping along upright layers also requires the maximum stress to be south dipping, and such a stress configuration can also be extrapolated for steeply (e.g. > 75°) south-dipping strata. In agreement, we consider that the stress rotation was not a discontinuous process, but instead it has continuously operated during folding.

As documented in Tavani et al. (2011), the layer-parallel shortening pattern in the Sant Corneli-Bóixols anticline indicates a sub-horizontal maximum stress before folding and during the early stages of folding. As evidenced by data presented here, the stress was in a configuration not allowing faulting in the Vallcarga Group layers during almost the entire folding process. Such stress configuration was able to produce slipping along the bedding surfaces and along the unconformity though, with faulting in the Vallcarga and Areny Groups being almost negligible. Apart from those structures associated with the layer-parallel slipping, the few additional deformation structures point out a maximum stress oriented at low angle to bedding. During the late stage of folding, when the unconformity angle exceeded 120–130°, the maximum stress was south dipping, with an angle higher than 30°. Then, the stress attained a state allowing faulting within the Vallcarga Group layers. Contextually, bedding surfaces of the Vallcarga Group continued to be sheared, while shearing along the unconformity in the central limb arrested.

6.3 Flexural slipping and stress reorientation

The information discussed above argues for a syn-folding maximum stress rotation and/or reorientation within the growing anticline, from sub-horizontal early-folding layer-parallel shortening in Tavani et al. (2011) to south-dipping maximum stress in overturned strata documented here. As schematically illustrated in Fig. 7d, we infer that the sub-horizontal maximum stress applied to the leading syncline of the growing Sant Corneli-Bóixols anticline (i.e. the remotely applied stress has an Andersonian compressive configuration; Anderson, 1951) progressively rotated as it was transmitted across folding rock volumes affected by widespread flexural slipping. In agreement with the hypothesis outlined in the introduction, this process of stress deflection is interpreted here as associable with the flexural-slipping mechanism. In fact, as largely documented, slipping along low-friction faults produces the perturbation of the remotely applied stress field (e.g. Pollard and Segall, 1987; Soliva et al., 2010) and, in particular, reduction of the fault-parallel shear stress component causes the orientation of principal stress to locally rotate towards a fault-parallel direction. Consistently with this, the coupling between flexural slip and maximum stress reorientation documented in other structures has been attributed to the fact that slipping along closely spaced low-friction bedding surface imposes the maximum stress to orient at low angle to the slipping bedding surface over a wide area (i.e. the flexural-slip folded area), as mentioned in the introduction (Wiltschko et al., 1985; Ohlmacher and Aydin, 1997; Tavani et al., 2012). This concept fully applies to the data presented in this work until the strata attain a strongly overturned attitude. The close link between flexural slipping and stress reorientation also implies that the amount of deflection of the maximum compressive stress scales with the amount of flexural slipping. Accordingly, if the growth of an anticline occurs in a discontinuous fashion, the orientation of maximum compressive stress is expected to rotate repeatedly during the repeated pulses of flexural slip. In the case documented here, the absence of any indicators of a sub-

horizontal maximum stress could be related to the fact that Andersonian stress configuration would characterise stages in which the maximum stress is low and in a subcritical state, not allowing faulting and folding. Repeated pulses of maximum stress increase would instead cause the progressive slipping of bedding surfaces, with consequent maximum stress deflection.

7 Conclusions

This work has allowed the determination of the threshold dip value at which flexural slipping can be operative and of sufficient magnitude to deflect the maximum principal stress direction from its regional orientation. In the studied area, flexural slipping has been an active process, discontinuous at the local scale, but sufficiently dense in time and space at the scale of the growing fold for local stress deflection to occur. The mechanism operated up to 120° of dip (i.e. overturned bedding) and caused the maximum stress to progressively reorient at low angle to bedding until strata attained an overturned attitude.

Acknowledgements. The authors sincerely thank the reviewers, Ryan Shackleton, Richard Lisle, Juliet Crider, Hugo Ortner, and the editors for their constructive comments and suggestions. The cross sections presented in this work were constructed using 3-D Move software. This work is a contribution of the following research institutions: DISTAR from the Università degli Studi di Napoli "Federico II", the Institut de Recerca Geomodels and the Geodinàmica i Analisi de Conques research group (2014SGR467SGR) from the Agència de Gestió d'Ajuts Universitaris i de Recerca (AGAUR) and the Secretaria d'Universitats i Recerca del Departament d'Economia i Coneixement de la Generalitat de Catalunya.

Edited by: B. Grasemann

Competing interests. The authors declare that they have no conflict of interest.

References

Alonso, J. L.: Fold reactivation involving angular unconformable sequences: theoretical analysis and natural examples from the Cantabrian Zone (Northwest Spain), Tectonophysics, 170, 57–77, 1989.

Anderson, E. M.: The Dynamics of Faulting, Oliver and Boyd, Edinburgh, 1951.

Arbués, P., Pi, E., and Berástegui, X.: Relaciones entre la evolución sedimentaria del Grupo de Arén y el cabalgamiento de Bóixols (Campaniense terminal-Maastrichtiense del Pirineo meridional-central), Geogaceta, 20, 446–449, 1996.

Balsamo, F., Clemenzi, L., Storti, F., Mozafari, M., Solum, J., Swennen, R., Taberner, C., and Tueckmantel, C.: Anatomy and paleofluid evolution of laterally restricted extensional fault zones in the Jabal Qusaybah anticline, Salakh arch, Oman, Geol. Soc. Am. Bull., 128, 957–972, 2016.

Beamud, E., Muñoz, J. A., Fitzgeral, P. G., Baldwin, S. L., Garcés, M., Cabrera, L., and Metcalf, J. R.: Magnetostratigraphy and detrital apatite fission track thermochronology in syntectonic conglomerates: constraints on the exhumation of the South-Central Pyrenees, Basin Res., 23, 309–331, 2011.

Belayneh, M. and Cosgrove, J. W.: Fracture-pattern variations around a major fold and their implications regarding fracture prediction using limited data: an example from the Bristol Channel Basin, Geological Society, London, Special Publications, 231, 89–102, 2004.

Bellahsen, N., Fiore, P. E., and Pollard, D. D.: From spatial variation of fracture patterns to fold kinematics: A geomechanical approach, Geophys. Res. Lett., 33, L02301, doi:10.1029/2005GL024189, 2006.

Berástegui, X., García-Senz, J. M., and Losantos, M.: Tectosedimentary evolution of the Organya extensional basin (central south Pyrenean unit, Spain) during the Lower Cretaceous, Bulletin de la Societe Geologique de France, 8, 251–264, 1990.

Bond, R. M. G. and McClay, K. R.: Inversion of a Lower Cretaceous extensional basin, south central Pyrenees, Spain, Geological Society, London, Special Publications, 88, 415–431, 1995.

Brandes, C. and Tanner, D. C.: Fault-related folding: A review of kinematic models and their application, Earth-Sci. Rev., 138, 352–370, 2014.

Burbank, D., Meigs, A., and Brozović, N.: Interactions of growing folds and coeval depositional systems, Basin Res., 8, 199–223, 1996.

Burbank, D. W. and Anderson, R. S.: Tectonic geomorphology, Blackwell, Oxford, 2011.

Callot, J. P., Robion, P., Sassi, W., Guiton, M. L. E., Faure, J.-L., Daniel, J. M., Mengus, J.-M., and Schmitz, J.: Magnetic characterisation of folded aeolian sandstones: Interpretation of magnetic fabrics in diamagnetic rocks, Tectonophysics, 495, 230–245, 2010.

Choukroune, P., Roure, F., and Pinet, B.: Main results of the ECORS Pyrenees profile, Tectonophysics, 173, 411–423, 1990.

Cowie, P. A. and Scholz, C. H.: Displacement-length scaling relationship for faults: data synthesis and discussion, J. Struct. Geol., 14, 1149–1156, 1992.

Cuevas, J. L.: Estratigrafía del "Garumniense" de la Conca de Tremp. Prepirineo de Lérida, Acta geológica hispánica, 27, 95–108, 1992.

Dahlstrom, C. D. A.: Balanced cross sections, Can. J. Earth Sci., 6, 743–757, 1969.

Donath, F. A. and Parker, R. B.: Folds and folding, Geol. Soc. Am. Bull., 75, 45–62, 1964.

Engelder, T. and Geiser, P.: On the use of regional joint sets as trajectories of paleostress fields during the development of the Appalachian plateau, New York, J. Geophys. Res., 85, 6319–6341, 1980.

Erslev, E. A. and Mayborn, K. R.: Multiple geometries and modes of fault-propagation folding in the Canadian thrust belt, J. Struct. Geol., 19, 321–335, 1997.

Etchecopar, A., Vasseur, G., and Daignieres, M.: An inverse problem in microtectonics for the determination of stress tensors from fault striation analysis, J. Struct. Geol., 3, 51–65, 1981.

Fischer, M. P. and Wilkerson, M. S.: Predicting the orientation of joints from fold shape: results of pseudo-three-dimensional modcling and curvature analysis, Geology, 28, 15–18, 2000.

Ford, M., Williams, E. A., Artoni, A., Vergés, J., and Hardy, S.: Progressive evolution of a fault-related fold pair from growth strata geometries, Sant Llorenç de Morunys, SE Pyrenees, J. Struct. Geol., 19, 413–441, 1997.

Fowler, T. J.: Flexural-slip generated bedding-parallel veins from central Victoria, Australia, J. Struct. Geol., 18, 1399–1415, 1996.

García-Senz, J.: Cuencas extensivas del Cretácico Inferior en los Pirineos centrales. Formación y subsecuente inversión, PhD Thesis, 310 pp., Universitat de Barcelona, 2002.

Garrido, A.: Estudio geologico y relacion entre teclonica y sedimentacion del Secundario y Terciario de la vertiente meridional Pirenaica en su zona central, PhD Thesis, 395 pp., Universidad de Granada, 1973.

Gutiérrez, F., Carbonel, D., Kirkham, R. M., Guerrero, J., Lucha, P., and Matthews, V.: Can flexural-slip faults related to evaporite dissolution generate hazardous earthquakes? The case of the Grand Hogback monocline of west-central Colorado, Geol. Soc. Am. Bull., 126, 1481–1494, 2014.

Hancock, P. L.: Brittle microtectonics: principles and practice, J. Struct. Geol., 7, 437–457, 1985.

Lacombe, O.: Do fault slip data inversions actually yield "paleostresses" that can be compared with contemporary stresses? A critical discussion, C. R. Geosci., 344, 159–173, 2012.

Lanaja, J. M.: Contribución de la exploración petrolífera al conocimiento de la geología de España, 465 pp., IGME, Madrid, 1987.

Laubach, S. E.: Paleostress directions from the preferred orientation of closed microfractures (fluid-inclusion planes) in sandstone, East Texas basin, U.S.A., J. Struct. Geol., 11, 603–611, 1989.

Li, T., Chen, J., Thompson, J. A., Burbank, D. W., and Yang, X.: Active flexural-slip faulting: A study from the Pamir-Tian Shan convergent zone, NW China, J. Geophys. Res.-Sol. Ea., 120, 4359–4378, 2015.

Lisle, R. J.: Detection of zones of abnormal strains in structures using Gaussian curvature analysis, AAPG Bull., 78, 1811–1819, 1994.

Medwedeff, D. W.: Growth fault-bend folding at southeast Lost Hills, San Joaquin Valley, California, Bulletin of the American Association of Petroleum Geologists, 73, 54–67, 1989.

Mencos, J.: Metodologies de reconstrucció i modetització 3D d'estructures geològiques: anticlinal de Sant Corneli-Bóixols (Pirineus centrals), PhD Thesis, 277 pp., Universitat de Barcelona, 2011.

Mencos, J., Carrera, N., and Muñoz, J. A.: Influence of rift basin geometry on the subsequent postrift sedimentation and basin inversion: The Organyà Basin and the Bóixols thrust sheet (south central Pyrenees), Tectonics, 34, 1452–1474, doi:10.1002/2014TC003692, 2015.

Mitra, S.: Fault-propagation folds: geometry, kinematic evolution, and hydrocarbon traps, AAPG Bull., 74, 921–945, 1990.

Muñoz, J. A.: Evolution of a continental collision belt: ECORS-Pyrenees crustal balanced cross-section, in: Thrust tectonics, edited by: McClay, K. R., 235–246, Chapman & Hall, London, 1992.

Muñoz, J. A.: The Pyrenees, in: The Geology of Spain, edited by: Gibbons, W. and Moreno, T., 370–385, Geological Society, London, 2002.

Muñoz, J. A., Carrera, N., Mencos, J., Beamud, B., Perea, H., Arbués, P., Rivas, G., Bausà J., and Garcia-Senz, J.: Cartografia geològica del substrat prequaternari, Map 252-2-2 (66–22), Aramunt scale 1 : 25000, Institut Cartogràfic de Catalunya (ICC), Servei Geològic de Catalunya (IGC), Barcelona, 2010.

Nagtegaal, P. J. C.: Depositional history and clay minerals of the Upper Cretaceous basin in the South-Central Pyrenees, Spain, Leidse Geologische Mededelingen, 47, 251–275, 1972.

Nicol, A. and Nathan, S.: Folding and the formation of bedding-parallel faults on the western limb of Grey Valley Syncline near Blackball, New Zealand, New Zeal. J. Geol. Geop., 44, 127–135, 2001.

Ohlmacher, G. C. and Aydin, A.: Mechanics of vein, fault and solution surface formation in the Appalachian Valley and Ridge, northeastern Tennessee, USA: implications for fault friction, state of stress and fluid pressure, J. Struct. Geol., 19, 927–944, 1997.

Poblet, J., Storti, F., McClay, K., and Muñoz, J. A.: Geometries of syntectonic sediments associated with single-layer detachment folds, J. Struct. Geol., 19, 369–381, 1997.

Pollard, D. D. and Segall, P.: Theoretical displacements and stresses near fractures in rock: with applications to faults, joints, veins, dikes, and solution surfaces, Fracture mechanics of rock, Academic Press, London, 1987.

Price, N. J. and Cosgrove, J. W.: Analysis of Geological Structures, Cambridge, New York, Cambridge University Press, 511 pp., ISBN 0521265819, 1990.

Ramsay, J. G.: Folding and Fracturing of Rocks, McGraw-Hill, New York, 1967.

Roma, M., Arbués, P., Granado, P., Gratacós, O., and Muñoz, J. A.: The Sallent growth strata: an example of complex fold amplification mechanisms interacting with deepwater to continental sedimentation (South-Central Pyrenees, Spain), Abstracts with Poster, T4b, Sedimentation and Tectonics, 28th IAS Meeting of Sedimentology, 2011.

Rosell, J.: Sobre la existencia de la discordancia precenomaniense en el Pirineo de la provincia de Lerida, Notas y Comunicaciones del Instituto Geológico y Minero de España, 72, 71–80, 1963.

Shackleton, J. R., Cooke, M. L., and Sussman, A. J.: Evidence for temporally changing mechanical stratigraphy and effects on joint-network architecture, Geology, 33, 101–104, 2005.

Shackleton, J. R., Cooke, M. L., Vergés, J., and Simó, T.: Temporal constraints on fracturing associated with fault-related folding at Sant Corneli anticline, Spanish Pyrenees, J. Struct. Geol., 33, 5–19, 2011.

Simó, A.: Carbonate platform depositional sequences, Upper Cretaceous, South Central Pyrenees (Spain), Tectonophysics, 129, 205–231, 1986.

Soliva, R., Maerten, F., Petit, J. P., and Auzias, V.: Field evidences for the role of static friction on fracture orientation in extensional relays along strike-slip faults: Comparison with photoelasticity and 3-D numerical modeling, J. Struct. Geol., 32, 1721–1731, 2010.

Suppe, J.: Geometry and kinematics of fault–bend folding, Am. J. Sci., 283, 684–721, 1983.

Suppe, J. and Medwedeff, D. A.: Geometry and kinematics of fault-propagation folding, Eclogae Geol. Helv., 83, 409–454, 1990.

Suppe, J., Chou, G. T., and Hook, S. C.: Rates of folding and faulting determined from growth strata, in: Thrust tectonics, edited by: McClay, K. R., 105–121, Chapman & Hall, London, 1992.

Suppe, J., Sàbat, F., Muñoz, J. A., Poblet, J., Roca, E., and Vergés, J.: Bed-by-bed fold growth by kink-band migration: Sant Llorenç de Morunys, eastern Pyrenees, J. Struct. Geol., 19, 443–461, 1997.

Tanner, P. W.: The flexural-slip mechanism, J. Struct. Geol., 11, 635–655, 1989.

Tavani, S., Storti, F., Fernández, O., Muñoz, J. A., and Salvini, F.: 3-D deformation pattern analysis and evolution of the Anisclo anticline, southern Pyrenees, J. Struct. Geol., 28, 695–712, 2006.

Tavani, S., Mencos, J., Bausà, J., and Muñoz, J. A.: The fracture pattern of the Sant Corneli Bóixols oblique inversion anticline (Spanish Pyrenees), J. Struct. Geol., 33, 1662–1680, 2011.

Tavani, S., Storti, F., Bausà, J., and Muñoz, J. A.: Late thrusting extensional collapse at the mountain front of the northern Apennines (Italy), Tectonics, 31, TC4019, doi:10.1029/2011TC003059, 2012.

Tavani, S., Storti, F., Lacombe, O., Corradetti, A., Muñoz, J. A., and Mazzoli, S.: A review of deformation pattern templates in foreland basin systems and fold-and-thrust belts: Implications for the state of stress in the frontal regions of thrust wedges, Earth-Sci. Rev., 141, 82–104, 2015.

Teixell, A.: Crustal structure and orogenic material budget in the west central Pyrenees, Tectonics, 17, 395–406, 1998.

Tesei, T., Collettini, C., Vitic, C., and Barchi, M. R.: Fault architecture and deformation mechanisms in exhumed analogues of seismogenic carbonate-bearing thrusts, J. Struct. Geol., 55, 167–181, 2013.

Tugend, J., Manatschal, G., Kusznir, N. J., Masini, E., Mohn, G., and Thinon, I.: Formation and deformation of hyperextended rift systems: Insights from rift domain mapping in the Bay of Biscay-Pyrenees, Tectonics, 33, 1239–1276, 2014.

Vergés, J., Fernàndez, M., and Martínez, A.: The Pyrenean orogen: pre-, syn-, and post-collisional evolution, Journal of the Virtual Explorer, 8, 57–76, 2002.

Vitale, S., Zaghloul, M. N., Tramparulo, F. D. A, and El Ouaragli, B.: Deformation characterization of a regional thrust zone in the northernRif (Chefchaouen, Morocco), J. Geodyn., 77, 22–38, 2014.

Wiltschko, D. V., Medwedeff, D. A., and Millson, H. E.: Distribution and mechanisms of strain within rocks on the northwest ramp of Pine Mountain block, southern Appalachian foreland: a field test of theory, Geol. Soc. Am. Bull., 96, 426–435, 1985.

Zapata, T. R. and Allmendinger, R. W.: Growth stratal records of instantaneous and progressive limb rotation in the Precordillera thrust belt and Bermejo basin, Argentina, Tectonics, 15, 1065–1083, 1996.

2

A new theoretical interpretation of Archie's saturation exponent

Paul W. J. Glover

School of Earth and Environment, University of Leeds, Leeds, UK

Correspondence to: Paul W. J. Glover (p.w.j.glover@leeds.ac.uk)

Abstract. This paper describes the extension of the concepts of connectedness and conservation of connectedness that underlie the generalized Archie's law for n phases to the interpretation of the saturation exponent. It is shown that the saturation exponent as defined originally by Archie arises naturally from the generalized Archie's law. In the generalized Archie's law the saturation exponent of any given phase can be thought of as formally the same as the phase (i.e. cementation) exponent, but with respect to a reference subset of phases in a larger n-phase medium. Furthermore, the connectedness of each of the phases occupying a reference subset of an n-phase medium can be related to the connectedness of the subset itself by $G_i = G_{\mathrm{ref}} S_i^{n_i}$. This leads naturally to the idea of the term $S_i^{n_i}$ for each phase i being a fractional connectedness, where the fractional connectednesses of any given reference subset sum to unity in the same way that the connectednesses sum to unity for the whole medium. One of the implications of this theory is that the saturation exponent of any phase can be now be interpreted as the rate of change of the fractional connectedness with saturation and connectivity within the reference subset.

1 Introduction

Currently, there is no well-accepted physical interpretation of the saturation exponent other than qualitatively as some measure of the efficiency with which electrical flow takes place within the water occupying a partially saturated rock. Some might say that the meaning is not important as long as one can reliably obtain the water saturation of reservoir rocks with sufficient accuracy to calculate reserves. According to the 2016 BP Statistical Review of World Energy (BP, 2016), the world had proven oil reserves at the end of 2015 of 1.6976 trillion (million million) barrels (Tbbl.), slightly

down on the value at the end of 2014 (1.7 Tbbl.) and significantly above the respective values at the end of 1995 (1.1262 Tbbl.) and 2005 (1.3744 Tbbl.). The same source lists proven natural gas reserves of 186.9 trillion cubic metres (Tcm) at the end of 2015, slightly lower than at the end of 2014 (187.0 Tcm) and significantly and progressively higher than the values at the end of 1995 (119.9 Tcm) and 2005 (157.3 Tcm). This represents combined oil and gas reserves of approximately USD 78.4 trillion at end December 2015 prices (using WTI crude and Henry Hub).

Even a tiny uncertainty of, say, 0.01 in a saturation exponent of 2 (i.e. 0.5 % or 2 ± 0.01) would result in an error in the reserves of about USD ± 254.36 billion; the equivalent of 82 *Queen Elizabeth* class aircraft carriers or one mission to Mars. This calculation has been carried out by calculating the percentage change in hydrocarbon saturation resulting from an error of 2 ± 0.01 in the value of the saturation exponent. Since the calculated change in hydrocarbon saturation also depends on other parameters in Archie's equations, typical representative values for these parameters have been used; $R_T = 500\,\Omega\mathrm{m}$, $R_w = 1\,\Omega\mathrm{m}$, $\phi = 0.1$, and $m = 2$. When these values are used with $n = 2 \pm 0.01$, a change of ± 0.3245 % was calculated for the hydrocarbon saturation, allowing the change in global reserves to be calculated. However, the degree to which we can carry out the real calculations does not match this precision. Uncertainties in input parameters – over how representative seismic and petrophysical parameters are and difficulties with heterogeneity and anisotropy, to name but a few – result in the real calculations having uncertainties in the order of ± 20–40 %.

Within the hydrocarbon industry it is extremely common to assume that the saturation exponent is about 2 for most rocks. However, it is worthwhile thinking about the USD 254 billion global shortfall in revenue if it really is equal to 2.01 instead. These frightening, large financial values make it extremely important that the physical interpretation of the saturation exponent in the classical Archie's law is well understood. This paper attempts to provide a new theoretical and physical interpretation.

The classical Archie's laws (Archie, 1942) link the electrical resistivity of a rock to its porosity, to the resistivity of the water saturating its pores, and to the fractional saturation of the pore space with the water. They have been used for many years to calculate the hydrocarbon saturation of the reservoir rock and hence hydrocarbon reserves. The classical Archie's laws contain two exponents, m and n, which Archie called the cementation exponent and the saturation exponent, respectively. The conductivity of the hydrocarbon-saturated rock is highly sensitive to changes in either exponent.

Like the cementation exponent, and despite its importance to reserves calculations, the physical meaning of the saturation exponent is difficult to understand from a physical point of view, which leads to petrophysicists not giving it the respect it deserves. It is common, for example, to hear that, in the absence of laboratory measurements, the saturation exponent has been taken to be equal to 2, which it has just been noted is bound to lead to gross errors. While it is true that there seems to be a strong preference for values of saturation exponent near 2 ± 0.5 for most water-wet rocks, oil-wet rocks show much higher values (4–5) (Montaron, 2009; Sweeney and Jennings, 1960), and there is evidence that the saturation exponent changes with saturation, with the type of rock microstructure, and with saturation history, leading to hysteresis in the plot of resistivity index as a function of water saturation.

When a saturation exponent is derived from laboratory measurements, it is commonly done by fitting a straight line to resistivity data where the y axis is the logarithm of the resistivity index and the x-axis is the logarithm of the water saturation. The resistivity index is the ratio of the measured rock resistivity at a given water saturation S_w divided by the resistivity of the same rock when the pore space is completely saturated with water (i.e. $S_w = 1$). The problem is that the saturation exponent varies with water saturation, becoming significantly smaller at low saturations, leading to an uncertainty in which value to use. This observation also gives us the first hint that it is the connectedness of the water phase that is controlling the saturation exponent just as it did the phase exponent in the generalized Archie's law.

It is clear that the physical understanding of the saturation exponent needs to be improved. The purpose of this paper is to investigate the elusive physical meaning of the saturation exponent, where it is shown that the saturation exponents are intimately linked to the phase exponents in the generalized Archie's model.

2 Traditional interpretations

Considering the classical form of Archie's laws; the first Archie's law relates the formation factor F, which is the ratio of the resistivity of a fully saturated rock $\rho_o (R_o)$ to the resistivity of the fluid occupying its pores $\rho_f (R_w)$, to the rock porosity ϕ and a parameter he called the cementation exponent m, where the symbols in parentheses are those traditionally used in the hydrocarbon industry. Archie's first law can be expressed as $F = \rho_o / \rho_f = \phi^{-m}$ using resistivities (Archie, 1942) or as $G = \sigma_o / \sigma_f = \phi^{+m}$ using conductivities. In the latter case, G is called the conductivity formation factor or the connectedness (Glover, 2009). It can easily be seen that the effective resistivity and effective conductivity of the fully saturated rock can be expressed as $\rho_o = \rho_f \phi^{-m}$ and $\sigma_o = \sigma_f \phi^{+m}$ using resistivities or conductivities, respectively. It should be noted that this work does not consider the form of Archie's law which includes the so-called "tortuosity factor" a, which was developed by Winsauer et al. (1952). The role of this parameter is discussed fully in Glover (2016).

Archie's second law considers that the rock is not fully saturated with a conductive fluid but is partially saturated with a fractional water saturation S_w. It relates the resistivity index I, which is the ratio of the resistivity of a partially saturated rock ρ_{eff} to the resistivity of the fully saturated rock ρ_o, to the water saturation S_w and a parameter he called the saturation exponent n. Archie's second law can be expressed as $I = \rho_{\text{eff}} / \rho_o = S_w^{-n}$ using resistivities or $1/I = \sigma_{\text{eff}} / \sigma_o = S_w^{+n}$ using conductivities.

The two laws may be combined to give $\rho_{\text{eff}} = \rho_f \phi^{-m} S_w^{-n}$ using resistivities and $\sigma_{\text{eff}} = \sigma_f \phi^{+m} S_w^{+n}$ if conductivities are used. In reserves calculations, the resistivity of the partially saturated rock, the resistivity of the pore water, the porosity of the rock, and the two exponents are "known" from logging or laboratory measurements. This enables the water saturation S_w and hence the hydrocarbon saturation $S_h = (1 - S_w)$ and, consequently, the reserves to be calculated.

Archie's laws require that both the rock matrix and all but one of the fluid phases that occupy the pores have infinite resistivity. Hence, it is a model for the distribution of one conducting phase (the pore water) within a rock sample consisting of a non-conducting matrix and other fluids which also have zero or negligible conductivity. Problems arise when there are other conducting phases in the rock, such as clay minerals. These problems have generated a huge amount of research in the past (e.g. Waxman and Smits, 1968; Clavier et al., 1984), which is reviewed in Glover (2015). The classical Archie's laws were based upon experimental determinations. However, there has been progressive theoretical work (Sen et al., 1981; Mendelson and Cohen, 1982) showing that for at least some values of cementation exponent, Archie's law has a theoretical pedigree, while hinting that the law may be truly theoretical for all physical values of cementation exponent. A study has recently shown that the Winsauer et al. (1952) modification to Archie's law is only needed to compensate for systematic errors in the measurement of its input parameters and has no theoretical basis (Glover, 2016). Meanwhile, independent modifications to the original Archie's law have allowed it to be used when both the pore fill and the matrix have significant electrical conductivities (Glover et al., 2000a; Glover, 2009), such as the case when a rock melt occupies spaces between a solid matrix in the lower crust (Glover et al., 2000b). This has culminated in a generalized Archie's law which is valid for any number of conductive phases in the three-dimensional medium and which was published in 2010 (Glover, 2010).

3   The generalized Archie's law

The generalized Archie's law (Glover, 2010) extends the classical Archie's law to a porous medium containing n phases. It is based on the same concept of connectedness that was introduced in the present author's previous interpretation of the cementation exponent (Glover, 2009). It should be noted that from this point in this paper the symbol ϕ refers not just to the porosity of the rock but to the volume fraction of a particular phase, whether it be the matrix, the water, hydrocarbon or whatever other phase may be present. It will either be used for a specific phase such as water (e.g. ϕ_f) or for a set of phases (e.g. ϕ_i). The unsubscripted symbol continues to refer to conventional porosity, where $\phi = \sum_i \phi_i - \phi_m$; ϕ_m is the phase fraction of the rock matrix (conventionally equal to $1 - \phi$). Occasionally, the unsubscripted symbol will also be used when the general properties of phase fractions are being discussed, such as in the following two equations.

In the 2009 paper the connectedness was defined as

$$G \equiv \frac{\sigma_o}{\sigma_w} = \frac{1}{F} = \phi^m, \tag{1}$$

where F is the formation factor. The connectedness of a given phase is a physical measure of the availability of pathways for conduction through that phase. The connectedness is the ratio of the measured conductivity to the maximum conductivity possible with that phase (i.e. when that phase occupies the whole sample). This implies that the connectedness of a sample composed of a single phase is unity. Connectedness is not the same as connectivity. The connectivity is defined as the measure of how the pore space is arranged in its most general sense as that distribution in space which makes the contribution of the specific conductivity of the material express itself as a different conductance (see Glover, 2010). The connectivity is given by $\chi = \phi^{m-1}$ and depends upon the porosity and the classical Archie's cementation exponent m. It should be noted that the connectedness is also given by

$$G = \phi\chi, \tag{2}$$

and then it becomes clear that the connectedness depends both upon the amount of pore space (given by the porosity)

and the arrangement of that pore space (given by the connectivity).

The generalized Archie's law was derived by Glover (2010) and is given by

$$\sigma = \sum_i \sigma_i \phi_i^{m_i} \text{ with } \sum_{i=1} \phi_i = 1, \tag{3}$$

where there are n phases, each with a conductivity σ_i, a phase volume fraction ϕ_i, and an exponent m_i. The porosity and cementation exponent in the classical Archie's law are the same as the pore space phase volume fraction and pore space phase exponent in the generalized Archie's law, respectively. However, the pore space and the matrix may be subdivided into any number of other phases as required. Indeed, the generalized Archie's law will not contain a term that represents the pore space unless the pore space is only occupied by a single phase.

In the generalized law the phase exponents can take any value from 0 to ∞. Values less than unity represent a phase with an extremely high degree of connectedness, such as that for the solid matrix of a rock. Connectedness decreases as the phase exponent increases. Phase exponents that tend towards 1 are associated with a highly connected phase which is analogous to the low cementation exponents occurring in the traditional Archie's law for networks of high aspect ratio cracks. Phase exponents about 2 represent the degree of connectedness that one might find when the phase is partially connected in a similar way to which the pore network in a sandstone is connected and which is, again, analogous to that scenario in the traditional Archie's law. By extension, higher values of phase exponents represent lower phase connectedness, such as that in the traditional Archie's law for the pores in a vuggy limestone.

It is clear that the classical and generalized laws share the property that the exponents modify the volume fraction of the relevant phase with respect to the total volume of the rock. However the exponents in the generalized law differ from the classical exponent because some of them have values which are not measurable because their phases are composed of materials with negligible conductivity. Despite this, each phase has a well-defined exponent providing (i) it has a non-zero volume fraction and (ii) the other phases are well-defined.

It should be noted that higher phase exponents tend to be related to lower phase fractions, although this relationship is not implicit in the generalized Archie's law as it is currently formulated.

The generalized Archie's law as formulated by Glover (2010) hinges upon the proposal that the sum of the connectednesses of the phases in a three-dimensional n-phase medium is given by

$$\sum_i \phi_i^{m_i} = \sum_i G_i = 1. \tag{4}$$

It is important to consider Eqs. (1) and (4) together to develop a fuller understanding of the model. There is an infinite number of solutions to Eq. (4) even in the most restrictive two-phase system. However, there is only a small subset of solutions if both Eqs. (1) and (4) are to be fulfilled together, as the model requires. The problem of having enough degrees of freedom is not problematic for three phases or more and is trivial for one phase. Consequently, if there is to be a problem with the Glover (2010) model, it should be clearest for a two-phase system.

Considering a two-phase system, Eq. (1) gives $\phi_1 = 1 - \phi_2$ while Eq. (4) can be written as $\phi_1^{m_1} + \phi_2^{m_2} = 1$. Substituting, we obtain either $(1 - \phi_2)^{m_1} + \phi_2^{m_2} = 1$ or $(1 - \phi_1)^{m_2} + \phi_1^{m_1} = 1$. These equations are formally the same. They each have trivial solutions when each of the volume fractions tends to unity, the other volume fraction consequently tending to 0. Another solution occurs when $m_1 = m_2 = 1$, which is the simple parallel conduction model. Only one other solution exists for the general case where the volume fractions are variable, and that requires $m_1 > 1$ when $m_2 < 1$ or vice versa. Consequently, the non-trivial solution for a two-phase medium falls into one of the following classes:

i. $m_1 = m_2 = 1$. The phases, whatever their volume fractions, are arranged in parallel and both have a unity exponent.

ii. $m_1 > 1$ and $m_2 < 1$. This implies that Phase 1 has a path across the 3-D medium that is less connected than a parallel arrangement of that phase. Since we have a two-phase medium, Phase 2 must have a path across the medium which is more connected than a parallel arrangement, hence forcing $m_2 < 1$.

iii. $m_1 < 1$ and $m_2 > 1$. Since the system is symmetric. This scenario is formally the same as (ii) above, but with the phase numbers switched around.

Consequently, for a two-phase medium, defining the porosity and connectedness (or exponent) of one of the phases immediately fully defines the other phase. For higher numbers of phases, there are more solutions, but if the porosity and connectedness (or exponent) of $n - 1$ of the phases is known, the nth phase is also fully defined in the same way. The logical extension of this idea is that both the sum of the volume fractions of the n phases is unity and the sum of the connectednesses of the n phases is also unity or that both volume fraction and connectedness are conserved in a three-dimensional n-phase mixture.

Another, more intuitive way of looking at this is as follows. It has already been shown that the connectedness of a system that contains only one phase is unity as a result of Eq. (1); i.e. if there is one phase, $\phi = 1$ and hence $G = 1$. Let us imagine that a second phase is introduced. Intuitively, it seems reasonable that as the phase fraction of the new phase increases, its connectedness will increase and that when this happens both the volume fraction and connectedness of the first phase will decrease. The same would be true if any number of new

phases were introduced – all the phases would compete for a fixed amount of connectedness, its increase for one phase being balanced by a decrease in at least one of the other phases. In other words there is a fixed maximum amount of connectedness possible in a three-dimensional sample, expressed by Glover (2010) as Eq. (4).

Figure 1 is an illustrative example of the idea of a fixed amount of connectedness, using a 2-D slice for simplicity and clarity. Hence, Fig. 1 shows a two-dimensional slice through a 3-D four-phase water-wet medium composed of detrital quartz grains, a string of clay, and a porosity that is partially filled with water, at near irreducible saturation and oil. The figure should be read in two columns. The left-hand column shows an arbitrary arrangement of the four phases that together completely make up the medium (Fig. 1a). In this case I have chosen to represent the detrital quartz as sub-angular detrital grains with a grain size distribution, the clay as a stringer, the near-irreducible water as covering the quartz grain surfaces and the oil as occupying the centre parts of the pores as these geometries can be found in typical water-wet shaly sandstone reservoirs. It should be noted, however, that the equations make no such distinction and what follows is true for any geometrical set of four phases composing the 3-D medium completely. Reading downwards, panels (c), (e), (g), and (i) show each of the quartz, clay, water, and oil phases alone and respectively. One can imagine that each phase has a certain phase fraction and a certain connectedness. Some of the phases look disconnected in the figure, but it should be remembered that there will be a greater connectedness in reality because there will be connection in the third dimension that is not shown in the figure. If we imagine hydraulic flow or electrical flow from the bottom to the top of the medium, the quartz seems to have a relatively high phase fraction and a moderate connectedness, the clay seems to have a moderate phase fraction and a high connectedness, the water seems to have a low phase fraction but a relatively high connectedness due to the multiple pathways formed by the thin "ribbons" of water, and the oil has a moderate phase fraction but a relatively low connectedness as the patches of oil are relatively isolated. The right-hand part of the figure represents the same medium but with the small addition of a quartz grain, labelled "Q", and its accompanying thin film of surface water. The addition of this makes a minuscule increase in the phase fractions of the detrital quartz and water phase fractions, and, literally, an equally small decrease in the phase fractions of the clay and oil. Reading the distributions for the quartz, clay, water, and oil phases alone (panels (d), (f), (h), and (j)) shows that the addition has made a significant increase in the connectedness of the quartz as well as some increase in that of the water, which was well connected anyway. The low connectedness of the oil will have changed little, but the addition has blocked the main pathway through the clay, leaving only a minor secondary pathway and consequently resulting in a significant decrease in the clay connectedness. Consequently, Fig. 1 shows the prin-

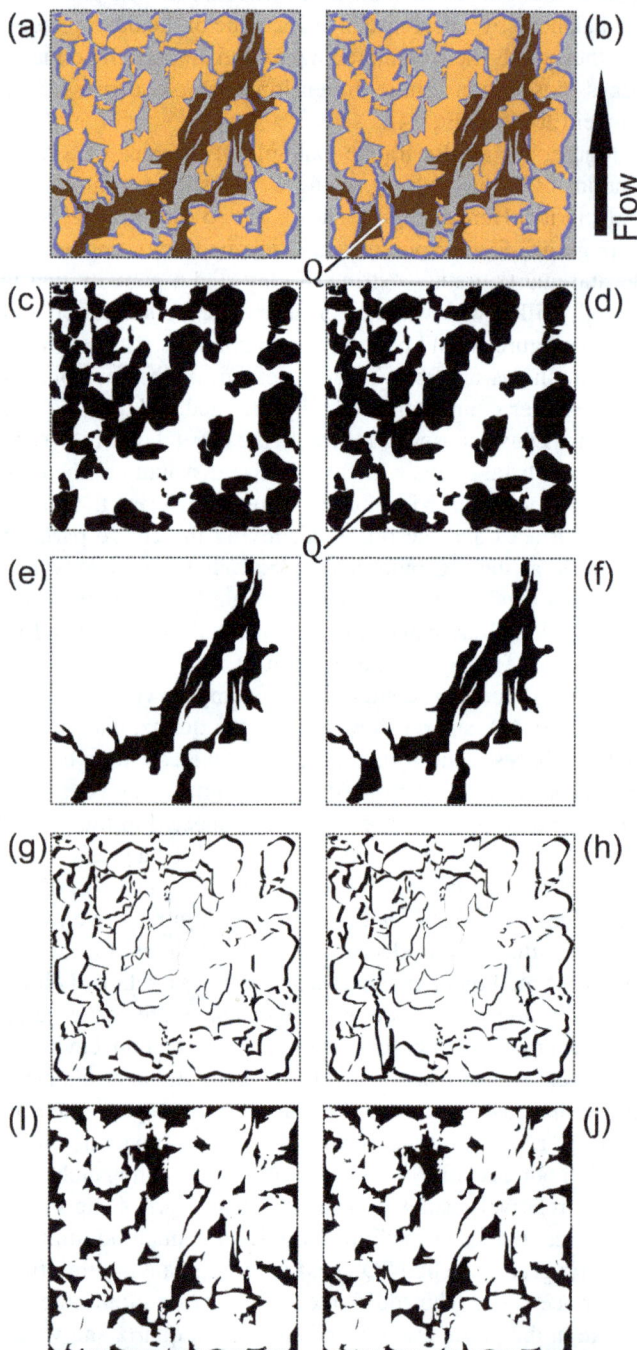

Figure 1. Distribution of a four-phase clay-rich, water-wet sandstone saturated with water and oil (quartz – orange; clay – brown; water – blue; oil – grey) represented by a 2-D slice through a 3-D medium. The left-hand column differs from the right-hand column by the addition of a single grain of quartz with its associated surface water, labelled Q. Consequently, the figure should be read vertically comparing the two columns: **(a, b)** complete medium; **(c, d)** quartz distribution; **(e, f)** clay distribution; **(g, h)** water distribution; and **(i, j)** oil distribution.

ciple behind the idea of the conservation of connectedness given in Eq. (4) but not a proof, the latter of which is considered in Glover (2010).

In summary, both the sum of the volume fractions and the sum of the connectednesses of the phases composing a 3-D medium is equal to unity. The corollary is that connectedness is conserved; if the connectedness of one phase diminishes, there must be an increase in the connectedness of one or more of the other phases to balance it.

It is interesting to consider the role of percolation effects within the generalized model (see Glover, 2010, for a full treatment). In percolation theory, the bulk value of a given transport property is only perturbed by the presence of a given phase with a well-defined phase conductivity after a certain phase volume fraction has been attained. This critical volume fraction is called the percolation threshold. This works well for a two-phase system when one phase is non-conductive, with a percolation threshold occurring near the 0.3316 to 0.342 (Montaron, 2009). For such a system, consisting of one non-conducting and one conducting phase, the effective conductivity of the medium depends only on the conductivity of the conducting phase, its volume fraction, and how connected it is. It is intuitive, therefore, that there may exist a phase volume fraction below which the conducting phase is not connected and for which the resulting effective conductivity will be zero. The concept of a percolation factor becomes unclear if the matrix phase has a non-zero conductivity or one or more additional, either solid or fluid conducting phases are added. Under these circumstances a percolation threshold may not exist. Glover (2010) went further than this claiming that Eq. (4) in this work (which is Eq. 26 in Glover, 2010) contains enough information to make the explicit inclusion of percolation effects unnecessary.

4 Origin of the saturation exponent

Within the framework of the classical Archie's laws, it is possible to envisage the cementation exponent as controlling how the porosity is connected within the rock sample volume and to envisage the saturation exponent as controlling how the water is connected within that porosity. The cementation exponent is defined relative to the total volume of the rock, while the saturation exponent is defined relative to the pore space, which is a subset of the whole rock. This is an important concept for what follows.

The water is one of two phases within the porosity, while that porosity is one of two phases within the rock. Hence, there exists a three-phase system to which the generalized Archie's law can be applied. In fact, the generalized Archie's law can be used to show that the saturation exponents arise naturally and have a physical meaning: they are defined in the same way as the phase exponents but are expressed relative to the pore space instead of the whole rock.

By writing the generalized law (Eq. 4) for three defined phases – let us say matrix, water, and hydrocarbon gas – and assuming that neither the matrix nor the gas is conductive, i.e. $\sigma_m = 0$ and $\sigma_h = 0$, but allowing the pore space to be partially saturated with water such that $\phi_h \neq 0$, it is possible to obtain $\sigma_{\text{eff}} = \sigma_f \phi_f^{m_f}$. This is a re-expression of Eq. (4), which is the sum of three terms, one for each phase, two of which are 0 because the conductivity of the material which makes up each of those is 0 (i.e. the matrix and hydrocarbon). The exponent m_f is the phase exponent of the fluid phase, which is the only phase contributing to the effective conductivity of the three-phase medium. Since $\phi_h \neq 0$, the pore space is partially saturated with hydrocarbon and partially saturated with water. It is also possible to write $\phi_f = \phi S_w$ and hence obtain

$$\sigma_{\text{eff}} = \sigma_f \phi^{m_f} S_w^{m_f}. \tag{5}$$

Comparison with the classical Archie's laws, which can be written as $\sigma_{\text{eff}} = \sigma_f \phi^m S_w^n$ (Tiab and Donaldson, 2004), shows structural similarity. However, the exponent m_f in Eq. (5) is expressed relative to the whole rock because it is the phase exponent for the fluid that appears in Eq. (4). By contrast, although the cementation exponent m in the classical first Archie's law is expressed relative to the whole rock, the saturation exponent n is related to the pore space, which is a subset of the whole rock. The distinction between whether the exponent is expressed relative to the whole rock or relative to a subset of the rock, such as the pore space, can be made easily by imagining whether the saturation exponent is independent of any changes one might make to the rock matrix. In this case, it is possible to see that the saturation exponent is independent of the rock matrix and is only sensitive to changes occurring within the pore space. Consequently, it is expressed relative to the pore space rather than the whole rock.

Accordingly, both equations provide a valid measure of the effective rock conductivity, so they may be equated as $\sigma_f \phi^{m_f} S_w^{m_f} = \sigma_{\text{eff}} = \sigma_f \phi^m S_w^n$, hence resulting in $\phi^m = \phi^{m_f} S_w^{(m_f - n)}$. It can be recognized that the classical Archie's saturation exponent refers to saturation with water and is hence renamed as n_f, giving

$$\phi^m = \phi^{m_f} S_w^{(m_f - n_f)}. \tag{6}$$

It is important to realize that the exponent n_f is a "saturation" exponent that refers to the arrangement of the water phase within the pore space. In other words it is expressed with respect to the pore space, not the whole rock, and is found experimentally by varying the saturation of the water in the pore space, the latter of which is assumed to always remain unchanged.

Now it is possible to write Eq. (6) in terms of connectednesses. The left-hand side of Eq. (6) is simply the connectedness of the pore space, as defined by Eq. (1). It is the phase

volume fraction of the pore space, i.e. the classical porosity, raised to the power of the phase exponent that contains the information about how that pore space is distributed, which is the classical cementation exponent m. Consequently, we can write $G_{\text{pore}} = \phi^m$, and Eq. (6) becomes

$$G_{\text{pore}} = \phi^{m_f} S_w^{(m_f - n_f)}. \tag{7}$$

The right-hand side of the equation may be rewritten as $(\phi S_w)^{m_f} / S_w^{n_f}$, which allows Eq. (7) to be written as

$$G_{\text{pore}} S_w^{n_f} = (\phi S_w)^{m_f}. \tag{8}$$

The term in brackets is simply the phase fraction of the water with respect to the whole rock, i.e. $\phi_f = \phi S_w$, and the exponent m_f is simply the phase exponent of the fluid phase with respect to the whole rock. Consequently, Eq. (1) can be applied for the fluid phase leading to

$$G_f = \phi_f^{m_f} = (\phi S_w)^{m_f}, \tag{9}$$

which, when substituted into Eq. (8) and rearranged, gives

$$G_f = G_{\text{pore}} S_w^{n_f}. \tag{10}$$

This equation is for one fluid phase, i.e. water, occupying the pore space. Since the system is symmetric, Eq. (10) can be generalized for any of the fluid phases occupying the pore space

$$G_j = G_{\text{pore}} S_j^{n_j}, \tag{11}$$

where G_j is the connectedness of fluid j, S_j is its saturation, and the exponent n_j is a saturation exponent that refers to the arrangement of the water phase within the pore space. In other words n_j is expressed with respect to the pore space, not the whole rock.

However, there is nothing geometrically special about the entity we call the pore space or any distinction between solid and fluid phases that compose the whole rock. Consequently, Eq. (11) is only a partial generalization, and it is possible to extend the result in Eq. (10) to any phase of i phases composing a three-dimensional medium each of which partially or fully occupies a saturation S_i of a subset of the medium whose connectedness is given as G_{ref}, according to

$$G_i = G_{\text{ref}} S_i^{n_i}. \tag{12}$$

The pore connectedness is relabelled as the reference connectedness because the equation is valid not only for multiple phases that fill the porosity but multiple phases composing any other phase.

Equation (12) gives the connectedness of the ith phase in an n-phase 3-D medium as depending on both its fractional saturation S_i within a larger volume which has a connectedness G_{ref} and that reference connectedness. The distribution of that saturation is taken into account by the exponent n_i, which will have a general functional form.

If one considers the whole 3-D n-phase medium (i.e. one where $\sum_i \phi_i = 1$), Eq. (1) states that the connectedness of each phase is the volume fraction of that phase raised to the value of its phase exponent, and Eq. (4) states that the sum of those connectednesses is unity.

If a subset of a whole n-phase medium (i.e. one where $\sum_i \phi_i < 1$) is considered and labelled the reference subset, the reference subset will have a connectedness $G_{\mathrm{ref}} = \phi_{\mathrm{ref}}^{m_{\mathrm{ref}}}$ relative to the whole rock, and the connectedness of any phase which partially occupies the reference subset (e.g. water within the pore space, clay within the rock matrix) is equal to the connectedness of the reference phase multiplied by the volume fraction of the phase within the reference subset (i.e. the saturation relative to the reference subset) raised to the value of its saturation exponent.

The definition above is somewhat complex due to the requirement to be both completely general and precise and due to the fact that there are two reference frames here. The first is the whole 3-D n-phase medium. The second is the 3-D reference subset which may contain between two and $n-1$ phases. Conversion between the two reference frames can be carried out using the relationship

$$\phi_i^{m_i - n_i} = \phi_{\mathrm{ref}}^{m_{\mathrm{ref}} - n_i}. \tag{13}$$

It can also be shown that (Glover, 2010)

$$\sum_i S_i^{n_i} = 1, \tag{14}$$

where the sum is carried out over all the phases within the reference subset.

It should be noted that Eq. (14) is formally the same as Eq. (4) except that Eq. (14) is valid for the reference subset of phases, while Eq. (4) is valid for the whole n-phase medium. Hence, it is possible to use $S_i = \phi_i / \phi_{\mathrm{ref}}$ to write both Eqs. (4) and (14) as

$$\sum_i \left(\frac{\phi_i}{\phi_{\mathrm{ref}}} \right)^{m_i} = 1. \tag{15}$$

For a whole n-phase medium, $\phi_{\mathrm{ref}} = 1$ and Eq. (15) becomes equal to Eq. (4). For a subset of the n-phase medium, $\phi_{\mathrm{ref}} < 1$ and Eq. (15) becomes equal to Eq. (14).

The distinction between the phase exponent and saturation exponent becomes trivial; they each control how connected the phase is relative to the reference volume fraction. In other words, the transformation

$$1 \leftrightarrow \phi_{\mathrm{ref}} \text{ leads to } \phi_i \leftrightarrow S_i \text{ and } m_i \leftrightarrow n_i. \tag{16}$$

Figure 2 illustrates the concept of a subset of an n-phase medium using a 2-D slice from a 3-D medium. Figure 2a shows a simple two-phase situation, where Phase 1 is brown and Phase 2 is yellow. Both phases are connected across the

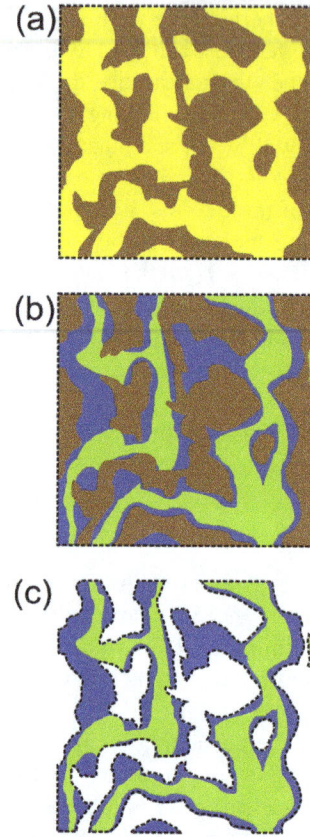

Figure 2. Sets and subsets of a three-phase medium using a 2-D slice to represent the whole 3-D medium. **(a)** Two phases: Phase 1, brown, representing solid matrix; Phase 2, yellow, represents pore space, with unspecified fill. Phase fractions and connectednesses can be defined for each phase with respect to the whole medium (dotted box). **(b)** Three phases created by filling (replacing) the porosity with two phases: Phase 1, brown, representing solid matrix as before; Phase 2, blue, representing water; Phase 3, green, representing oil. Phase fractions and connectednesses can be defined for each of the three phases with respect to the whole medium (dotted box). **(c)** If only the pore space is considered by considering Phase 1 to be unchanging, what remains is a two-phase subset of the three-phase situation. Phase fractions and connectednesses can be defined for the two fluid phases with respect to the subset which is the porosity (inside the dotted interface).

medium from top to bottom, and were they not in the 2-D slice, they would likely be connected through the third dimension. Phase 1 (brown) can be considered to be the solid matrix of a rock, and Phase 2 (yellow) is considered to be the pore spaces in the rock for the purposes of this illustration, but the distinction is arbitrary. The rock matrix has a phase fraction ϕ_1 and a connectedness $G_1 = \phi_1^{m_1}$ and the pore space has a phase fraction ϕ_2 and a connectedness $G_2 = \phi_2^{m_2}$ (Eq. 1). Both of these are expressed with respect to the whole medium, which is bounded in the figure by the dotted box. Consequently, $\phi_1 + \phi_2 = 1$ and $G_1 + G_2 = 1$ (Eqs. 3 and 4).

The pore space may be occupied by any number of miscible or immiscible fluids. Let us assume there are two immiscible fluids completely occupying the pores, which are water and oil and which we will assign the names Phase 3 and Phase 4. Figure 2b shows this situation. Once again, the phase fraction and connectedness of each of the three phases that compose the medium can be defined as phase fractions ϕ_1, ϕ_3, and ϕ_4 and $G_1 = \phi_1^{m_1}$, $G_3 = \phi_3^{m_3}$, and $G_4 = \phi_4^{m_4}$ for the solid matrix, water, and oil, respectively. Since these parameters are being considered with respect to the whole medium, it is possible to write $\sum_{i=1,3,4} \phi_i = 1$ and $\sum_{i=1,3,4} G_i = 1$.

However, it is possible to use a different reference medium for calculations. For example, the classical Archie's second law is expressed in terms of saturations and uses the pore space as a reference space in order to express the amount of water and hydrocarbons not with respect to the total volume of the rock but as a fraction of the pore space. Let us, therefore, also take the pore space as a convenient reference sub-space of the whole medium. This situation is shown in Fig. 2c, where the dotted line delineated the extent of the reference space. In this space, (i) what was the whole medium, represented by unity in the transform given in Eq. (16), becomes the volume fraction of the reference space $1 \leftrightarrow \phi_{ref}$ (i.e. the pore space in this example), (ii) the volumes of the different phases are more efficiently described using saturations Si with respect to the reference space (i.e. the pore space) than using phase volume fractions which are defined relative to the whole medium $\phi_i \leftrightarrow S_i$, and (iii) the whole-medium connectednesses $G_i = \phi_i^{m_i}$ are replaced by the entity $S_i^{n_i}$, which uses the saturation exponent in place of the phase exponent $m_i \leftrightarrow n_i$. It will be seen that the entity $S_i^{n_i}$ has its own properties in the next section and will be labelled the fractional connectedness. Topologically, the occupation of the fluids within the pore space (Fig. 2c) is identical to the occupation of the whole medium by the matrix and pore space (Fig. 2a), which leads to the symmetry in the mathematical equations.

The transformation given in Eq. (16) is perhaps not immediately clear when expressed in these most general terms. Let us take an illustrative example. Imagine a three-dimensional five-phase medium where the phases are (i) detrital quartz (dq), (ii) calcite cement (cc), (iii) distributed clay (dc), (iv) saline water (sw), and (v) hydrocarbon gas (hg), where the subscripts that will be used for each phase are given in parentheses. First let us consider the whole medium (i.e. $\phi_{ref} = 1$). Each of the phase volume fractions is given by $\phi_{dq}, \phi_{cc}, \phi_{dc}, \phi_{sw}$, and ϕ_{hg}, respectively. Each of their connectednesses is equal to their phase volume fraction raised to the power of their phase exponents (according to Eq. 1), where the phase exponents contain the information about how each of the five phases is distributed in the medium. The connectednesses are $G_{dq} = \phi_{dq}^{m_{dq}}$, $G_{cc} = \phi_{cc}^{m_{cc}}$, $G_{dc} = \phi_{dc}^{m_{dc}}$, $G_{sw} = \phi_{sw}^{m_{sw}}$, and $G_{hg} = \phi_{hg}^{m_{hg}}$. Equation (15) can be used,

setting $\phi_{ref} = 1$, to give

$$\phi_{dq}^{m_{dq}} + \phi_{cc}^{m_{cc}} + \phi_{dc}^{m_{dc}} + \phi_{sw}^{m_{sw}} + \phi_{hg}^{m_{hg}} = 1. \tag{17}$$

This is the same result as applying Eq. (4) directly. It is expressed in terms of the parameters (i) $\phi_{ref} = 1$ (i.e. the whole medium), (ii) individual phase fractions (ϕ_i), and (iii) individual phase exponents (m_i); the latter two are expressed relative to the whole medium. These are the conditions and parameters expressed by the left-hand components of the transformation given by Eq. (16).

Now consider the subset of the whole medium which comprises just its solid parts. The reference fraction ϕ_{ref} is the sum of the solid-phase fractions (i.e. $\phi_{dq}^{m_{dq}} + \phi_{cc}^{m_{cc}} + \phi_{dc}^{m_{dc}}$), which is less than unity. Rewriting Eq. (15) for the reference subset gives

$$\left(\frac{\phi_{dq}^{m_{dq}}}{\phi_{dq}^{m_{dq}} + \phi_{cc}^{m_{cc}} + \phi_{dc}^{m_{dc}}} \right)^{n_{dq}} + \left(\frac{\phi_{cc}^{m_{cc}}}{\phi_{dq}^{m_{dq}} + \phi_{cc}^{m_{cc}} + \phi_{dc}^{m_{dc}}} \right)^{n_{cc}}$$
$$+ \left(\frac{\phi_{dc}^{m_{dc}}}{\phi_{dq}^{m_{dq}} + \phi_{cc}^{m_{cc}} + \phi_{dc}^{m_{dc}}} \right)^{n_{dc}} = 1, \tag{18}$$

which can be written in terms of "saturations" (i.e. fractional volumes of the reference subset) as

$$S_{dq}^{n_{dq}} + S_{cc}^{n_{cc}} + S_{dc}^{n_{dc}} = 1, \tag{19}$$

because $S_{dq} = \phi_{dq}^{m_{dq}} / \left(\phi_{dq}^{m_{dq}} + \phi_{cc}^{m_{cc}} + \phi_{dc}^{m_{dc}} \right)$, etc.

There are two important aspects to note about Eq. (19). First, there are no terms for the saline water and hydrocarbon gas in the equation because these phases are not present in the reference subset. Second, the phase exponents that were used when considering the whole medium have been replaced by saturation exponents because we are now considering the distribution of each of the phases within the reference subset rather than within the whole medium. Third, both Eqs. (17) and (19) are simultaneously true and may be equated.

Equation (19) is clearly the same as Eq. (14). Under the transformation that considers a subset of the whole medium (in this case the solid fractions only) where $1 \leftrightarrow \phi_{ref}$, the individual phase fractions relating to the whole medium are replaced by saturations relative to the subset (i.e. $\phi_i \leftrightarrow S_i$) and the original phase exponents, which were related to the whole medium, are now saturation exponents that are related only to the reference subset (i.e. $m_i \leftrightarrow n_i$).

Both the phase (cementation) exponent and the saturation exponent control how the phase is connected. The phase exponent does this with reference to the whole rock, while the saturation exponent does it with reference to a subset of the whole rock. The underlying physical meaning of the saturation exponent is the same as that of the phase (cementation) exponent; it is only the reference frame that changes. The implication is that the general Archie's law replaces both of the classical Archie's laws. For an application to a sandstone gas reservoir, one would use a three-phase generalized Archie's law.

Equation (12) is easily transformed to provide a calculable value for the saturation exponent by taking the logarithm of both sides of Eq. (12) and rearranging the result before substituting Eq. (1) for the relevant connectednesses and using the relationship $S_i = \phi_i / \phi_{ref}$ to obtain

$$n_i = \frac{\log(G_i) - \log(G_{ref})}{\log(S_i)}$$
$$= \frac{m_i \log(\phi_i) - m_{ref} \log(\phi_{ref})}{\log(\phi_i) - \log(\phi_{ref})}. \quad (20)$$

This equation may be illustrated using a three-phase medium. Imagine a reservoir rock with a 20 % porosity. The pore space contains only oil and water with a water saturation of 0.25. We want to calculate the saturation exponent of the water if the phase exponents of the matrix and the oil are 0.2 and 1.68, respectively. It is simple to calculate the volume fractions of matrix, oil, and water to be 0.8, 0.15, and 0.05, respectively. The connectednesses of matrix and oil can be calculated using Eq. (1) to be 0.956 and 0.0413, respectively. Using Eq. (4) we obtain the connectednesses of the pores and water as 0.0436 and 0.00236, respectively. In this case the reference subset is the pore space, so $G_{ref} = G_{pore} = 0.0436$. Equation (20) can now be used with $G_{water} = 0.00236$, $G_{ref} = 0.0436$, and $S_w = 0.25$ to give $n_w = 2.105$. The saturation exponent of the oil can also be calculated as $n_o = 0.1931$. There is no value for the matrix as the matrix is not included in the pore space reference subset.

There is a reiterative symmetry in this transformation where both the whole-medium phase fractions and the reference subset saturations are both volume fractions with respect to the whole medium and the reference subset, respectively. Similarly, the phase exponents and the saturation exponents are also defined with respect to the whole medium and the reference subset, respectively. This would, therefore, allow the calculation of a reference subset of a subset of a whole medium if required, and so on. There is of course the possibility that the whole n-phase medium is itself a subset of a larger medium with more phases. In this case Eq. (15) still holds, but with $\phi_{ref} > 1$. The implication is that the definition of the original whole medium is arbitrary and can be defined to make the solution of the problem more tractable.

5   Physical interpretation of the saturation exponent

This section provides a physical interpretation for the saturation exponent in a perfect analogy to that derived for the cementation exponent by Glover (2009).

The connectedness G is the inverse of the Archie's formation factor and is central to the generalized Archie's law. The inverse of the Archie's resistivity (saturation) index $1/I = S_w^n$ is also rather important. It relates the connectednesses

of each phase with respect to the whole rock to the connectedness of the reference subset in Eq. (12), and when summed over all the phases that occupy the reference subset, it produces unity as in Eq. (14). In this paper the inverse of the Archie's resistivity (saturation) index has been given the symbol H_i and defined as

$$H_i \equiv S_i^{n_i}. \quad (21)$$

Just as the saturation of any given phase S_i is the ratio of the volume fraction of the phase to that of all the phases making up any reference set of phases, H_i is the ratio of the connectedness of the phase to that of the all the phases making up any reference set of phases. The parameter H_i is in fact a fractional connectedness.

We follow the approach of Glover (2009) in the analysis of the physical interpretation of the cementation exponent. In this work Glover (2009) showed that the cementation exponent was the differential of the connectedness with respect to both porosity and pore connectivity. Following the same methodology, differentiating the fractional connectedness with respect to the phase saturation S_i gives

$$\frac{\partial H_i}{\partial S} = n_i S_i^{n_i - 1}. \quad (22)$$

By analogy we recognize that $S_i^{n_i-1}$ represents the connectivity of Phase i with respect to the reference subset and define this connectivity as

$$\psi_i = S_i^{n_i - 1}, \quad (23)$$

to give

$$\frac{\partial H_i}{\partial S} = n_i \psi_i. \quad (24)$$

A further differentiation, this time with respect to the connectivity ψ_i allows us to obtain

$$n_i = \frac{\partial}{\partial \psi}\left(\frac{\partial H_i}{\partial S}\right). \quad (25)$$

Consequently, the saturation exponent is the rate of change of fractional connectedness with respect to both phase saturation and phase connectivity in a similar way that Glover (2009) found that the physical interpretation of the cementation exponent was the rate of change of connectedness with respect to phase fraction (porosity) and its connectivity. This shows once again the symmetry between phase fractions and saturations and between phase exponents and saturation exponents.

The fractional connectedness is also the product of the saturation and the connectivity with respect to the reference subset

$$H_i = S_i \psi_i. \quad (26)$$

Table 1. Comparison of all the parameters in the classical and generalized Archie's laws.

Parameter	Generalized Archie's law		Classical Archie's law	
	With respect to the whole medium	With respect to a reference subset of the whole medium	First law	Second Law
Phase volume fraction	ϕ_i $\phi_i = \phi_{ref} S_i$	S_i $S_i = \phi_i/\phi_{ref}$	ϕ $V_f = V_{pore} S_w$	S $S_w = V_f/V_{pore}$
Exponent	$m_i = \frac{d}{d\chi}\left(\frac{dG_i}{d\phi}\right)$ $m_i = \frac{\log(\sigma_i)-\log(\sigma_f)}{\log(\phi_i)}$	$n_i = \frac{d}{d\psi}\left(\frac{dH_i}{dS}\right)$ $n_i = $ $\frac{m_i \log(\phi_i)-m_{ref}\log(\phi_{ref})}{\log(\phi_i)-\log(\phi_{ref})}$	m $m = \frac{\log(\sigma_{eff})-\log(\sigma_f)}{\log(\phi)}$	n $n = \frac{\log(\sigma_{eff})-\log(\sigma_{100})}{\log(S_w)}$
Connectedness	$G_i \equiv \phi_i^{m_i}$ $G_i = \phi_i \chi_i$ $G_i = 1/F_i$ $G_i = G_{ref} H_i$	$H_i \equiv S_i^{n_i}$ $H_i = S_i \psi_i$ $H_i = 1/I_i$ $H_i = G_i/G_{ref}$	Undefined	Undefined
Connectivity	$\chi = \phi_i^{m_i-1}$	$\psi = S_i^{n_i-1}$	$\chi = \phi^{m-1}$	Undefined
Rate of change of connectedness	$\frac{dG_i}{d\phi_i} = m_i \chi_i$	$\frac{dH_i}{dS_i} = n_i \psi_i$	Undefined	Undefined
Sum of phases	$\sum_{i=1} \phi_i = 1$ $\sum_i S_i > 1$	$\sum_{i=1} \phi_i < 1$ $\sum_i S_i = 1$	$\phi_{pore} + \phi_{matrix} = 1$	$S_w + S_o + S_g = 1$
Sum of connectednesses	$\sum_i \phi_i^{m_i} = \sum_i G_i = 1$	$\sum_i S_i^{n_i} = \sum_i H_i = 1$	Undefined	Undefined

$$\sum_i \left(\frac{\phi_i}{\phi_{ref}}\right)^{m_i} = 1$$

The transformation $1 \leftrightarrow \phi_{ref}$ leads to $\phi_i \leftrightarrow S_i$ and $m_i \leftrightarrow n_i$

Parameter	Generalized Archie's law		Classical Archie's law	
Effective conductivity	$\sigma_{eff} = \sum_i \sigma_i \phi_i^{m_i}$	$\sigma_{eff} = \sum_i \sigma_i \phi_{ref}^{m_i} S_i^{m_i}$	$\sigma_{eff} = \sigma_f \phi^m$	$\sigma_{eff} = \sigma_f \phi^m S_w^n$

Hence, the saturation exponents obey the same laws as the phase (cementation) exponents, but whereas the phase exponents are defined relative to the whole rock, the saturation exponents are defined relative to some subset of the rock. Table 1 shows the relationships of the generalized Archie's law expressed relative to the whole rock and with respect to a reference subset of the whole rock.

For petrophysicists the reference subset has been the porosity, and there has only been one conducting phase that partially saturates that porosity – the pore water. Now we are not restricted to that model. The reference subset could be, for example, the solid matrix, in which a number of separate mineral phases can be defined, one of which might be, say, a target ore or a clay phase. Let us take a four-phase medium as an example. Imagine a four-phase medium composed of 65 % quartz matrix with a phase volume exponent of 0.3 and 15 % clay. Consequently, the medium's porosity is $\phi = 0.2$. The porosity is occupied by gas and saline water with saturations $S_g = 0.625$ and $S_w = 0.375$, respectively, and the

classical cementation exponent $m = 1.8$ and the classical saturation exponent is $n = 2.05$. Imagine needing to calculate the resistivity of the rock if the resistivity of the clay and the water are known; $\rho_{clay} = 50\ \Omega m$ and $\rho_{water} = 5\ \Omega m$, say. Equation (1) can be used to calculate $G_{quartz} = 0.8788$ and $G_{pore} = 0.0552$. Using Eq. (4) provides $G_{clay} = 0.0660$, with no need to consider the various saturations of the fluids occupying the pores. The phase exponent of the clay can be found to be $m_{clay} = 1.43$. The contribution of the clay to resistivity can be calculated as $\rho_{clay} = 757\ \Omega m$ using Eq. (3), rewritten as $\rho_{contclay} = \rho_{clay}\phi_{clay}^{-m_{clay}} = \rho_{clay}/G_{clay}$, noting that this value takes full account of its volume fraction and its geometrical distribution. Now we must consider the relative distributions of water and gas in the medium. Calculations can be carried out in terms of connectednesses G or fractional connectednesses H. In this case we use the connectednesses G. Equations (11) or (12) can be used to calculate $G_{water} = 0.00739$ and Eq. (4) applied to give $G_{gas} = 0.0478$. Once again, Eq. (1) may be applied, but this time in the rearranged form $m_i = \log G_i / \log \phi_i$ in order to calculate the respective phase exponents $m_{\text{water}} = 1.895$ and $m_{\text{gas}} = 1.462$. Now, the contribution of the saline water to the overall resistivity can be calculated as $\rho_{\text{water}} = 677$ Ωm using Eq. (3), rewritten as $\rho_{\text{contcwater}} = \rho_{\text{water}}\phi_{\text{water}}^{-m_{\text{water}}} = \rho_{\text{water}}/G_{\text{water}}$, noting that this value takes full account of its volume fraction and its geometrical distribution. The resistivity of the rock can now be calculated by simply summing the contributions to conductivity as implied by Eq. (3) to give $\rho_{\text{eff}} = 357$ Ωm. In this particular example, the conductivity of the medium is controlled by the clay and water fractions in approximately equal measure. It should also be noted that there are a number of different pathways for obtaining the same result using the equations contained in this paper.

6   Conclusions

The main conceptual steps in this paper are summarized as follows:

- The classical Archie's saturation exponent arises naturally from the generalized Archie's law.

- The saturation exponent of any given phase can be thought of as formally the same as the phase (i.e. cementation) exponent, but with respect to a reference subset of phases in a larger n-phase medium.

- The connectedness of each of the phases occupying a reference subset of an n-phase medium can be related to the connectedness of the subset itself by $G_i = G_{\text{ref}}S_i^{n_i}$.

- The sum of the connectednesses of a 3-D n-phase medium is given by $\sum_i \phi_i^{m_i} = 1$, mirroring the relationship for phase volumes $\sum_i \phi_i = 1$.

- Connectedness is conserved in a 3-D n-phase medium. If one phase increases in connectedness, the connectedness of one or more of the other phases must decrease to compensate for it, just as phase volumes are conserved with the decrease in one leading to the increase in another phase.

- The sum of the fractional connectednesses (saturations) of an n-phase medium is given by $\sum_i S_i^{n_i} = 1$.

- Fractional connectedness is conserved in a 3-D n-phase medium.

- The saturation exponent may be calculated using the relationship $n_i = \frac{m_i \log(\phi_i) - m_{\text{ref}} \log(\phi_{\text{ref}})}{\log(\phi_i) - \log(\phi_{\text{ref}})}$.

- The connectivity of any phase with respect to the reference subset is given by $\psi_i = S_i^{n_i - 1}$.

- The connectedness of a phase with respect to a reference subset (also called the fractional connectedness) is given by $H_i = S_i \psi_i$ and depends upon the fractional volume of the phase divided by that of the reference subset (i.e. its saturation) and the arrangement of the phase within the reference subset (i.e. its connectivity with respect to the reference subset).

- The rate of change of fractional connectedness with saturation $\frac{dH_i}{dS_i} = n_i \psi_i$ depends upon the connectivity with respect to the reference subset ψ_i and the saturation exponent n_i.

- Hence, the saturation exponent is interpreted as being the rate of change of the fractional connectedness with saturation and connectivity within the reference subset, $n_i = \frac{d^2 H_i}{d\psi_i dS_i}$.

While this paper represents a theoretical treatment of the saturation exponent and attempts to develop a theoretical interpretation that should offer insight into the physical meaning of the saturation exponent, it does not contain a physical proof of these equations. That can only come from targeted experimental work on multiphase media, which is difficult to carry out and represents one of our research goals.

Competing interests. The author declares that he has no conflict of interest.

Acknowledgements. The author would like to thank Harald Milsch, Graham Heinson, and one anonymous reviewer for their detailed reading and constructive comments on the initial submission of this paper.

Edited by: Charlotte Krawczyk

References

Archie, G. E.: The electrical resistivity log as an aid in determining some reservoir characteristics, T. AIME, 146, 54–67, 1942.

British Petroleum: BP Statistical Review of World Energy, June 2016, British Petroleum, London, UK, 48 pp., 2016.

Clavier, C., Coates, G., and Dumanoir, J.: Theoretical and experimental bases for the dual-water model for interpretation of shaly sands, SPE J., 24, 153–168, 1984.

Glover, P. W. J.: What is the cementation exponent? A new interpretation, The Leading Edge, 28, 82–85, https://doi.org/10.1190/1.3064150, 2009.

Glover, P. W. J.: A generalised Archie's law for n phases, Geophysics, 75, E247–E265, https://doi.org/10.1190/1.3509781, 2010.

Glover, P. W. J.: Geophysical properties of the near surface Earth: Electrical properties, Treatise on Geophysics, 11, 89–137, 2015.

Glover, P. W. J.: Archie's law – a reappraisal, Solid Earth, 7, 1157–1169, https://doi.org/10.5194/se-7-1157-2016, 2016.

Glover, P. W. J., Hole, M. J., and Pous, J.: A modified Archie's law for two conducting phases, Earth Planet. Sc. Lett., 180, 369–383, 2000a.

Glover, P. W. J., Pous, J., Queralt, P., Muñoz, J.-A., Liesa, M., and Hole, M. J.: Integrated two dimensional lithospheric conductivity modelling in the Pyrenees using field-scale and laboratory measurements, Earth Planet. Sc. Lett., 178, 59–72, 2000b.

Mendelson, K. S. and Cohen, M. H.: The effects of grain anisotropy on the electrical properties of sedimentary rocks, Geophysics, 47, 257–263, 1982.

Montaron, B.: A quantitative model for the effect of wettability on the conductivity of porous rocks, Texas, USA, SPE 105041, 2009.

Sen, P. N., Scala, C., and Cohen, M. H.: Self-similar model for sedimentary rocks with application to the dielectric constant of fused glass beads, Geophysics, 46, 781–795, 1981.

Sweeney, S. A. and Jennings, H. Y.: The electrical resistivity of preferentially water-wet and preferentially oil-wet carbonate rock, Producers Monthly, 24, 29–32, 1960.

Tiab, D. and Donaldon, E. C.: Petrophysics: theory and practice of measuring reservoir rock and fluid transport properties, 4th Edn., golf professional publishing, 918 pp., eBook ISBN: 9780128031896, Hardcover ISBN: 9780128031889, 2004.

Waxman, M. M. and Smits, L. J. M., Electrical conductivity in oil-bearing shaly sand, Soc. Pet. Eng. J., 8, 107–122, 1968.

Winsauer, W. O., Shearin, H. M., Masson, P. H., and Williams, M.: Resistivity of brine-saturated sands in relation to pore geometry, AAPG Bulletin, 36, 253–277, 1952.

3

Soil erosion evolution and spatial correlation analysis in a typical karst geomorphology using RUSLE with GIS

Cheng Zeng[1,2,3], **Shijie Wang**[1,3], **Xiaoyong Bai**[1,3], **Yangbing Li**[2], **Yichao Tian**[1,3], **Yue Li**[4], **Luhua Wu**[1,3], **and Guangjie Luo**[3,5]

[1]State Key Laboratory of Environmental Geochemistry, Institute of Geochemistry, Chinese Academy of Sciences, 99 Lincheng West Road, Guiyang 550081, Guizhou Province, PR China
[2]School of Geographyical and Environmental Sciences, Guizhou Normal University, Guiyang 550001, China
[3]Puding Karst Ecosystem Observation and Research Station, Chinese Academy of Sciences, Puding 562100, Guizhou Province, PR China
[4]Key Laboratory of State Forestry Administration on Soil and Water Conservation, Beijing Forestry University, Beijing 100083, China
[5]Institute of Agricultural Ecology and Rural Development, Guizhou Normal College, Guiyang 550018, China

Correspondence to: Xiaoyong Bai (baixiaoyong@126.com)

Abstract. Although some scholars have studied soil erosion in karst landforms, analyses of the spatial and temporal evolution of soil erosion and correlation analyses with spatial elements have been insufficient. The lack of research has led to an inaccurate assessment of environmental effects, especially in the mountainous area of Wuling in China. Soil erosion and rocky desertification in this area influence the survival and sustainability of a population of 0.22 billion people. This paper analyzes the spatiotemporal evolution of soil erosion and explores its relationship with rocky desertification using GIS technology and the revised universal soil loss equation (RUSLE). Furthermore, this paper analyzes the relationship between soil erosion and major natural elements in southern China. The results are as follows: (1) from 2000 to 2013, the proportion of the area experiencing micro-erosion and mild erosion was at increasing risk in contrast to areas where moderate and high erosion are decreasing. The area changes in this time sequence reflect moderate to high levels of erosion tending to convert into micro-erosion and mild erosion. (2) The soil erosion area on the slope, at 15–35°, accounted for 60.59 % of the total erosion area, and the corresponding soil erosion accounted for 40.44 %. (3) The annual erosion rate in the karst region decreased much faster than in the non-karst region. Soil erosion in all of the rock outcrop areas indicates an improving trend, and dynamic changes in soil erosion significantly differ among the various lithological distribution belts. (4) The soil erosion rate decreased in the rocky desertification regions, to below moderate levels, but increased in the severe rocky desertification areas. The temporal and spatial variations in soil erosion gradually decreased in the study area. Differences in the spatial distribution between lithology and rocky desertification induced extensive soil loss. As rocky desertification became worse, the erosion modulus decreased and the decreasing rate of annual erosion slowed.

1 Introduction

Soil erosion is one of the most serious environmental problems that affect the environment and human development worldwide (Higgitt, 1993; Martínez-Casasnovas et al., 2016; Borrelli et al., 2016). It not only causes a loss of soil nutrients and land degradation, but also exacerbates the occurrence of droughts, floods, landslides and other disasters (Munodawafa, 2007; Park et al., 2011; Rickson, 2014; Arnhold et al., 2014). Severe soil erosion directly influences the development, application, and protection of regional resources (Cai and Liu, 2003; Ligonja and Shrestha, 2015). In particular, soil erosion threatens ecological security patterns at regional and even global scales.

Many factors affect the evolution of soil erosion in karst areas (Karamesouti et al., 2016; Krklec et al., 2016; Y. B. Li et al., 2016; Wang et al., 2016; Wu et al., 2016) because of the complicated natural conditions (Bai et al., 2013a, b; Tian et al., 2016). Therefore, it is necessary for ecology and soil erosion research in karst areas to explore the spatial evolution characteristics of soil erosion and their influencing factors in a karst area. In the context of global soil erosion and land degradation, traditional methods, such as runoff plots and watershed hydrological stations, are inapplicable for the study of soil erosion in karsts. This has caused fundamental research on soil erosion to lag behind that on soil and water conservation in karst areas.

China possesses the most concentrated, widely distributed, and complex areas of karst landforms worldwide. Guizhou Province is in the center of the karst landform, which is a typical representation of southern China. Due to the slow soil formation rate, mismatched water and soil space, specific geological and hydrological background, and underground structure (Wang and Li, 2007) in the karst zone, soil erosion in the area is more complex and unique than in the non-karst zone. Soil erosion in the karst area exhibits a complex relationship with topography, lithology, and rocky desertification. In addition to surface soil loss, underground leakage has been observed in the area. The karst area has minimal environmental capacity and low restorability of the ecological system (Wallbrink et al., 2002). As such, soil erosion in the area leads to serious consequences that may restrict the sustainable development of the local economy in the region.

Many scholars have studied soil erosion and determined its causes and spatial evolution. Erosion force (Bai and Wan, 1998; Feng et al., 2011), erosion processes (Edgington et al., 1991; Cao et al., 2012), soil degradation (M. Feng et al., 2016; Gao et al., 2015; Guo et al., 2015), and erosion mechanisms (Hancock et al., 2014) have also been explored. Studies on soil erosion have been mainly concentrated in non-karst areas or basins (Fernández and Vega, 2016; Park et al., 2011), whereas few studies have investigated the fragile ecological–geological environment within the karst zone. Some scholars have also conducted preliminary studies on soil erosion in the karst landform areas. For example, Y. Li et al. (2016) evaluated soil erosion in a typical karst basin by using the RUSLE model and explored the influence of slope on the temporal and spatial evolution laws of soil erosion in a karst area. The results indicated that the main erosion on the slope section in the basin was within 8–25°. Yang et al. (2014) analyzed soil erosion in Chaotiangong County in Guilin using an analytic hierarchy and fuzzy model; they found that the risk of soil erosion was very high in the southeastern study area but relatively low in the northwest. Biswas and Pani (2015) studied soil erosion in the Barakar River basin in eastern India using the RUSLE model combined with GIS technology; the soil erosion rate is more than

$100\,\mathrm{t\,km^{-2}\,a^{-1}}$, which accounts for only 0.08 % of the total study area. T. Feng et al. (2016) compared the soil erosion rate between two karst peak-cluster depression basins in northwestern Guangxi, China, using ^{137}Cs and RUSLE models. Runoff discontinuity and underground seepage on the karst slope are significant factors to consider in the RUSLE model because they reduce the effect of slope length. However, previous research exhibits some deficiencies and limitations. Most studies are conducted in karst basins or mountain areas (Shi et al., 2004; Terranova et al., 2009) and analyze the effect of terrain, rainfall, vegetation cover, and other factors on soil erosion (M. Feng et al., 2016; Ganasri and Ramesh, 2016; Liu et al., 2016). The effects of soil erosion on rocky desertification and lithology have been ignored. Few scholars have analyzed the soil erosion evolution in a karst valley area on a long time sequence or determined the effect of spatial factors on evolution. Therefore, the available data on the correlation between soil erosion evolution and spatial factors in the karst zone are limited, particularly for the mountainous area of Wuling, China. This lack of knowledge leads to an inaccurate assessment of the environmental effects in the region; soil erosion and rocky desertification in this area influence the survival and development of 0.22 billion people. Studying the evolution of the temporal and spatial distribution of soil erosion in the karst area and analyzing its correlation with spatial factors remains challenging. Studies have rarely been conducted worldwide because of a lack of supporting data, insufficient experience, and lack of applicable technical methods.

This paper evaluated typical karst areas in southern China and combined current surveys on soil types with calculation results from a soil erodibility test. Soil erosion was analyzed in different periods using a revised universal soil loss equation (RUSLE) model. The specific aims of this study were as follows: (1) to identify the evolution of the temporal and spatial distribution of soil erosion in typical karst areas in southern China; (2) to explore the relationship between soil erosion and rocky desertification; and (3) to determine the correlation between soil erosion and major natural elements and evaluate their ecological effect. This study improved upon existing research methods and proposes suggestions for additional research. It provides a basis for macro-decision-making by government policy makers and environmental managers as well as relevant data on methodology and references for research into soil erosion in karst landform areas.

2 Study area

Yinjiang County is located on the northeastern Guizhou Plateau (China); the geographical position of the study area is 108° 17' to 108°48' N, 27°35' to 28°28' E and the land area is 196 900 hm². Mount Fanjing, the main peak in the Wuling Mountains, is located in the east of Yinjiang. The topogra-

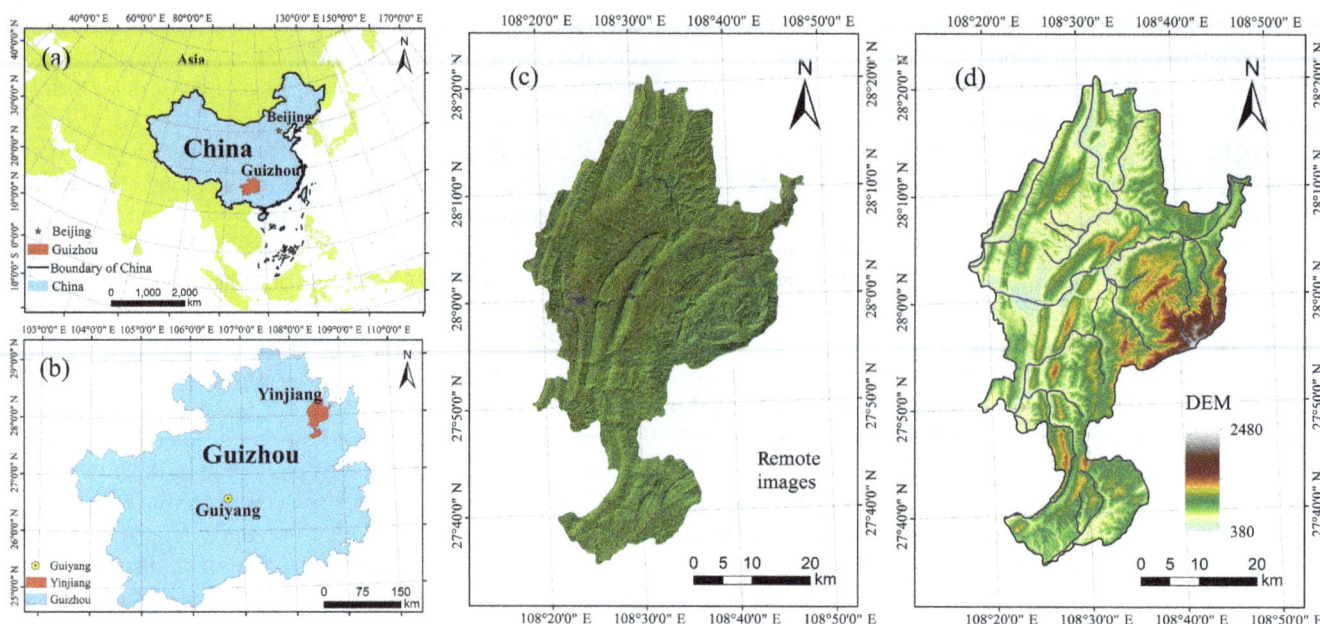

Figure 1. Study area in Guizhou, China (**a, b**). Study area remote images (**c**) and topography (**d**).

phy is such that the east is at a high elevation and the west is at a low elevation, sloping from southeast to northwest. Yinjiang County has a relative elevation difference of 2000 m and an average altitude of 2480 m (Fig. 1). The study area has a subtropical monsoonal climate with annual precipitation of 1100 mm. Rainfall occurs mainly between April and August. The temperature in this area ranges from −3.1 to 29.8 °C with an annual average of 16.8 °C. The highest monthly temperature occurs in July, and the lowest occurs in January. The vegetation is primarily composed of evergreen broad-leaved forest, coniferous forest, evergreen deciduous broad-leaved mixed forest, and temperate coniferous mixed forest. The vegetation coverage increased from 49.1 to 58.5 % during the study period.

Carbonate rocks are widely distributed in Yinjiang County, accounting for 60.06 % of the total area (Fig. 2b). During karst activity, the mantle rock is discontinuous with underground fissures and karst development. Widely distributed soil erosion led to a thin soil layer in the study area and a fragile ecology. Yinjiang County has suffered from different degrees of rocky desertification, accounting for 57.69 % of the total area of the whole county (Fig. 2c). Rocky desertification has been mainly caused by soil erosion due to unsustainable land use. According to the classification of soil zonality, the zonal soil is yellow soil in the study area, but a large area is distributed with lime soil. Moreover, based on the site survey, mountain shrub meadow soil, soil mud, a purple mud field, a tidal sand mud field, and other soil types are distributed in Yinjiang (Fig. 1a). All of these factors are dominant in a typical karst area.

3 Materials and methods

3.1 Data sources

The related data collected based on the RUSLE model mainly include the following: (1) monthly rainfall data in the study area for 2000, 2005, and 2013 from the Tongren Meteorological Bureau (http://tongren04264.11467.com). (2) A soil database was established according to a current survey of soil types, particle size, and the content of organic substances in various soil types that are mainly based on Chinese soil records. (3) A digital elevation model (DEM) was obtained from a Chinese remote-sensing satellite ground station at the Chinese Academy of Sciences (http://www.cas.cn), with a spatial resolution of 30 m. (4) ArcGIS 10.0 was used to determine the three study periods of the NDVI data from the Chinese geospatial data cloud platform (http://www.gscloud.cn). (5) Landsat 7 OLI and Landsat 8 OLI remote sensing images (P126, R40 and P126, R41) were synthesized in ArcGIS 10.0 for stitching and cutting using the data from the Chinese geospatial data cloud platform, with a spatial resolution of 30 m; based on these data, a land-use map was drawn in ArcGIS 10.0 software. The Albers equal-area conic projection was used for a geographic coordinate system.

3.2 RUSLE model

The RUSLE model (Renard et al., 1997) is an empirical model revised from the USLE model for predicting soil erosion. The calculation is as follows:

$$A = R \times K \times L \times S \times C \times P, \tag{1}$$

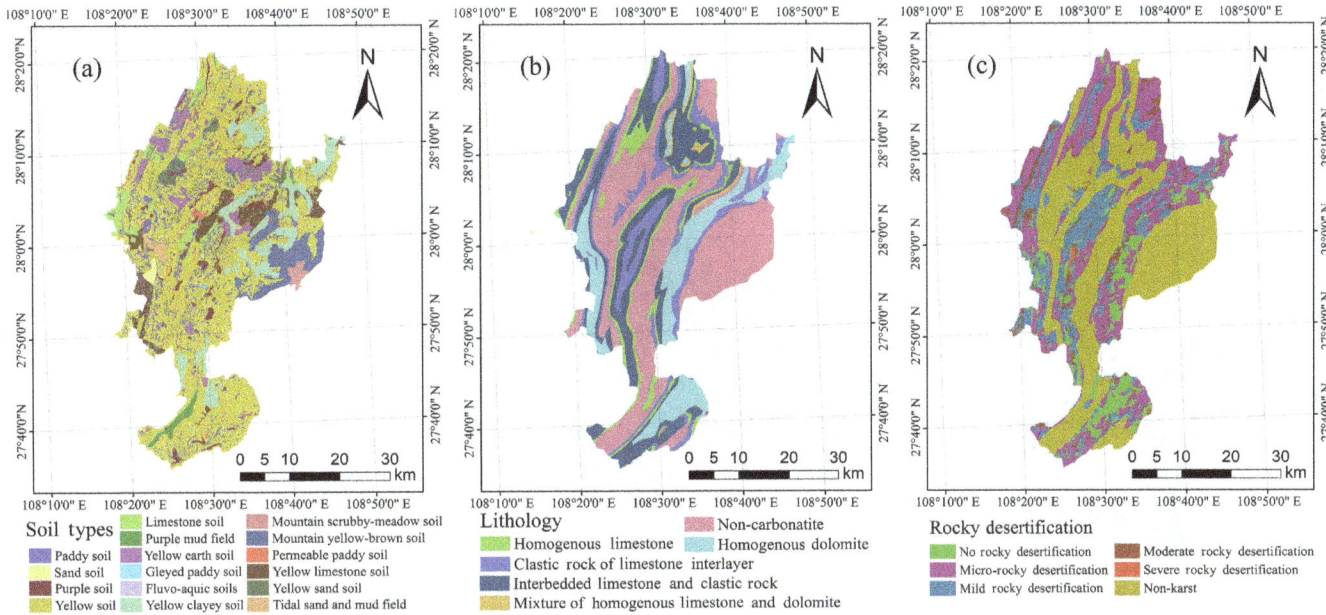

Figure 2. Study area geological background: soil map **(a)**, lithology **(b)**, and rocky desertification **(c)**.

where A [US unit $t\,km^{-2}\,a^{-1}$] refers to the amount of soil loss per unit area in time and space and depends on the K and R units. R [$MJ\,mm\,(hm^2\,h\,a)^{-1}$] refers to the rainfall erosivity factor in consideration of the erosion of snowmelt runoff. K [$t\,hm^2\,h\,(hm^2\,MJ\,mm)^{-1}$] refers to the soil erodibility factor, which is the soil loss rate of specific soil rainfall erosivity per unit measured in a standard plot. L and S refer to the slope aspect factor. C refers to the coverage factor for vegetation. P refers to the conservation measure factor, which includes engineering and tillage measure factors.

3.2.1 Rainfall erosivity factor (R)

Rainfall erosivity represents the potential ability of rainfall to induce erosion. Rainfall erosivity is the primary factor that should be considered in the soil loss equation and is related to rainfall, duration of rainfall, and rainfall energy. This factor reflects the effect of rainfall characteristics on soil erosion. Rainfall erosivity is difficult to directly measure. Most studies use rainfall parameters, including rainfall intensity and precipitation, to estimate rainfall erosivity. Given the relatively fragmented surface, concentrated precipitation, and strong water erosion in the study area, this paper adopts a simple monthly rainfall formula developed by Zhou et al. (1995) to estimate rainfall erosivity (R) in Yinjiang by comparing various algorithms and determining the accuracy of the acquired climate data. The formula is as follows:

$$R = \sum_{i=1}^{12} (-1.5527 + 0.7297 P_i), \qquad (2)$$

where P_i refers to the rainfall in month i (mm). The unit of the calculated R is $100\,ft\,t\,in\,ac^{-1}\,h^{-1}\,a^{-1}$. If R is changed to the international unit $MJ\,mm\,hm^{-2}\,h^{-1}\,a^{-1}$, then a coefficient of 17.02 should be the multiplier (Table 1).

3.2.2 Soil erodibility factor (K)

Soil erodibility is an important indicator that reflects the rainfall infiltration capacity of soil and the sensitivity of soil to rainfall and runoff erosion. This internal factor affects soil loss. The size of K is related to the soil texture and organic material content. In this paper, soil erodibility and soil mechanical composition are used to form a calculation and are closely related to the organic carbon content (Sharpley and Williams, 1990):

$$
K = \left\{ 0.2 + 0.3 \exp\left[-0.0256 SAN \left(1 - \frac{SIL}{100} \right) \right] \right\}
$$
$$
\times \left(\frac{SIL}{CLA - SIL} \right)^{0.3}
$$
$$
\times \left(1 - \frac{0.7 SN1}{SN1 + \exp(-5.51 + 22.9 SN1)} \right), \qquad (3)
$$

where K refers to the soil erodibility [$(t\,acre\,h)\,(100\,acre\,ft\,tanf\,in)^{-1}$]. A conversion factor of 0.1317 should be multiplied to obtain the international unit $(t\,hm^2\,h)\,(hm^2\,MJ\,mm)^{-1}$. SAN, SIL, CLA, and C refer to the sand particles (0.050–2.000 mm), powder particles (0.002–0.050 mm), clay particles (< 0.002 mm), and organic material content (%); $SN1 = 1 - SN/100$. Different K (Fig. 3a) values are obtained from the different soil types on the soil type map (Fig. 2a).

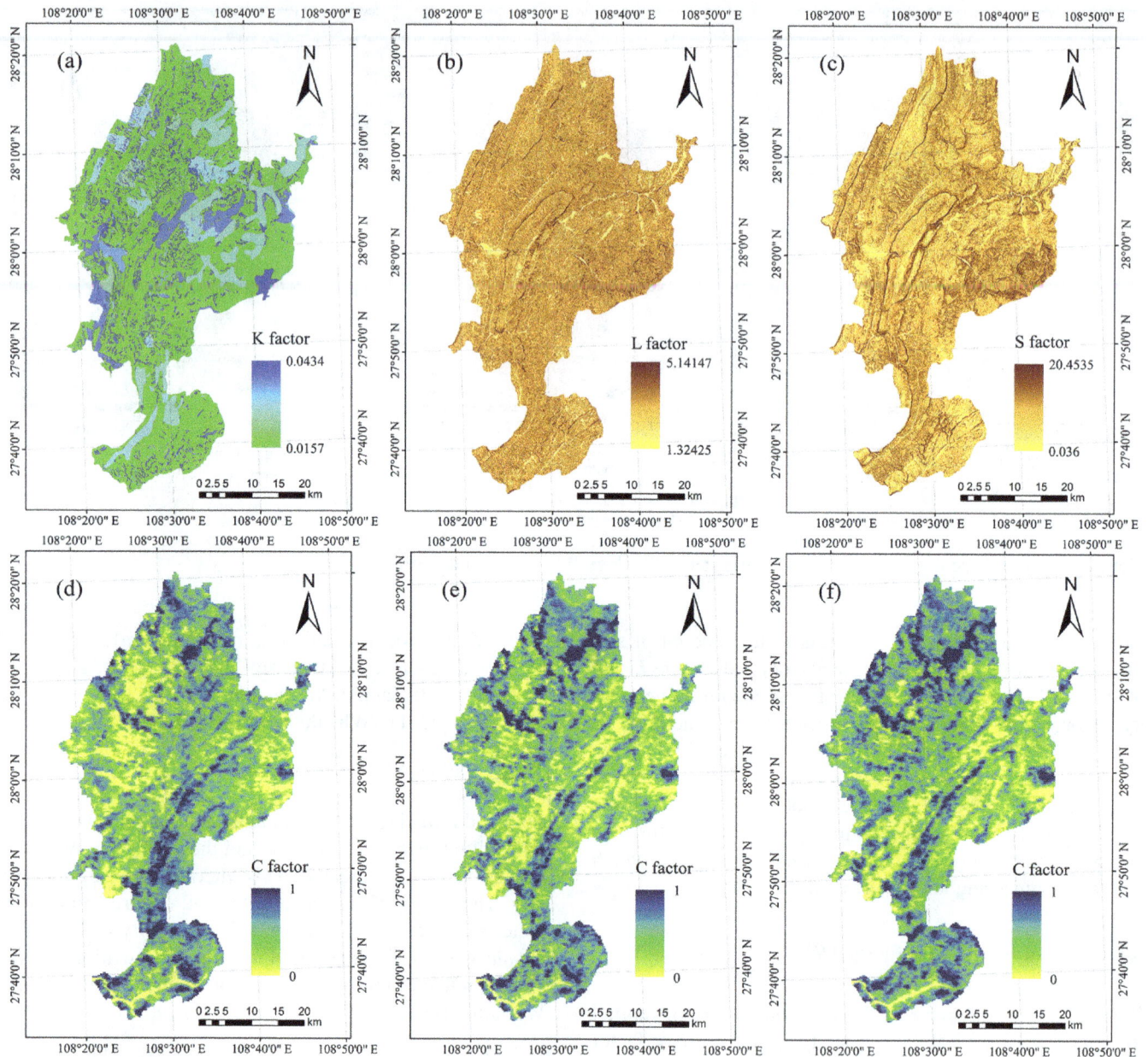

Figure 3. Soil erodibility map (**a**), slope length factor map (**b**), slope gradient factor map (**c**), 2000 vegetation cover factor map (**d**), 2005 vegetation cover factor map (**e**), and 2013 vegetation cover factor map (**f**).

3.2.3 Topographic factor (L)(S)

Slope length is a basic terrain factor that influences soil erosion. In this paper, the formula developed by Liu et al. (2000) is used to calculate slope length in Yinjiang County. The calculation is as follows:

$$S = \begin{cases} 10.8\sin\theta + 0.03 & \theta \lhd 5° \\ 16.8\sin\theta - 0.05 & 5° \leq \theta \lhd 10° \\ 21.9\sin\theta - 0.96 & \theta \geq 10°, \end{cases} \quad (4)$$

$$L = (\lambda/22.13)^m, \quad (5)$$

where S refers to the slope factor, θ refers to the slope value (°), L refers to the slope length factor, and λ refers to the slope length (m). To determine the slope and length, 30 m DEM data from ArcGIS are used and then placed in the formula to calculate L, S (Fig. 3b and c).

3.2.4 Vegetation cover factor (C)

The vegetation cover is correlated with C; hence, this paper used the NDVI of MODIS as a data resource for calculating the vegetation coverage factor C using the methods of Cai et

Table 1. The rainfall erosivity factor (R) in Yinjiang during the study period.

Year	Annual rainfall (mm)	Annual rainfall erosivity [MJ mm hm^{-2} h^{-1} a^{-1}]
2000	1121.03	3183.25
2005	884.23	2460.92
2013	734.39	2003.93

al. (2000). The vegetation coverage rate is also determined using the algorithm established by Tan et al. (2005) with the following equations:

$$C = \begin{cases} 1 & f_c = 0 \\ 0.6508 - 0.3436 \lg f_c & 0 \lhd f_c \lhd 0.783 \\ 0 & f_c = 0.783 \end{cases} \quad (6)$$

$$f_c = (\text{NDVI} - \text{NDVI}_{\text{soil}}) / (\text{NDVI}_{\text{veg}} - \text{NDVI}_{\text{soil}}), \quad (7)$$

$$\text{NDVI} = (\rho_{\text{NIR}} - \rho_{\text{R}}) / (\rho_{\text{NIR}} + \rho_{\text{R}}), \quad (8)$$

where C refers to the vegetation coverage factor, f_c refers to the vegetation coverage (%), and NDVI refers to the normalized differential vegetation index. In this paper, the cumulative percentages of 5 and 95 % are used as the confidence intervals to determine the corresponding pixel values and the effective NDVI$_{\text{soil}}$ and NDVI$_{\text{veg}}$ in the study area. ρ_{NIR} refers to the near-infrared band, and ρ_{R} refers to the red band. The above formula is used to calculate the vegetation coverage distribution map in different periods (Fig. 3d, e, f).

3.2.5 Conservation practice factor (P)

The soil and water conservation factor P refers to the percentage of soil loss to planting down the slope after adopting soil and water conservation measures. The obtained value is within 0–1. If the value is 0, then the area is not affected by soil erosion; if the value is 1, the area has not been subjected to any soil or water conservation measures (Table 2).

3.3 Calculation of the soil erosion and evaluation methods

The above factor layers are converted into raster layers in 30×30 m equal coordinates with ArcGIS 10.0 software. All of the layers are multiplied to obtain the spatial distribution of the soil erosion modulus in the study area. Reference SL190-2007 criteria are used for the classification and grading of soil erosion intensity relative to water erosion grading standards for Yinjiang County (Fig. 4). On this basis, the spatial and temporal evolution of soil erosion in the study area was analyzed and evaluated.

4 Results

4.1 Evolution of soil erosion

From 2000 to 2013, the total amount of soil erosion in Yinjiang decreased from 477.48×10^4 t a^{-1} in 2000 to 366.56×10^4 t a^{-1} in 2005 and 314.64×10^4 t a^{-1} in 2013, with a total reduction of 34.11 % (Table 3).

The area of micro-erosion accounts for 28.97, 30.27, and 34.21 % of the total erosion area in the three study periods, with a total increase of 5.24 %. The area of mild erosion accounts for 39.99, 43.90, and 44.29 % of the total erosion area; the area decreased by 1860 hm^2 overall within the study period, but mild erosion conversion led to an increase of 4.30 %. The total of micro-erosion and mild erosion in the three periods was more than 65 %, and the moderate to higher levels for 2000 to 2013 are declining. The decreased amplitudes of moderate erosion areas, strong erosion areas, pole strong erosion areas, and violent erosion areas were 24, 49, 63, and 89 %, respectively. Yinjiang County exhibited a transformation from moderate erosion, strong erosion, pole strong erosion, and violent erosion to micro-erosion and mild erosion.

The percentages of micro-erosion, mild erosion, and moderate erosion to the total erosion amount increased during the study period. Mild and moderate erosion amounts contributed to the total erosion amount in Yinjiang County. The total percentage of erosion increased from 57.14 % in 2000 to 71.63 % in 2013, whereas the percentages of strong, pole strong, and violent erosion significantly decreased. The total percentage of strong and pole strong erosion decreased from 36.15 to 24.33 %.

In summary, Yinjiang County was mainly affected by mild and moderate erosion. The total percentage of soil erosion increased by 12.57 % from 2000 to 2013. In the entirety of Yinjiang County, a large portion of land experienced micro-erosion and mild erosion in 2000, 2005, and 2013. The total erosion was more than 65 %. The corresponding soil erosion accounted for 28.21, 34.66, and 40.78 % of the total erosion. Although the total area affected by erosion increased to 2374 hm^2, the areas with more than micro-erosion levels decreased. The erosion amount decreased yearly, and the erosion level significantly changed from high to low over a large area.

4.2 Grade shifting of soil erosion intensity

From 2000 to 2005, the percentages of areas with unchanged soil erosion intensity, increased soil erosion intensity, and decreased erosion intensity were 22.76, 33.68, and 43.56 %, respectively. Hence, the soil erosion level was transformed from moderate and high levels to low levels during the study period (Fig. 5).

From 2005 to 2013, the percentage of area with unchanged soil erosion intensity was 23.19 %, which increased by 0.43 % relative to 2000–2005. The percentage of areas

Table 2. Soil and water conservation factors in Yinjiang County.

Land use types	Forest	Grassland	Cropland	Paddy field	Town	Village	Road	Water	Unused land
p	1	1	0.4	0.15	0	0	0	0	1

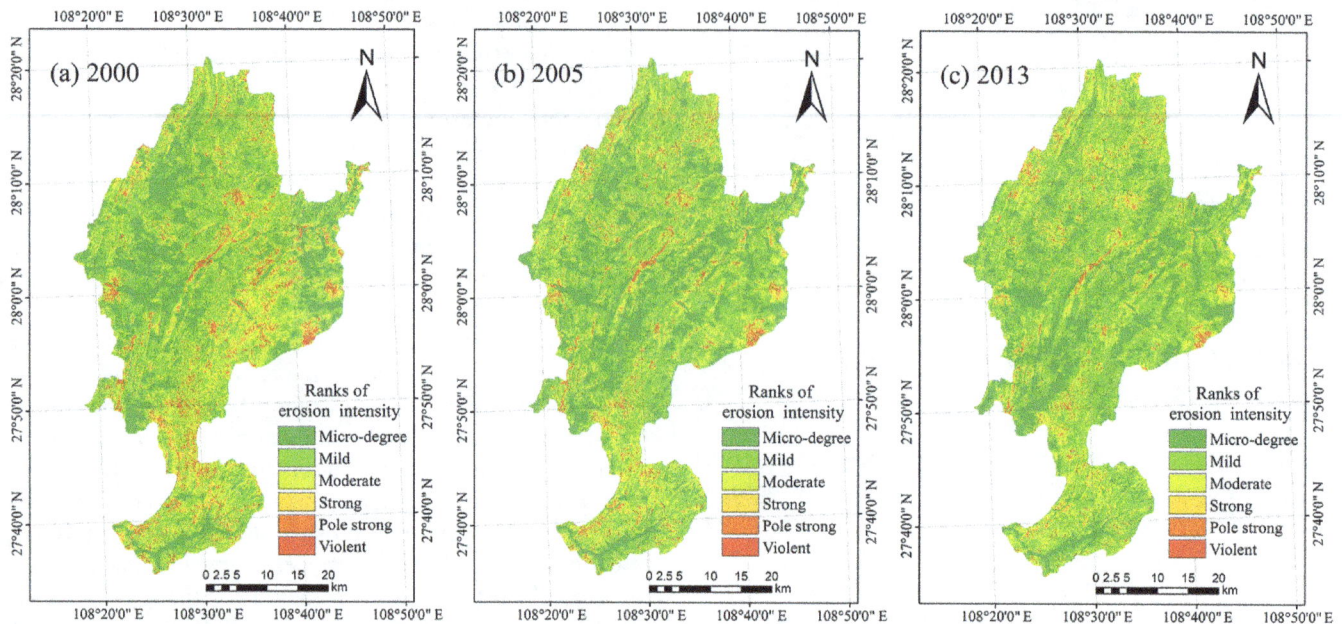

Figure 4. Spatial distribution of soil erosion in Yinjiang in different periods.

with increased and decreased soil erosion intensity slightly increased and attained values of 40.2 % and 36.59 %, respectively.

From 2000 to 2013, the percentages of the total area with increased and decreased erosion intensity were 31.6 and 48.66 %, respectively. This finding reveals that soil erosion intensity has an improving trend.

4.3 Spatial variation in soil erosion

4.3.1 Different slope zones

Slope is the most important terrain factor that influences soil erosion, and it is related to the soil erosion modulus; the modulus in Yinjiang County gradually increased with increasing slope. Hence, slope exhibits a significantly positive correlation with the soil erosion modulus. High-slope areas possess a high mean soil erosion modulus but a small erosion area and erosion amount (Fig. 6).

The soil erosion area of 33.31 % represents the largest area within 15–25° slope bands, followed by 25–35° slope bands (area of 27.28 %). The 25–35, 15–25, 8–15, and 5–8° slope bands account for 20.71, 19.68, 18.09, and 17.32 % of the total erosion. The band with a slope < 5° represents the lowest erosion amount, accounting for 10.85 % (Table 4). All of the

slope bands exhibit a slight erosion level in terms of the mean erosion modulus.

4.3.2 Outcrop area of different rocks

The karst surface is broken and contains peak clusters, needle karst, and isolated peaks. The area with carbonate rock distribution accounts for 60.06 % of the total study area. From 2000 to 2013, the annual erosion rate decreased by $8.22\,\mathrm{t\,(hm^2\,a)^{-1}}$ with a decreased amplitude of 30.82 %. In non-carbonate rock areas, the annual erosion rate from 2000 to 2013 decreased by $6.19\,\mathrm{t\,(hm^2\,a)^{-1}}$ with a decreased amplitude of 24.29 %, which is smaller than in carbonate rock areas (Fig. 7).

The annual erosion rate in the carbonate rock area from 2000 to 2013 demonstrated the following trends: erosion was reduced by $12.24\,\mathrm{t\,(hm^2\,a)^{-1}}$ with a decreased amplitude of 40.40 % in the homogenous dolomite (HD) area (soil loss tolerance in the area $T = 20$). It was reduced by $3.8\,\mathrm{t\,(hm^2\,a)^{-1}}$ with a decreased amplitude of 15.99 % in the homogenous limestone (HL) area. It was reduced by $1.28\,\mathrm{t\,(hm^2\,a)^{-1}}$ with a decreased amplitude of only 5.26 % in the mixed area of homogenous limestone and homogenous dolomite (MHLD). It was reduced by $4.38\,\mathrm{t\,(hm^2\,a)^{-1}}$ with a decreased amplitude of 20.11 % in the clastic rock area of limestone interlayer

Table 3. The soil erosion estimates for different periods in Yinjiang.

	Erosion rating	Erosion area (hm^2)	Total soil loss ($\times 10^4$ t)	Average modulus (t hm^{-2} a^{-1})	Area ratio (%)	Erosion ratio (%)
2000	Micro-degree	36 187	8.47	2.30	28.97	1.77
	Mild	87 470	126.25	126	39.99	26.44
	Moderate	40 506	146.58	36.11	19.27	30.70
	Strong	15 719	98.88	62.88	7.78	20.71
	Pole strong	7153	73.73	103.30	3.46	15.44
	Violent	1244	23.57	184.80	0.54	4.94
2005	Micro-degree	56 529	9.74	2.35	30.27	2.66
	Mild	84 898	117.30	13.92	43.90	32.00
	Moderate	34 362	120.91	35.23	17.76	32.99
	Strong	10 929	67.95	62.17	5.65	18.54
	Pole strong	4352	44.67	102.70	2.25	12.19
	Violent	338	5.99	177.59	0.17	1.64
2013	Micro-degree	63 544	10.57	2.32	34.21	3.36
	Mild	85 610	117.63	13.83	44.29	37.42
	Moderate	30 801	107.54	34.97	15.92	34.21
	Strong	8010	49.73	62.11	4.14	15.82
	Pole strong	2663	26.76	100.52	1.38	8.51
	Violent	125	2.11	168.55	0.06	0.67

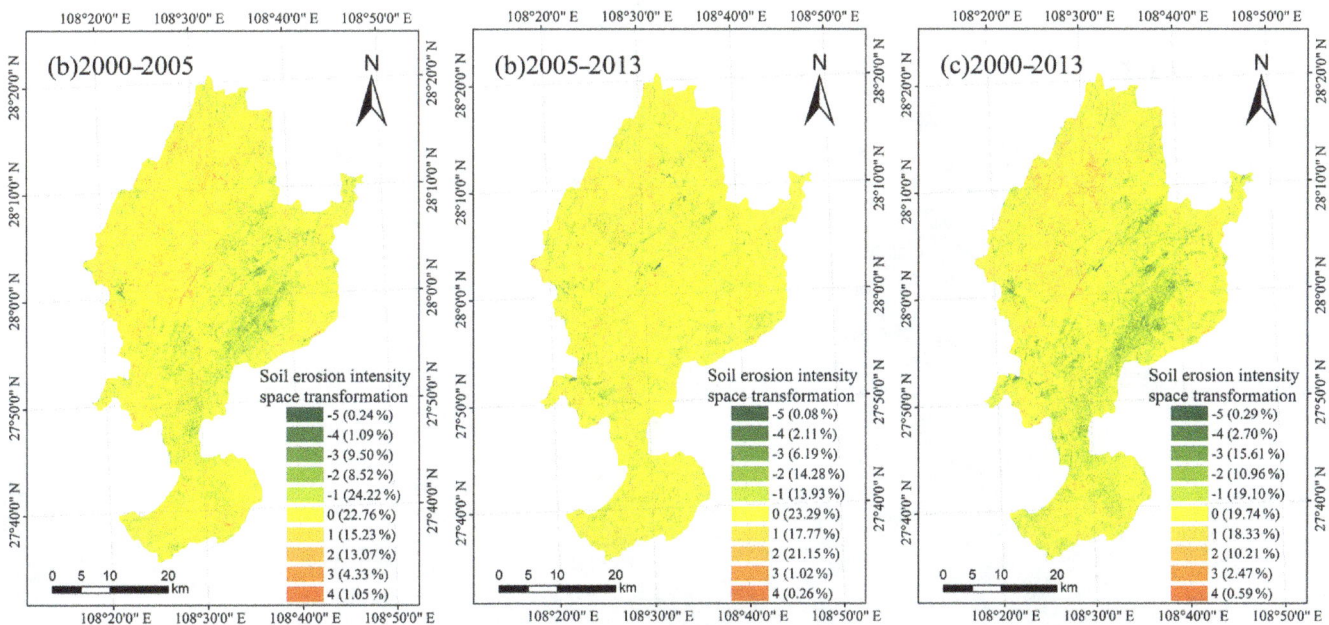

Figure 5. The intensity variation map of the soil erosion in Yinjiang. Note: 0 refers to unchanged soil erosion intensity; 1 refers to the soil erosion intensity increasing by one level; 2 refers to the soil erosion intensity increasing by two levels; 3 refers to the soil erosion intensity increasing by three levels; 4 refers to the soil erosion intensity increasing by four levels; 5 refers to the soil erosion intensity increasing by five levels; −1 refers to the soil erosion intensity decreasing by one level; −2 refers to the soil erosion intensity decreasing by two levels; −3 refers to the soil erosion intensity decreasing by three levels; −4 refers to the soil erosion intensity decreasing by four levels; and −5 refers to the soil erosion intensity decreasing by five levels.

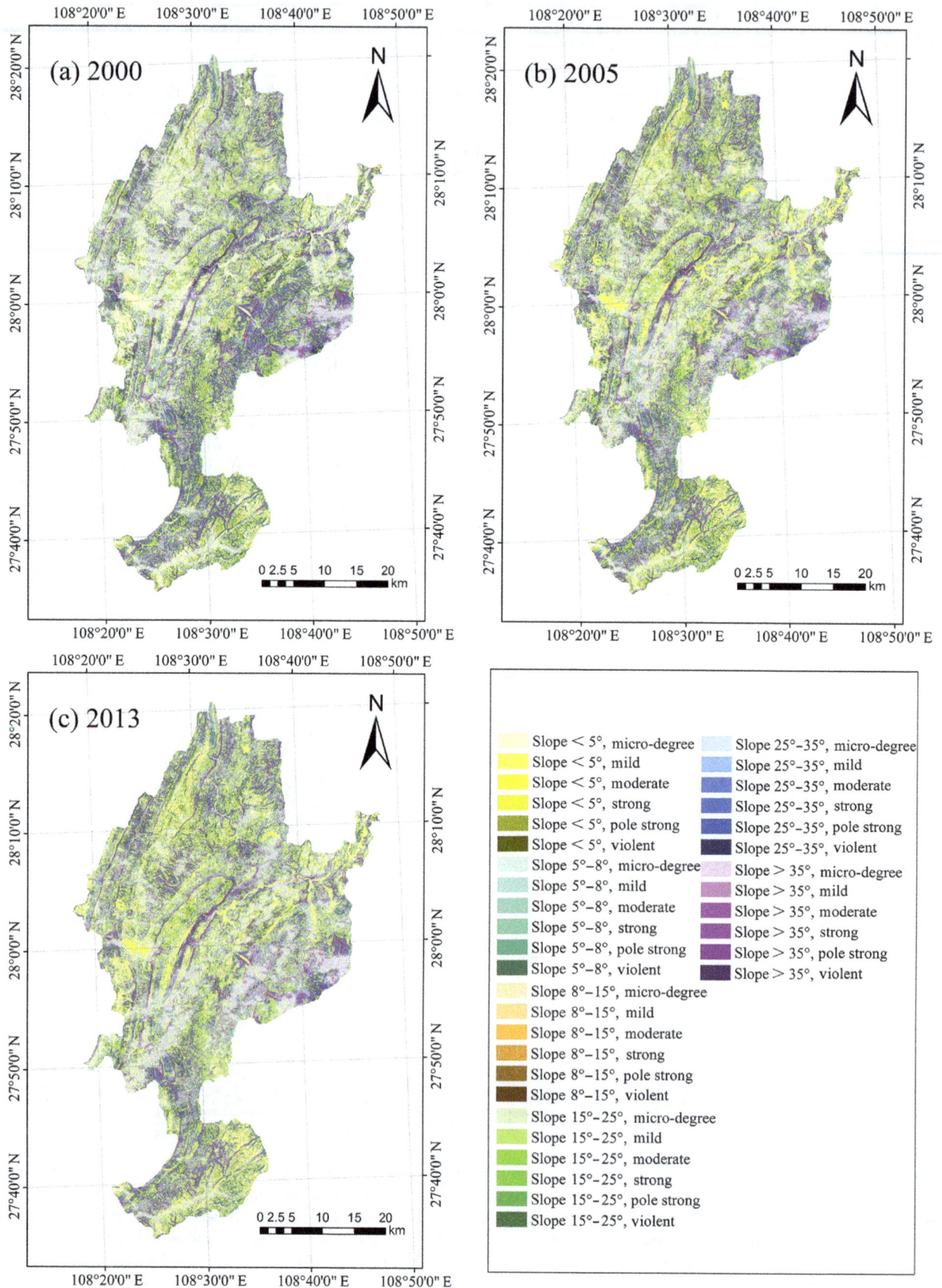

Figure 6. Spatial distribution of soil erosion in different slope bands.

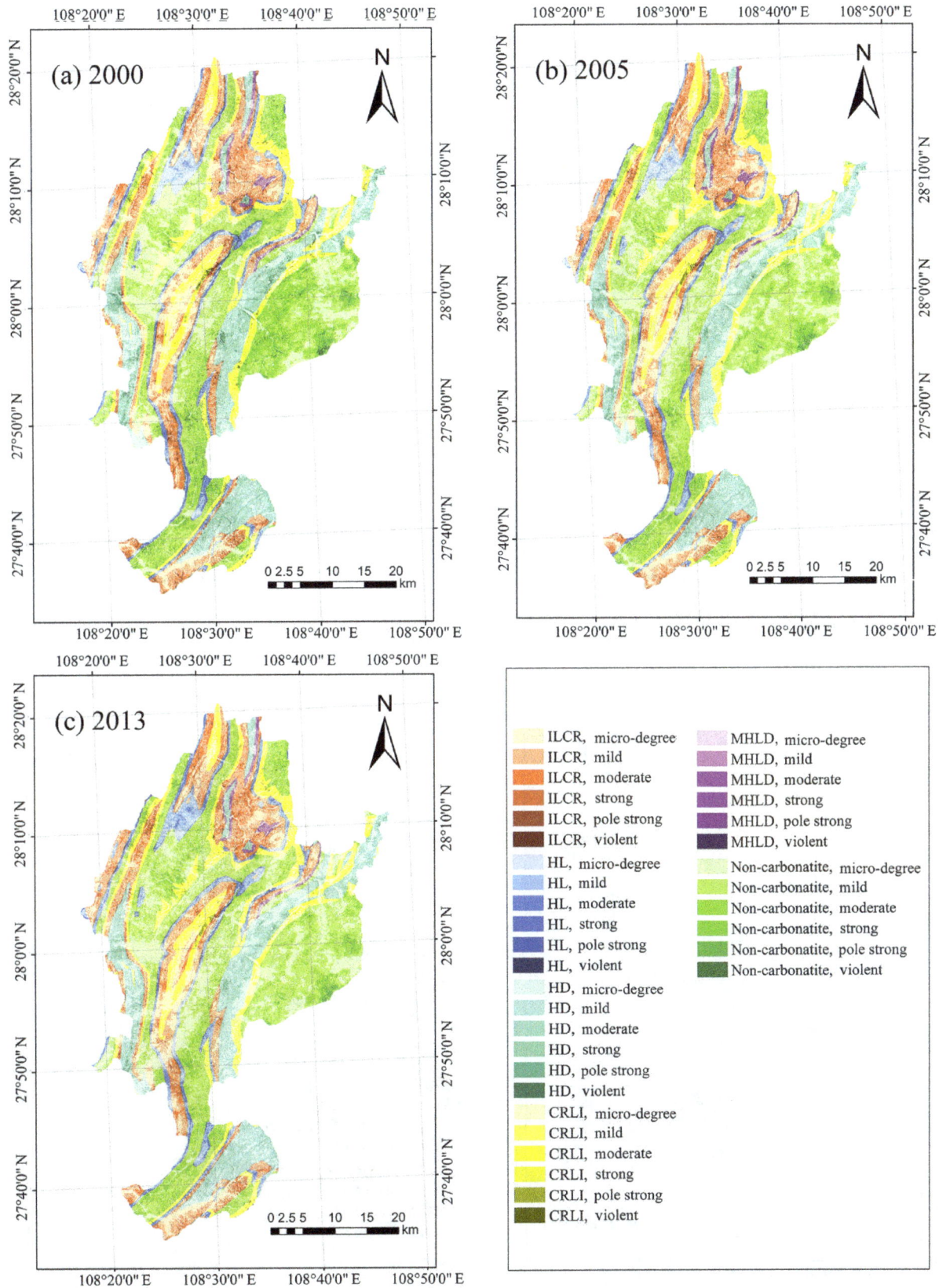

Figure 7. Spatial distribution of soil erosion in different rock outcrop areas.

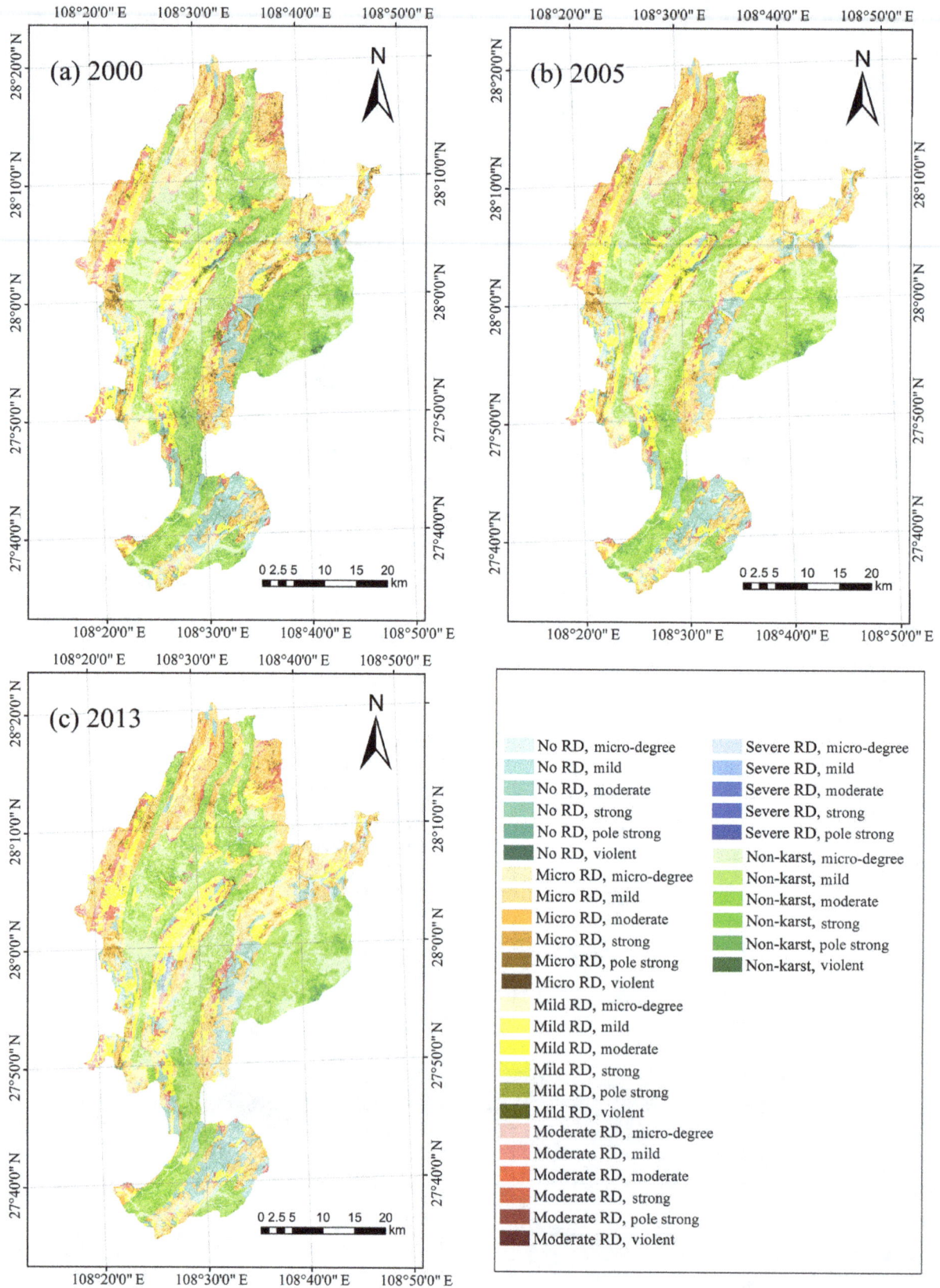

Figure 8. Spatial distribution of soil erosion on different rocky desertification grades.

Table 4. Soil erosion conditions on different slope grades.

Slope	Average modulus $(t\,hm^{-2}\,a^{-1})$	Area ratio (%)	Erosion ratio (%)
<5°	15.32	9.68	10.85
5–8°	13.31	4.76	17.32
8–15°	15.33	12.94	18.09
15–25°	17.56	33.31	19.68
25–35°	18.54	27.28	20.72
>35°	20.15	12.03	13.33

(CRLI; soil loss tolerance in the area $T = 100$), and it was reduced by $4.31\,t\,(hm^2\,a)^{-1}$ with a decreased amplitude of 17.07 % in the interbedded area of limestone and clastic rock (ILCR; soil loss tolerance in the area $T = 250$) (Table 5).

The relationship of the changes in the decreased amplitude in the study period was as follows: homogenous dolomite ($T = 20$) > clastic rock of limestone interlayer ($T = 100$) > interbedded of limestone and clastic rock ($T = 250$) > homogenous limestone > mixture of homogenous limestone and dolomite.

4.3.3 Different grades of rocky desertification

Different degrees of rocky desertification are distributed in approximately 57.69 % of the study area. In the karst area, interference and destruction from invasive social and economic activities caused severe soil erosion, leading to soil particle loss, a thin soil layer, and outcropped base rock in the desertification area (Fig. 8).

From 2000 to 2013, the annual erosion rate in Yinjiang County exhibited the following trend: erosion was reduced by $11.99\,t\,(hm^2\,a)^{-1}$ with a decreased amplitude of 39.36 % in the non-rocky desertification area. It was reduced by $6.23\,t\,(hm^2\,a)^{-1}$ with a decreased amplitude of 24.53 % in the micro-rocky desertification area. It was reduced by $3.2\,t\,(hm^2\,a)^{-1}$ with a decreased amplitude of 14.90 % in the mild rocky desertification area. It was reduced by $1.68\,t\,(hm^2\,a)^{-1}$ with a decreased amplitude of 9.06 % in the moderate rocky desertification area. It increased by $1.86\,t\,(hm^2\,a)^{-1}$ with an increased amplitude of 19.16 % in the severe rocky desertification area, and it was reduced by $7.42\,t\,(hm^2\,a)^{-1}$ with a decreased amplitude of 28.62 % in the non-rocky desertification area (Table 6).

The relationship of the decreasing amplitude of erosion rates in karst areas during the study period was as follows: non-rocky desertification area > micro-rocky desertification area > mild rocky desertification area > moderate rocky desertification area > severe rocky desertification area. The soil erosion amounts decreased in the non-rocky desertification area, micro-rocky desertification area, mild rocky desertification area, and moderate rocky desertification area; they increased in the severe rocky desertification area. The micro-rocky desertification zone occupied the largest soil erosion

area (47.55 % of the total area) and had the highest erosion amount (48.86 % of the total erosion amount). The mean erosion modulus was a mild level of erosion.

5 Discussion

5.1 Spatiotemporal evolution characteristics of soil erosion

The overall soil erosion conditions in Yinjiang County improved annually. The erosion area and erosion amount were distinguished by conversion from strong, pole strong, and violent erosion to moderate and lower levels of erosion. This phenomenon occurred because rainfall and vegetation coverage mainly affect the dynamic changes in soil erosion in Yinjiang County. On the one hand, rainfall decreased yearly from 1121.03 mm in 2000 to 734.39 mm in 2013 in the study period, which led to a weakening of rainfall erosion (Mohamadi and Kavian, 2015). On the other hand, Yinjiang County has a wide range of farmland returning to forests and closed forest projects, so vegetation management and soil and water conservation measures in the study area correspondingly changed. The improved vegetation coverage plays a role in the prevention and control of soil and water erosion. Soil and water conservation measures have a large-scale active effect and cause significant results.

Slope determines the speed of surface runoff. If other factors remain unchanged, the surface runoff impacts on soil in an area with a slope below 35° become stronger and the soil erosion amount increases with increasing slope. When the slope reaches 35°, the erosion amount decreases and is weakly influenced by the increasing slope. The band with a slope of 15–35° accounts for 60.59 % of the total erosion area and 40.44 % of the total erosion amount. This band is the main erosion slope section in the study area. This phenomenon is the result of artificial reclamation in the slope area. Based on the current results, as has been reported in previous studies (Xu et al., 2008; Chen et al., 2012), the slope is about 25° in areas prone to soil erosion. The 15–35° slope area in Yinjiang County must have enhanced prevention and control measures for soil erosion.

5.2 Influence of spatial factors on soil erosion

5.2.1 Influence of lithology on soil erosion

The decreasing amplitude of the soil erosion rate in the carbonate area was larger than in the non-carbonate area. This finding is related to the widely distributed rocky desertification in the karst area, soil formation rate, soil type, and other factors. After the carbonate rock is dissolved in the study area, soluble matter is removed by water, and insoluble matter forms soil. The content of insoluble matter in carbonate rock in the southwest is 1–9 % and is generally less than 5 %. The soil-forming efficiency is low. After erosion and weath-

Table 5. Annual erosion rates in different rock outcrop areas.

	Average soil erosion rate ($t\,hm^{-2}\,a^{-1}$)						
	Non-carbonatite	carbonatite	HD	HL	MHLD	CRLI	ILCR
2000	26.67	25.48	30.30	23.77	24.34	21.78	25.25
2005	21.79	21.82	22.26	21.86	27.44	19.10	23.03
2013	18.45	19.29	18.06	19.97	23.06	17.40	20.94

Table 6. Annual erosion rates in different rocky desertification grades.

	Average soil erosion rate ($t\,hm^{-2}\,a^{-1}$)					
	No RD	Micro RD	Mild RD	Moderate RD	Severe RD	Non-karst
2000	30.46	25.40	21.48	18.54	9.71	25.93
2005	22.17	21.79	20.09	18.57	8.98	21.74
2013	18.47	19.17	18.28	16.86	11.56	18.51

ering, 630–7880 ka of carbonate is required to form a 1 m thick soil layer. The soil-forming rate is 10–40 times slower than in the general non-karst area (Chen, 1997). Moreover, the soil-forming rate and soil thickness are higher in non-carbonate areas than in carbonate areas. The formation time of runoff is short after rainfall, and the surface water storage capacity is low in the karst area. Rainfall forms underground runoff; hence, underground soil loss is high and the vegetation coverage is lower than in the non-karst area.

In the study period, only 10–22.37 % of the areas are within the allowable loss amount. These areas are mainly distributed in the valley zone, with low altitudes in the south of Yinjiang and the smooth zone in the southwestern and Fanjingshan areas. These areas are mostly located in non-karst zones with a wide distribution of non-carbonates. The soil formation is rapid, the underground soil loss is low, and the vegetation coverage is high.

Soil erosion exhibited an improving trend in different outcrop areas. However, the dynamic changes in soil erosion in various lithological distribution belts were significant. The decreasing amplitude of the annual erosion rate in homogenous dolomite, limestone intercalated with clastic rock, and the interbedded region of limestone and clastic rock gradually decreased with decreasing carbonate content. This phenomenon occurred because of the mineral composition and chemical characteristics of the parent rock, which directly affect the speed and direction of soil formation. The weathering degree of different lithologies, the speed and direction of soil formation, and the erosion type, intensity, and rate are also different. If the carbonate content is high, then the soil formation rate is slow and the soil layer is shallow. Therefore, the decreasing amplitude of the annual erosion rate is low. The homogenous limestone region and the mixed region of homogenous dolomite and limestone are mainly distributed in an area of low altitude with a slope of less than 8°. There is

therefore a specific soil thickness, resulting in a large erosion model and small decreasing amplitude of the annual erosion rate. Moreover, the lithology controls the spatial distribution and development of soil erosion. Li et al. (2006) reported that the allowable soil loss is $6.75\,t\,(km^2\,a)^{-1}$ in the carbonate area and $7.08\,t\,(km^2\,a)^{-1}$ in the homogenous limestone area and homogenous dolomite area. The rank of allowable loss amounts is as follows: homogenous dolomite composition distribution area > homogenous limestone composition distribution area. In the present study, the rank of calculated loss amounts (homogenous dolomite area > homogenous limestone area) is consistent with a previous study. The allowable soil loss amounts are $45.40\,t\,(km^2\,a)^{-1}$ in limestone intercalated with clastic rock and $103.\,38\,t\,(km^2\,a)^{-1}$ in the interbedded region of limestone and clastic rock. The relationship of the allowable loss amount is as follows: interbedded region of limestone and clastic rock > limestone intercalated with clastic rock. The allowable loss is positively correlated with the amount of loss calculated in areas of $T = 100$ (limestone intercalated with clastic rock) and $T = 250$ (interbedded layer of limestone and clastic rock).

5.2.2 Effects of rocky desertification on soil erosion

In terms of soil erosion intensity in the study area, the decreasing amplitude in the annual soil erosion rate gradually decreases with the aggravation of rocky desertification. When the degree of rocky desertification is high, the erosion modulus is low and the decreasing amplitude of the annual erosion rate is small. The decreasing amplitude of the annual erosion rate in non-rocky desertification areas is higher than in rocky desertification areas. This finding could be because the non-rocky desertification areas are mainly distributed in valleys and low-altitude regions with sufficient soil thickness and good vegetation coverage. Currently, the soil erosion rate in the severe rocky desertification region of the study area is

Table 7. Soil erosion data obtained in previous studies in typical karst areas.

Reference	Study area	Timescale	Average modulus $(\mathrm{t\,hm^{-2}\,a^{-1}})$	Total soil loss $(\times 10^4\,\mathrm{t})$
Zeng et al. (2014)	Hongfeng Lake watershed	1960–1986	38.35	610.53
		1987–1997	52.80	839.90
		1998–2004	40.24	640.18
Xu and Peng (2008)	Maotiao River watershed	2002	28.70	875.65
Y. Wang et al. (2014)	Wujiang River basin	1980–1989	26.78	133.36
		1990–1999	23.13	115.18
This paper	Yinjiang County	2000	25.09	477.49
		2005	21.53	366.56
		2013	18.84	314.64

increasing and the loss intensity is large; however, they were not obvious in general (the total amount of soil erosion is small and very low). As these areas, which are concentrated in the Langxi valley, are small areas with poor conditions for growing vegetation, are in a soil accumulation environment, or are on negative terrain, there are specific soil thicknesses causing high erosion rates.

The decrease in the erosion rate in other rocky desertification bands reveals that soil erosion in the rocky desertification area improved during the study period. The soil loss in the karst rocky desertification areas could be due to the particular geology (wide distribution of carbonate rocks), topography (presence of underground space), vegetation, and climate conditions, which lead to a low soil formation rate and shallow soil layer in the study area. Abundant rainfall in the study area provides a dynamic potential for soil and water loss. Furthermore, underground pores, cracks, and pipes are widely distributed in the karst area. In addition to surface loss, soil loss also occurs through karst caves, underground rivers, and other means (Peng and Wang, 2012; J. Wang et al., 2014).

The current study method exhibits certain limitations in a typical karst area. In future studies, underground soil and water loss in the karst area should be calculated. The localization of the model calculation factor in the karst area should also be considered for calculating soil erosion using the proposed model. Based on the specificity of soil erosion in the karst area, improving the method and exploring erosion indicators can improve and enrich the study of soil erosion in karst areas.

5.3 Modulus of different soil erosion statistics in karst areas

The RUSLE model is a classical model for evaluating soil erosion and is widely used in various countries and regions worldwide. Although the RUSLE model is a mature and classical model, its application in karst areas is relatively scarce. Several scientists have conducted research on different parts of the karst areas in Guizhou Province. Different results have

been derived; thus, a simple control should be adopted. The results are given in Table 7.

6 Conclusions

The temporal and spatial variation in soil erosion gradually declined in the study area and exhibited a changing trend from moderate and higher levels to lower levels. Slope was the most important topographic factor that affected different spatial and temporal distributions of soil erosion. The band with a slope of 15–35° was the main erosion slope section in the study area. The soil erosion in all rocky outcrop areas exhibited an improving trend, but the dynamic changes in soil erosion in each lithological distribution zone varied greatly. As rocky desertification worsens, the erosion modulus lowers and the decreasing rate of annual erosion will slow.

In karst areas, lithology and rocky desertification are the most important natural factors that cause different temporal and spatial variations in soil erosion. Lithology is the geological basis of soil erosion, and rocky desertification is widely distributed in karst valley areas. Different spatial distributions of lithology and rocky desertification lead to a large area of soil loss. Lithological and rocky desertification factors introduced in the soil erosion model can accurately reflect and predict soil erosion conditions and spatial distribution characteristics in karst areas. This finding will help promote research into soil erosion in karst areas worldwide.

In karst areas, underground space is complicated and consists of multiple geological and geomorphological features. In addition to surface loss, soil loss occurs through karst caves, underground rivers, and other means, causing differences between the measured soil loss and the calculated value in the model. Most of the time, soil erosion study methods and indicators that are used for non-karst areas cannot reflect the actual conditions of karst areas.

Competing interests. The authors declare that they have no conflict of interest.

Acknowledgements. This research was supported by the National Key Research Program of China (nos. 2016YFC0502300, 2016YFC0502102, 2013CB956700, and 2014BAB03B02), the UNESCO Research Center on Karst (no. U1612441), international cooperation research projects of the National Natural Science Fund Committee (nos. 41571130074 and 41571130042), the Science and Technology Plan of Guizhou Province of China (nos. 2012-6015, 2013-3190, and 2017-2966), and science and technology cooperation projects (no. 2014-3).

Edited by: Antonio Jordán

References

Arnhold, S., Lindner, S., Lee, B., Martin, E., Kettering, J., and Nguyen, T. T.: Conventional and organic farming: soil erosion and conservation potential for row crop cultivation, Geoderma, 219–220, 89–105, https://doi.org/10.1016/j.geoderma.2013.12.023, 2014.

Bai, X. Y., Zhang, X. B., Long, Y., Liu, X., and Zhang, S.: Use of ^{137}cs and ^{210}pb ex, measurements on deposits in a karst depression to study the erosional response of a small karst catchment in southwest china to land-use change, Hydrol. Process., 27, 822–829, https://doi.org/10.1002/hyp.9530, 2013a.

Bai, X. Y., Wang, S. J., and Xiong, K. N.: Assessing spatial-temporal evolution processes of karst rocky desertification land: indications for restoration strategies, Land Degrad. Dev., 24, 47–56, https://doi.org/10.1002/ldr.1102, 2013b.

Bai, Z. G. and Wan, G. J.: Study on watershed erosion rate and its environmental effects in Guizhou Karst region, Journal of Soil Erosion and Soil and Water Conservation, 4, 1–7, 1998.

Biswas, S. S. and Pani, P.: Estimation of soil erosion using rusle and gis techniques: a case study of barakar river basin, jharkhand, india, Modeling Earth Systems and Environment, 1, 1–13, https://doi.org/10.1007/s40808-015-0040-3, 2015.

Borrelli, P., Panagos, P., Märker, M., Modugno, S., and Schütt, B.: Assessment of the impacts of clear-cutting on soil loss by water erosion in italian forests: first comprehensive monitoring and modelling approach, Catena, 149, 770–781, https://doi.org/10.1016/j.catena.2016.02.017, 2016.

Cai, C. F., Ding, S. W., Shi, Z. H., Huang, L., and Zhang, G. Y.: Study of applying USLE and geographical information system IDRISI to predict soil erosion in small watershed, J. Soil Water Conserv., 14, 19–24, https://doi.org/10.3321/j.issn:1009-2242.2000.02.005, 2000.

Cai, G. Q. and Liu, J. G.: Evolution of soil erosion models in China, Progress in Geography, 22, 242–250, https://doi.org/10.3969/j.issn.1007-6301.2003.03.003, 2003.

Cao, J., Yuan, D., Groves, C., Huang, F., Hui, Y., and Qian, L. U.: Carbon fluxes and sinks: the consumption of atmospheric and soil CO_2 by carbonate rock dissolution, Acta Geol. Sin.-Engl., 86, 963–972, https://doi.org/10.1111/j.1755-6724.2012.00720.x, 2012.

Chen, L., Xie, G. D., Zhang, C. S., Li, S. M., Fan, N., Zhang, C. X., Pei, S., and Ge, L. Q.: Spatial distribution characteristics of soil erosion in Lancang river basin, Resources Science, 34, 1240–1247, 2012.

Chen, X. P.: Research on characteristics of soil erosion in Karst mountainous region environment, Journal of Soil Erosion and Soil and Water Conservation, 3, 31–36, 1997.

Edgington, D. N., Klump, J. V., Robbins, J. A., Kusner, Y. S., Pampura, V. D., and Sandimirov, I. V.: Sedimentation rates, residence times and radionuclide inventories in lake baikal from 137cs and 210pb in sediment cores, Nature, 350, 601–604, https://doi.org/10.1038/350601a0, 1991.

Fernández, C. and Vega, J. A.: Evaluation of rusle and pesera models for predicting soil erosion losses in the first year after wildfire in nw spain, Geoderma, 273, 64–72, https://doi.org/10.1016/j.geoderma.2016.03.016, 2016.

Feng, M., Wang, Q., Hao, Q., Yin, Y., Song, Z., Wang, H., and Liu, H.: Determinants of soil erosion during the last 1600 years in the forest–steppe ecotone in northern china reconstructed from lacustrine sediments, Palaeogeogr. Palaeocl., 449, 79–84, 2016.

Feng, T., Chen, H. S., and Wang, K. G.: 137Cs profile distribution character and its implication for soil erosion on Karst slopes of Northwest Guangxi, Chinese Journal of Applied Ecology, 22, 593–599, https://doi.org/10.13287/j.1001-9332.2011.0123, 2011.

Feng, T., Chen, H., Polyakov, V. O., Wang, K., Zhang, X., and Zhang, W.: Soil erosion rates in two karst peak-cluster depression basins of northwest guangxi,china:comparison of the rusle model with 137cs measurements, Geomorphology, 253, 217–224, https://doi.org/10.1016/j.geomorph.2015.10.013, 2016.

Ganasri, B. P. and Ramesh, H.: Assessment of soil erosion by RUSLE model using remote sensing and GIS-A case study of Nethravathi Basin, Geoscience Frontiers, 7, 953–961, 2016.

Gao, X., Xie, Y., Liu, G., Liu, B., and Duan, X.: Effects of soil erosion on soybean yield as estimated by simulating gradually eroded soil profiles, Soil Till. Res., 145, 126–134, https://doi.org/10.1016/j.still.2014.09.004, 2015.

Guo, Q., Hao, Y., and Liu, B.: Rates of soil erosion in china: a study based on runoff plot data, Catena, 124, 6–76, https://doi.org/10.1016/j.catena.2014.08.013, 2015.

Hancock, G. J., Wilkinson, S. N., Hawdon, A. A., and Keen, R. J.: Use of fallout tracers 7 be, 210 pb and 137 cs to distinguish the form of sub-surface soil erosion delivering sediment to rivers in large catchments, Hydrol. Process, 28, 3855–3874, https://doi.org/10.1002/hyp.9926, 2014.

Higgitt, D.: Soil erosion and soil problems, Prog. Phys. Geog., 17, 461–472, 1993.

Karamesouti, M., Petropoulos, G. P., Papanikolaou, I. D., Kairis, O., and Kosmas, K.: Erosion rate predictions from pesera and rusle at a mediterranean site before and after a wildfire: comparison and implications, Geoderma, 261, 44–58, https://doi.org/10.1016/j.geoderma.2015.06.025, 2016.

Krklec, K., Domínguez-Villar, D., Carrasco, R. M., and Pedraza, J.: Current denudation rates in dolostone karst from central spain: implications for the formation of unroofed caves, Geomorphology, 264, 1–11, https://doi.org/10.1016/j.geomorph.2016.04.007, 2016.

Li, Y., Bai, X. Y., Zhou, Y., Qin, L., Tian, X., Tian, Y., and Li, P. L.: Spatial–temporal evolution of soil erosion in a typical mountainous karst basin in sw china, based on gis and rusle, Arab. J. Sci. Eng., 41, 1–13, https://doi.org/10.1007/s13369-015-1742-6, 2016.

Li, Y. B., Wang, S. J., Wei, C. F., and Long, J.: The spatial distribution of soil loss tolerance in carbonate area in Guizhou province, Earth of Environment, 4, 36–40, 2006.

Li, Y. B., Li, Q. Y., Luo, G. J., Bai, X. Y., Wang, Y. Y., Wang, S. J., Xie, J., and Yang, G. B.: Discussing the genesis of karst rocky desertification research based on the correlations between cropland and settlements in typical peak-cluster depressions, Solid Earth, 7, 741–750, https://doi.org/10.5194/se-7-741-2016, 2016.

Ligonja, P. J. and Shrestha, R. P.: Soil erosion assessment in kondoa eroded area in tanzania using universal soil loss equation, geographic information systems and socioeconomic approach, Land Degrad. Dev., 26, 367–379, https://doi.org/10.1002/ldr.2215, 2015.

Liu, B. Y., Nearing, M. A., Shi, P. J., and Jia, Z. W.: Slope length effects on soil loss for steep slopes, Soil Sci. Soc. Am. J., 64, 1759–1763, 2000.

Liu, Q. J., An, J., Zhang, G. H., and Wu, X. Y.: The effect of row grade and length on soil erosion from concentrated flow in furrows of contouring ridge system, Soil Till. Res., 160, 92–100, 2016.

Martínez-Casasnovas, J. A., Ramos, M. C., and Benites, G.: Soil and water assessment tool soil loss simulation at the sub-basin scale in the alt penedÈs–anoia vineyard region (ne spain) in the 2000s, Land Degrad. Dev., 27, 160–170, https://doi.org/10.1002/ldr.2240, 2016.

Mohamadi, M. A. and Kavian, A.: Effects of rainfall patterns on runoff and soil erosion in field plots, International Soil and Water Conservation Research, 3, 273–281, 2015.

Munodawafa, A.: Assessing nutrient losses with soil erosion under different tillage systems and their implications on water quality, Phys. Chem. Earth Pt. A/B/C, 32, 1135–1140, https://doi.org/10.1016/j.pce.2007.07.033, 2007.

Park, S., Oh, C., Jeon, S., Jung, H., and Choi, C.: Soil erosion risk in Korean watersheds, assessed using the revised universal soil loss equation, J. Hydrol., 399, 263–273, https://doi.org/10.1016/j.jhydrol.2011.01.004, 2011.

Peng, T. and Wang, S. J.: Effects of land use, land cover and rainfall regimes on the surface runoff and soil loss on karst slopes in southwest china, Catena, 90, 53–62, 2012.

Renard, K. G., Foster, G. R., Weesies, G. A., Mccool, D. K., and Yoder, D. C.: Predicting soil erosion by water: a guide to conservation planning with the revised universal soil loss equation (rusle), Agriculture Handbook, United States Department of Agriculture (USDA), USA, 1997.

Rickson, R. J.: Can control of soil erosion mitigate water pollution by sediments?, Sci. Total Environ., 468–469, 1187–1197, https://doi.org/10.1016/j.scitotenv.2013.05.057, 2014.

Sharpley, A. N. and Williams, J. R.: Epic-erosion/productivity impact calculator: 1. model documentation, Technical Bulletin – United States Department of Agriculture, 4, 206–207, 1990.

Shi, Z. H., Cai, C. F., Ding, S. W., Wang, T. W., and Chow, T. L.: Soil conservation planning at the small watershed level using RUSLE with GIS: a case study in the Three Gorge Area of China, Catena, 55, 33–48, 2004.

Tan, B. X., Li, Z. Y., Wang, Y. H., Yu, P. T., and Liu, L. B.: Estimation of vegetation coverage and analysis of soil erosion using remote sensing data for Guishuihe drainage basin, Remote Sensing Technology and Application, 20, 215–220, 2005.

Terranova, O., Antronico, L., Coscarelli, R., and Iaquinta, P.: Soil erosion risk scenarios in the Mediterranean environment using RUSLE and GIS: an application model for Calabria (southern Italy), Geomorphology, 112, 228–245, 2009.

Tian, Y., Wang, S., Bai, X., Luo, G., and Xu, Y.: Trade-offs among ecosystem services in a typical karst watershed, sw china, Sci. Total Environ., 566–567, 1297–1308, https://doi.org/10.1016/j.scitotenv.2016.05.190, 2016.

Wallbrink, P. J., Roddy, B. P., and Olley, J. M.: A tracer budget quantifying soil redistribution on hillslopes after forest harvesting, Catena, 47, 179–201, https://doi.org/10.1016/S0341-8162(01)00185-0, 2002.

Wang, J., Zou, B., Liu, Y., Tang, Y., Zhang, X., and Yang, P.: Erosion-creep-collapse mechanism of underground soil loss for the karst rocky desertification in chenqi village, puding county, guizhou, China, Environ. Earth Sci., 72, 2751–2764, 2014.

Wang, S. J. and Li, Y. B.: Problems and development trends about researches on karst rocky desertification, Adv. Earth Sci., 22, 573–582, 2007.

Wang, X., Zhao, X., Zhang, Z., Yi, L., Zuo, L., Wen, Q., Liu, F., Xu, J., Hu, S., and Liu, B.: Assessment of soil erosion change and its relationships with land use/cover change in china from the end of the 1980s to 2010, Catena, 137, 256–268, https://doi.org/10.1016/j.catena.2015.10.004, 2016.

Wang, Y., Cai, Y. L., and Pan, M.: Soil erosion simulation of the Wujiang River Basin in Guizhou Province Based on GIS, RUSLE and ANN, Geology in China, 41, 1735–1747, 2014.

Wu, L., Liu, X., and Ma, X.: Application of a modified distributed-dynamic erosion and sediment yield model in a typical watershed of a hilly and gully region, Chinese Loess Plateau, Solid Earth, 7, 1577–1590, https://doi.org/10.5194/se-7-1577-2016, 2016.

Xu, Y. Q. and Peng, J.: Effects of simulated land use change on soil erosion in the Maotiao River watershed of Guizhou Province, Resources Science, 30, 1218–1225, 2008.

Xu, Y. Q., Shao, X. M., Kong, X. B., Peng, J., and Cai, Y. L.: Adapting the rusle and gis to model soil erosion risk in a mountains karst watershed, guizhou province, china, Environ. Monit. Assess., 141, 275–286, https://doi.org/10.1007/s10661-007-9894-9, 2008.

Yang, Q., Xie, Y., Li, W., Jiang, Z., Li, H., and Qin, X.: Assessing soil erosion risk in karst area using fuzzy modeling and method of the analytical hierarchy process, Environ. Earth Sci., 71, 287–292, https://doi.org/10.1007/s12665-013-2432-x, 2014.

Zeng, L. Y., Wang, M. H., and Li, C. M.: Study on soil erosion and its spatio-temporal change at Hongfeng Lake watershed based on RUSLE Model, Hydrogeology and Engineering Geology, 38, 113–118, https://doi.org/10.16030/j.cnki.issn.1000-3665.2011.02.003, 2014.

Zhou, F. J., Chen, M. H., and Liu, F. X.: The rainfall erosivity index in Fujian Province, Soil Water Conserv., 9, 13–18, 1995.

4

Microscale and nanoscale strain mapping techniques applied to creep of rocks

Alejandra Quintanilla-Terminel[1], Mark E. Zimmerman[1], Brian Evans[2], and David L. Kohlstedt[1]

[1]Department of Earth Sciences, University of Minnesota, Minneapolis, MN 55455, USA
[2]Earth, Atmospheric and Planetary Sciences, Massachusetts Institute of Technology, Cambridge, MA 02139, USA

Correspondence to: Alejandra Quintanilla-Terminel (aqt@alum.mit.edu)

Abstract. Usually several deformation mechanisms interact to accommodate plastic deformation. Quantifying the contribution of each to the total strain is necessary to bridge the gaps from observations of microstructures, to geomechanical descriptions, to extrapolating from laboratory data to field observations. Here, we describe the experimental and computational techniques involved in microscale strain mapping (MSSM), which allows strain produced during high-pressure, high-temperature deformation experiments to be tracked with high resolution. MSSM relies on the analysis of the relative displacement of initially regularly spaced markers after deformation. We present two lithography techniques used to pattern rock substrates at different scales: photolithography and electron-beam lithography. Further, we discuss the challenges of applying the MSSM technique to samples used in high-temperature and high-pressure experiments. We applied the MSSM technique to a study of strain partitioning during creep of Carrara marble and grain boundary sliding in San Carlos olivine, synthetic forsterite, and Solnhofen limestone at a confining pressure, P_c, of 300 MPa and homologous temperatures, T/T_m, of 0.3 to 0.6. The MSSM technique works very well up to temperatures of 700 °C. The experimental developments described here show promising results for higher-temperature applications.

1 Introduction

During plastic deformation of a crystalline material, strain is accommodated by a variety of deformation mechanisms, including atomic diffusion, mechanical twinning, grain boundary sliding, and dislocation glide, climb, and cross slip, which may operate separately or in combination. Deformation mechanism maps are conceptual tools that divide stress, temperature, and strain rate space into different fields with a particular deformation mechanism prevalent in each. Simplified flow laws are established by assuming specific micromechanical models corresponding to the prevalent mechanisms and using experimental data to fit the material constants in the theoretical flow laws (Ashby, 1972; Frost and Ashby, 1982). Such simplified flow laws are a useful first step for the extrapolation of laboratory results to larger-scale geomechanical problems, but it is also possible that the relative activity of the various deformation processes might change significantly when extended to natural conditions of rate, mean pressure, chemical environment, and temperature. Thus, the quantification of the strain due to each process and the determination of the interactions among the processes is crucial for establishing robust microphysical models that can be used to interpret observations of natural microstructures and nanostructures and that can describe natural deformation at larger spatial and temporal scales.

Quantification of the strain accommodated by each mechanism is not an easy task. Ideally, the microstructural evolution of a polycrystalline material would be monitored continuously as a sample is strained. However, high pressures and temperatures are required when deforming rocks through plastic or viscous processes in the laboratory, making microscopic observations of deformation in situ a significant experimental challenge. Fortunately, even stepwise observations of the amount of strain accommodated by individual intracrystalline mechanisms could significantly enhance our understanding of high-temperature creep.

In this paper, we describe a technique for microscale strain measurement (MSSM), which permits strain mapping at micrometer and submicrometer scales, and discuss several observations of strain partitioning in rocks deformed at high pressures and temperatures. To quantify the deformation, the coordinates of the material points (at the microscale) must be identified before and after deformation. Then, the strain tensor can be computed from their relative displacements. We used two lithography techniques, described in Sect. 2, to print a grid of points onto the polished surface of a rock. After deformation, we computed strain following the formulation described in Sect. 3. We applied the MSSM technique to four different high-temperature experiments described in Sect. 4.

2 Lithographic techniques

The literal meaning of lithography, writing on stones, is very appropriate for our application. More generally, lithography refers to different printing techniques that allow a pattern to be transferred onto a substrate. Microlithograpy and nanolithography are widely used in the semiconductor industry (Razegui, 2006; Mack, 2006) to pattern micrometric and nanometric circuits onto silicon wafers. Applying these techniques to nonconductive, heterogeneous materials is challenging, but we developed two protocols using photolithography and electron-beam (e-beam) lithography to produce patterns on the surfaces of Carrara marble, Solnhofen limestone, San Carlos olivine, and synthetic forsterite. The patterns had features with spatial dimensions on the order of 1 µm down to tens of nanometers. In this section, we will first describe both lithographic processes and then provide the specific protocols used for our applications. The communities developing and using microfabrication and nanofabrication techniques are extremely active. Undoubtedly, new processes and improvements will allow for other innovative applications to geomaterials.

2.1 Photolithography and e-beam lithography

Figure 1 summarizes the steps needed for the lithographic process. Both photolithography and electron-beam lithography require special equipment, which is usually present in standard clean laboratories used to fabricate solid-state devices. The process described here is commonly referred as "lift-off" in the microfabrication community. It uses a sacrificial layer of polymer, which is a film that can be spread uniformly on the sample, to create a pattern on the sample surface (Razeghi, 2006). Note that the microfabrication processes used in solid-state engineering require extra steps to ensure the functionalities of the devices (conductivity and reflectivity, among others). We present the steps required to deposit micrometric and nanometric patterns on a rock surface with the main objective of printing a particular

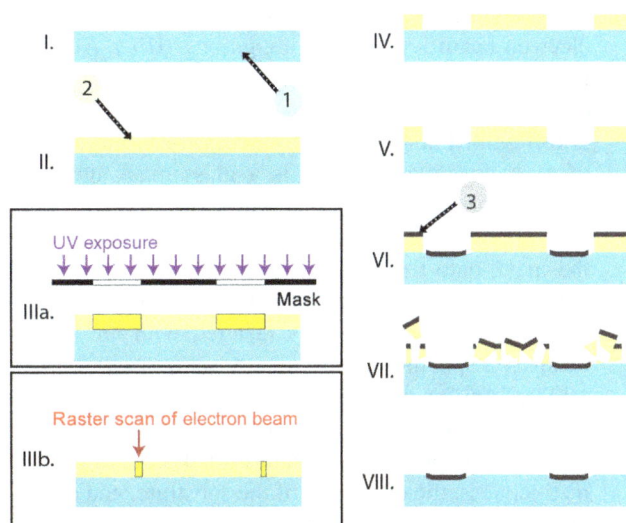

Figure 1. Lithographic process: (1) rock, (2) photoresist or PMMA, and (3) metal layer (sputtered or evaporated). Each step is described in more detail in the text: (I) sample preparation, (II) coating with a polymer (photoresist for photolithography, PMMA for e-beam lithography), (III) exposure to UV rays for photolithography (IIIa) or an electron beam for e-beam lithography (IIIb), (IV) development of the coating, (V) etching, (VI) metal deposition, (VII) dissolution of the remaining polymer, and (VIII) patterned sample.

grid designed to track strain at a granular level. Quintanilla-Terminel (2014) describes in more detail the different exploratory steps that were taken before developing the following protocols.

  I. Sample preparation: the surface to be patterned needs to be polished to a roughness of less than 0.5 µm. The details of the polishing technique will depend on the composition, porosity, and grain size of the rock. For e-beam lithography, the sample must be flat with less than 1 µm of variation in height for each 1 mm^2 region.

 II. Coating with a polymer: the surface then needs to be uniformly coated with a polymer sensitive to the energy used to transfer the pattern. A photoresist is sensitive to UV light (photolithography) and poly(methyl methacrylate), or PMMA, is sensitive to an electron beam (e-beam lithography). The coating method will depend on the sample geometry. Spin coating is the standard method for flat samples. The sample is held on a rotating device by a vacuum chuck and the polymer is spread onto the surface to uniform thickness by spinning the sample at a controlled speed, typically around 3000 rpm. For more challenging geometries, such as half cylinders or curved surfaces, a paintbrush or an aerosol spray can be used, but care must be taken to ensure that the polymer thickness is relatively uniform from sample to sample.

III. Exposure to UV rays for photolithography (IIIa) or an electron beam for e-beam lithography (IIIb). Although both techniques allow for the transfer of a custom-designed pattern, the process by which this is achieved is different. Photolithography necessitates the creation of a shadow mask, which is a glass mask imprinted with the wanted design analogous to a negative in the photographic process. UV light is then shined through the mask onto the sample coated with the photoresist. E-beam lithography does not necessitate a mask; instead the pattern is directly rastered with an electron beam at a specific dose (typically set at around 500 to $1000\,\mu C\,cm^{-2}$) onto the PMMA covering the sample. The correct exposure time and the dose of the electron beam depend on the nature of the polymer used, its thickness, the chemistry of the substrate, and the resolution of the pattern. For both photolithography and e-beam lithography, the pattern can be overexposed or underexposed. Overexposed features or patterns will appear larger than designed, while those underexposed will be smaller or will not appear at all.

IV. Development of the coating: the structure of the polymer exposed to the UV light or to the electron beam changes, and a chemical bath is used to dissolve this sacrificial layer. The pattern is now exposed on the sample surface. The development time depends on the exposure time and on the resolution of the pattern. Some play exists between the exposure time and the development time; ultimately, the lithographer must determine the optimum match for the intended application.

V. Etching: the exposed surface is slightly etched before proceeding to the metal deposition. This step was found to be important in geomaterials to allow for a good adhesion of the metal layer. The etching step can be dry (via reactive or nonreactive gas) or wet (typically in a diluted HCl or HF bath). If the sample cannot be in contact with water, a dry etch would be preferable.

VI. Metal deposition: a physical vapor deposition is used to coat the sample surface with a thin film of a chosen metal. There is latitude in the choice of the metal sputtered. Our applications require a metal that can be seen before and after deformation and that has a minimum interaction with the rock. Therefore, the parameters that are important in the choice of metal include adhesion, melting temperature, visibility, and interactions with the rock.

VII. Dissolution of the remaining polymer: the remaining polymer is dissolved, leaving only the metal deposited onto the sample surface.

VIII. Patterned sample: the patterned sample is now ready to be deformed.

Both photolithography and e-beam lithography can be used for decorating rocks; a combination of both would also be possible. The main factors to take into account are the size of the area on which the pattern will be deposited, the design of the pattern (shape and resolution), and the sample geometry. Photolithography is a good choice when many samples are required, when large areas are to be patterned, and when the sample geometry is more complicated. For applications requiring a pattern with submicron features, e-beam lithography will be more appropriate. Because e-beam lithography does not necessitate a mask, it would be faster to implement for single applications. In the following sections we provide a detailed protocol for each technique. With some adaptation, particularly regarding the exposure, developing steps, and the choice of the metal to be deposited, these protocols might be used on other geomaterials.

2.2 Photolithography applied to Carrara marble

Photolithography was carried out at the MIT Microsystems Technology Laboratory (MTL), where half cylinders of Carrara marble were patterned. First, each half cylinder was polished with a succession of aluminum oxide suspensions down to a 0.5 μm particle size to produce a flat, mirror-like surface. After polishing, the pattern was produced using the following steps.

I. The polished half cylinders were dried in a convection oven at 130 °C for 20 min.

II. The sample surfaces were manually coated with a light-sensitive photoresist, Fujifilm OCG 825 g-line, using a paintbrush; the thickness of the polymer was determined to be ~400 nm using a profilometer. Each half cylinder was then baked for 20 min in a convection oven at 90 °C.

III. The pattern inscribed in a photomask was transferred onto the polymer coating by exposing the coating to ultraviolet light in a mask aligner for 60 s. The exposure time depends on the substrate, the photoresist, and photoresist thickness. Our photomask was created using computer-aided design (CAD) software to create the pattern and a laser to draw the pattern onto a glass plate covered with chrome and photoresist, followed by a typical etch-back technique (Mack, 2006). Photomasks are also available from various commercial sources in the microfabrication industry.

IV. The nonexposed polymer was dissolved using a developer, Fujifilm OCG 934, for 30 to 60 s.

V. The samples were then very lightly etched with an acidic solution (a 1 mass % dilute solution of HCl for 1 s).

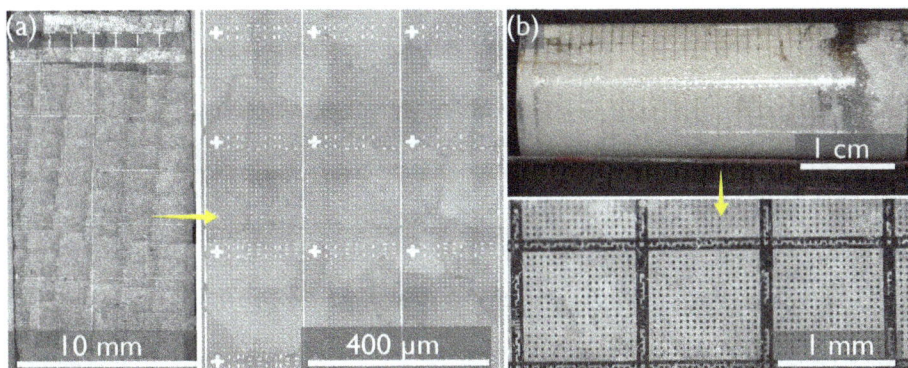

Figure 2. Patterned Carrara marble using photolithography. (**a**) The optical light micrograph on the left shows the patterned half cylinder with more than 1.5 million markers, while the micrograph on the right presents a zoomed-in view of an area revealing the 2 µm wide markers with centers spaced every 10 µm and the embedded numbering system that allows us to locate each marker before and after deformation. (**b**) The photograph on top illustrates the structure of a patterned cylinder, while the optical light micrograph on the bottom shows a closer view of the patterned curved surface revealing 20 µm wide markers with centers spaced every 50 µm.

VI. The surface of the samples was sputtered with a double layer composed of 20 nm of chromium and 30 nm of gold using a plasma sputterer.

VII. Finally, the protective layer of photoresist was removed with 2-(2-aminoethoxy)ethanol, N-methyl-2-pyrrolidine at 100 °C for 30 min.

Photolithography is a very flexible patterning technique. Patterns can be deposited on large surfaces with a customized design, little limitation exists regarding the geometry of the sample, and a wide selection of metals can be used. For example, the 10×20 mm patterned area in Fig. 2a contains roughly 1.5 million markers spaced every 10 µm and is composed of a double layer of chromium and gold. The polished surface of the outside of a Carrara marble cylinder in Fig. 2b is patterned with markers spaced every 50 µm and composed of a double layer of titanium and gold. The main disadvantages of photolithography are related to pattern resolution. Although the resolution of photolithography has been pushed to submicrometric scales (Dong et al., 2014), the resolution is limited to a few micrometers for nonconventional substrates like rocks and to hundreds of micrometers for patterns on curved surfaces. Furthermore, a photomask is needed, and its fabrication adds an additional step to the protocol. More details on the photolithography process and the development of the described protocol are given by Quintanilla-Terminel (2014) and Quintanilla-Terminel and Evans (2016).

2.3 Electron-beam lithography applied to olivine and Solnhofen limestone

The e-beam lithography process was carried out at the Minnesota Nano Center (MNC). The forsterite and San Carlos olivine samples were polished using diamond-lapping film down to a 0.5 µm particle size. The Solnhofen limestone samples were polished with aluminum oxide powder following the same procedure used on the Carrara marble samples. The electron-beam lithography steps are described below.

I. A wafer of San Carlos olivine, Solnhofen limestone, or synthetic forsterite was initially left in a vacuum oven at 130 °C for 20 min to dry.

II. The sample was first coated with PMMA diluted in 4 vol % chlorobenzene. The coating was done with a precision spin coater spinning at 3000 rpm for 60 s to produce a thickness of the PMMA film of 350 nm (evaluated with ellipsometry). The sample was then baked on a hot plate at 180 °C for 10 min. Afterward, the sample was coated with 15 nm of gold in a plasma sputterer to avoid charging.

III. The patterning was achieved using a Vistec EBPG5000+ e-beam lithography system. First, the design was created using CAD software. In e-beam lithography, a dose array is typically used in order to find the optimum exposure dose for each material and each resolution. To find the optimum exposure for San Carlos olivine, forsterite, and Solnhofen limestone, patterns of different resolution were exposed with doses ranging from 300 to 1200 µC cm^{-2} in increments of 50 µC cm^{-2}. The optical microscope image in Fig. 3a illustrates the results of a dose array performed on a sample of San Carlos olivine. The visible radial lines forming the spokes of a wheel are 10 µm thick and were exposed to 900 µC cm^{-2}. Inside each rectangle along the radial lines, different doses were tested for grids lines of 50, 100, and 200 nm in width. Examples of good exposure and overexposure are shown in the secondary electron micrographs in Fig. 3a and b, respectively. Based on these results, we selected optimal doses of 900 µC cm^{-2} for submicron features, 750 µC cm^{-2} for micrometer features on San Carlos

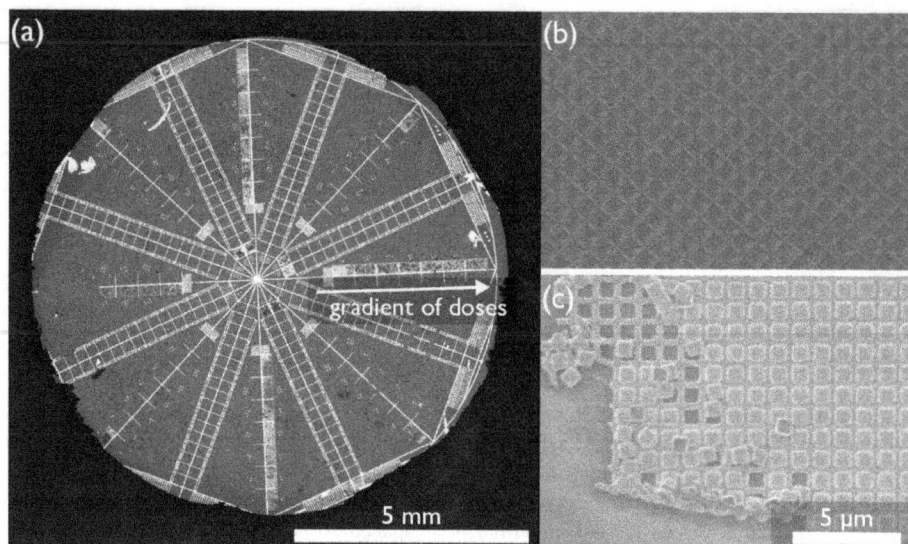

Figure 3. (a) Light micrograph of a patterned San Carlos olivine crystal. Electron-beam lithography was used with varying doses along the radial lines. All radial lines 10 μm thick and exposed to a 900 μC cm^{-2} dose are visible; inside each rectangle different doses were tested: examples of a well-exposed and an overexposed area are seen in panel **(b)** and **(c)**, respectively. In the overexposed area, the pattern appears blurred, and causes the rounding of the corners in the grid. An underexposed area will typically show no signs of the pattern. The dose test allowed for the identification of the optimum dose to achieve the desired resolution: 700 μC cm^{-2} for 200 nm wide lines.

olivine and synthetic forsterite, and 800 μC cm^{-2} for the submicron features on Solnhofen limestone. A variety of layouts were designed to meet the needs of the type of application; for example, the wheel design (Fig. 3a) was developed to evaluate the inelastic strain distribution produced during a torsion experiment.

IV. The first step in the developing process required the dissolution of the gold layer using a solution of KI, routinely called a gold etcher. The sample was first submerged in the KI solution for 30 s, then rinsed in deionized (DI) water for 10 s, and finally rinsed in isopropanol for 10 s. The sample was then developed. Two developing processes were explored. The typical developing solution for PMMA is a 3 : 1 mixture of isopropanol and methyl isobutyl ketone. The coated sample was submerged in this solution for 60 s. If a higher resolution is required during developing, a mixture of 3 : 1 isopropanol and DI water can be used. This second mixture was used with samples that were exposed to about half of the dose previously described in order to avoid overdeveloping the features (Yasin et al., 2002). In any case, the choice of the particular developer, exposure, and developing time should be adapted to the patterned substrate and the resolution needed.

V. Two etching processes were explored; a wet etch using HF diluted to 1 vol % applied for 1 s and a chlorine-gas etch using an Oxford plasma etcher for 1 min. Although the two processes produced similar results, the wet etch was simpler to implement.

VI. A 110 nm thick layer of Cr was deposited in a plasma evaporator.

VII. To dissolve the remaining PMMA, the sample was immersed in 2-(2-aminoethoxy)ethanol, N-methyl-2-pyrrolidine at 120 °C for at least 4 h. Gentle, 1 s bursts of sonication were sometimes necessary to completely remove the PMMA.

The resulting patterned surfaces of San Carlos olivine and Solnhofen limestone are illustrated in Fig. 4. Notice the difference in scale compared to the patterned marble surface in Fig. 2; the images of the patterned Carrara marble in Fig. 2 are optical light micrographs, whereas the images of the San Carlos and Solnhofen limestone in Fig. 4 are scanning secondary-electron micrographs. The high resolution, the sharpness of lines, and the pattern profile can be better appreciated in images of the patterns deposited on San Carlos olivine obtained by atomic-force microscopy (AFM; Fig. 5). It is apparent that electron-beam lithography permits much higher resolution than photolithography; in theory, nanometer scales can be achieved (Vieu et al., 2000; Manfrinato et al., 2013). For geomaterials, a resolution of 20 nm is easily attained. The pattern is rastered directly onto the sample without the need for a mask, which facilitates tests with different designs. However, because the rastering process has to be repeated for each sample, e-beam lithography is much slower and more costly than photolithography. Furthermore, samples have to fit into the electron-beam lithography system, limiting the geometry of the sample that can be patterned. In the Vistec EBPG5000+ used at the MNC at the

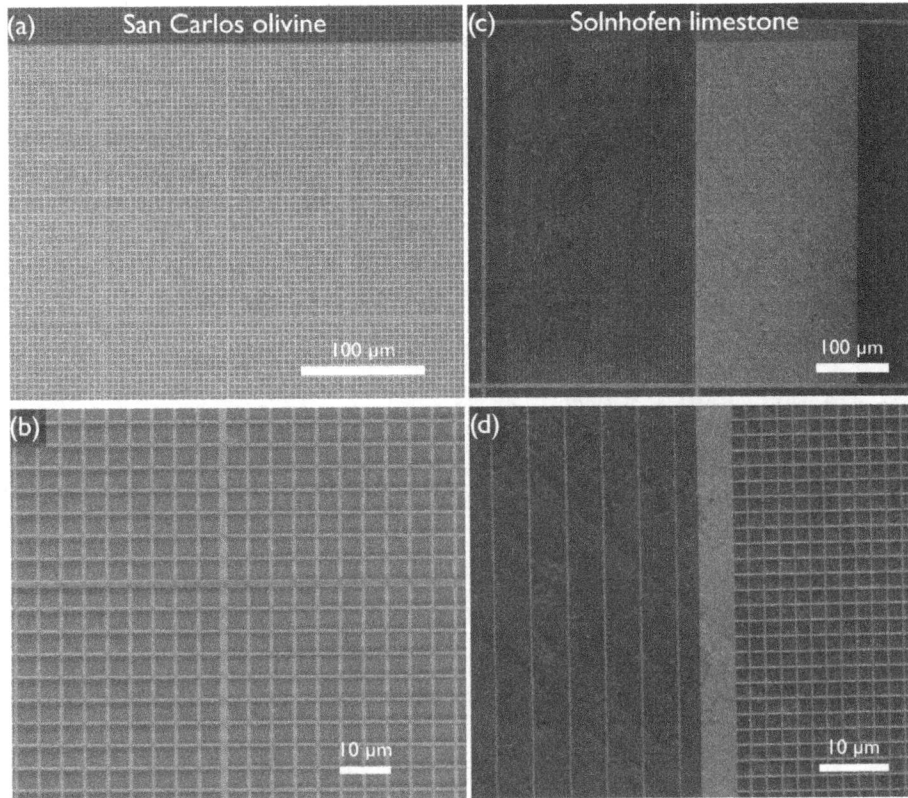

Figure 4. Secondary electron images of patterned San Carlos olivine (**a, b**) and Solnhofen limestone (**c, d**). The grid in the San Carlos olivine sample is composed of 500 μm lines spaced every 4.5 μm (**a**); a zoom in is seen in panel (**b**). For Solnhofen limestone, we combined 200 nm straight lines with a regular grid of 200 μm lines spaced every 2 μm (**c**); a zoom into both areas is seen in panel (**d**).

University of Minnesota, the sample has to be as thin and flat as a silicon wafer (i.e., no more than 1 mm thick), which is a geometric constraint that has to be considered when designing deformation experiments.

3   Strain analysis

The regular grid that is printed onto the sample is used to track the strain accommodated by the sample. Indeed, the ultimate goal of using the MSSM technique is to understand which micromechanical processes accommodate strain in a deforming rock and to relate the observed microstructure to the imposed macroscopic deformation conditions. For this purpose, it is necessary to calculate the local strains using microscale gauge lengths within larger regions of interest.

3.1   Need for a regular grid

Different techniques allow for the computation of strain at a local scale; most are image-based particle tracking techniques and therefore use a Lagrangian description in which strain is calculated by following a material point before and after deformation (Reddy, 2013; Malvern, 1969). The more widely used strain analysis technique, digital image corre-

lation (DIC), has been successfully applied in a variety of rock mechanics applications. For instance, DIC was used to map the localization of damage in a heterogeneous carbonate (Dautriat et al., 2011), to quantify the role of crystal slip and grain boundary sliding during creep of synthetic halite (Bourcier et al., 2013), and to better understand creep of ice (Chauve et al., 2015; Grennerat et al., 2012). DIC is an excellent tool for extracting displacement and strain measurements at a microscopic scale (Bruck et al., 1987; Sutton et al., 2009) by comparing digital micrographs of the sample at different stages of deformation. The strain is computed using various algorithms that allow the observer to track blocks of pixels and cross correlate them between deformation steps (Bornert et al., 2008). Each block of pixels therefore needs to be unique and recognizable between deformation steps.

High-temperature and high-pressure deformation often changes the reflectivity of the material to a point at which blocks of pixels are no longer recognizable between deformation steps. Because our experimental setup makes it difficult to analyze incrementally strained samples, we require a technique that allows us to identify a material point independently of the amount of deformation the material has experienced. We therefore rely on the analysis of an initially regular grid characterized after deformation (Allais et al., 1994;

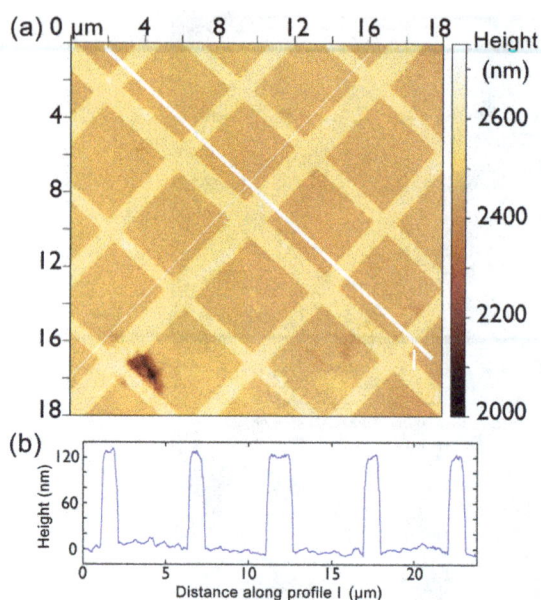

Figure 5. Atomic force micrographs of patterned San Carlos olivine **(a)** and the height profile **(b)** of the surface along profile 1. The 500 nm wide chromium lines are spaced every 4.5 μm and are 110 nm high. The sharpness of the lines can be appreciated in the AFM data.

Ghadbeigi et al., 2012; Karimi, 1984; Moulart et al., 2007; Sharpe, 2008; Wu et al., 2006; Martin et al., 2014). Because our experiments are performed at high temperature and high pressure with a metal jacket surrounding the sample, the grid is introduced in the middle of the sample, following different variations of a split-cylinder assembly (Raleigh, 1965). The initial reference grid is formed by deposition using either photolithography or e-beam lithography. Other deposition methods have been used, such as sputtering through a commercial screen (Xu and Evans, 2010) or introducing a metal mesh between the sample halves (Spiers, 1979). Recently, Hiraga (2015) used a focused ion beam to groove lines on samples composed of synthetic forsterite and diopside in order to track grain displacement and rotation during diffusion creep. In all the techniques, the main challenge is to accurately identify each individual marker or line before and after deformation. Various complications can arise; for instance, the markers can be unstable at high temperature, or the split-cylinder surfaces can weld together at high pressures, making the recovery of the patterned surface challenging or impossible. Nevertheless, if the identification of individual markers after deformation is accomplished, then spatial variations in relative displacement can be used to compute the strain field across the reference surface. Lithography has the advantages that higher resolutions can be achieved, the patterns are more robust and they survive high-temperature, high-pressure deformation, and the specific patterns can be custom designed to account for the grain size of the material or the research questions to be answered. Furthermore, because lithographic

techniques allow the researcher to create extremely regular patterns with known dimensions, the strain can be computed by locating the markers only after deformation. It is, however, advisable to image the sample before deformation as well, as this step could provide additional information (particularly if the grain boundaries are visible) and eliminate the potential error introduced by irregularities in the undeformed grid.

3.2 Strain computation

3.2.1 Marker location

Different image processing techniques can be used for locating the markers before and after deformation, such as a convolution (Biery et al., 2003) or a Hough transform algorithm (Quintanilla-Terminel and Evans, 2016). Depending on the application, the user could rely on the regularity of the grid and not identify the markers before deformation. Often, some manual input is required to ensure that the marker is correctly identified after deformation.

3.2.2 The *n*-point analysis

The *n*-point analysis is an inversion technique that allows the user to probe the strain at different scales. The positions of *n* material points are registered before and after deformation, and the best fit of the deformation gradient tensor transforming the undeformed to the deformed material lines related to these *n* material points is computed. This inversion assumes that the deformation is homogenous over the area spanned by the *n*-point ensemble. In practice, the technique requires the following steps. (1) Imaging the region of interest (ROI) before deformation. Based on the size and spacing of the markers, this step can be performed with light or electron-beam microscopy. Note that the strain can also be computed without imaging the area before deformation, relying on the regularity of the grid. (2) Registration of the coordinates of all markers in the ROI before deformation. (3) Deformation experiments under pressure and temperature. (4) Imaging of the same ROI after deformation. (4) Registration of the coordinates of all markers in the ROI after deformation. (5) Determination of deformation gradient tensor **F** using the *n*-point inversion technique. To eliminate rigid-body translations, the material displacements are calculated relative to a moving centroid for every set of *n* points. That is, the coordinates of each point *i* before and after deformation, X_i and x_i, respectively, are referenced to the centroid of the set of *n* points before and after deformation, C_n and c_n. The 2-D deformation gradient tensor **F** is determined from the relation between material displacement vectors before, $dX_i = X_i - C_n$, and after deformation, $dx_i = x_i - c_n$ using a least-squares estimate.

The deformation gradient tensor, **F**, is determined by calculating the least-squares fit for the ensemble of *n* material vectors, i.e., minimizing the sum of the squares of the differ-

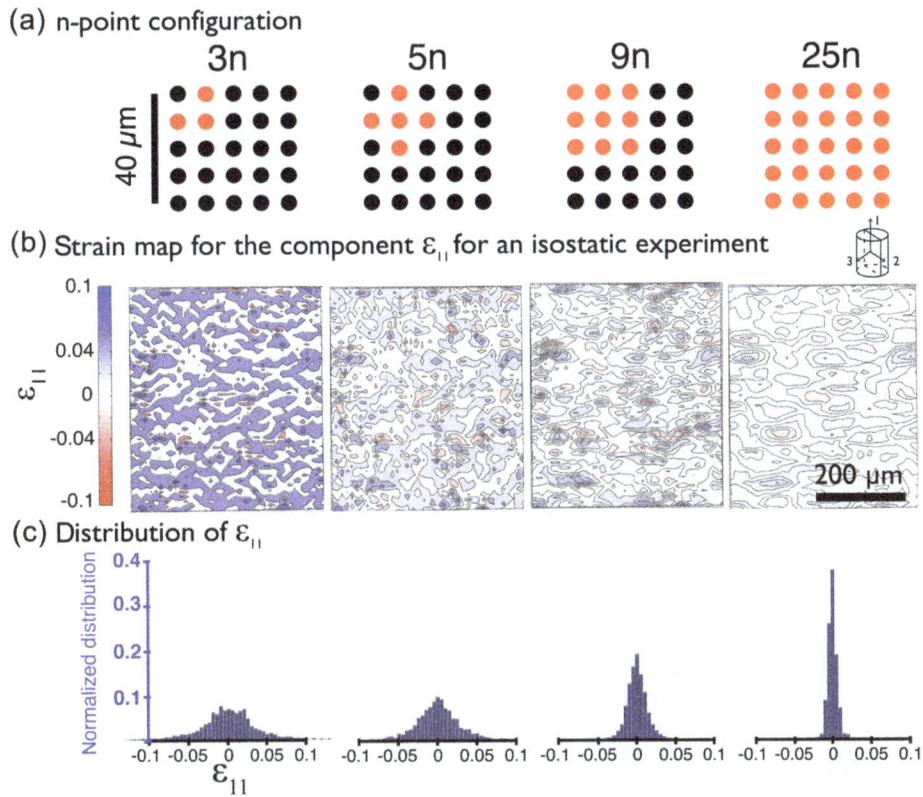

Figure 6. (a) The n-point configurations for $n = 3, 5, 9$, and 25. **(b)** Corresponding strain maps and strain distributions for strain along axis 1 in the sample reference (see insert for reference frame) for a Carrara marble cylinder isostatically annealed at $T = 700\,°C$ and $P_c = 300\,MPa$ for 3 h. **(c)** Corresponding strain distribution for each strain map; it can be observed that the mean is always 0 but the spread of the distribution becomes smaller with increasing n. Modified after Quintanilla-Terminel and Evans (2016).

ence between the modeled and the measured material lines, $dx_i^* = \mathbf{F} \cdot d\mathbf{X}$ and dx_i, respectively. Consequently, the tensor $\mathbf{F}$ describes the homogeneous deformation that best fits the displacement field of n material points. Four different configurations of n points are illustrated in Fig. 6a.

3.2.3   Computation of the strain tensor

The deformation gradient tensor $\mathbf{F}$ can be decomposed into a product of a rotation tensor $\mathbf{R}$ and either $\mathbf{U}$, a right-stretch tensor, or $\mathbf{V}$, a left-stretch tensor. $\mathbf{U}$ can be diagonalized following Eq. (1):

$$\mathbf{U} = \mathbf{Q}^T \mathbf{D} \, \mathbf{Q}. \tag{1}$$

The Hencky strain tensor, $\boldsymbol{\varepsilon}$, is defined in Eq. (2):

$$\boldsymbol{\varepsilon} = \mathbf{Q}^T \ln \mathbf{D} \, \mathbf{Q}. \tag{2}$$

An optical or electron micrograph allows for a 2-D strain inversion. The strain tensor, $\boldsymbol{\varepsilon}$, is a 2 by 2 tensor with three independent components that can be computed in the sample reference as seen in the insert in Fig. 7, one along the 1 axis, a second along the 2 axis, and a third shear component. In Fig. 7, the three components for an area of Carrara

marble deformed to 11 % compressive strain at $T = 600\,°C$, $P_c = 300\,MPa$, and a strain rate of $3 \times 10^{-5}\,s^{-1}$ are contoured over a map of the grain boundaries of the analyzed area. The geological convention is used in which positive strains correspond to shortening strains and negative strains correspond to lengthening strains. Topographic information would be required to compute the full 3-D strain tensor. However, in the split-cylinder setup, the out-of plane component of the strain tensor can be inferred from the 2-D tensor if the deformation is isochoric and axisymmetric as explained in Quintanilla-Terminel and Evans (2016).

3.3   Resolution of the technique

The resolution of the strain analysis technique depends on two main factors that should be assessed for each application before interpreting the strain results. The main sources of error are related to the location of the markers and the strain induced by the sample preparation. Any error in the marker locations will propagate into the strain calculation. This error is a function of the resolution of the images and the image analysis technique used to locate the markers. The second source of error comes from actual local strains produced by

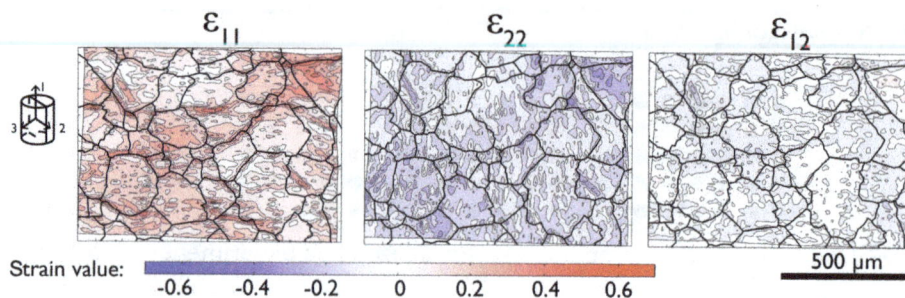

Figure 7. Strain maps for the three components of the 2-D strain tensor for a sample of Carrara marble deformed at $T = 600\,°C$ and $P_c = 300\,MPa$ to 11 % shortening strain at $3 \times 10^{-5}\,s^{-1}$. The strain tensor was computed using the n-point technique with $n = 9$. The grain boundaries are overlaid on the strain maps. Positive strains correspond to shortening.

the preparation of the sample. This error has to be estimated with a "zero strain" experiment by measuring the local strain in a sample left under the pressure and temperature conditions for a time length typical of the experiment.

The n-point inversion technique assumes homogeneous deformation for all n points, and it therefore introduces a characteristic spatial dimension for the local inelastic strain. In practice, the inversion is realized point by point; for each point, its $n-1$ neighbors are located. Thus, when n is larger, the computation error is reduced, providing that strain is actually homogenous within the ROI. However, when n is larger, the characteristic length scale of the strain heterogeneity is also larger. Thus, a compromise exists between the resolution of spatial variations in local strain and the errors involved in its calculation (Quintanilla-Terminel and Evans, 2016; Xu and Evans, 2010; Bourcier et al., 2013).

Figure 6 illustrates the different n configurations and their corresponding strain map and strain distribution along the 1 axis (in the sample reference) for a "zero strain" experiment. This experiment was performed on a half cylinder of Carrara marble at $T = 700\,°C$ and $P_c = 300\,MPa$ in order to evaluate the resolution of the experimental series in Quintanilla-Terminel and Evans (2016). The mean strain is zero for all n point configurations, but the standard deviation is larger for smaller n. Both the "real" strain coming from experimental artifacts and the computational error coming from the marker locations are evaluated. In the Carrara marble series we chose the $9n$ point as a compromise, with a strain gauge length of $20\,\mu m$ and an experimentally estimated uncertainty in the strain of ± 0.001 (Quintanilla-Terminel and Evans, 2016).

4 Application to creep of rock

The choice of grid design and patterning technique should reflect the test geometry, physical characteristics of the target rocks, and research goals, including specific hypotheses regarding constitutive behavior. The design choices made for tests on Carrara marble and San Carlos olivine, synthetic forsterite, and Solnhofen limestone can be used for illustration. In the first case, Carrara marble deformation by power-

law creep (Renner and Evans, 2002; Renner et al., 2002), we wished to investigate the partitioning of the strain between different deformation mechanisms to determine which were dominant and to identify the internal state variables needed for a more accurate flow law (Evans, 2005). Patterns in this study were made using photolithography. In the second case, we wished to observe microstructures produced during dislocation creep in San Carlos olivine (Hansen et al., 2011; Hirth and Kohlstedt, 2003) and during diffusion creep in synthetic forsterite (Dillman, 2016) and Solnhofen limestone (Schmid et al., 1977). Here, our goals included determining the strain contribution of grain boundary sliding in different creep regimes and testing current constitutive models. For this work, we produced patterns with e-beam lithography.

4.1 Grid design and sample geometry

The average grain size of our Carrara marble samples was $130\,\mu m$ (Xu and Evans, 2010; Quintanilla-Terminel and Evans, 2016); therefore, patterns with micrometer resolution, as formed by photolithography, were optimal. Our designs included circular markers $2\,\mu m$ in diameter with centers spaced at $10\,\mu m$ intervals, complemented by a printed numbering system. Thus, even though there were about 1.5 million markers, each could be assigned a unique address. Strain was mapped at a scale of $20\,\mu m$ across the sample. For MSSM in samples of San Carlos olivine, synthetic forsterite, and Solnhofen limestone, all of which have grain sizes $\leq 10\,\mu m$, higher-resolution patterns were necessary. We therefore used electron-beam lithography to print two intersecting grid patterns. In the first, the lines were $500\,nm$ thick and spaced by $4.5\,\mu m$; in the second, they were $200\,nm$ thick and spaced by $2\,\mu m$. We chose lines rather than circular markers to emphasize localized offsets at grain boundaries.

Photolithography is a flexible technique that can be used to mark surfaces with varying geometry, including split cylinders or cylindrical surfaces (Figs. 2a, b, 8a). E-beam lithography has better spatial resolution but limits the surface geometry. As seen in Sect. 2, the sample has to be flat and not more than 1 mm thick. Two different composite cylinder

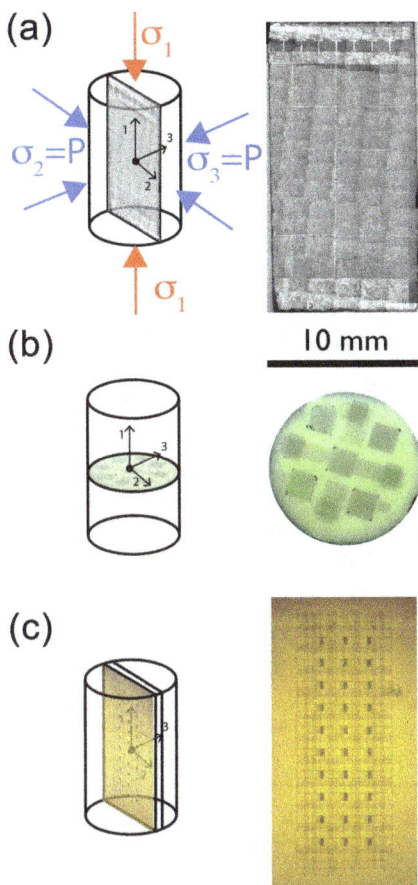

Figure 8. Experimental setup for studying creep in Carrara marble, San Carlos olivine, and fine-grained Solnhofen limestone at high temperatures and $P_c = 300\,\text{MPa}$ in a Paterson Instruments gas apparatus. **(a)** The composite half cylinder is composed of two pieces; one is gridded using photolithography, and the other is polished and sputtered with a window-shaped metal layer. Panels **(b)** and **(c)** show the composite setups used when the patterning involved e-beam lithography. San Carlos olivine was deformed using the setup in panel **(b)**: a patterned 1 mm thick disk was introduced between two short olivine samples. Solnhofen limestone was deformed using the setup in panel **(c)**: a patterned 1 mm thick rectangular slab was introduced between two half cylinders.

sample configurations were used to introduce a 1 mm thick patterned sample. In the first, a 1 mm thick disk of San Carlos olivine disk was placed between two short olivine cylinders (Fig. 8b). In the second, a 1 mm thick rectangular slab of patterned Solnhofen limestone was inserted between two half cylinders (Fig. 8c).

4.2 Application of photolithography: creep of Carrara marble

We used a split-cylinder setup following Raleigh (1965) and Spiers (1979). Half cylinders of "Lorano Bianco" Carrara marble, a standard material for deformation experiments

(Molli and Heilbronner, 1999; Heege et. al., 2002), were polished down to a 0.5 µm grit size using aluminum oxide and patterned by photolithography. Optical micrographs of the surfaces were made before and after deformation. Matlab™ was used for the strain computation. The circular markers were identified using a Hough transform and registered with the embedded coordinate system. The 2-D strain tensor was computed using the n-point algorithm. The MSSM technique allowed for an inversion of strain at different scales up to shortening strains of 36 % at a temperature of 700 °C. Notice that strains measured over an ROI at sample scales, macroscales, and microscales have the same average values but larger standard deviations as the spatial scale of measurement decreases (Fig. 9). The markers belonging to each grain are identified and classified; maps of the different strain tensor components can be used to determine strain accommodation owing to slip along grain boundaries, twinning, and intragranular deformation (Quintanilla-Terminel, 2014). Figure 10 shows samples shortened to 11 % at $T = 400$, 500, 600, and 700 °C and $P_c = 300\,\text{MPa}$; Fig. 11 illustrates samples shortened to 11, 22, and 36 % at 600 °C. In all experiments, the means of the local strains agreed remarkably well with those measured during the mechanical tests. Importantly, the spatial heterogeneity varied with strain and temperature. For more detail, see Quintanilla-Terminel and Evans (2016).

4.3 Application of electron-beam lithography: creep of San Carlos olivine and synthetic forsterite

The greatest challenge in studying strain distributions during creep in olivine rocks is the separation of the surfaces after deformation. Because creep tests must be carried out at $T > 1000$ °C and $P_c = 300\,\text{MPa}$, the metal of the grid and the rock surface strongly adhere. Nonetheless, the results described here are promising and suggest that, with further work, strain distributions during grain boundary sliding might be successfully measured (Figs. 12–13).

4.3.1 Compression experiments on San Carlos olivine at 300 MPa and 1150 °C

We used electron-beam lithography to pattern monophase polycrystalline samples prepared from powders of San Carlos olivine, Fo$_{90}$. First, samples were fabricated by cold pressing powders into a nickel can, followed by hot-pressing at $P_c = 300\,\text{MPa}$ and $T = 1250$ °C for 3 h (Hansen et al., 2011). A 1 mm thick slice was cut from the hot-pressed cylinder and one side was ground and polished in a final step using lapping films with 0.5 µm diamond grit. The olivine wafers were then patterned following the protocol described in Sect. 2. The rest of the cylinder was cut in half, one was end polished, and the three pieces were assembled as shown in Fig. 8b. This composite sample was deformed at $P_c = 300\,\text{MPa}$, $T = 1150$ °C, and a constant strain rate of

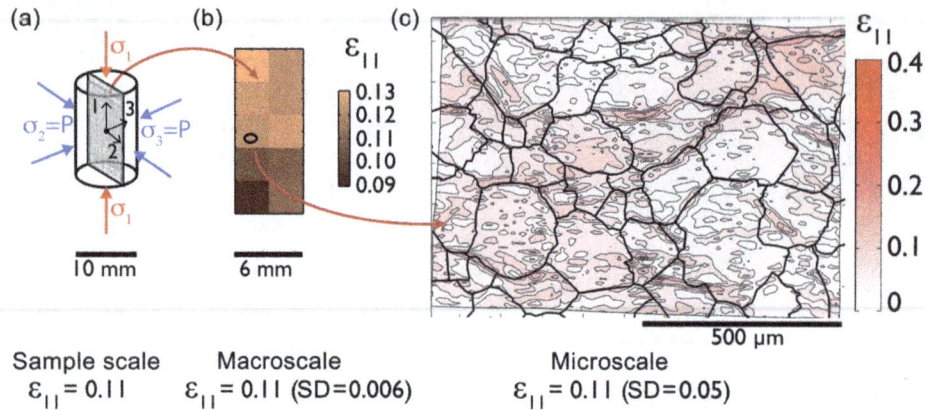

Figure 9. Strain maps over different regions of interest for a Carrara marble split cylinder deformed at $T = 600\,°C$ and $P_c = 300\,MPa$ to 11 % shortening strain at $3 \times 10^{-5}\,s^{-1}$ **(a)**. The component along the shortening direction of the strain tensor, ε_{11}, was computed at different scales. In panel **(b)**, ε_{11} was computed over the whole sample with a gauge length of 3 mm (macroscale). In panel **(c)**, ε_{11} was computed over a smaller region of interest with a 20 μm gauge length (microscale) and is mapped with the overlaying grain boundaries. Positive strains correspond to shortening. Modified after Quintanilla-Terminel and Evans (2016).

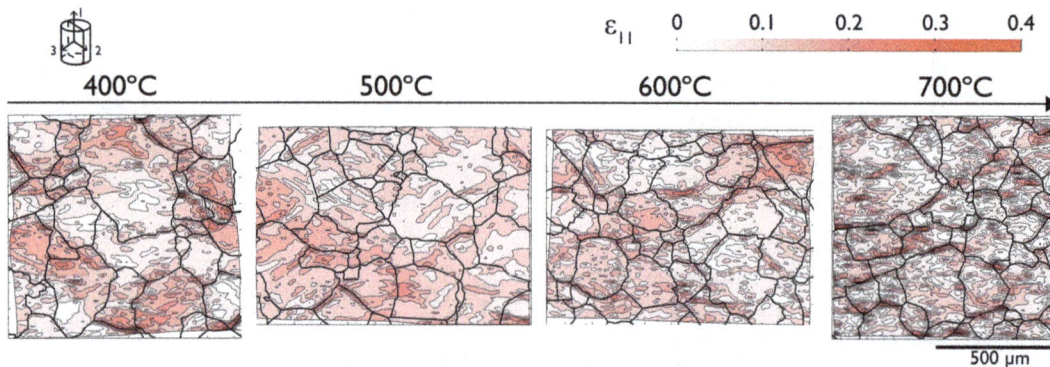

Figure 10. Strain maps for ε_{11} computed with an n-point technique (with $n = 9$) for samples of Carrara marble deformed at $T = 400, 500,$ 600, and 700 °C, $P_c = 300\,MPa$, and $3 \times 10^{-5}\,s^{-1}$ to 11 % shortening strain. The strain component along the shortening direction is mapped with the overlaying grain boundaries. Shortening is positive. Note the concentration of strain along grain boundaries and intragranular features as T increases.

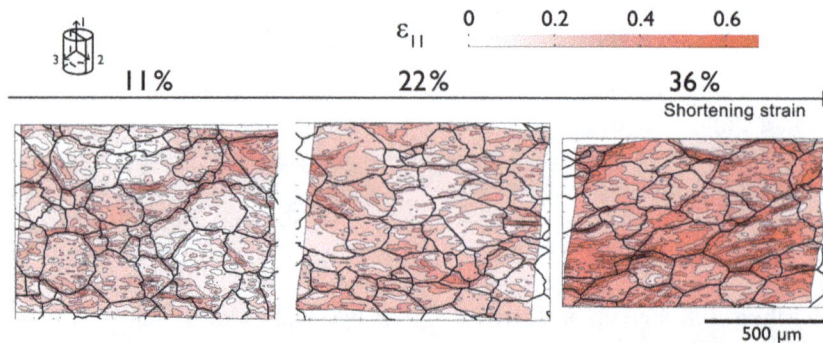

Figure 11. Strain maps for ε_{11} computed with an n-point technique (with $n = 9$) for samples of Carrara marble deformed at $T = 600$ °C, $P_c = 300\,MPa$, and $3 \times 10^{-5}\,s^{-1}$ to 11, 22, and 36 % shortening strain. The strain component along the shortening direction is mapped with the overlaying grain boundaries. Positive strains correspond to shortening strains.

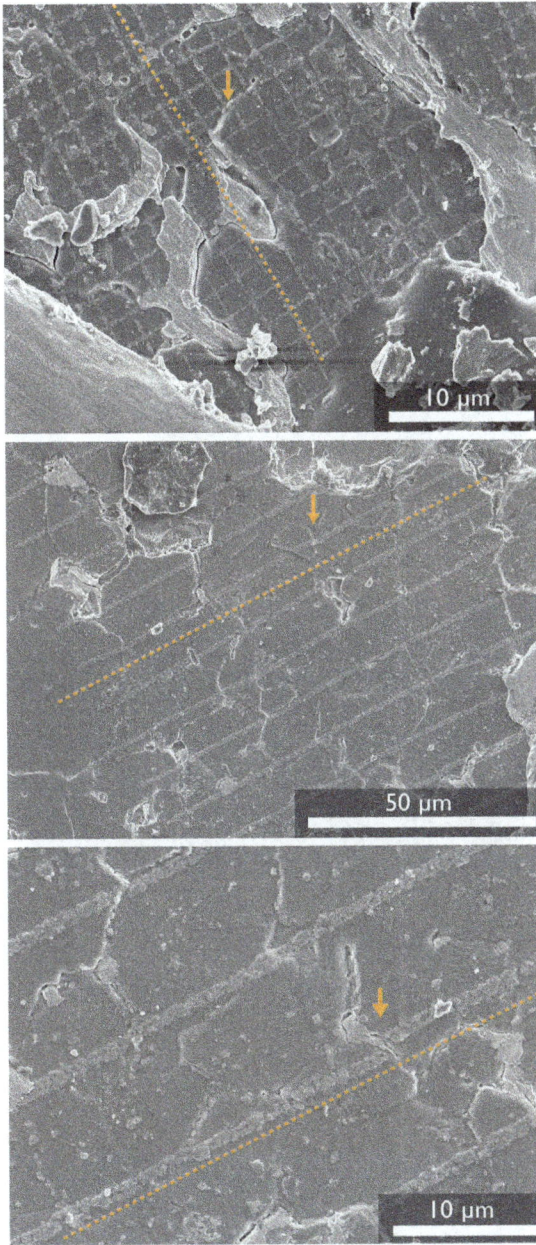

Figure 12. Secondary electron images of a gridded fine-grained San Carlos olivine after deformation at $T = 1100\,°C$ and $P_c = 300\,MPa$ to 5 % shortening at $1 \times 10^{-5}\,s^{-1}$. The deformed lines show clear offsets at grain boundaries. The dotted orange lines show evidence of the deformation of the lines in the grid. The orange arrows show grain boundary displacement.

$1 \times 10^{-5}\,s^{-1}$ to 15 % shortening strain in a gas-medium high-pressure apparatus (Hansen et al., 2011). The strength at a steady state of the composite, 245 MPa, is in excellent agreement with that determined from earlier tests with intact cylinders (Hansen et al., 2011).

Recovering the grid after deformation was more challenging because of the high temperatures and normal loads on the marked surface. We experimented with different metals and separation techniques. Sputtering a window shape with a high-melting-temperature metal, such as tungsten, on the facing sample (a technique similar to the one used for Carrara marble) was unsuccessful, and the two faces still could not be separated. Inserting a thin (0.01 mm), ring-shaped, nickel foil between the gridded surface and the facing surface was still not sufficient, but the gridded surface was successfully recovered if a full nickel foil was placed between both surfaces. Secondary electron micrographs of grids with lines 200 nm and 1 µm thick show offsets of 0.5 to 1 µm along the grain boundaries (Fig. 12a, b). On a recovered area (Fig. 13a), the strain was inverted using the n-point algorithm described above; the component ε_{22} is mapped in Fig. 13b. Note that the gridded surface is perpendicular to the shortening axis compared to the Carrara marble tests in which the gridded surface was parallel to the shortening axis (Figs. 10–11). Because the size of the recovered area is small, we cannot draw meaningful conclusions regarding the partitioning between intragranular strain and grain boundary sliding, but the technique offers further opportunities for success.

4.3.2 Compression experiments on forsterite at 1 atm and 1100 °C

Samples of synthetic forsterite were synthesized following Koizumi et al. (2010). After gridding, they were deformed at $T = 1250\,°C$ and $P_c = 1\,atm$ (Dillman, 2016) with the grid on an outer surface of a sample with a square cross section. The tests used a grid 500 nm thick and were designed to be compared with previous measurements of creep (Dillman, 2016) using a line offset caused by grain boundary sliding (Langdon, 2006). Grid recovery for these samples was much easier because deformation occurred under atmospheric pressure. We used an AFM to measure the vertical displacement of grains and to map the deformed grid (Fig. 14). The state of the initially regular grid after deformation highlights the deformation occurring along grain boundaries. 2-D offsets of $\sim 500\,nm$ are evident. More details on the 1 atm compression experiments on synthetic forsterite can be found in Dillman (2016).

4.3.3 Creep of fine-grained Solnhofen limestone

Three-piece samples of Solnhofen limestone were prepared (Fig. 8c), gridded using e-beam lithography, and deformed in compression in gas-medium apparatus (Paterson, 1990) at $T = 700\,°C$ and $P_c = 300\,MPa$ to a shortening strain of 9 % at a strain rate of $3 \times 10^{-4}\,s^{-1}$. The patterned surfaces could be separated without the addition of a metal foil, and the deformed pattern was easily found using electron microscopy. In Fig. 15 with the two secondary electron micrographs of the same area before and after deformation, the grid and the deformed lines are clearly visible; however, the grain structure is not. Additional characterization permitting the identi-

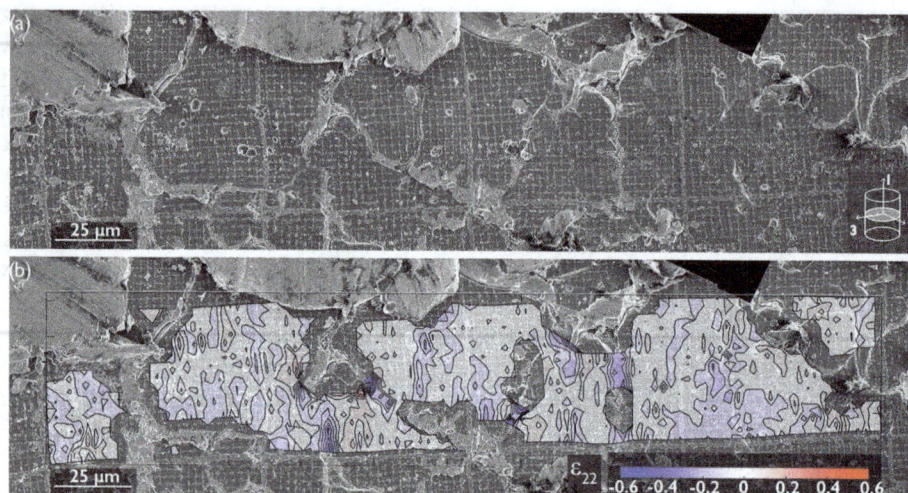

Figure 13. Strain inversion on a recovered grid in a fine-grained sample of San Carlos olivine deformed at $T = 1100\,^{\circ}\mathrm{C}$ and $P_{\mathrm{c}} = 300\,\mathrm{MPa}$ to 5 % deformation at $1 \times 10^{-5}\,\mathrm{s}^{-1}$. **(a)** Secondary electron image of the recovered area and **(b)** a superposed strain map of ε_{22}, a component of the strain tensor that is perpendicular to the shortening direction.

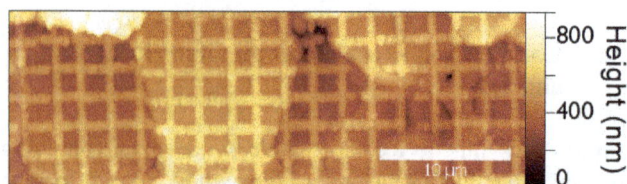

Figure 14. Atomic force micrograph of a deformed grid in synthetic forsterite deformed at $T = 1110\,^{\circ}\mathrm{C}$ and $P_{\mathrm{c}} = 1$ atm to 5 % shortening. The chromium lines are about 110 nm thick and are clearly visible in the height map. The grain structure is also clearly visible and both the height difference and the offset of the lines show evidence of grain displacement. Amanda Dillman performed the deformation experiment and the AFM observations for this sample.

fication of the grain boundaries (possibly imaging with electron backscatter diffraction) will be necessary to evaluate the contribution of the grain boundary sliding to the total strain. Nonetheless, this application demonstrates that the deformation of the lines and grid is visible and that the electron-beam lithography provides the required resolution to quantify the strain at the granular level.

5   Conclusions

Patterning using photolithography and e-beam lithography can provide maps of strain calculated over spatial scales of 10–0.5 µm. The experimental preparation is technologically demanding and labor intensive, but MSSM is a unique characterization tool that yields a detailed description of strain accommodation within intragranular and intergranular regions. The experimental protocols described here can be adapted to many other rocks, and the pattern can be designed for

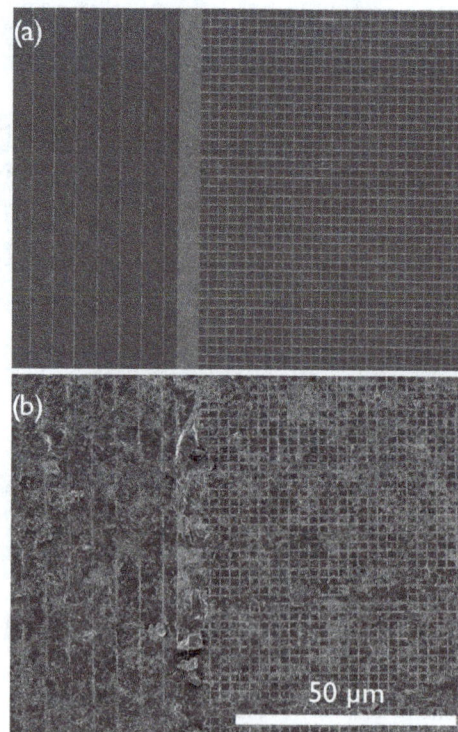

Figure 15. Secondary electron micrographs of a patterned Solnhofen limestone sample **(a)** before deformation and **(b)** after deformation at $T = 700\,^{\circ}\mathrm{C}$ and $P_{\mathrm{c}} = 300\,\mathrm{MPa}$ to 9 % shortening strain at $3 \times 10^{-4}\,\mathrm{s}^{-1}$. Before deformation, the pattern consisted of 200 nm wide lines spaced every 5 µm and a grid of 200 nm lines spaced every 2 µm. The lines were made of 110 nm thick chromium. After deformation, the lines are clearly deformed but the grain structure is not apparent.

specific purposes. The MSSM technique has the potential to improve the interpretation of microstructural observations of rocks deformed in nature and in the lab, constrain the partitioning of strain among several deformation mechanisms, and improve the constitutive modeling of creep mechanisms in the Earth.

Competing interests. The authors declare that they have no conflict of interest.

Special issue statement. This article is part of the special issue "Analysis of deformation microstructures and mechanisms on all scales". It is a result of the EGU General Assembly 2016, Vienna, Austria, 17–22 April 2016.

Acknowledgements. Matej Pec is acknowledged for his help in acquiring the SEM pictures. Amanda Dillman is acknowledged for providing the data related to the forsterite sample and for her help in acquiring the AFM data. The lithography techniques were developed at the Minnesota Nano Center (MNC) of the University of Minnesota and at the MIT Micro Technology Laboratories (MTL). AQT thanks Bryan Cord at the MNC and Kurt Broderick at the MTL for their guidance through the fabrication processes. We benefitted from enriching discussions with Matej Pec, Amanda Dillman, Renée Heilbronner, Holger Stünitz, and William Durham and thoughtful reviews by Lars Hansen and an anonymous reviewer. Support through NSF grants EAR-1520647 (UMN) and 145122 (MIT) is gratefully acknowledged.

Edited by: Florian Fusseis

References

Allais, L., Bornert, M., Bretheau, T., and Caldemaison, D.: Experimental characterization of the local strain field in a heterogeneous elastoplastic material, Acta Metallurgica et Materialia, 42, 3865–3880, https://doi.org/10.1016/0956-7151(94)90452-9, 1994.

Ashby, M. F.: A First Report on Deformation-Mechanism Maps, Acta Metallurgica, 20, 887–897, https://doi.org/10.1016/0001-6160(72)90082-X, 1972.

Biery, N., De Graef, M., and Pollock, T. M.: A Method for Measuring Microstructural-Scale Strains Using a Scanning Electron Microscope: Applications to γ-Titanium Aluminides, Metall. Mater. Trans. A, 34, 2301–2313, 2003.

Bornert, M., Brémand, F., Doumalin, P., Dupré, J.-C., Fazzini, M., Grédiac, M., Hild, F., Mistou, S., Molimard, J., Orteu, J.-J., Robert, L., Surrel, Y., Vacher, P., and Wattrisse, B.: Assessment of Digital Image Correlation Measurement Errors: Methodology and Results, Exp. Mech., 49, 353–370, https://doi.org/10.1007/s11340-008-9204-7, 2008.

Bourcier, M., Bornert, M., Dimanov, A., Héripré, E., and Raphanel, J. L.: Multiscale Experimental Investigation of Crystal Plasticity and Grain Boundary Sliding in Synthetic Halite Using Digital Image Correlation, J. Geophys. Res.-Solid Earth, 118, 511–526, 2013.

Bruck, H. A., McNeill, S. R., Sutton, M. A., and Peters, W. H.: Digital Image Correlation Using Newton-Raphson Method of Partial Differential Correction, Exp. Mech., 29, 261–267, https://doi.org/10.1007/BF02321405, 1987.

Chauve, T., Montagnat, M., and Vacher, P.: Strain Field Evolution during Dynamic Recrystallization Nucleation; A Case Study on Ice, Acta Materialia, 101, 116–124, 2015.

Dautriat, J., Bornert, M., Gland, N., Dimanov, A., and Raphanel, J.: Localized Deformation Induced by Heterogeneities in Porous Carbonate Analysed by Multi-Scale Digital Image Correlation, Tectonophysics, 503, 100–116, 2011.

Dillman, A.: Influence of Grain Boundaries and Their Composition on the Deformation Strength of High-Purity, Synthetic Forsterite, available at: http://conservancy.umn.edu/handle/11299/182257 (last access: July 2017), 2016.

Dong, J., Liu, J., Kang, G., Xie, J., and Wang, Y.: Pushing the Resolution of Photolithography down to 15 nm by Surface Plasmon Interference, Scientific Reports, 4, 5618, https://doi.org/10.1038/srep05618, 2014.

Evans, B.: Creep constitutive laws for rocks with evolving structure, in: High-Strain Zones: Structure and Physical Properties, edited by: Bruhn, D. and Burlini, L., 329–346, Geological Society of London, London, UK, https://doi.org/10.1144/GSL.SP.2005.245.01.16, 2005.

Frost, H. J. and Ashby, M. F.: Deformation-Mechanism Maps: The Plasticity and Creep of Metals and Ceramics, Pergamon Press, 1982.

Ghadbeigi, H., Pinna, C., and Celotto, S.: Quantitative strain analysis of the large deformation at the scale of microstructure: Comparison between digital image correlation and microgrid techniques, Exp. Mech., 52, 1483–1492, https://doi.org/10.1007/s11340-012-9612-6, 2012.

Grennerat, F., Montagnat, M., Castelnau, O., Vacher, P., Moulinec, H., Suquet, P., and Duval, P.: Experimental Characterization of the Intragranular Strain Field in Columnar Ice during Transient Creep, Acta Materialia, 60, 3655–3666, https://doi.org/10.1016/j.actamat.2012.03.025, 2012.

Hansen, L. N., Zimmerman, M. E., and Kohlstedt, D. L.: Grain Boundary Sliding in San Carlos Olivine: Flow Law Parameters and Crystallographic-Preferred Orientation, J. Geophys. Res.-Solid Earth (1978–2012), 116, B08201, https://doi.org/10.1029/2011JB008220, 2011.

Hiraga, T.: Grain rotation during diffusion creep of forsterite + diopside, Abstract T51I-03 presented at 2015 Fall Meeting, AGU, San Francisco, Calif., 14–18 December, 2015.

Hirth, G. and Kohlstedt, D.: Rheology of the Upper Mantle and the Mantle Wedge: A View from the Experimentalists, Inside the Subduction Factory, 138, 83–105, 2003.

Karimi, R.: Plastic flow study using the microgrid technique, Mater. Sci. Eng., 63, 267–276, https://doi.org/0025-5416/84, 1984.

Koizumi, S., Hiraga, T., Tachibana, C., Tasaka, M., Miyazaki, T., Kobayashi, T., Takamasa, A., Ohashi, N., and Sano, S.: Synthesis of Highly Dense and Fine-Grained Aggregates of Mantle Composites by Vacuum Sintering of Nano-Sized Mineral Powders, Phys. Chem. Miner., 37, 505–518, https://doi.org/10.1007/s00269-009-0350-y, 2010.

Langdon, T. G.: Grain Boundary Sliding Revisited: Developments in Sliding over Four Decades, J. Mater. Sci., 41, 597–609, https://doi.org/10.1007/s10853-006-6476-0, 2006.

Mack, C. A.: Field Guide to Optical Lithography, Spi edition. Bellingham, Wash: SPIE Publications, 2006.

Malvern, L. E.: Introduction to the Mechanics of a Continuous Medium, Prentice-Hall, 154–183, 1969.

Manfrinato, V. R., Zhang, L., Su, D., Duan, H., Hobbs, R. G., Stach, E. A., and Berggren, K. K.: Resolution Limits of Electron-Beam Lithography toward the Atomic Scale, Nano Letters, 13, 1555–1558, https://doi.org/10.1021/nl304715p, 2013.

Martin, G., Sinclair, C. W., and Lebensohn, R. A.: Microscale Plastic Strain Heterogeneity in Slip Dominated Deformation of Magnesium Alloy Containing Rare Earth, Mater. Sci. Eng.: A, 603, 37–51, https://doi.org/10.1016/j.msea.2014.01.102, 2014.

Molli, G. and Heilbronner, R.: Microstructures Associated with Static and Dynamic Recrystallization of Carrara Marble (Alpi Apuane, NW Tuscany, Italy), Geologie En Mijnbouw, 78, 119–126, https://doi.org/10.1023/A:1003826904858, 1999.

Moulart, R., Rotinat, R., Pierron, F., and Lerondel, G.: On the realization of microscopic grids for local strain measurement by direct interferometric photolithography, Opt. Laser. Eng., 45, 1131–1147, https://doi.org/10.1016/j.optlaseng.2007.06.009, 2007.

Paterson, M. S.: Rock deformation experimentation, in Brittle-Ductile Transition in: Rocks, Heard Volume, edited by: Duba, A. G., Durham, W. B., Handin, J. W., and Wang, H. F., 187–194, Am. Geophys. Union, Washington, 1990.

Quintanilla-Terminel, A.: Strain Heterogeneity during Creep of Carrara Marble, Thesis, Massachusetts Institute of Technology, available at: http://dspace.mit.edu/handle/1721.1/95556 (last access: July 2017), 2014.

Quintanilla-Terminel, A. and Evans, B.: Heterogeneity of Inelastic Strain during Creep of Carrara Marble: Microscale Strain Measurement Technique, J. Geophys. Res.-Solid Earth, 121, 2016JB012970, https://doi.org/10.1002/2016JB012970, 2016.

Quintanilla-Terminel, A. and Evans, B.: Data set of microscale strain for creep of Carrara marble [Data set], Zenodo, available at: https://doi.org/10.5281/zenodo.822501 (last access: July 2017), 2017.

Quintanilla-Terminel, A., Zimmerman, M. E., Evans, B., and Kohsltedt, D. L.: Data set for Microscale and nanoscale strain mapping techniques applied to creep of rocks [Data set], Zenodo, available at: https://doi.org/10.5281/zenodo.822490 (last access: 7 July 2017), 2017.

Raleigh, C. B.: Glide Mechanisms in Experimentally Deformed Minerals, Science, 150, 739–741, https://doi.org/10.1126/science.150.3697.739, 1965.

Razegui, M.: Fundamentals of Solid State Engineering, Springer US, 615–664, 2006.

Reddy, J. N.: An Introduction to Continuum Mechanics, Cambridge University Press, 68–84, 2013.

Renner, J. and Evans, B.: Do Calcite Rocks Obey the Power-Law Creep Equation?, Geological Society, London, Special Publications, 200, 293–307, https://doi.org/10.1144/GSL.SP.2001.200.01.17, 2002.

Renner, J., Evans, B., and Siddiqi, G.: Dislocation Creep of Calcite, J. Geophys. Res., 107, 2364, https://doi.org/10.1029/2001JB001680, 2002.

Schmid, S. M., Boland, J. N., and Paterson, M. S.: Superplastic Flow in Finegrained Limestone, Tectonophysics, 43, 257–291, https://doi.org/10.1016/0040-1951(77)90120-2, 1977.

Sharpe, W. N.: Springer handbook of experimental solid mechanics, Springer, Berlin, 2008.

Spiers, C.: Fabric Development in Calcite Polycrystals Deformed at 400 °C, Bull. Mineral., 102, 282–289, 1979.

Sutton, M. A., Orteu, J. J., and Schreier, H.: Image Correlation for Shape, Motion and Deformation Measurements: Basic Concepts, Theory and Applications, 2009 edition, New York, N.Y: Springer, 2009.

Ter Heege, J. H., De Bresser, J. H. P., and Spiers, C. J.: The Influence of Dynamic Recrystallization on the Grain Size Distribution and Rheological Behaviour of Carrara Marble Deformed in Axial Compression, Geological Society, London, Special Publications, 200, 331–553, https://doi.org/10.1144/GSL.SP.2001.200.01.19, 2002.

Vieu, C., Carcenac, F., Pepin, A., Chen, Y., Mejias, M., Lebib, A., Manin-Ferlazzo, L., Couraud, L., and Launois, H.: Electron Beam Lithography: Resolution Limits and Applications, Appl. Surf. Sci., 164, 111–117, 2000.

Wu, A., de Graef, M., and Pollock, T. M.: Grain-scale strain mapping for analysis of slip activity in polycrystalline b2rual, Philosophical Magazine, 86, 3995–4008, 2006.

Xu, L. and Evans, B.: Strain Heterogeneity in Deformed Carrara Marble Using a Microscale Strain Mapping Technique, J. Geophys. Res.-Solid Earth, 115, B04202, https://doi.org/10.1029/2009JB006458, 2010.

Yasin, S., Hasko, D. G., and Ahmed, H.: Comparison of MIBK/IPA and Water/IPA as PMMA Developers for Electron Beam Nanolithography, Microelectronic Engineering, Micro- and Nano-Engineering, 2001, 61–62 (July), 745–753, https://doi.org/10.1016/S0167-9317(02)00468-9, 2002.

5

Global patterns in Earth's dynamic topography since the Jurassic: the role of subducted slabs

Michael Rubey[1], **Sascha Brune**[2,3], **Christian Heine**[4], **D. Rhodri Davies**[5], **Simon E. Williams**[1], and **R. Dietmar Müller**[1]

[1]Earthbyte Group, School of Geosciences, the University of Sydney, Sydney, Australia

[2]Helmholtz Centre Potsdam, GFZ German Research Centre for Geosciences, Potsdam, Germany

[3]Institute of Earth and Environmental Science, University of Potsdam, Potsdam, Germany

[4]Specialist Geosciences, Shell Projects & Technology, Rijswijk, the Netherlands

[5]Research School of Earth Sciences, Australian National University, Canberra, Australia

Correspondence to: Michael Rubey (michael.rubey@sydney.edu.au)

Abstract. We evaluate the spatial and temporal evolution of Earth's long-wavelength surface dynamic topography since the Jurassic using a series of high-resolution global mantle convection models. These models are Earth-like in terms of convective vigour, thermal structure, surface heat-flux and the geographic distribution of heterogeneity. The models generate a degree-2-dominated spectrum of dynamic topography with negative amplitudes above subducted slabs (i.e. circum-Pacific regions and southern Eurasia) and positive amplitudes elsewhere (i.e. Africa, north-western Eurasia and the central Pacific). Model predictions are compared with published observations and subsidence patterns from well data, both globally and for the Australian and southern African regions. We find that our models reproduce the long-wavelength component of these observations, although observed smaller-scale variations are not reproduced. We subsequently define "geodynamic rules" for how different surface tectonic settings are affected by mantle processes: (i) locations in the vicinity of a subduction zone show large negative dynamic topography amplitudes; (ii) regions far away from convergent margins feature long-term positive dynamic topography; and (iii) rapid variations in dynamic support occur along the margins of overriding plates (e.g. the western US) and at points located on a plate that rapidly approaches a subduction zone (e.g. India and the Arabia Peninsula). Our models provide a predictive quantitative framework linking mantle convection with plate tectonics and sedimentary basin evolution, thus improving our understanding of how subduction and mantle convection affect the spatio-temporal evolution of basin architecture.

1 Introduction

At short spatial scales, Earth's topography depends on interactions between crustal structures, tectonic deformation and surface processes. However, at length scales exceeding the flexural strength of the lithosphere, topography is an expression of the force balance acting at Earth's free surface. This balance has (i) an *isostatic* component, which depends on the composition and thickness of the crust and lithospheric mantle, and (ii) a *dynamic* component, which is the deflection of Earth's surface resulting from vertical stresses generated by convection in the underlying mantle. Their relative importance varies according to the tectonic setting. Most orogens show patterns of thickened continental crust and hence isostatically compensated topography (e.g. Frisch et al., 2011), while the anomalous elevation of the stable southern African craton has been attributed to dynamic topography generated by deep-mantle upwellings (e.g. Lithgow-Bertelloni and Silver, 1998; Gurnis et al., 2000).

Dynamic topography generates significant surface deflections in both continental and oceanic regions (e.g. Ricard et al., 2006). The connection to sea level and continental flooding is direct: as continents migrate over mantle upwellings (regions of positive dynamic topography)

and downwellings (negative dynamic topography), vertical motions may be sufficiently large to control the emergence and submergence of continents (e.g. Mitrovica et al., 1989), thus respectively generating surface depressions in which sediments accumulate or erosional surfaces. Dynamic topography is identified in Earth's present-day relief via deviations from isostasy, yielding the so-called residual topography field (e.g. Kaban et al., 1999; Doin and Fleitout, 2000; Crosby et al., 2006; Winterbourne et al., 2009; Flament et al., 2013; Hoggard et al., 2016). It is recorded via anomalous vertical motion events in the basin infill (e.g. Mitrovica et al., 1989; Heine et al., 2008) or long-wavelength uplift in non-basinal areas, as seen in plateau exhumation and geomorphic features (e.g. Roberts and White, 2010). It is, therefore, a fundamental component of Earth's topographic expression and an improved understanding of the history of Earth's surface topography can only be achieved by successfully accounting for its spatial and temporal variations.

There are two principal ways to retrieve the component of dynamic support: (i) the "top-down" inverse approach, which is based on geophysical and geological observations and subtracts isostatic topography from observed surface elevations, and (ii) the "bottom-up" strategy, which infers dynamic support by determining mantle flow from Earth's internal buoyancy and viscosity distribution. In this study, we follow the bottom-up strategy and discuss our findings in relation to results from the top-down approach (e.g. Panasyuk and Hager, 2000; Kaban et al., 2003, 2004; Steinberger, 2007; Heine et al., 2010; Flament et al., 2013; Hoggard et al., 2016). Discrepancies between these two basic strategies lead to much discussion in the literature (e.g. Flament et al., 2013; Molnar et al., 2015; Colli et al., 2016; Hoggard et al., 2016). Molnar et al. (2015) argue that only small free-air gravity anomalies (< 30–$50\,\text{mGal}$) are caused by long-wavelength topography, interpreting most of the signal as isostatic. They suggest that a maximum of $300\,\text{m}$ of dynamic support is indicated by long-wavelength gravity anomalies, which is in contrast to many published geodynamic models (e.g. Flament et al., 2013). However, Colli et al. (2016) demonstrate that mantle convection with a depth-dependent viscosity generates small perturbations (geoid $37\,\text{m}$; gravity $8\,\text{mGal}$), considerable dynamic topography ($1.1\,\text{km}$) and a gravitational admittance of mostly less than $10\,\text{mGal\,km}^{-1}$. This supports the idea that relatively small long-wavelength free-air gravity anomalies do not preclude dynamic topography at amplitudes substantially larger than $300\,\text{m}$. It also helps to underpin the results of Hoggard et al. (2016) on variations of $1\,\text{km}$ in dynamically induced topography inferred from observations.

Numerous bottom-up studies have been undertaken in recent years in an attempt to constrain the amplitude, spatial pattern and time dependence of dynamic topography using numerical modelling. These studies examined the temporal evolution of dynamic support by forward modelling (e.g.

Ricard et al., 1993; Zhang et al., 2012; Flament et al., 2013), backward advection (e.g. Conrad and Gurnis, 2003; Moucha et al., 2008; Heine et al., 2010), adjoint schemes (e.g. Bunge et al., 2003; Liu et al., 2008; Spasojevic et al., 2009; Colli et al., 2017) or a back-and-forth iterative method (Glišović and Forte, 2016, 2017). Present-day dynamic topography is discussed on a global scale by Ricard et al. (1993), Steinberger (2007), Zhang et al. (2012) and Flament et al. (2013), but relatively few studies (e.g. Zhang et al., 2012; Flament et al., 2013) which use the forward modelling approach have examined the temporal evolution of dynamic support globally.

In this paper, we also use numerical forward modelling to compute the transient evolution of dynamic topography. The distribution of heterogeneity in our models is generated by solving the equations relevant to mantle convection inside a global spherical shell, with plate velocities derived from a global plate kinematic model (Seton et al., 2012, with modifications by Shephard et al., 2013) and featuring topological plate boundaries extending back to $200\,\text{Ma}$ imposed as a surface boundary condition. We compare our predicted present-day dynamic topography to previous numerical studies before discussing the temporal evolution of dynamic topography within selected regions. Finally, we propose a categorisation of regions on Earth's surface in terms of their exposure to the effects of dynamic support over the past $200\,\text{Myr}$, which is corroborated by an automated classification technique.

2   Numerical model

2.1   Model set-up

We use the global mantle convection code TERRA (e.g. Baumgardner, 1985; Bunge et al., 1997; Davies and Davies, 2009; Wolstencroft et al., 2009; Davies et al., 2013; Wolstencroft and Davies, 2017), which solves the Stokes and energy equations relevant to mantle convection inside a spherical shell, with appropriate material properties and boundary conditions (Fig. 1). Calculations are performed on a mesh with ~ 80 million nodal points (lateral nodal spacing of $\sim 28\,\text{km}$ at the surface and $\sim 14\,\text{km}$ at the core–mantle boundary (CMB); radial spacing of $\sim 22\,\text{km}$ across the entire domain), thus providing the resolution necessary to explore mantle flow at Earth-like convective vigour. Our models incorporate compressibility via the anelastic liquid approximation (ALA), with radial reference values represented through a Murnaghan equation of state (Murnaghan, 1944). Isothermal boundary conditions are specified at the surface ($273\,\text{K}$) and CMB (models M1 and M3 $4000\,\text{K}$, model M2 $3500\,\text{K}$), with the mantle also heated internally at roughly chondritic rates. The present-day surface heat flux of model M1 is $34\,\text{TW}$, which is consistent with current estimates for Earth's surface heat flux

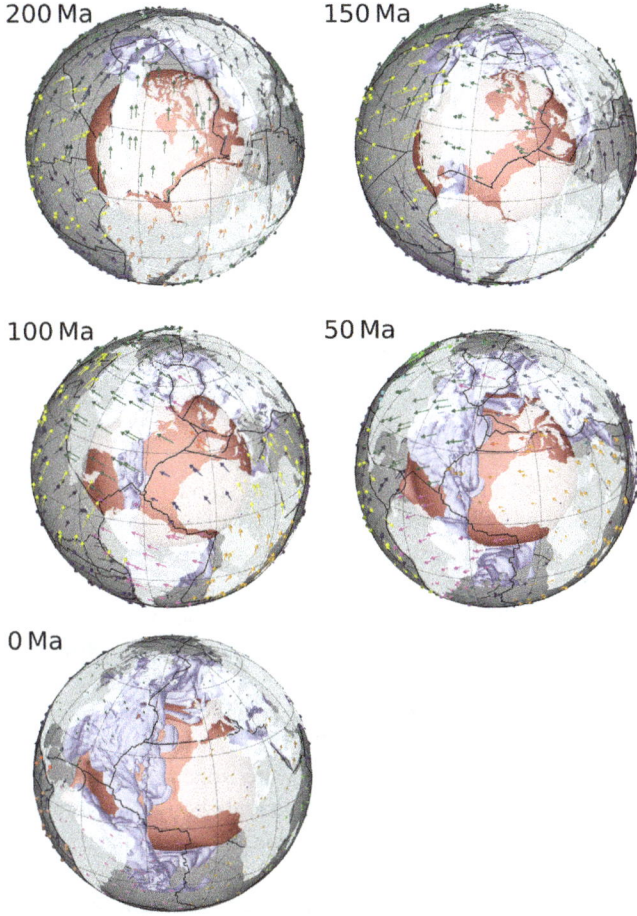

Figure 1. Schematic depiction of model run showing the evolution of the model from 200 Ma to present day. The velocity field as denoted by the coloured arrows is generated by the plate tectonic reconstruction software GPlates based on the Earthbyte 2013 plate model (see text). This velocity field is used as a time-dependent velocity boundary condition for the TERRA model runs. The model temperature field within the lower mantle is shown in terms of the 2000 and the 2800 K isotherm depicted as blue and red isosurfaces, respectively.

Table 1. Physical parameters common to all model runs. See Table 2 for the parameters varied between models.

Model parameters	
T_{surf} (surface temperature)	300 K
T_{CMB} (CMB temperature)	(varied)
H (internal heating rate)	5.5×10^{-12} W kg^{-1}
η_{ref} (reference viscosity)	1.0×10^{20} Pa s
ρ_S (surface density)	3500 kg m^{-3}
ρ_{CMB} (CMB density)	5568 kg m^{-3}
α_S (surface thermal expansivity)	3.8×10^{-5} K^{-1}
α_{CMB} (CMB thermal expansivity)	1.2×10^{-5} K^{-1}
k (thermal conductivity)	4.0 W m^{-1} K^{-1}
c (specific heat)	1081.0 J kg^{-1} K^{-1}

Table 2. Key parameters of models M1, M2 and M3. Only the viscosity profile (Fig. 2) and the temperature at the core–mantle boundary (CMB) have been varied. The viscosity structure refers to a reference viscosity η_{ref} of 1020 Pa s and comprises different viscosities for the following radial mantle layers: lithosphere (L), upper mantle (UM), transition zone (TZ) and lower mantle (LM).

Model	Viscosity structure (w.r.t η_{ref})				T_{CMB} [K]
	L	UM	TZ	LM	
M1	1000	1	10	1000	4000
M2	1000	1	10	1000	3500
M3	1000	1	10	300	4000

$$\eta = \eta_{ref}(r)\, e^{4.61(0.5 - T^*)}, \quad \text{with} \quad T^* = \frac{T - T_{surf}}{T_{CMB} - T_{surf}}, \qquad (1)$$

where T^* represents the non-dimensionalised temperature. For initial conditions, we approximate the unknown Jurassic mantle heterogeneity structure by applying a 70 Myr pre-conditioning stage during which global plate configurations are fixed to the oldest available reconstruction at 200 Ma. From here, models run forward for 200 Myr toward the present day. Note that for simplicity our models assume no chemical heterogeneity. The distribution of heterogeneity is principally dictated by the imposed plate motion histories, the material properties of the model and the heating mode. Purely thermal models, as in the present study, have previously been shown to provide a good match to seismic observations (Schuberth et al., 2009; Davies et al., 2012, 2015) and dynamic topography (Colli et al., 2017). Key model parameters are listed in Table 1 with reference values also illustrated in Fig. 2. Three separate cases are examined with the variations between these cases summarised in Table 2.

($\sim$ 47 TW) minus the contributions from crustal radioactivity (Davies and Davies, 2010). Of this 34 TW, $\sim$ 30 % is transferred across the CMB with the remainder introduced through internal heating.

A free-slip boundary condition is specified at the CMB, whilst surface velocities are assimilated via 200 Myr of plate motion histories (Seton et al., 2012; Shephard et al., 2013) generated by the plate tectonic reconstruction software GPlates (Boyden et al., 2011; Fig. 1) at discrete 1 Myr intervals. In our reference model (M1), we prescribe a radial viscosity profile with four layers, namely the lithosphere (10^{23} Pa s), the sub-lithospheric upper mantle (10^{20} Pa s), the transition zone (10^{21} Pa s) and the lower mantle (10^{23} Pa s). Viscosity also varies with temperature T, following the relation:

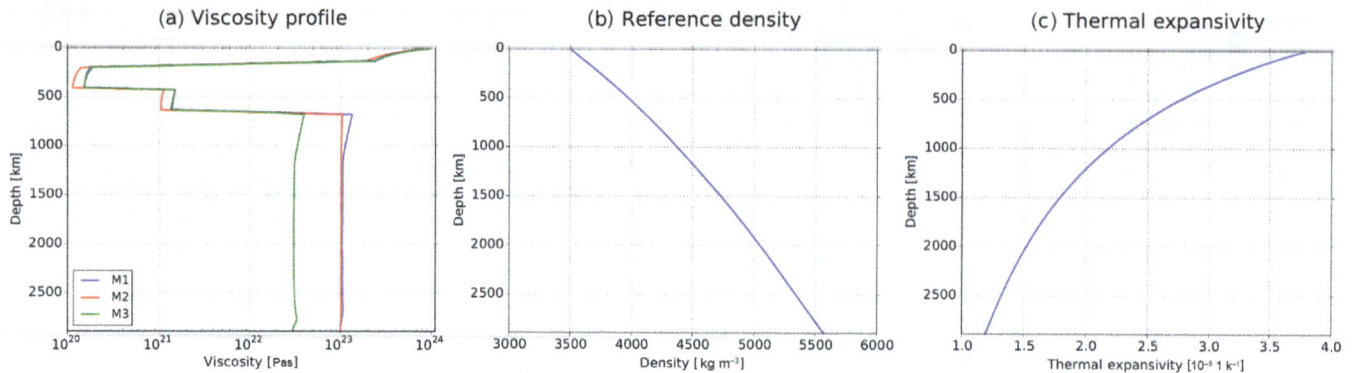

Figure 2. Radial profiles of (**a**) viscosity, (**b**) density and (**c**) thermal expansivity in our models.

Dynamic topography is computed from normal stresses at the model surface using the following formula:

$$h = \frac{1}{\Delta \rho g} \sigma_{\mathrm{rr}}, \tag{2}$$

where σ_{rr} is the normal stress, $\Delta \rho$ is the density difference between surface and air and g is the acceleration due to gravity. This is likely to yield elevations that are too high, given that we are restricted in our choice of the asthenospheric viscosity, which we limit to a minimum of 10^{20} Pa s. Estimates for the viscosity of Earth's asthenosphere are between 10^{19} and 10^{21} Pa s (e.g. Mitrovica and Forte, 2004; Steinberger and Calderwood, 2006; Petersen et al., 2010). Whilst we cannot compensate for the weaker coupling of tectonic plates to the mantle that this entails, we can account for the increased elevations predicted in our models through the introduction of a scaling. Thus, we assume simple linear scaling and introduce a modified formula for dynamic topography calculations:

$$h = \frac{1}{\Delta \rho g} \sigma_{\mathrm{rr}} \frac{\eta_0}{\eta_{\mathrm{ref}}}, \tag{3}$$

where $\eta_0 = 3 \times 10^{19}$ Pa s is an estimate of the asthenospheric viscosity, while η_{ref} is the model reference viscosity (see Table 2). This is less than the lowest reference viscosity used in our model (10^{20} Pa s) and subsequently acts to scale our predicted dynamic topography to the observed amplitudes. The scaling factor thus encapsulates our assumption that the stresses scale linearly with the ratio of asthenosphere to model viscosity found at the isostatic compensation depth of the lithosphere.

Note that to calculate dynamic topography, we focus on long-wavelength subduction-driven global flow below the lithosphere, as this is the contribution to dynamic topography that does not depend on the specific structure of the lithosphere and the depth of the lithosphere–asthenosphere boundary. Following these considerations, we replace the model thermal structure above 225 km of depth with the constant median mantle temperature of

2150 K. We subsequently re-run the model for one time step employing a rigid (zero-slip) boundary condition at the model surface and use the resulting surface normal stresses to evaluate dynamic topography. The depth above which we applied the blanking procedure has been chosen such that all lithospheric structures are encompassed. However, we varied this limiting depth between 300 and 150 km and the impact on the results was negligible. We also varied the blanking temperature of 2150 K, but since the chosen temperature is applied to the entire spherical shell, it is always neutral with respect to the mantle buoyancy structure and the amplitude of dynamic topography does not depend on the specific temperature choice. Similar approaches to discard the lithospheric buoyancy structure from the dynamic topography signal have been applied by Lithgow-Bertelloni and Silver (1998), Steinberger (2007), Conrad and Husson (2009) and Flament et al. (2013) for structures above 325, 220, 300 and 350 km of depth, respectively.

2.2 Model robustness

To assess the robustness of our results, we vary two key parameters in our reference model, M1, and compare the temporal evolution of dynamic topography at selected locations. These synthetic sample points have been picked to illustrate the effect on regions where either the influence of dynamic support at present day or through time is debated, or the geological setting of which prompts that question. Model M2 employs the same viscosity profile as M1 (Fig. 2 and Table 2) but differs in CMB temperature; a temperature of 3500 K is prescribed instead of 4000 K. Model M3, on the other hand, utilises the same CMB temperature of 4000 K, but its viscosity profile differs from model M1: the lower mantle viscosity is reduced from 1×10^{23} to 3×10^{22} Pa s.

The computed dynamic topography is not strongly affected by these variations, as demonstrated by the elevation histories of chosen sample locations in Fig. 3. Decreasing the CMB temperature has a negligible effect, with models M1 and M2 predicting very similar dynamic topography amplitudes and trends. Decreasing the lower mantle viscosity

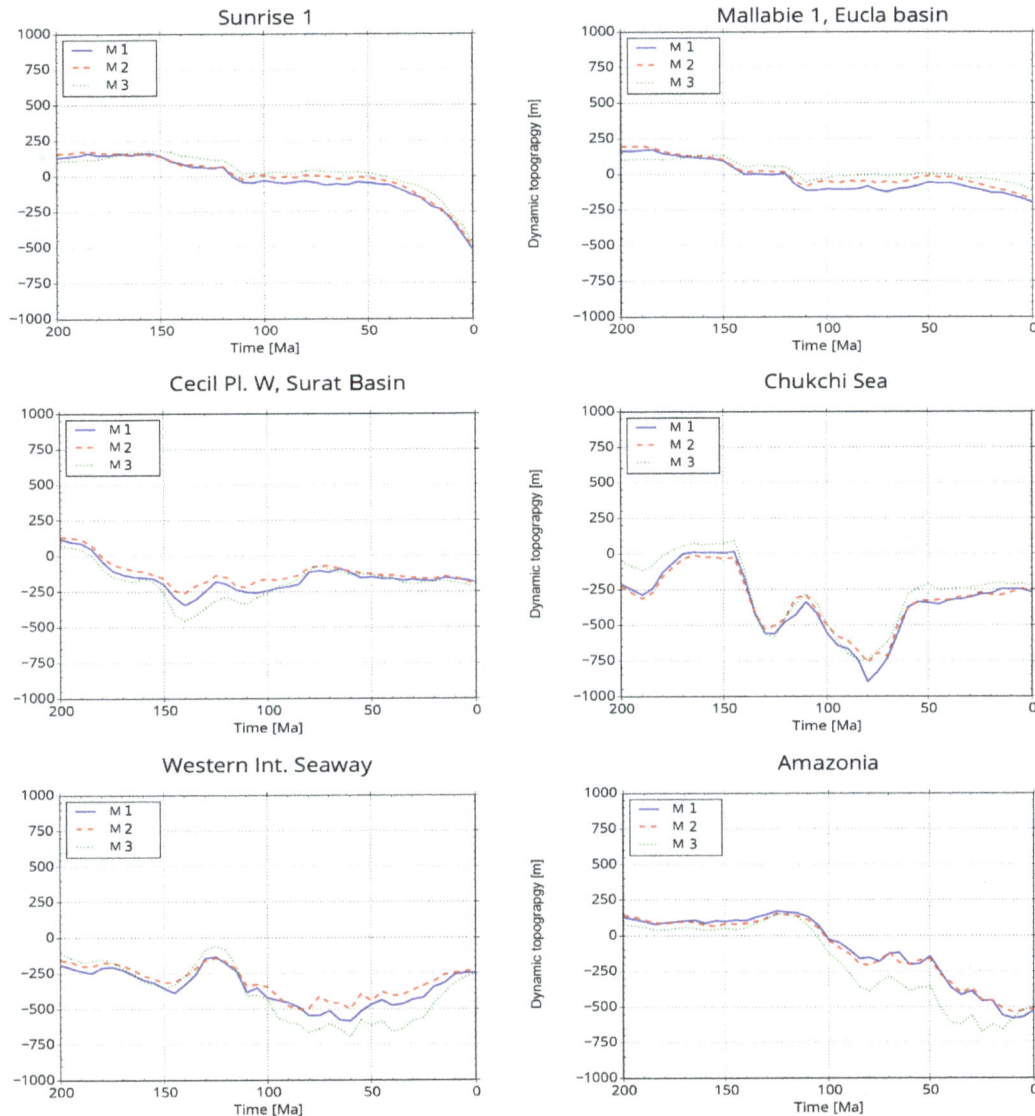

Figure 3. Dynamic topography time series of selected locations for all three models (M1–M3). Vertical motions are dominated by the plate tectonic history, while variations in core–mantle boundary temperature (M2) and lower mantle viscosity (M3) have only marginal effects on dynamic topography. For locations of sample points, see Figs. 11 and 12.

has a more pronounced influence, as slab penetration into the lower mantle is facilitated and slab-sinking rates are increased. This has two principal effects: (i) due to viscous coupling, faster-sinking slabs increase the amplitude of negative dynamic topography, and (ii) the suction effect of a faster-sinking detached slab fades quicker after subduction slows down or ceases. Nevertheless, the first-order characteristics of all three models (Fig. 3) are in good agreement, indicating that long-wavelength dynamic topography is principally controlled by the plate tectonic history and the location of subduction zones. Accordingly, for the remainder of this paper, we focus solely on our reference model, M1.

3 Model results and implications

3.1 Predicted mantle structure

Snapshots of the mantle structure predicted in model M1 are presented in Fig. 1, illustrating that the upper mantle planform is dominated by strong downwellings in regions of present-day plate convergence. In the mid-mantle, cold downwellings are prominent beneath North America (the subducted Farallon slab) and southern Eurasia (the former Neo-Tethys ocean), whilst remnants of older subduction (those encircling the former supercontinent Pangaea) are visible above the CMB. In our model M1, slabs sink at rates of $1.5\,\mathrm{cm\,year^{-1}}$, which is comparable to those

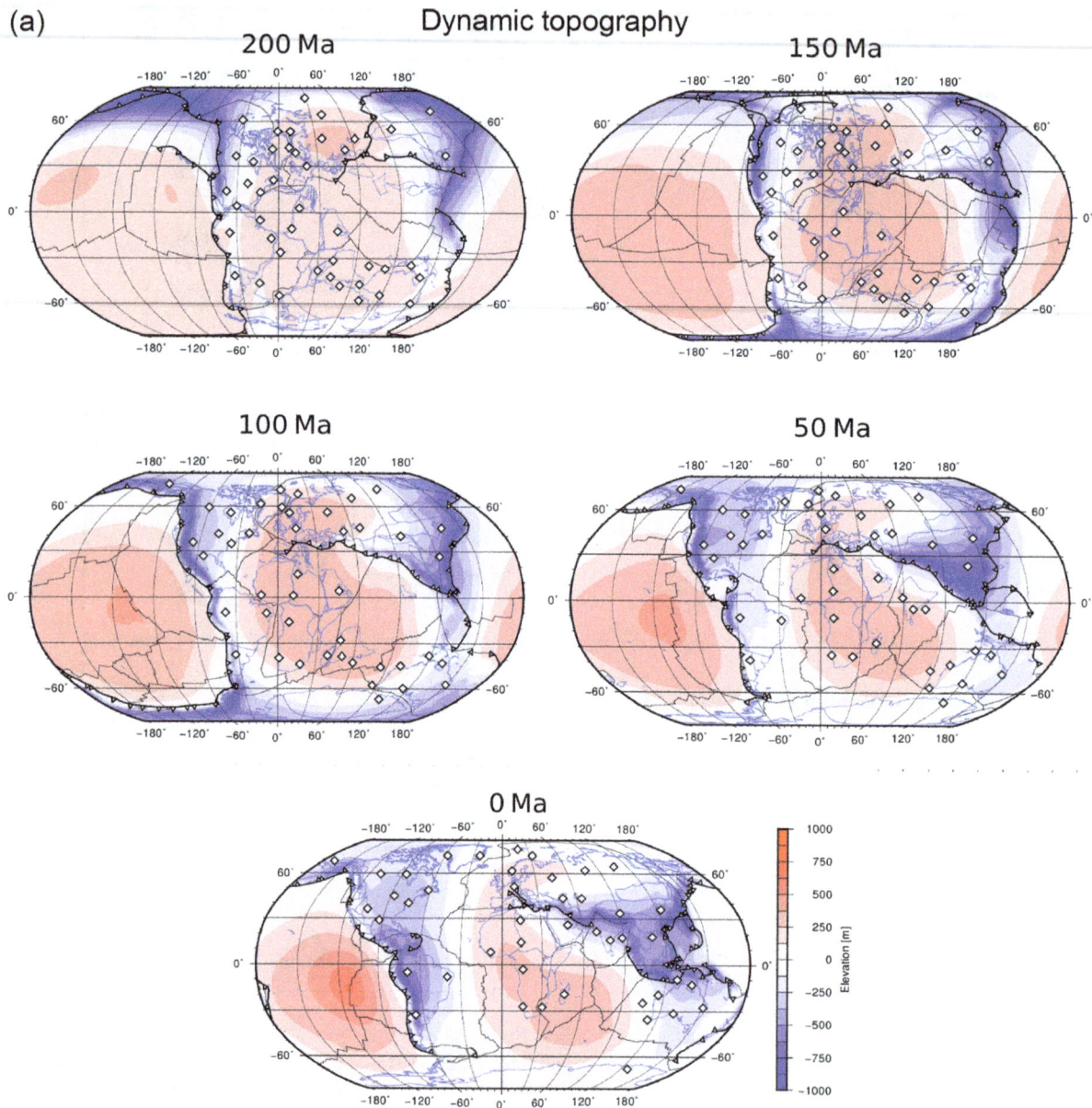

Figure 4.

inferred from seismic tomography (e.g. Li. et al., 2008; Simmons et al., 2010). These downwellings modulate the location of upwelling flow, such that it becomes concentrated beneath Africa and the Pacific (e.g. McNamara and Zhong, 2005; Davies et al., 2012, 2015). The Pacific anomaly is approximately circular, whilst the African anomaly is a NW–SE trending structure, which to the north curves eastward under Europe and to the south extends into the Indian Ocean. Note that due to (i) the short evolution times of our models and (ii) the limited temperature dependence of viscosity, plumes play a negligible role in the convective planform. Accordingly, the regions of upwelling flow beneath Africa and the Pacific discussed above principally represent passive

return flow from subduction. Nevertheless, we also examined models with different initial conditions, under which plumes were able to form. We found that dynamic topography is modified in regions of active upwelling. However, these dynamic topography highs constitute shorter wavelength features (Moucha and Forte, 2011; Hoggard et al., 2016) and, on a global scale, are less significant than the role of subducting slabs; this is why we do not examine plume-related dynamic topography in this study.

3.2 Global and regional dynamic topography

The spatio-temporal evolution of dynamic topography from case M1 is illustrated in Fig. 4a in the mantle reference frame

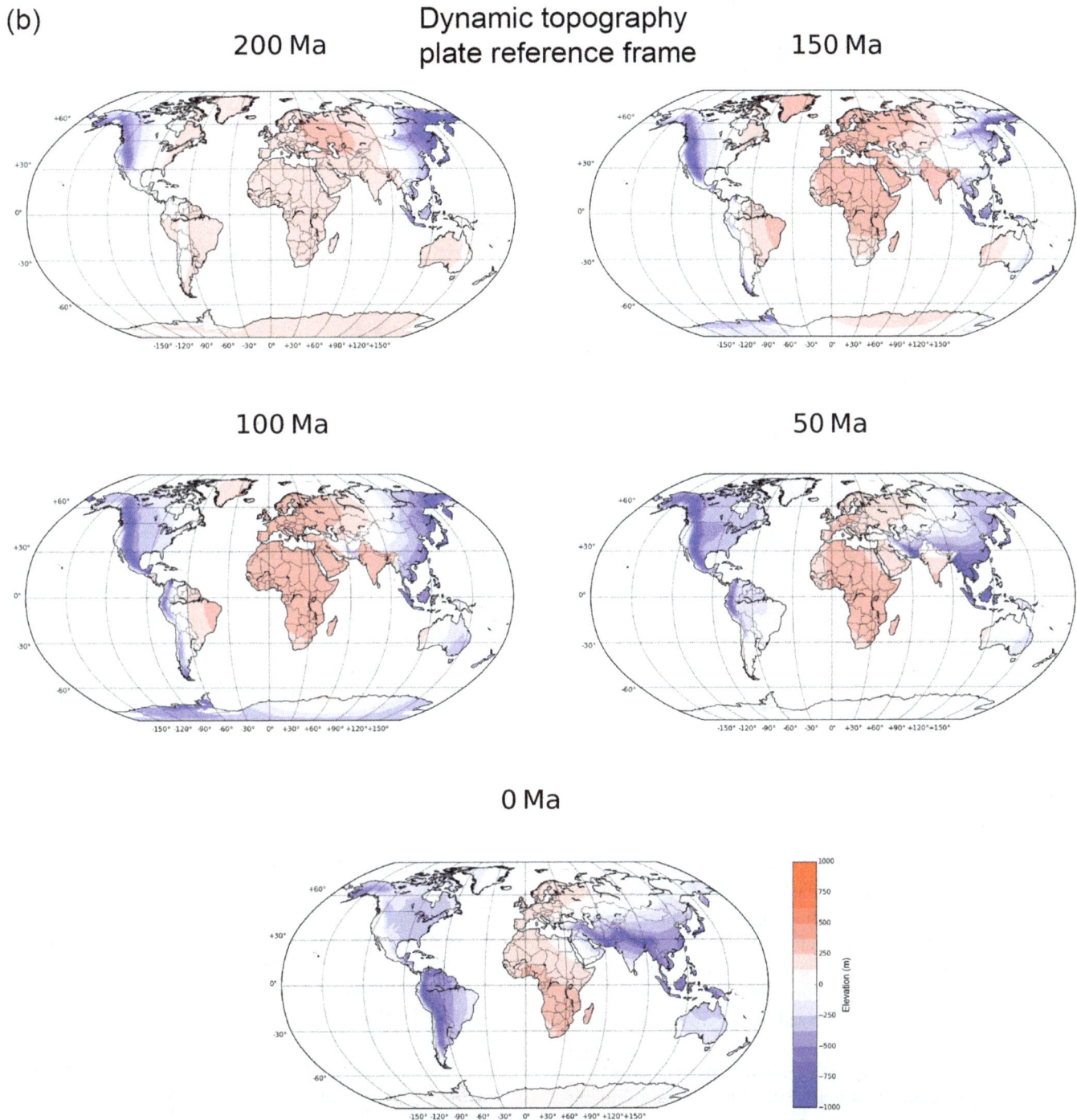

Figure 4. Evolution of global dynamic topographic elevation from 200 Ma to present. Blue colours denote negative dynamic topography, and red colours mark positive dynamic topography. (**a**) Dynamic topography in the absolute reference frame. The indicated points show synthetic sample locations as discussed in the text. Note that sample points move along with their associated tectonic plates. The evolution of dynamic topography in 10 Myr intervals can be found in the Supplement. (**b**) Dynamic topography in the plate frame of reference. The evolution of dynamic topography in 10 Myr intervals can be found in the Supplement.

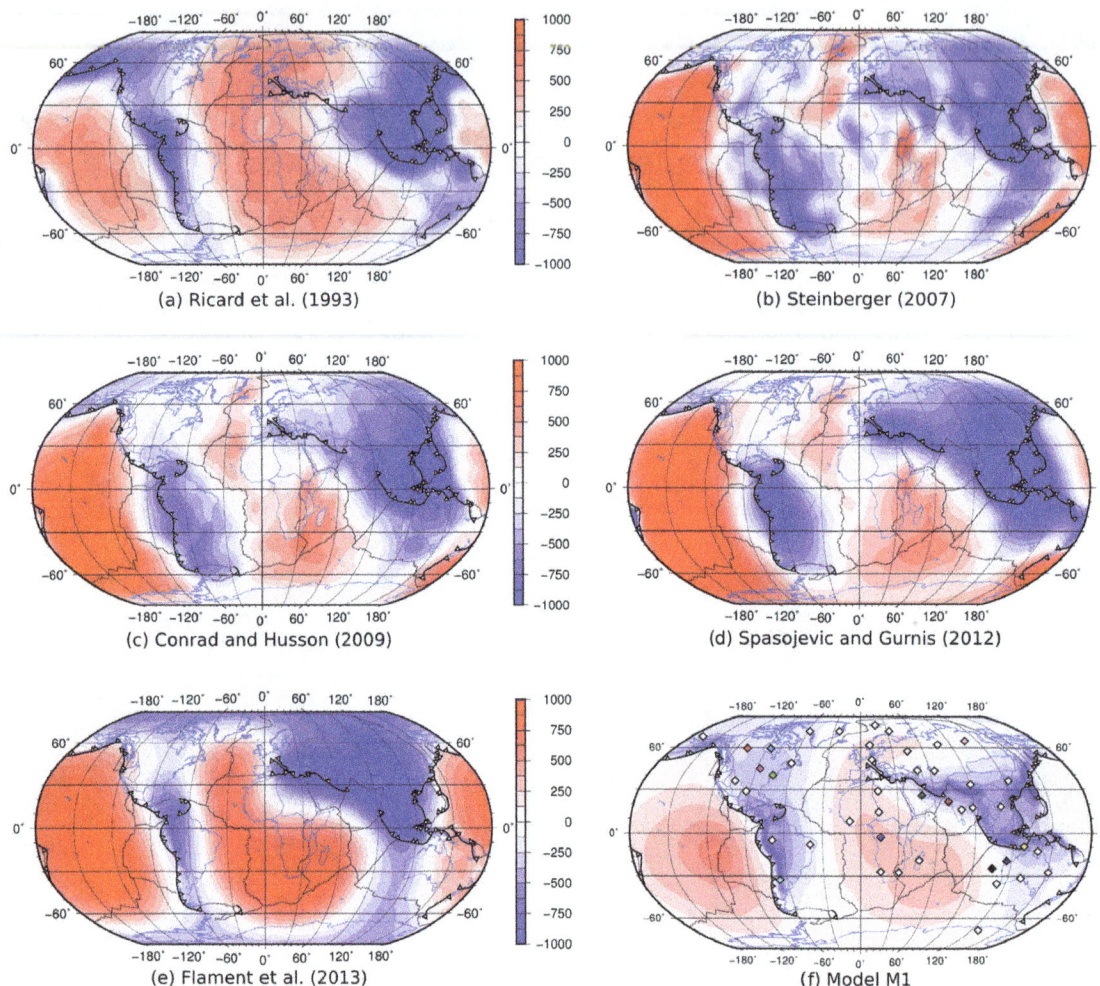

Figure 5. Comparison of present-day global dynamic topography as inferred from published mantle convection models: **(a)** Ricard et al. (1993), **(b)** Steinberger (2007), **(c)** Conrad and Husson (2009), **(d)** Spasojevic and Gurnis (2012), **(e)** Flament et al. (2013) and **(f)** our model M1. All grids display dynamic support that has been corrected for air loading.

and in Fig. 4b in a plate frame of reference. Note the strong correspondence of the upper mantle thermal heterogeneity distribution with negative dynamic topography in regions of present-day and former subduction, and positive dynamic topography elsewhere. Before we further interpret our results we compare them with previous geodynamic modelling studies and with selected observations on a regional scale. We evaluate our model against well data at Australian locations, which have experienced two marked periods of anomalous subsidence and uplift since the Jurassic. Finally we consider observations that constrain the uplift history of southern Africa. Due to its particular geological setting, southern Africa has been the target of many studies on constraining its uplift history and assessing to what extent and when it experienced dynamic support.

3.2.1 Global models of dynamic topography

To place our results in the context of previous studies, we compare our predictions of present-day dynamic topography (Fig. 4a) to those of Ricard et al. (1993), Steinberger (2007), Conrad and Husson (2009), Spasojevic and Gurnis (2012) and Flament et al. (2013) as shown in Fig. 5. All models display a dominant degree-2 pattern, whereby the Pacific part is most prominent in all studies. In Steinberger (2007), Conrad and Husson (2009) and Spasojevic and Gurnis (2012), the African and Indian Ocean region consists of a cluster of upwellings with moderate amplitudes as opposed to a more coherent, strong high in the other models. In those models, mantle heterogeneities derive from subduction history, in which upwellings feature less internal heterogeneity. In a very recent study, Guerri et al. (2016) examine combinations of 11 different crustal models (Guerri et al., 2016) with 10

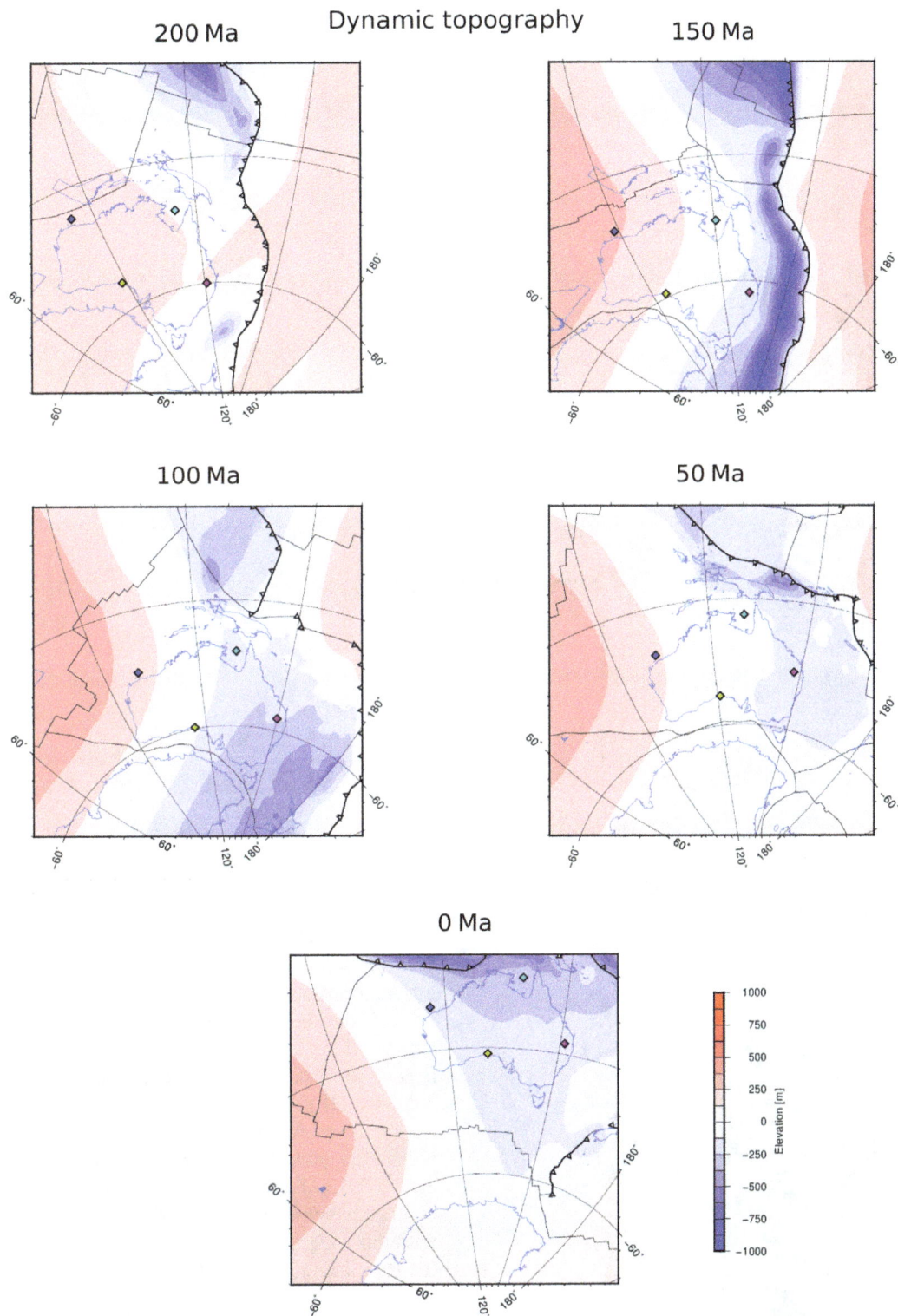

Figure 6. Dynamic topography evolution for Australia, analogous to Fig. 4a. Between 200 and 120 Ma, Australia's eastern margin experiences westward-directed subduction of the Phoenix plate. From 50 Ma onward Australia's dynamic topography is dominated by the northward-directed subduction of the Australian plate under the Indonesian archipelago. See the Supplement for dynamic topography evolution in 10 Myr time steps.

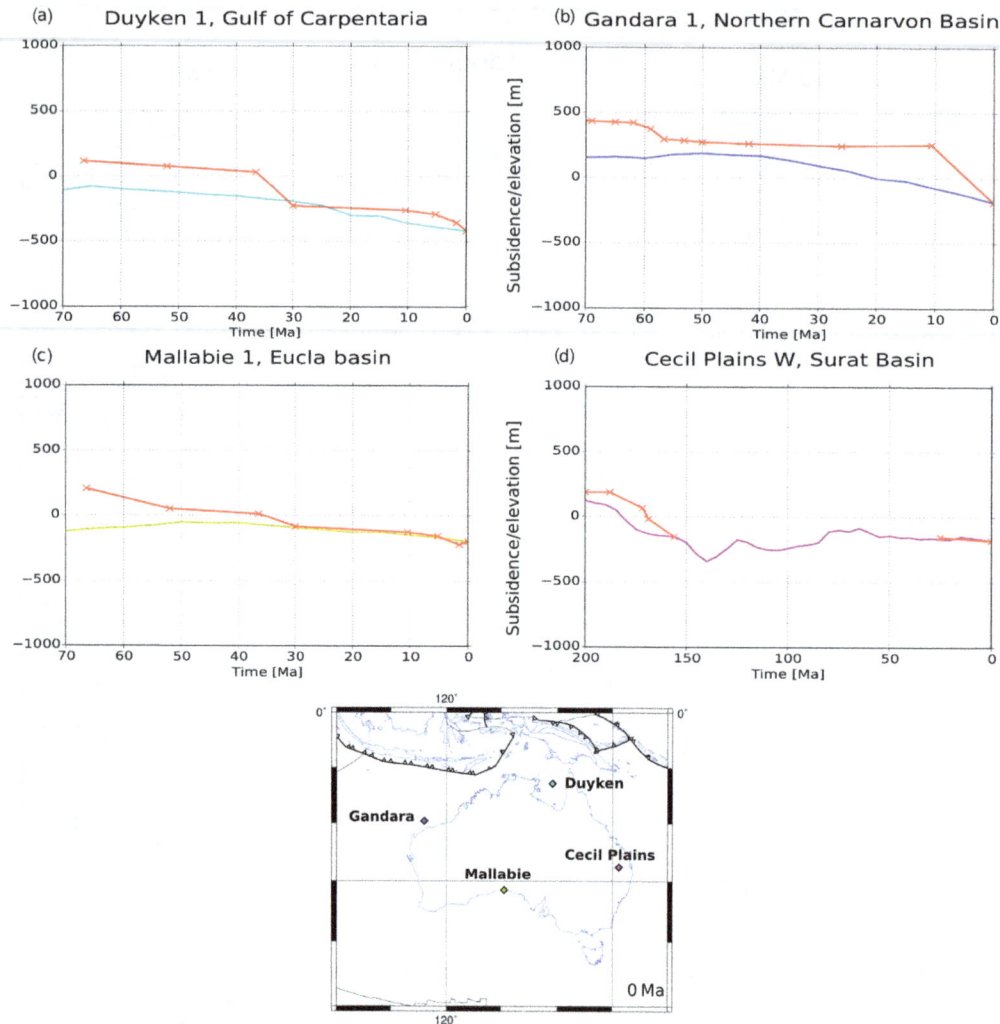

Figure 7. Tectonic subsidence curves for selected wells in Australia (red) compared to our model output (coloured). Well data curves (accordingly coloured) are offset to model elevation at present day.

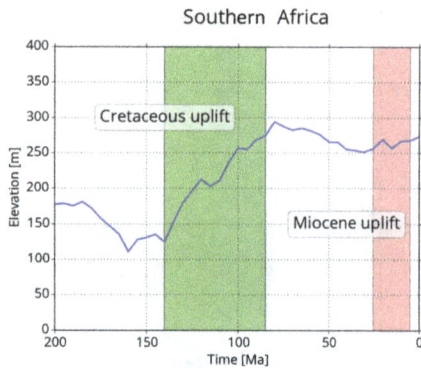

Figure 8. Southern African (sample point southern Africa east) uplift resulting from our models. While it successfully reproduces the Cretaceous uplift as described in previous work, uplift in the Miocene period remains comparatively small (see text).

different tomography models after obtaining mantle density through the thermodynamic modelling of various chemical compositions in the mineralogical assemblage. This yields a peak-to-peak dynamic topography amplitude exceeding 3 km for all models. Other studies suggest that amplitudes for long-wavelength (10^3 km) dynamic topography should be smaller than ± 1 km (Wheeler and White, 2002; Molnar et al., 2015; Forte et al., 2015; Hoggard et al., 2016). In contrast to some previous models, our model M1 agrees well with this suggestion, mapping deep mantle return flow above the Pacific and African LLSVPs at amplitudes not exceeding ± 1 km.

3.2.2 Australia

The tectonic stability of the Australian continent and its relatively large displacement relative to the underlying mantle since the Cretaceous makes it an ideal location

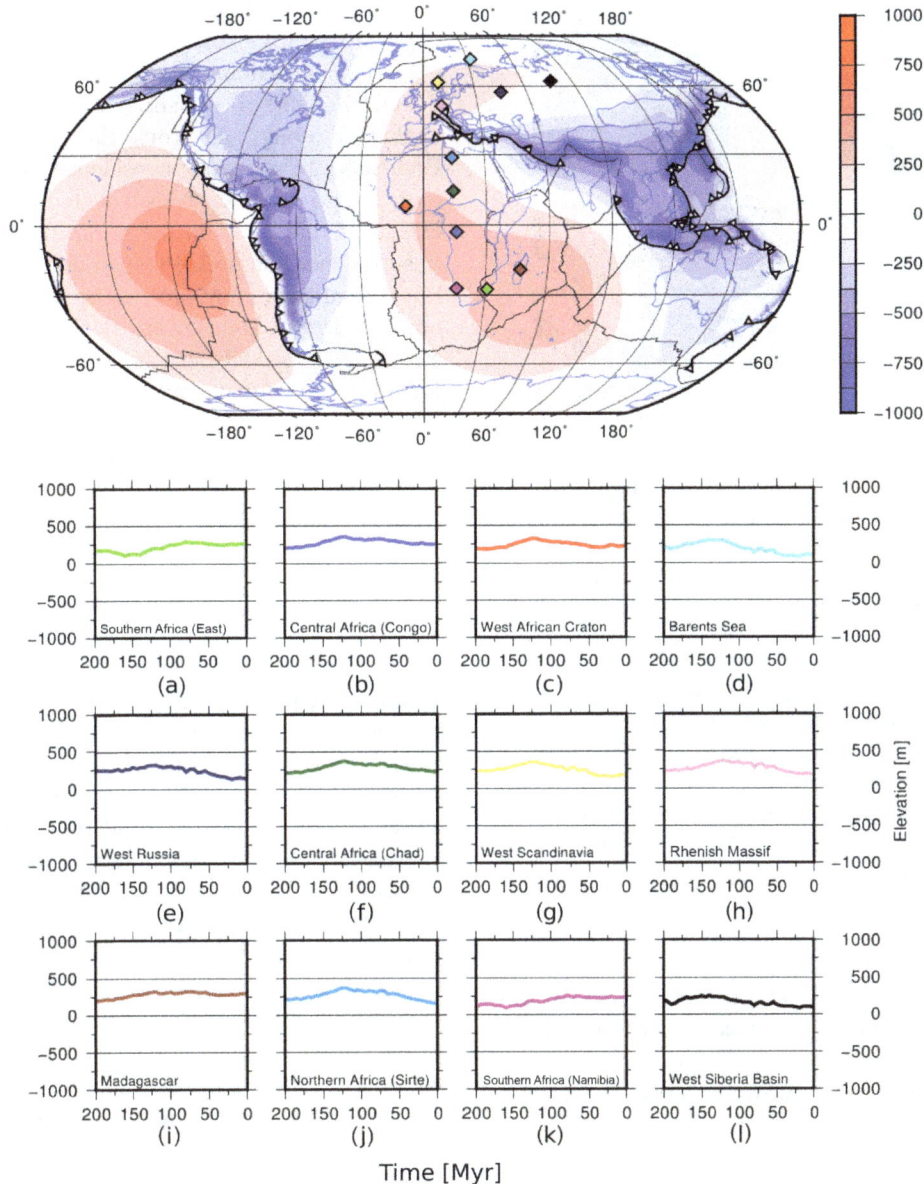

Figure 9. Overview map (top) and time series (bottom) of dynamic topography for Category I, "topographic stable areas".

for investigating mantle-induced topography changes. It is the only continent to have undergone two distinct episodes of dynamic topography during the last 200 Myr: (i) a pronounced west–east-dominated warping during the middle and late Jurassic and early Cretaceous (Gurnis et al., 1998) caused by subduction along its eastern margin as the Phoenix plate subducted westward under East Gondwana, and (ii) a south–north tilt due to subduction of the Australian plate underneath the Eurasian and Pacific plates (Fig. 6) that started in the Miocene and continued through to the present day (Sandiford, 2007). The effect of transient mantle-convection-induced vertical Australian motions has been the subject of many observation- and model-based research

efforts (Veevers, 1984; DiCaprio et al., 2009; Heine et al., 2010; Czarnota et al., 2014).

To ground-truth our predicted dynamic topography we focus on tectonic uplift and subsidence histories from four selected Australian exploration wells located in opposite corners of the Australian continent. The well data were retrieved from the Geoscience Australia PIMS system online (http:// dbforms.ga.gov.au/pls/www/npm.pims_web.search; Fig. 7).

For the well on the North West Shelf we find good agreement of the long-term Cenozoic elevation history (Fig. 7b; Müller et al., 2000). The modelled vertical motion history shows that the area was subjected to gradually increasing subsidence during the Cenozoic, which is in good agreement with observations (Czarnota et al., 2013). Also for

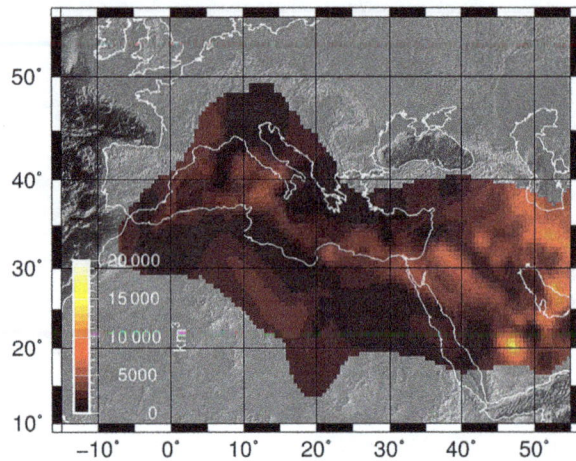

Figure 10. Volume of subducted oceanic lithosphere (in km^3) for the western Tethyan region for the past 200 Myr. The amount of material was calculated based on paleo-age grids with a 6 arcmin grid cell size for the subducted oceanic lithosphere (see Müller et al., 2008, for methodology) and convergence velocities derived from the input plate kinematic model. We assume a simple half-space cooling model to compute the thickness of the oceanic lithosphere for ages < 69.9 Myr as $z = 11.36\sqrt{age}$, where age is the age of the oceanic lithosphere. Older lithosphere is set to a default thickness of 95 km (Kanamori and Press, 1970). Volumes are derived by multiplying thickness with computed trench-normal convergence in 1 Myr time steps. Grid cell size for the subduction volume is 30 arcmin. Map background is ETOPO1 topography (Amante and Eakins, 2009).

the remaining wells (Fig. 7a, c and d), our modelled vertical motion histories follow the backstripped subsidence patterns generally quite well.

The Gandara 1 well in the north-western part of Australia (Fig. 7b) shows a distinctly different temporal subsidence behaviour from the most eastern one in the Surat Basin, Cecil Plains West 1 (Fig. 7d). We explain this through the evolution of dynamic topography shown in Fig. 6. The former well is not affected by subducted, cold, dense slab material from the Proto-Western Pacific/East Gondwana subduction zone. When, from around 70 Ma onward, Australia starts progressively moving north, this area experiences the increasing influence of downward mantle flow related to the subduction of the Australian lithosphere along the Indonesian subduction zone system. The Cecil Plains well, on the other hand, mostly displays the Jurassic and early Cretaceous East Gondwana subduction setting, while Australia's Cenozoic tilting only has a moderate effect: this location indicates phases of accelerating subsidence until about the latest Jurassic, followed by a phase of non-deposition and erosion and a recent phase of gentle subsidence. Our models can reproduce the initial Jurassic phase of subsidence. We attribute the time of the depositional hiatus ($\sim$ 155 to 25 Ma) to the rebounding of the west–east

tilt, while the renewed phase of subsidence seems to be related to the Cenozoic warping of the Australian continent.

For the remaining bore holes, the modelled Duyken 1 and Mallabie 1 dynamic topography histories can be explained along analogous lines of argument (Fig. 6). Both reflect Cretaceous subsidence, with the former being more pronounced as the subducted material progressively extends farther west in the north of Australia. A decrease in subsidence can be seen when that influence fades, just to be affected even more profoundly later when the Australian north is drawn down by the negative dynamic topography due to sinking slabs in the Indonesian subduction zone. Keeping in mind that dynamic topography varies only slowly over time, we observe a good correlation between model-predicted vertical motions and reconstructed vertical motions from well data. Hence, these data corroborate the dynamic topography amplitudes from our study and are in line with other studies (e.g. DiCaprio et al., 2011; Heine et al, 2010; Czarnota et al., 2013).

3.2.3 Southern Africa

Southern Africa's unusually high elevation relative to other continents, despite being surrounded by passive margins, makes it an area of particular interest. Braun et al. (2014) analysed sedimentary flux data, which confirmed a major phase of erosion on the southern African plateau in the late Cretaceous. By examining paleoprecipitation indicators across the plateau, they ruled out climate variability as the main driver of increased erosional activity. Thus, Braun et al. (2014) concluded that the continent migrated over a source of positive dynamic topography at that time. Tinker et al. (2008) quantified exhumation using apatite fission track dating. While denudation rates during the Cenozoic indicate that mantle processes did not primarily cause the uplift of the region, the authors suggest that mantle buoyancy during the Cretaceous was responsible for the enhanced erosion across southern Africa, which in turn generated up to 3 km of exhumation (Tinker et al., 2008). Stanley et al. (2013) performed a detailed thermochronometry study across the southern African plateau by employing (U-Th)/He dating. They suggest that a mid-Cretaceous buoyancy increase in the lithospheric mantle caused coeval denudation. Our model predicts dynamic support equivalent to $\sim$ 200 m underneath southern Africa gradually developing in the Cretaceous period (Fig. 8), which is in line with previous modelling studies (Zhang et al., 2012; Flament et al., 2014). However, it does not reproduce a significant post-30 Ma uplift phase as revealed by the inversion of river profiles (Paul et al., 2014), which may reflect the lack of active upwellings in our convection model or convective processes at (sub-) lithospheric levels.

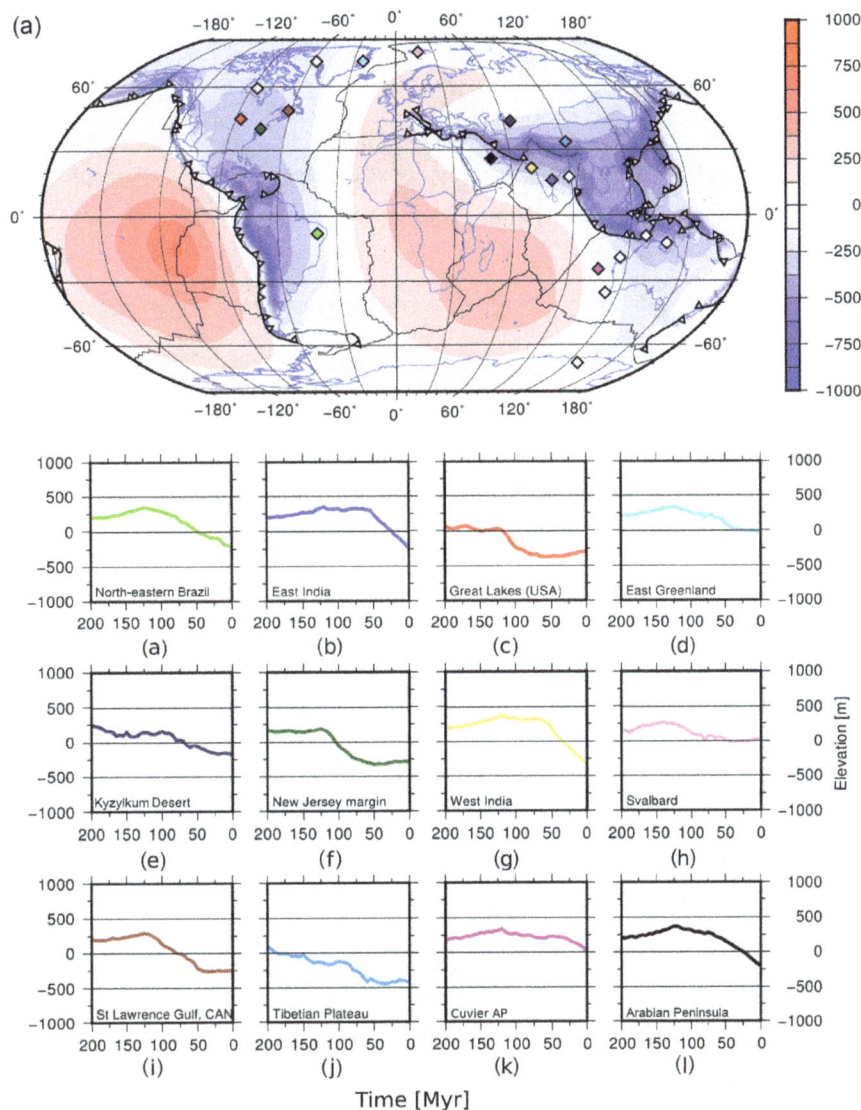

Figure 11.

3.3 Modes of long-term dynamic surface topography evolution

In this section we use our model results to establish a set of "geodynamic patterns" for how different continental regions are affected by underlying mantle processes. We define three categories that are exemplified by selected locations using two alternative approaches. First, we plot time series of their dynamic topography and visually classify them into these categories based on the overall temporal trends in elevation. Secondly, we apply a k-means-based cluster analysis to the time series of dynamic topography as an independent means to validate our interpretation and explore the consequences of classifying the time series into different numbers of groups.

3.3.1 Category I – Topographic stable areas

The first category is comprised of locations in topographically stable regions. These areas experience only positive dynamic support over the 200 Myr time period examined, generating stable positive topography of a few hundred metres. As can be seen from Fig. 9, these regions lie on a curved band that stretches from the Eurasian to the African plate and extends into the Somali plate, encompassing locations in Africa, Madagascar, Western and Northern Europe and Western Siberia. This category generally consists of regions with no adjacent subduction zone. Where points are located adjacent to a subduction zone (for example the Rhenish Massif next to the western Tethyan subduction zones; Fig. 9h), there is insufficient cold, dense slab material in the mantle to cause negative dynamic

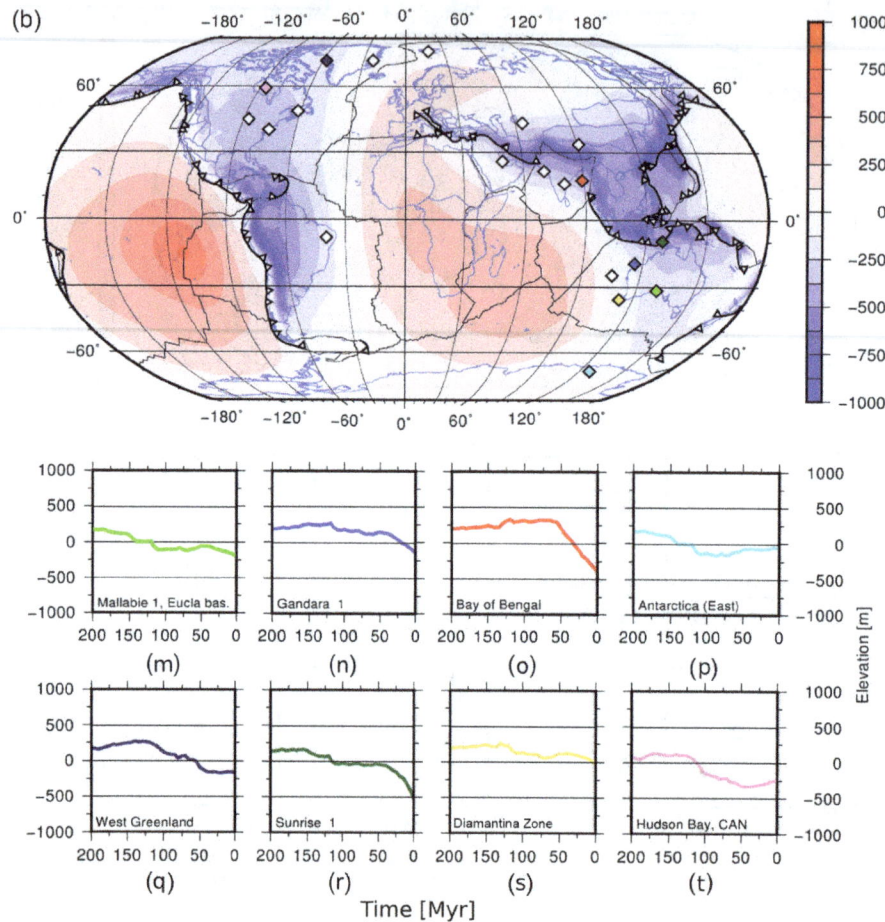

Figure 11. (a) Overview (top) and time series (bottom) plot of dynamic topography for Category II, "dynamically subsiding areas". **(b)** Overview (top) and time series (bottom) plot of dynamic topography for Category II, "dynamically subsiding areas", continued.

topography, as evidenced by a computation of subducted oceanic lithosphere volumes (Fig. 10) for the past 200 Myr. The total volume of subducted material in the western Tethyan and present-day Mediterranean region is on the order of 10 000–15 000 km^3 per 30 arcmin grid cell, yielding very low average volume fluxes of around 50–75 km^3 Myr^{-1}.

3.3.2 Category II – Dynamically subsiding areas

Locations that fall into the second category experience a long-term continuous decrease in dynamic support. Examples include East and Western India, the Arabian Peninsula, north-eastern Brazil, the north-eastern United States, north-eastern Canada and Western Australia (Fig. 11a and b). These regions subside as they approach sites of active or recent subduction. Note that Australia is a special case in which two subduction zones exert a sequential influence (Sect. 3.2.2). First the East Gondwana and subsequently the Indonesian subduction zone give rise to a decrease in elevation, as described above.

3.3.3 Category III – Fluctuating areas

Locations in this category (Fig. 12) exhibit a non-monotonic temporal evolution with mostly negative dynamic topography. This is likely caused by two factors: (i) the influence of multiple subduction zones or (ii) the influence of a single subduction zone that exhibits strongly time-dependent characteristics.

Multiple subduction zones affected the dynamic topography evolution of Australia, as discussed above: two lows in dynamic subsidence can be identified when the particular location is influenced by one of these. The Chukchi Sea region (Fig. 12a) is a second example where three local minima in dynamic topography correlate with sequential activity in Arctic subduction zones: the Koni-Taigonos subduction zone at ∼ 200 Ma, the Koyukuk and Nutesyn subduction between 140 and ∼ 125 Ma and the Alaskan subduction zone since ∼ 110 Ma (Shephard et al., 2013; see Fig. 4a and the corresponding Supplement). A third example is the Central Siberian Plateau (Fig. 12g), which was first affected by the East Gondwana subduction

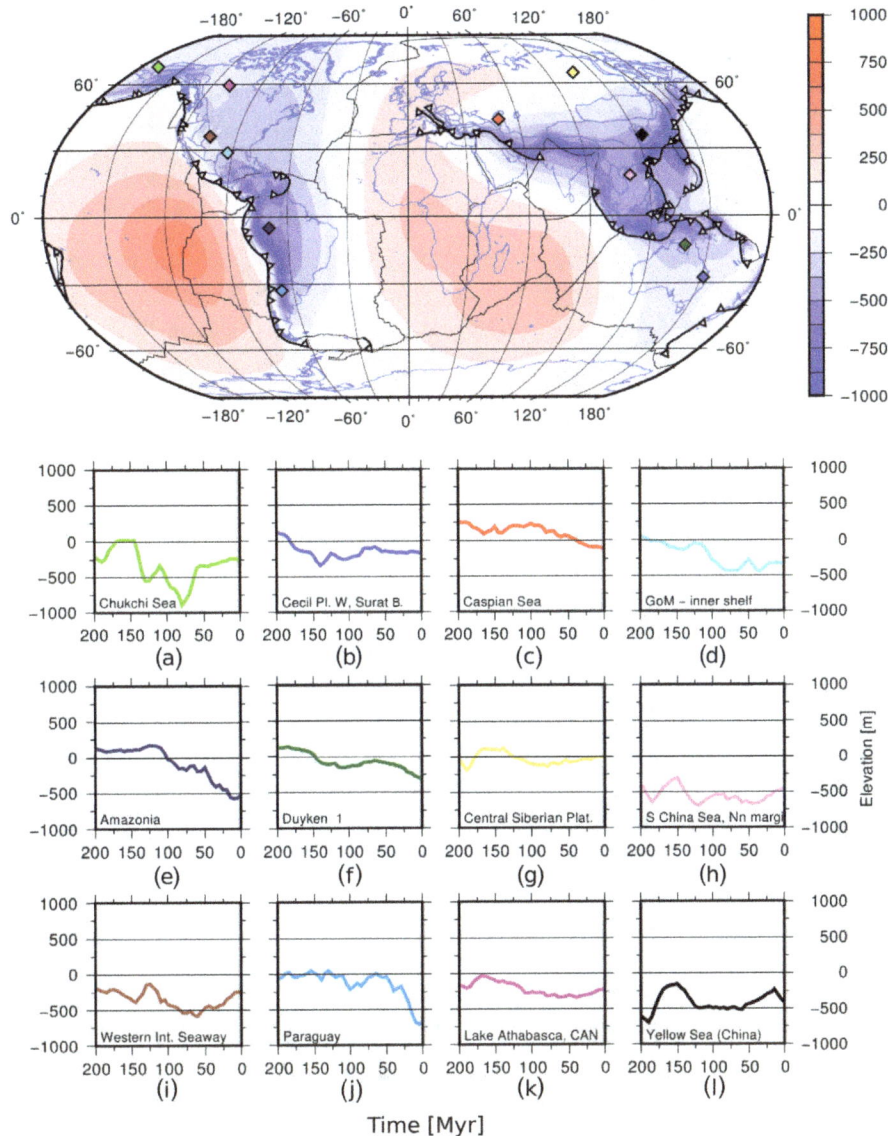

Figure 12. Overview (top) and time series (bottom) plot of dynamic topography for Category III, "fluctuating areas".

zone (~ 180 Ma) and subsequently (~ 120 Ma onward) influenced by the East Asian subduction system.

A key time-dependent parameter of a subduction zone is the subduction velocity. Rapid convergence tends to introduce more cold, dense material into the mantle, which leads to a more pronounced negative dynamic topography. To isolate the role of subduction velocity, we compare dynamic topography time series at three different subduction zones with the orthogonal component of the relative velocity between the subducting and overriding plate at the closest subduction zone (Fig. 13). For instance, the Chukchi Sea location exhibits a significant increase in negative dynamic topography about 0 m at 145 Ma to about -500 m at 120 Ma. This modelled dynamic subsidence phase is followed by a short period of rebound ($+150$ m over 15 Myr) before the

area is again modelled as experiencing dynamic subsidence of more than 500 m between 110 and 80 Ma. Here, we can correlate the modelled trench-normal convergence velocities from our input plate kinematic model (Seton et al., 2012; Shephard et al., 2013) directly with predicted dynamic topography evolution over time. Similarly, accelerating convergence from 200 Ma until ca. 150 Ma at the Western Interior Seaway sample point corresponds to decreasing dynamic topography. This is followed by a short uplift period due to slow convergence before elevation decreases again, driven by the increased convergence velocity. An analogous situation occurs in the South American Amazonian region, where the general trend of orthogonal convergence correlates well with the evolution of dynamic topography.

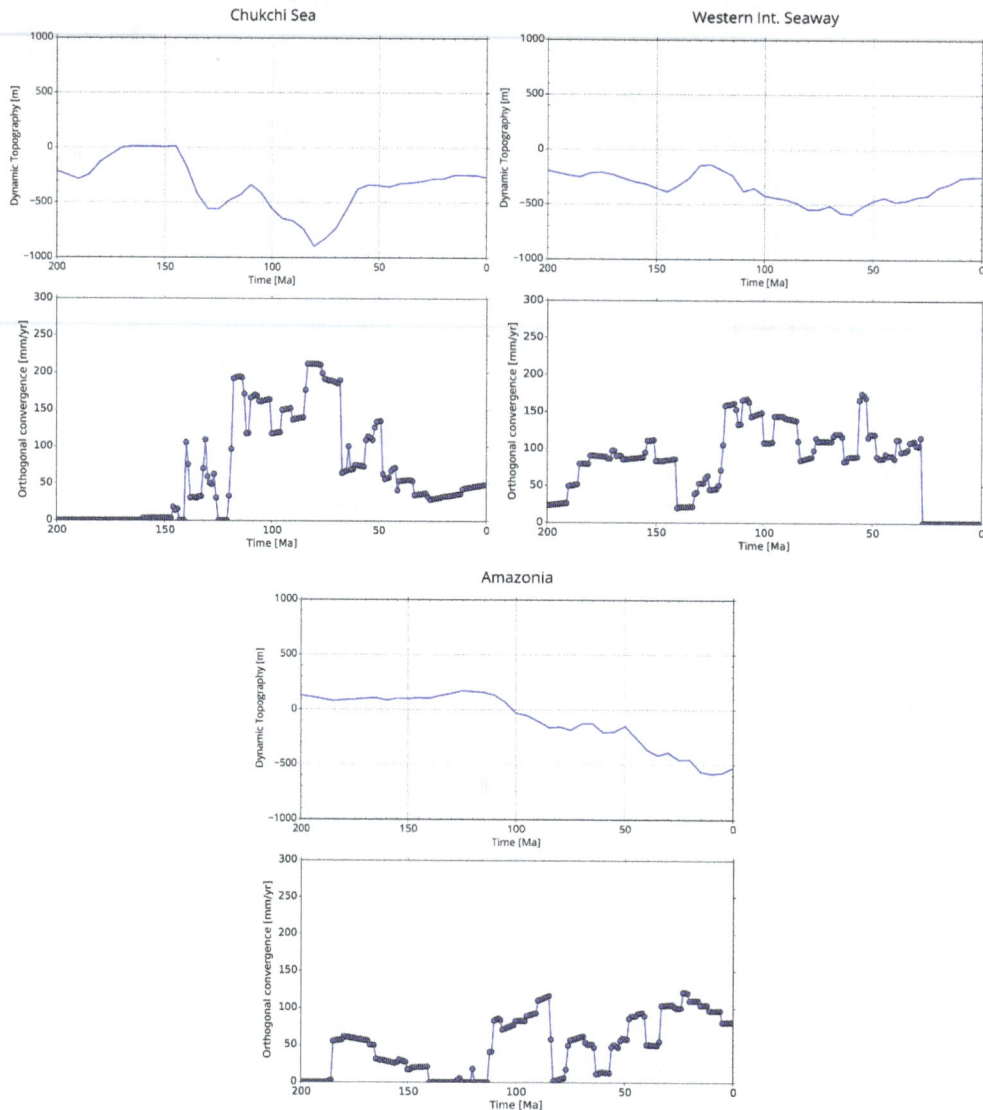

Figure 13. Comparison of dynamic topography time series at three sample locations with the relative orthogonal velocity of the down-going and overriding plate at the nearest point on the subduction zone. For locations of sample points, see Fig. 12.

It should be noted that the slabs in our models are generally weaker than slabs on Earth due to our simplistic representation of a primarily depth- and temperature-dependent rheology (see Garel et al., 2014, for comparison). As a consequence, upon slab descent, slabs often tear and break, which leads to rapid, localised changes in regional dynamic topography (e.g. Fig. 12c–e; h–l). Similar tears or breaks occur due to abrupt changes in plate motions. The closer a point lies to a subduction zone where this occurs, the larger the associated amplitude of dynamic topography change.

3.3.4   Cluster analysis

In the previous section, we isolated three characteristic groups which represent distinct trends in vertical motion. This classification was based on a simultaneous consideration of absolute elevations, changes in elevation and the geodynamic context. Here we use cluster analysis, based on the k-means algorithm (e.g. Lekic and Romanowicz, 2011), as an independent means to validate our classifications (see the Supplement for a detailed description). To this end, we search for clusters of similarity within our dynamic topography time series at individual surface points using a grid of nearly 2×10^5 equidistantly spaced nodes on the sphere. This provides a resolution of ~ 28 km on the surface sampled at 5 Myr intervals in time. The cluster analysis was

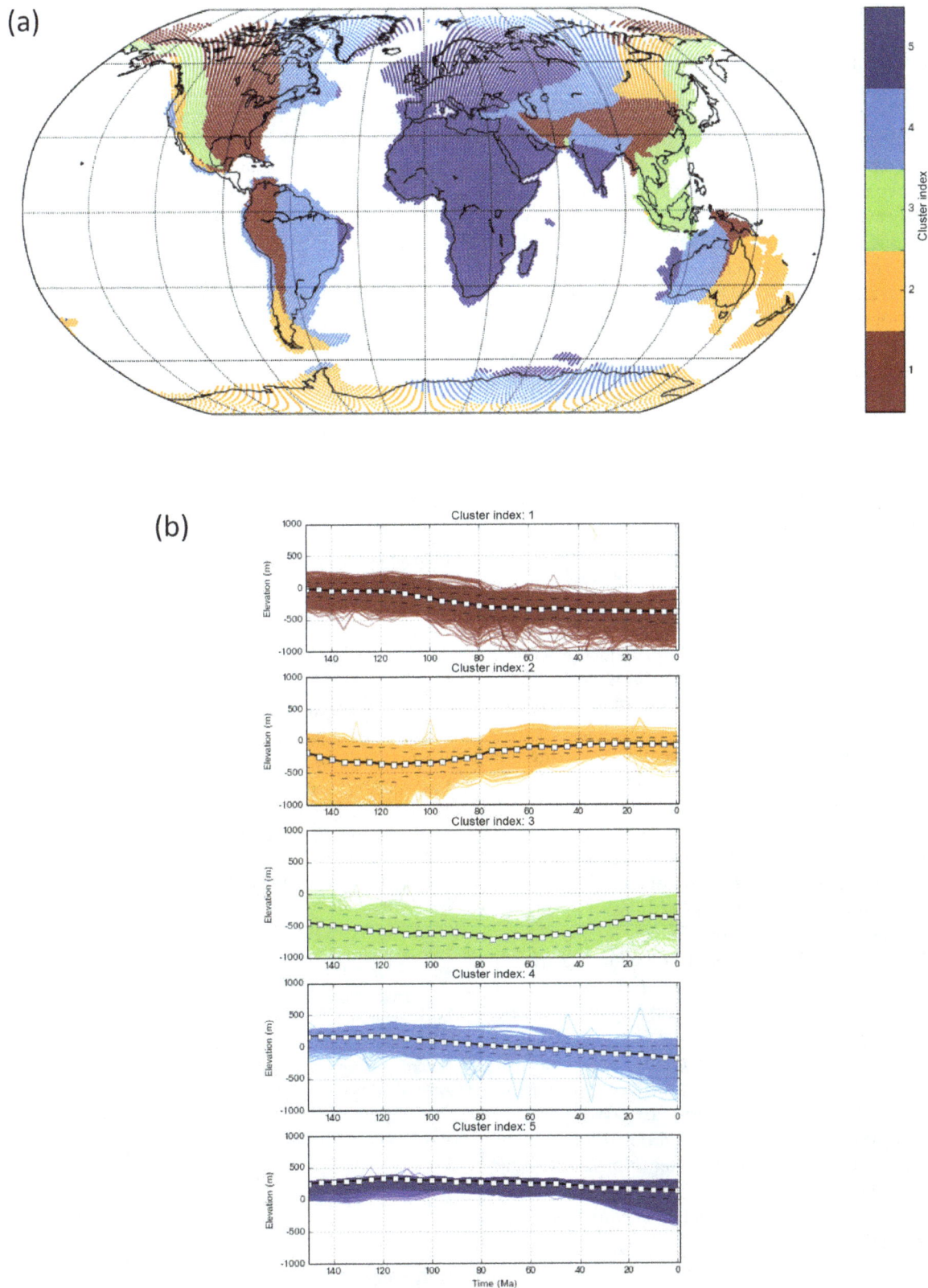

Figure 14. The cluster analysis underpins the suggested categorisation; $k = 5$. **(a)** The dark blue cluster consists of locations in stable areas, light blue corresponds to (long-term) dynamically subsiding areas and the fluctuating category is encompassed by three clusters (green, orange and red). **(b)** Different clusters of elevation time series. The average trend is marked in black. See the Supplement for alternative choices of k.

applied to the dynamic topography time series from 150 Ma to present; this excludes the first 50 Myr of the model run in which our dynamic topography predictions will be strongly sensitive to the initial condition.

The k-means algorithm partitions a given dataset into k clusters such that the sum of the squares of the deviation from the cluster mean is minimised. The number of clusters k is an input parameter for the algorithm. Note that this approach only evaluates elevations at the particular grid point at a certain time and does not account for relative variations (i.e. uplift or subsidence), nor does it take into account the history of elevations. For a detailed description of the k-means algorithm and the results for all clusters (k), see the Supplement.

We vary k between 2 and 6, which allows us to highlight a variety of spatio-temporal patterns in dynamic topography. Here we discuss $k = 5$ because this is the minimal number of clusters that reflects the underlying geodynamics to a first order (Fig. 14). Cluster 5 can be directly associated with our Category I ("topographic stable regions"), while Cluster 4 is an approximate representation of our Category II regions ("dynamically subsiding areas"). To first order, the map in Fig. 14 mirrors our dynamic topography maps in that the topographically high regions (dark blue areas) are concentrated in a region that stretches from Russia over Europe to Africa, mostly far away from subduction zones. Cluster 4 (light blue) corresponds to subsiding areas which approach subduction zones. Also, North India falls into this category (as opposed to $k < 5$; see the Supplement), which describes the history of the subcontinent starting out as a part of the highly elevated Pangaea supercontinent to the plunging of its northern edge as it collides with Asia (Fig. 11a, b, g). The Arabian Peninsula, however, has been determined not to reside in this group, but in the "topographic high" group instead; on average for most of the observed time period, this location displays positive elevations (Fig. 11a, l). The same holds true for the borderline case of the Cuvier Abyssal Plain (Fig. 11a, k). Category III ("fluctuating areas") disintegrates into three distinct clusters (red, orange and green): Cluster 1 with points that remain at negative dynamic topography throughout the entire model run, and Clusters 2 and 3 with locations that experience an initial phase of long-term dynamic subsidence followed by a subsequent uplift. Higher cluster numbers k serve to further partition our categories, yet do not reveal any additional information (see the Supplement). To summarise, the k-means scheme provides an independent instrument for the quantitative analysis of our model data and we find good correspondence to our classification.

4 Conclusions

We have demonstrated that the long-wavelength influence of dynamic topography in different continental regions can be classified via their spatio-temporal vicinity to a subduction zone. (i) Locations that are far away from subduction zones are generally stable and dynamically supported; these lie on a curved band stretching from the Eurasian plate to the African plate and extending into the Somali plate. (ii) Locations that move toward subduction zones dynamically subside over time, and (iii) locations that are being influenced by numerous subduction zones or a subduction zone that shows strong time dependence (i.e. abrupt changes in plate motions of both the overriding and subducting plate) will display an oscillatory history of dynamic topography. While this classification is based on a geodynamic understanding, it is confirmed through an independent approach based on a k-means cluster analysis. As geosciences moves toward more integrated approaches that couple deep Earth and surface processes, our pattern analysis of dynamic topography provides a useful means to evaluate the dynamic relationships between basin infill and sediment-sourcing regions to reconstruct regional, spatio-temporal vertical motion histories and support resource exploration.

Competing interests. The authors declare that they have no conflict of interest.

Acknowledgements. This research was supported by resources provided by the Pawsey Supercomputing Centre with funding from the Australian Government and the Government of Western Australia and with the assistance of resources from the National Computational Infrastructure (NCI), which is supported by the Australian Government. Sascha Brune was funded by the Marie Curie International Outgoing Fellowship 326115 and the Helmholtz Young Investigators Group CRYSTALS. Christian Heine was supported by ARC Linkage Project LP0989312 with Shell E & P and TOTAL. D. Rhodri Davies is funded by an ARC Future Fellowship (FT140101262) and Simon Williams and R. Dietmar Müller are supported by ARC grants DP130101946 and IH130200012. Leonardo Quevedo is acknowledged for the numerical routines to compute subduction volumes and the implementation of the GPlates TERRA output routines, along with John Cannon. The authors thank the employees of both supercomputing centres for their generous support and Geoscience Australia for their vast, open and easily accessible database. Some figures were generated using the Generic Mapping Tools (GMT; Wessel et al., 2013).

Edited by: J. Huw Davies

References

Amante, C. and Eakins, B. W.: ETOPO1 1 Arc-Minute Global Relief Model: Procedures, Data Sources and Analysis, National Geophysical Data Center, NOAA, https://doi.org/10.7289/V5C8276M, 2009.

Baumgardner, J. R.: Three-dimensional treatment of convective flow in the Earth's mantle, J. Stat. Phys., 39, 501–511, https://doi.org/10.1007/BF01008348, 1985.

Boyden, J. A., Müller, R. D., Gurnis, M., Torsvik, T. H., Clark, J. A., Turner, M., Ivey-Law, H., Watson, R. J., and Cannon, J. S.: Next-generation plate-tectonic reconstructions using GPlates, in Geoinformatics, Cambridge University Press, 95–114, https://doi.org/10.1017/CBO9780511976308.008, 2011.

Braun, J., Guillocheau, F., Robin, C., Baby, G., and Jelsma, H.: Rapid erosion of the Southern African Plateau as it climbs over a mantle superswell, J. Geophys. Res.-Sol. Ea., 119, 6093–6112, 2014.

Bunge, H.-P., Richards, M. A., and Baumgardner, J. R.: A sensitivity study of three-dimensional spherical mantle convection at 108 Rayleigh number: Effects of depth-dependent viscosity, heating mode, and an endothermic phase change, J. Geophys. Res.-Sol. Ea., 102, 11991–12007, https://doi.org/10.1029/96JB03806, 1997.

Bunge, H.-P., Hagelberg, C. R., and Travis, B. J.: Mantle circulation models with variational data assimilation: Inferring past mantle flow and structure from plate motion histories and seismic tomography, Geophys. J. Int., 152, 280–301, https://doi.org/10.1046/j.1365-246X.2003.01823.x, 2003.

Colli, L., Ghelichkhan, S., and Bunge, H.-P.: On the ratio of dynamic topography and gravity anomalies in a dynamic Earth, Geophys. Res. Lett., 43, 2510–2516, https://doi.org/10.1002/2016GL067929, 2016.

Colli, L., Ghelichkhan, S., Bunge, H.-P., and Oeser, J.: Retrodictions of Mid Paleogene mantle flow and dynamic topography in the Atlantic region from compressible high resolution adjoint mantle convection models: Sensitivity to deep mantle viscosity and tomographic input model, Gondwana Res., https://doi.org/10.1016/j.gr.2017.04.027, 2017.

Conrad, C. P. and Gurnis, M.: Seismic tomography, surface uplift, and the breakup of Gondwanaland: Integrating mantle convection backwards in time, Geochem. Geophy. Geosy., 4, 1031, https://doi.org/10.1029/2001GC000299, 2003.

Conrad, C. P. and Husson, L.: Influence of dynamic topography on sea level and its rate of change, Lithosphere, 1, 110–120, 2009.

Crosby, A. G., McKenzie, D., and Sclater, J. G.: The relationship between depth, age and gravity in the oceans, Geophys. J. Int., 166, 553–573, https://doi.org/10.1111/j.1365-246X.2006.03015.x, 2006.

Czarnota, K., Hoggard, M. J., White, N., and Winterbourne, J.: Spatial and temporal patterns of Cenozoic dynamic topography around Australia, Geochem. Geophys. Geosy., 14, 634–658, https://doi.org/10.1029/2012GC004392, 2013.

Czarnota, K., Roberts, G. G., White, N. J., and Fishwick, S.: Spatial and temporal patterns of Australian dynamic topography from River Profile Modeling, J. Geophys. Res.-Sol. Ea., 119, 1384–1424, https://doi.org/10.1002/2013JB010436, 2014.

Davies, D. R. and Davies, J. H.: Thermally-driven mantle plumes reconcile multiple hot-spot observations, Earth Planet. Sc. Lett., 278, 50–54, 2009.

Davies, J. H. and Davies, D. R.: Earth's surface heat flux, Solid Earth, 1, 5–24, https://doi.org/10.5194/se-1-5-2010, 2010.

Davies, D. R., Davies, J. H., Bollada, P. C., Hassan, O., Morgan, K., and Nithiarasu, P.: A hierarchical mesh refinement technique for global 3-D spherical mantle convection modelling, Geosci. Model Dev., 6, 1095–1107, https://doi.org/10.5194/gmd-6-1095-2013, 2013.

Davies, D. R., Goes, S., Davies, J. H., Schuberth, B. S. A., Bunge, H.-P., and Ritsema, J.: Reconciling dynamic and seismic models of Earth's lower mantle: The dominant role of thermal heterogeneity, Earth Planet. Sc. Lett., 353, 253–269, https://doi.org/10.1016/j.epsl.2012.08.016, 2012.

Davies, D. R., Goes, S., and Lau, H. C. P.: Thermally dominated deep mantle LLSVPs: a review, in: The Earth's heterogeneous mantle, edited by: Khan, A., Deschamps, F., and Kawai, K., Springer, 441–478, https://doi.org/10.1007/978-3-319-15627-9_14, 2015.

DiCaprio, L., Gurnis, M., and Müller, R. D.: Long-wavelength tilting of the Australian continent since the Late Cretaceous, Earth Planet. Sc. Lett., 278, 175–185, https://doi.org/10.1016/j.epsl.2008.11.030, 2009.

DiCaprio, L., Gurnis, M., Müller, R. D., and Tan, E.: Mantle dynamics of continentwide cenozoic subsidence and tilting of Australia, Lithosphere, 3, 311–316, https://doi.org/10.1130/L140.1, 2011.

Doin, M.-P. and Fleitout, L.: Flattening of the oceanic topography and geoid: thermal versus dynamic origin, Geophys. J. Int., 143, 582–594, https://doi.org/10.1046/j.1365-246X.2000.00229.x, 2000.

Flament, N., Gurnis, M., and Muller, R. D.: A review of observations and models of dynamic topography, Lithosphere, 5, 189–210, https://doi.org/10.1130/L245.1, 2013.

Flament, N., Gurnis, M., Williams, S., Seton, M., Skogseid, J., Heine, C., and Müller, R. D.: Topographic asymmetry of the South Atlantic from global models of mantle flow and lithospheric stretching, Earth Planet. Sc. Lett., 387, 107–119, 2014.

Forte, A. M., Simmons, N. A., and Grand, S. P.: Constraints on 3-D Seismic Models from Global Geodynamic Observables: Implications for the Global Mantle Convective Flow, in: Treatise of Geophysics, 2nd Edition, Volume 1, Deep Earth Seismology, 853–907, 2015.

Frisch, W., Meschede, M., and Blakey, R. C.: Plate Tectonics Continental Drift and Mountain Building, Springer Berlin Heidelberg, https://doi.org/10.1007/978-3-540-76504-2, 2011.

Garel, F., Goes, S., Davies, D. R., Davies, J. H., Kramer, S. C., and Wilson, C. R.: Interaction of subducted slabs with the mantle transition-zone: A regime diagram from 2-D thermo-mechanical models with a mobile trench and an overriding plate, Geochem. Geophy. Geosy., 15, 1739–1765, 2014.

Glišović, P. and Forte, A. M.: A new back-and-forth iterative method for time-reversed convection modeling: Implications for the Cenozoic evolution of 3-D structure and dynamics of the mantle, J. Geophys. Res.-Sol. Ea., 121, 4067–4084, https://doi.org/10.1002/2016JB012841, 2016.

Glišović, P. and Forte, A. M.: On the deep-mantle origin of the Deccan Traps, Science, 355, 613–616, https://doi.org/10.1126/science.aah4390, 2017.

Guerri, M., Cammarano, F., and Tackley, P. J.: Modelling Earth's surface topography: Decomposition of the static and dynamic components, Phys. Earth Planet. In., 261, 172–186, 2016.

Gurnis, M., Müller, R. D., and Moresi, L.: Cretaceous Vertical Motion of Australia and the Australian Antarctic Discordance, Science, 279, 1499–1504, https://doi.org/10.1126/science.279.5356.1499, 1998.

Gurnis, M., Mitrovica, J. X., Ritsema, J., and van Heijst, H.-J.: Constraining mantle density structure using geological evidence of surface uplift rates: The case of the African Superplume, Geochem. Geophy. Geosy., 1, 1020, https://doi.org/10.1029/1999GC000035, 2000.

Heine, C., Müller, R. D., Steinberger, B., and DiCaprio, L.: Integrating deep Earth dynamics in paleogeographic reconstructions of Australia, Tectonophysics, 483, 135–150, https://doi.org/10.1016/j.tecto.2009.08.028, 2010.

Heine, C., Müller, R. D., Steinberger, B., and Torsvik, T. H.: Subsidence in intracontinental basins due to dynamic topography, Phys. Earth Planet. Int., 171, 252–264, https://doi.org/10.1016/j.pepi.2008.05.008, 2008.

Hoggard, M. J., White, N. J., and Al-Attar, D.: Global dynamic topography observations reveal limited influence of large-scale mantle flow, Nat. Geosci., 9, 456–463, https://doi.org/10.1038/ngeo2709, 2016.

Kaban, M. K., Schwintzer, P., and Tikhotsky, S. A.: A global isostatic gravity model of the Earth, Geophys. J. Int., 136, 519–536, https://doi.org/10.1046/j.1365-246X.1999.00731.x, 1999.

Kaban, M. K., Schwintzer, P., Artemieva, I. M., and Mooney, W. D.: Density of the continental roots: compositional and thermal contributions, Earth Planet. Sc. Lett., 209, 53–69, 2003.

Kaban, M. K., Schwintzer, P., and Reigber, C.: A new isostatic model of the lithosphere and gravity field, J. Geodesy, 78, 368–385, 2004.

Kanamori, H. and Press, F.: How thick is the lithosphere?, Nature, 226, 330–331, 1970.

Lekic, V. and Romanowicz, B.: Tectonic regionalization without a priori information: A cluster analysis of upper mantle tomography, Earth Planet. Sc. Lett., 308, 151–160, https://doi.org/10.1016/j.epsl.2011.05.050, 2011.

Li, C., van der Hilst, R. D., Engdahl, E. R., and Burdick, S.: A new global model for P wave speed variations in Earth's mantle, Geochem. Geophy. Geosy., 9, Q05018, https://doi.org/10.1029/2007GC001806, 2008.

Lithgow-Bertelloni, C. and Silver, P. G.: Dynamic topography, plate driving forces and the African superswell, Nature, 395, 269–272, 1998.

Liu, L., Spasojević, S., and Gurnis, M.: Reconstructing Farallon plate subduction beneath North America back to the Late Cretaceous, Science, 322, 934–938, https://doi.org/10.1126/science.1162921, 2008.

McNamara, A. K. and Zhong, S.: Thermochemical structures beneath Africa and the Pacific Ocean, Nature, 437, 1136–1139, https://doi.org/10.1038/nature04066, 2005.

Mitrovica, J. X. and Forte, A. M.: A new inference of mantle viscosity based upon joint inversion of convection and glacial isostatic adjustment data, Earth Planet. Sc. Lett., 225, 177–189, 2004.

Mitrovica, J. X., Beaumont, C., and Jarvis, G. T.: Tilting of continental interiors by the dynamical effects of subduction, Tectonics, 8, 1079–1094, 1989.

Molnar, P., England, P. C., and Jones, C. H.: Mantle dynamics, isostasy, and the support of high terrain, J. Geophys. Res.-Sol. Ea., 120, 1932–1957, https://doi.org/10.1002/2014JB011724, 2015.

Moucha, R., Forte, A. M., Mitrovica, J. X., Rowley, D. B., Quéré, S., Simmons, N. A., and Grand, S. P.: Dynamic topography and long-term sea-level variations: There is no such thing as a stable continental platform, Earth Planet. Sc. Lett., 271, 101–108, https://doi.org/10.1016/j.epsl.2008.03.056, 2008.

Moucha, R. and Forte, A. M.: Changes in African topography driven by mantle convection, Nat. Geosci., 4, 707–712, https://doi.org/10.1038/ngeo1235, 2011.

Müller, R. D., Gaina, C., Tikku, A., Mihut, D., Cande, S. C., and Stock, J. M.: Mesozoic/Cenozoic tectonic events around Australia, The history and dynamics of Global Plate Motions, 161–188, https://doi.org/10.1029/GM121p0161, 2000.

Müller, R. D., Sdrolias, M., Gaina, C., and Roest, W. R.: Age, spreading rates, and spreading asymmetry of the world's ocean crust, Geochem. Geophy. Geosy., 9, Q04006, https://doi.org/10.1029/2007GC001743, 2008.

Murnaghan, F. D.: The compressibility of media under extreme pressures, P. Natl. Acad. Sci. USA, 30, 244–247, 1944.

Panasyuk, S. V. and Hager, B. H.: Models of isostatic and dynamic topography, geoid anomalies, and their uncertainties, J. Geophys. Res.-Sol. Ea., 105, 28199–28209, 2000.

Paul, J. D., Roberts, G. G., and White, N.: The African landscape through space and time, Tectonics, 33, 898–935, https://doi.org/10.1002/2013TC003479, 2014.

Petersen, K. D., Nielsen, S. B., Clausen, O. R., Stephenson, R., and Gerya, T.: Small-Scale Mantle Convection Produces Stratigraphic Sequences in Sedimentary Basins, Science, 329, 827–830, https://doi.org/10.1126/science.1190115, 2010.

Ricard, Y., Richards, M., Lithgow-Bertelloni, C., and Le Stunff, Y.: A geodynamic model of mantle density heterogeneity, J. Geophys. Res.-Sol. Ea., 98, 21895–21909, https://doi.org/10.1029/93JB02216, 1993.

Ricard, Y., Chambat, F., and Lithgow-Bertelloni, C.: Gravity observations and 3D structure of the Earth, C. R. Geosci., 338, 992–1001, 2006.

Roberts, G. G. and White, N.: Estimating uplift rate histories from river profiles using African examples, J. Geophys. Res.-Sol. Ea., 115, B02406, https://doi.org/10.1029/2009JB006692, 2010.

Sandiford, M.: The tilting continent: a new constraint on the dynamic topographic field from Australia, Earth Planet. Sc. Lett., 261, 152–163, https://doi.org/10.1016/j.epsl.2007.06.023, 2007.

Schuberth, B. S. A., Bunge, H. P., and Ritsema, J.: Tomographic filtering of high-resolution mantle circulation models: Can seismic heterogeneity be explained by temperature alone?, Geochem. Geophy. Geosy., 10, Q05W03, https://doi.org/10.1029/2009GC002401, 2009.

Seton, M., Müller, R. D., Zahirovic, S., Gaina, C., Torsvik, T., Shephard, G., Talsma, A., Gurnis, M., Turner, M., Maus, S., and Chandler, M.: Global continental and ocean basin reconstructions since 200Ma, Earth-Sci. Rev., 113, 212–270, https://doi.org/10.1016/j.earscirev.2012.03.002, 2012.

Shephard, G. E., Müller, R. D., and Seton, M.: The tectonic evolution of the Arctic since Pangea breakup: Integrating constraints from surface geology and geophysics with mantle structure, Earth-Sci. Rev., 124, 148–183, https://doi.org/10.1016/j.earscirev.2013.05.012, 2013.

Simmons, N. A., Forte, A. M., Boschi, L., and Grand, S. P.: GyPSuM: A joint tomographic model of mantle density and seismic wave speeds, J. Geophys. Res.-Sol. Ea., 115, B12310, https://doi.org/10.1029/2010JB007631, 2010.

Spasojevic, S. and Gurnis, M.: Sea level and vertical motion of continents from dynamic earth models since the Late Cretaceous, AAPG bulletin, 96, 2037–2064, 2012.

Spasojevic, S., Liu, L., and Gurnis, M.: Adjoint models of mantle convection with seismic, plate motion, and stratigraphic constraints: North America since the Late Cretaceous, Geochem. Geophy. Geosy., 10, Q05W02, https://doi.org/10.1029/2008GC002345, 2009.

Stanley, J. R., Flowers, R. M., and Bell, D. R.: Kimberlite (U-Th)/He dating links surface erosion with lithospheric heating, thinning, and metasomatism in the southern African Plateau, Geology, 41, 1243–1246, 2013.

Steinberger, B.: Effects of latent heat release at phase boundaries on flow in the Earth's mantle, phase boundary topography and dynamic topography at the Earth's surface, Phys. Earth Planet. In., 164, 2–20, https://doi.org/10.1016/j.pepi.2007.04.021, 2007.

Steinberger, B.: Topography caused by mantle density variations: observation-based estimates and models derived from tomography and lithosphere thickness, Geophys. J. Int., 205, 604–621, https://doi.org/10.1093/gji/ggw040, 2016.

Steinberger, B. and Calderwood, A. R.: Models of large-scale viscous flow in the Earth's mantle with constraints from mineral physics and surface observations, Geophys. J. Int., 167, 1461–1481, https://doi.org/10.1111/j.1365-246X.2006.03131.x, 2006.

Tinker, J., de Wit, M., and Brown, R.: Mesozoic exhumation of the southern Cape, South Africa, quantified using apatite fission track thermochronology, Tectonophysics, 455, 77–93, 2008.

Veevers, J. J. (Ed.): Phanerozoic Earth History of Australia, Oxford Monographs on Geology and Geophysics, No. 2, Vol. 21, p. 212, https://doi.org/10.1002/gj.3350210211, 1984.

Wessel, P., Smith, W. H. F., Scharroo, R., Luis, J., and Wobbe, F.: Generic mapping tools: Improved version released, Eos, Transactions American Geophysical Union, 94, 409–410, https://doi.org/10.1002/2013EO450001, 2013.

Wheeler, P. and White, N.: Measuring dynamic topography: An analysis of Southeast Asia, Tectonics, 21, 1040,, https://doi.org/10.1029/2001TC900023, 2002.

Winterbourne, J., Crosby, A., and White, N.: Depth, age and dynamic topography of oceanic lithosphere beneath heavily sedimented Atlantic margins, Earth Planet. Sc. Lett., 287, 137–151, 2009.

Wolstencroft, M., Davies, J. H., and Davies, D. R.: Nusselt–Rayleigh number scaling for spherical shell Earth mantle simulation up to a Rayleigh number of 109, Phys. Earth Planet. In., 176, 132–141, https://doi.org/10.1016/j.pepi.2009.05.002, 2009.

Wolstencroft, M. and Davies, H.: Breaking supercontinents; no need to choose between passive or active, Solid Earth Discuss., https://doi.org/10.5194/se-2017-14, in review, 2017.

Zhang, N., Zhong, S., and Flowers, R. M.: Predicting and testing continental vertical motion histories since the Paleozoic, Earth Planet. Sc. Lett., 317, 426–435, https://doi.org/10.1016/j.epsl.2011.10.041, 2012.

Soil Atterberg limits of different weathering profiles of the collapsing gullies in the hilly granitic region of Southern China

Yusong Deng[1], **Chongfa Cai**[1], **Dong Xia**[2], **Shuwen Ding**[1], **Jiazhou Chen**[1], **and Tianwei Wang**[1]

[1]Key Laboratory of Arable Land Conservation (Middle and Lower Reaches of Yangtze River) of the Ministry of Agriculture, College of Resources and Environment, Huazhong Agricultural University, Wuhan, 430070, People's Republic of China
[2]College of Hydraulic and Environmental Engineering, China Three Gorges University, Yichang 443002, People's Republic of China

Correspondence to: Chongfa Cai (chongfacai@126.com)

Abstract. Collapsing gullies are one of the most serious soil erosion problems in the tropical and subtropical areas of southern China. However, few studies have been performed on the relationship of soil Atterberg limits with soil profiles of the collapsing gullies. Soil Atterberg limits, which include plastic limit and liquid limit, have been proposed as indicators for soil vulnerability to degradation. Here, the soil Atterberg limits within different weathering profiles and their relationships with soil physicochemical properties were investigated by characterizing four collapsing gullies in four counties in the hilly granitic region of southern China. The results showed that with the fall of weathering degree, there was a sharp decrease in plastic limit, liquid limit, plasticity index, soil organic matter, cation exchange capacity and free iron oxide. Additionally, there was a gradual increase in liquidity index, a sharp increase in particle density and bulk density followed by a slight decline, a decrease in the finer soil particles, a noticeable decline in the clay contents, and a considerable increase in the gravel and sand contents. The plastic limit varied from 19.43 to 35.93 % in TC, 19.51 to 33.82 % in GX, 19.32 to 35.58 % in AX and 18.91 to 36.56 % in WH, while the liquid limit varied from 30.91 to 62.68 % in TC, 30.89 to 57.70 % in GX, 32.48 to 65.71 % in AX and 30.77 to 62.70 % in WH, respectively. The soil Atterberg limits in the sandy soil layers and detritus layers were lower than those in the surface layers and red soil layers, which results in higher vulnerability of the sandy soil layers and detritus layers to erosion and finally the formation of the collapsing gully. The regression analyses showed that soil Atterberg limits had significant and positive correlation with SOM, clay content, cationic exchange capacity and Fe_d, significant and negative correlation with sand content and no obvious correlation with other properties. The results of this study revealed that soil Atterberg limits are an informative indicator to reflect the weathering degree of different weathering profiles of the collapsing gullies in the hilly granitic region.

1 Introduction

In the early 20th century, Atterberg proposed the limits of consistency for agricultural purposes to get a clear concept of the range of water contents of a soil in the plastic state (Atterberg, 1911). These limits of consistency, namely plastic limit and liquid limit, are well known as soil Atterberg limits. Plastic limit is the boundary between semi-solid and plastic state, and liquid limit separates plastic state from liquid state (Campbell, 2001). The methods developed by Casagrande (1932, 1958) to determine the liquid and plastic limits are considered as standard international tests. The width of the plastic state (liquid limit minus plastic limit), the plasticity index, is very useful for characterization, classification and prediction of the engineering behavior of fine soils. Moreover, several researchers have identified the relationship between in situ water content and Atterberg limits, the liquidity index, which is an indicator of soil hardness under natural conditions (Shahminan et al.,2014; Rashid et al., 2014). Atterberg limits were used in early studies on the tillage of soils, with the plastic limit recommended as the highest possible soil water content for cultivation (Baver, 1930;

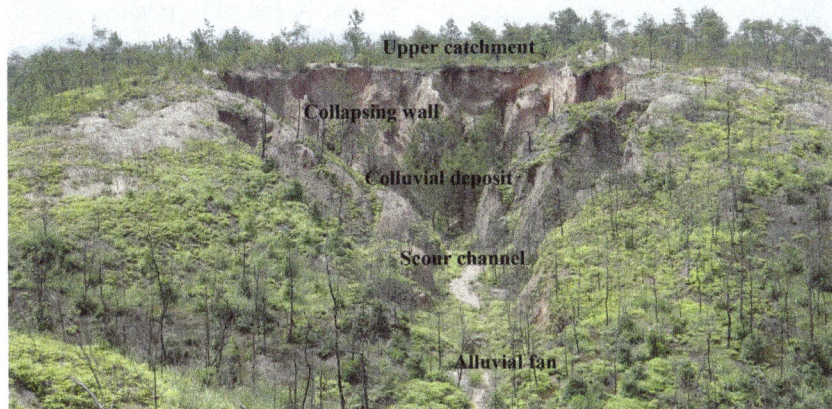

Figure 1. A typical collapsing gully in the hilly granitic region, Gan County, Jiangxi Province (photo: Yusong Deng).

Jong et al., 1990). Later on, Atterberg limits were mainly used in the classification of soils for engineering purposes. They also provide information for interpreting several soil mechanical and physical properties such as shear strength, bearing capacity, compressibility and shrinkage-swelling potential (Archer, 1975; Wroth and Wood, 1978; Cathy et al., 2008; McBride, 2008). Meanwhile, Atterberg limits are also essential for infrastructure design (e.g., construction of buildings and roads; Zolfaghari et al., 2015). These studies clearly show that there is a close relationship between Atterberg limits and certain properties of soils. More recently, Atterberg limits have been proposed as indicators for soil vulnerability to degradation processes of both natural and anthropogenic origin (Stanchi et al., 2015). Yalcin (2007) emphasized that, when subjected to water saturation, soils with limited cohesion are susceptible to erosion during heavy rainfall. Curtaz et al. (2014), Vacchiano et al. (2014) and Stanchi et al. (2012) have examined plastic limit and liquid limit in common soil types and proposed them as indicators to assess the soil vulnerability to erosion.

Soil degradation by processes such as soil erosion, shallow landslides and debris flows is a significant problem in mountainous areas, and is a crucial issue for natural hazard assessment in these areas (Jordán et al., 2014; Moreno-Ramón et al., 2014; Peng et al., 2015; Stanchi et al., 2015; Muñoz-Rojas et al., 2016a). A collapsing gully is a serious type of soil erosion widely distributed in the hilly granitic region of southern China, which is formed in the hill slopes covered by thick granite weathering mantle (Xu, 1996). The concept of a collapsing gully was first proposed by Zeng in 1960, which is a composite erosion formed by hydraulic scour and gravitational collapse (Zeng, 1960; Jiang et al., 2014; Xia et al., 2015; Deng et al., 2016b; Xia et al., 2016). These gul-

lies develop quickly and erupt suddenly, with an annual average erosion of over $50 \, \text{kt} \, \text{km}^{-2} \, \text{yr}^{-1}$ in these areas, more than 50-fold faster than the erosion on gentler slopes or on slopes with high vegetation cover (Zhong et al., 2013). The flooding, debris flows, and other disasters resulting from collapsing gullies can jeopardize sustainable development in the related regions. From 1950 to 2005, gully erosion affected $1220 \, \text{km}^2$ in the granitic red clay soil region, leading to the loss of more than 60 Mt of soil (Zhang, 2010). It is worth mentioning that the collapsing gullies in turn caused the loss of 360 000 ha of farmland, 521 000 houses, 36 000 km of road, 10 000 bridges, 9000 reservoirs, and 73 000 ponds, as well as an economic loss of USD 3.28 billion that affected 9.17 million residents (Jiang et al., 2014; Liang et al., 2009). According to a 2005 survey by the Monitoring Center of Soil and Water Conservation of China, collapsing gullies are widely distributed in the granitic red clay soil regions of southern China, which consist of Guangdong, Jiangxi, Hubei, Hunan, Fujian, Anhui, and Guangxi provinces, with the number of collapsing gullies up to 239 100 (Feng et al., 2009). A collapsing gully consists of five parts: (1) upper catchment, where a large amount of water is accumulated; (2) collapsing wall, where mass soil wasting, water erosion and gravity erosion are quite serious; (3) colluvial deposit, where residual material is deposited; (4) scour channel, where the sediment accumulation and transport is usually significantly deep and narrow; and (5) alluvial fan, the zone below the gully mouth where sediments transported by the collapse are deposited (Xu, 1996; Sheng and Liao, 1997; Xia et al., 2015; Deng et al., 2017; Fig. 1).

In a collapsing gully system, slumps and massive collapses of the collapsing wall are one of the main influential factors responsible for the collapsing gully enlargement and devel-

opment (Xia et al., 2015). Researchers have paid close attention to the damage of collapsing gully, and found that there
is a close relationship between the stability of the collapsing
wall, the amount of erosion and the development speed (Xu,
1996; Sheng and Liao, 1997; Luk et al., 1997a, b; Lan et al.,
2003). Qiu (1994) maintained that the mechanical composition of soil and the change in its action with water have an
important influence on the development of collapsing gully.
Li (1992) stated that there is an important relationship between the soil water content and critical height of collapsing
wall, with the height being 8–9 m at a low water content and
only 2–3 m in the saturated state. Zhang et al. (2013) pointed
out that granite soil (an Ultisol in the south of China) is easy
to disintegrate with increasing water content, and the process
is irreversible. Zhang et al. (2012) proposed that the cohesion
and internal friction angle of the soil showed a nonlinear attenuation trend with the increase in water content, and the
shear strength index showed a peak value when the soil water content was about 13 %. Liu and Zhang (2015) and Deng
et al. (2015) reported that the water content of the collapsing wall varied in different soil layers. Deng et al. (2016a)
proposed that the soil water characteristic curve of the layers
of granite is different, and the subsoil layers have greater de-
watering ability than the topsoil layers. From these studies,
we can find the soil water content is a common influencing
factor, and the stability of the collapsing wall will vary with
it. Wang et al. (2000) believe that the mechanical properties
of soil will change significantly when the rain is in full contact with the soil. Similar conclusions were reported by Luk
et al. (1997a), who revealed that the main cause for collapse
occurrence is the short-term rainfall intensity. The soil Atterberg limits refer to the highest and lowest water content
in the plastic state, which are of important significance in
predicting the influence of surface runoff and rainfall on the
collapsing gully. Several studies found that the soil Atterberg
limits are in general influenced by many soil properties, especially by organic matter and clay content (Hemmat et al.,
2010; Stanchi et al., 2015). However, few studies have been
performed on the relationship between Atterberg limits and
soil physicochemical properties and the occurrence of collapsing gully in the hilly granitic region of southern China.

The objectives of this study are (1) to evaluate the similarities and differences in soil Atterberg limits and soil physicochemical properties of different weathering profiles in the
four collapsing gullies, (2) to investigate the relationship between soil Atterberg limits and soil physicochemical properties by analyzing the status and variation in soil Atterberg
limits and (3) to explore the possibility of using soil Atterberg limits as an integrated index for quantifying collapsing
gully and soil weathering degree of different weathering profiles in the hilly granitic region.

2   Materials and methods

2.1   Study area

The sampling plots (22°58–29°24′ N, 110°51–118°17′) are
located in the hilly granitic region of southern China, including Tongcheng County (TC) in Hubei Province, Gan
County (GX) in Jiangxi Province, Anxi County (AX) in
Fujian Province and Wuhua County (WH) in Guangdong
Province, which are the most serious collapsing gully centers in southern China and thus were selected as the study
sites. These study areas are in a temperate monsoonal continental climate zone, with an average temperature of 15–22°
and an average annual precipitation of about 1500 mm with
high variability. The region is dominated by granitic red soil
(an Ultisol) that developed in the Yanshan period. There were
1102, 4138, 4744 and 22 117 collapsing gullies in TC, GX,
AX and WH, respectively. The control soil samples were collected from Xianning, Hubei.

2.2   Soil sampling

According to previous studies and the soil color and soil
structural characteristics, the weathering profiles of the collapsing gullies of the study area in the hilly granitic region
can be subdivided into four soil layers: surface layer, red soil
layer, sandy soil layer and detritus layer (Luk et al., 1997a;
Zhang et al., 2012; Xia et al., 2015).

The soil samples were collected in surface layer, red soil
layer, sandy soil layer and detritus layer. According to the
height of the collapsing gully wall, we collected 6, 8, 8 and
8 soil samples in four weathered layers, respectively, with a
total of 30 sampling sites. The detritus layer of the collapsing gully in Tongcheng County was not exposed, so the soil
samples were not collected. The information of soil sample
sites and soil sampling depth is presented in Tables 1 and 2.
The soil samples of control sites were collected from four
soil layers (A, B, C1, C2) in Xianning.

When collecting the samples of each soil layer, about 1–
2 kg soil sample was obtained by means of quartering and
transported to the laboratory for measurement of soil Atterberg limits (including plastic limit and liquid limit) and soil
physicochemical properties (including soil particle density,
organic matter, cation exchange capacity and free iron oxide). At each layer, six soil samples were obtained by using a
cutting ring to determine soil bulk density and calculate the
total porosity.

2.3   Soil analysis

The soil samples were air-dried and then sieved at the fraction < 0.452 mm for Atterberg limits determination, and at
< 2 mm for measurement of soil physical and chemical prop-
erties including particle density, particle-size distribution and
chemical analyses. Soil Atterberg limits (liquid limit and
plastic limit) were determined using the air-dried soil for

Table 1. Description of soil sampling sites (Xia et al., 2015).

Location	Collapsing gully code	Longitude and latitude	Altitude (m)	Height of collapsing gully wall (m)	Coverage of tree layer (%)	Coverage of surface layer (%)	Vegetation community
Tongcheng County	TC	29°12'39" N, 113°46'26" E	142	9	45	64	*Pinus massoniana + Cunninghamia lanceolata + Liquidambar formosana + Phyllostachys heterocycla − Rosa laevigata +Smilax china + Gardenia jasminoides +Vaccinium carlesii + Lespedeza bicolor − Dicranopteris linearis + Miscanthus floridulus*
Gan County	GX	26°11'22.2" N, 115°10'39.4" E	175	15	35	38	*P. massoniana + L. formosana + Schima superba − L. bicolor − D. linearis*
Anxi County	AX	24°57'14.3" N, 118°3'35.1" E	172	20	30	43	*P. massoniana + Eucalyptus robusta + Acacia confusa − Rhus chinensis + Rhodomyrtus tomentosa+ Loropetalum chinense − D. linearis + M. floridulus*
Wuhua County	WH	24°06'10.4" N, 115°34'57.1" E	157	35	28	35	*P. massoniana − R. tomentosa + Baeckea frutescens − D. linearis*

Table 2. Description of weathering profile, soil sampling depth and soil properties of different weathering profiles of the four collapsing gullies.

Soil layer code	Weathering profile	D (m)	PD (g cm^{-3})	BD (g cm^{-3})	TP (%)	SOM (g kg^{-1})	CEC (cmol kg^{-1})	Fe$_d$ (g kg^{-1})
TC1	Surface layer	0.3	2.58	1.29 ± 0.05d	49.03 ± 2.37a	23.37 ± 0.55a	16.39 ± 0.90a	21.38 ± 0.46bc
TC2	Red soil layer	0.8	2.64	1.47 ± 0.01a	44.11 ± 0.29c	6.81 ± 0.17b	8.37 ± 1.14b	27.37 ± 0.84a
TC3	Red soil layer	2	2.68	1.34 ± 0.05c	49.53 ± 1.79a	5.84 ± 0.20c	7.59 ± 0.27b	23.29 ± 1.29b
TC4	Red soil layer	4	2.65	1.39 ± 0.02b	47.26 ± 0.85b	2.68 ± 0.13d	3.32 ± 0.44c	19.42 ± 1.72c
TC5	Sandy soil layer	7	2.62	1.33 ± 0.02c	49.72 ± 0.83a	1.20 ± 0.11e	4.07 ± 0.61c	13.84 ± 0.93d
TC6	Sandy soil layer	9	2.65	1.35 ± 0.01c	48.63 ± 0.35ab	1.02 ± 0.06e	3.92 ± 0.34c	11.89 ± 1.00e
GX1	Surface layer	0.3	2.57	1.27 ± 0.05c	50.94 ± 2.34a	7.93 ± 0.11a	10.28 ± 0.17a	25.31 ± 1.45a
GX2	Red soil layer	0.8	2.67	1.40 ± 0.03ab	47.65 ± 1.50b	1.35 ± 0.08b	8.27 ± 0.44bc	26.59 ± 2.90a
GX3	Red soil layer	1.8	2.64	1.40 ± 0.02ab	46.79 ± 0.87bc	1.07 ± 0.12c	7.91 ± 0.60c	22.72 ± 0.57bc
GX4	Red soil layer	4	2.63	1.42 ± 0.02a	46.02 ± 0.95c	0.86 ± 0.07d	8.90 ± 0.69b	23.96 ± 1.11b
GX5	Sandy soil layer	7.5	2.62	1.41 ± 0.02ab	46.13 ± 1.06c	0.42 ± 0.06f	5.41 ± 0.86d	18.36 ± 0.77c
GX6	Sandy soil layer	9	2.69	1.37 ± 0.04bc	49.20 ± 1.59ab	0.72 ± 0.09e	5.98 ± 0.52d	13.30 ± 0.43d
GX7	Detritus layer	11	2.64	1.33 ± 0.06c	48.32 ± 1.27b	0.40 ± 0.06f	2.09 ± 0.19e	9.90 ± 0.78e
GX8	Detritus layer	13.5	2.59	1.38 ± 0.04ab	46.65 ± 1.96bc	0.71 ± 0.11e	3.43 ± 0.36e	9.41 ± 0.63e
AX1	Surface layer	0.3	2.54	1.31 ± 0.06c	44.40 ± 2.78d	44.06 ± 0.04a	22.18 ± 0.21a	31.03 ± 1.80a
AX2	Red soil layer	0.8	2.63	1.39 ± 0.06ab	54.24 ± 2.89a	11.23 ± 0.61b	14.63 ± 1.30b	27.53 ± 0.56b
AX3	Red soil layer	2	2.66	1.43 ± 0.03a	52.38 ± 1.73ab	6.33 ± 0.11c	9.20 ± 0.58c	26.35 ± 0.74b
AX4	Red soil layer	4	2.60	1.41 ± 0.01a	50.81 ± 0.45b	2.41 ± 0.11d	6.37 ± 0.61d	24.38 ± 1.11c
AX5	Sandy soil layer	8	2.65	1.37 ± 0.03b	48.39 ± 1.31bc	0.82 ± 0.03f	4.82 ± 0.18e	11.87 ± 1.04d
AX6	Sandy soil layer	10	2.54	1.35 ± 0.02bc	47.01 ± 0.88c	1.31 ± 0.09e	5.02 ± 0.27de	10.55 ± 1.23d
AX7	Detritus layer	12	2.62	1.32 ± 0.02c	49.50 ± 0.82bc	0.81 ± 0.07f	2.36 ± 0.32f	7.34 ± 0.56e
AX8	Detritus layer	15	2.53	1.31 ± 0.02c	48.12 ± 1.33bc	0.67 ± 0.09f	3.80 ± 0.71ef	7.30 ± 0.80e
WH1	Surface layer	0.3	2.52	1.33 ± 0.04d	48.19 ± 0.93a	15.17 ± 1.73a	13.84 ± 0.88a	28.40 ± 0.64a
WH2	Red soil layer	1	2.69	1.48 ± 0.01b	44.96 ± 0.29c	4.65 ± 0.29b	7.69 ± 0.39b	24.52 ± 0.54b
WH3	Red soil layer	2.5	2.72	1.47 ± 0.03b	45.68 ± 1.15bc	2.59 ± 0.14c	6.62 ± 0.51b	22.94 ± 0.91bc
WH4	Sandy soil layer	5	2.68	1.44 ± 0.02c	46.15 ± 0.83b	2.82 ± 0.03c	6.54 ± 0.45b	16.28 ± 1.10c
WH5	Sandy soil layer	9	2.63	1.40 ± 0.03cd	46.44 ± 1.64b	1.61 ± 0.10d	4.18 ± 0.50c	12.41 ± 0.27d
WH6	Sandy soil layer	11	2.62	1.49 ± 0.02b	43.01 ± 1.01c	0.57 ± 0.08f	2.28 ± 0.22d	14.23 ± 0.78cd
WH7	Detritus layer	14	2.59	1.54 ± 0.03a	40.34 ± 1.46d	0.74 ± 0.05e	3.91 ± 0.18cd	8.86 ± 0.40e
WH8	Detritus layer	17	2.61	1.37 ± 0.05d	46.41 ± 1.59b	0.23 ± 0.18g	1.93 ± 0.30e	8.37 ± 0.32e

Values with different letters are significantly different at the $P < 0.05$ level among the different soil layers of the same collapsing gully. SOM: soil organic matter; Fe$_d$: free iron oxide.

each layer according to the cone penetrometer and the thread roll method (Stanchi et al., 2015), which are reported in S.I.S.S (1997) after ASTM D 4318-10e1 (2010), i.e. the plasticity index and the liquidity index are obtained by the following Eqs. (1, 2):

$$\text{Plasticity index} = \text{liquid limit- plastic limit,} \qquad (1)$$

$$\text{Liquidity index} = (\text{WC}_{\text{insitu}} - \text{plastic limit})/$$

$$(\text{liquid limit} - \text{plastic limit}), \qquad (2)$$

where WC$_{\text{insitu}}$ is in situ water content.

The particle density (PD) was measured by the pycnometer method, the bulk density (BD) was determined by the cutting ring method, and the total porosity (TP) was calculated as TP = 1 − (BD/PD) (Anderson and Ingram, 1993; Cerdà and Doerr, 2010). The particle-size distribution (PSD) was determined by the sieve-and-pipette method (Gee and

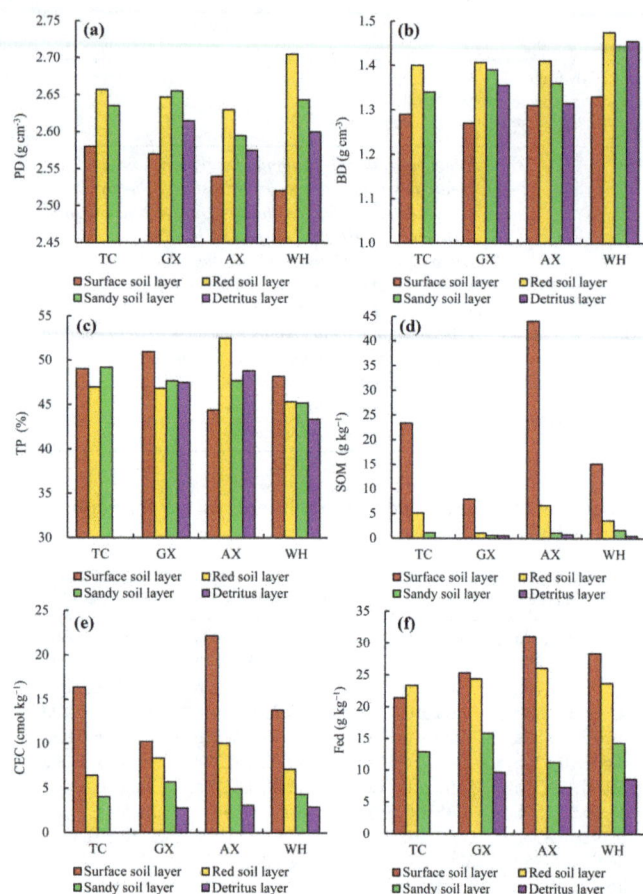

Figure 2. Averages of soil properties for different weathering profiles of the four collapsing gullies. (**a**) Particle density, (**b**) bulk density, (**c**) total porosity, (**d**) soil organic matter, (**e**) cation exchange capacity and (**f**) free iron oxide.

Bauder, 1986). Soil organic matter (SOM) was measured by the $K_2Cr_2O_7$–H_2SO_4 oxidation method of Walkey and Black (Nelson and Sommers, 1982). Cation exchange capacity (CEC) was measured after extraction with ammonium acetate (Rhoades, 1982). Free iron oxide (Fe_d) was extracted by dithionite–citrate–bicarbonate (DCB; Mehra and Jackson, 1958).

2.4   Statistical analysis

Statistical analyses were performed using SPSS 19.0 software (SPSS Inc., Chicago, IL, USA). A one-way analysis of variance (ANOVA) was performed to examine the effects of soil depth on soil Atterberg limits and soil physicochemical properties. The least squares difference (LSD) test (at $P < 0.05$) was used to compare means of soil variables when the results of ANOVA were significant at $P < 0.05$. Regression analysis was used to analyze the relationship between soil Atterberg limits and soil physicochemical properties.

3   Results and discussion

3.1   Soil physicochemical properties

The soil physical and chemical properties for the different weathering profiles in the four collapsing gullies (TC, GX, AX and WH) were described in terms of soil particle density (PD), soil bulk density (BD), total porosity (TP), soil organic matter (SOM), cation exchange capacity (CEC), free iron oxide (Fe_d) and particle-size distribution (PSD). The values for these properties are shown in Tables 2 and 3. Average values at varying soil layers including surface soil layer, red soil layer, sandy soil layer and detritus layer are given in Figs. 2 and 3.

3.1.1   Soil particle density (PD)

From Table 2, it can be seen that the soil PD was the highest in TC3 ($2.68\,\mathrm{g\,cm^{-3}}$), GX6 ($2.69\,\mathrm{g\,cm^{-3}}$), AX3 ($2.66\,\mathrm{g\,cm^{-3}}$) and WH3 ($2.72\,\mathrm{g\,cm^{-3}}$) of each collapsing gully, but the lowest in TC1 ($2.58\,\mathrm{g\,cm^{-3}}$), GX1 ($2.57\,\mathrm{g\,cm^{-3}}$), AX8 ($2.53\,\mathrm{g\,cm^{-3}}$) and WH1 ($2.52\,\mathrm{g\,cm^{-3}}$). Significant differences ($P < 0.01$) were observed in the average PD values of the different soil layers in TC, GX, AX and WH (Fig. 2a). The PD was the lowest in the surface soil layer, followed by the detritus layer. In addition, the highest PD was observed in the red soil layer of TC, AX and WH and the sandy soil layer of GX. Furthermore, as shown in Table 2, most of the soil PD values in all the four soil layers were less than $2.65\,\mathrm{g\,cm^{-3}}$, which are often used to calculate the value of soil BD (Lee et al., 2009; Sharma and Bora, 2015). The lower PD value may be due to the loose structure of granite soil (Luk et al., 1997a).

3.1.2   Bulk density (BD)

From Table 2, it can also be seen that soil BD values were the lowest in the surface layer of all the collapsing gullies (1.29, 1.27, 1.21 and $1.33\,\mathrm{g\,cm^{-3}}$ for TC, GX, AX and WH, respectively). However, relatively higher BD values were observed in the red soil layer (1.47, 1.42, 1.43 and $1.48\,\mathrm{g\,cm^{-3}}$ for TC, GX, AX and WH, respectively), followed by the sandy layer. The average soil BD values had significant difference ($p < 0.01$) in the different soil layers of TC, GX, AX and WH except in the surface layer of WH (Fig. 2b). Meanwhile, the bulk density first increased sharply ($p < 0.01$) and thus declined slightly from the surface layer to the sandy soil layer of TC and to the detritus layer of GX, AX and WH (Table 2), which are similar to the report by Perrin et al. (2014). The soil BD values of the surface layer were lower than those of the other layers, probably due to the higher content of SOM, more plant root distribution, and better soil structure and texture (Huang et al., 2014). The bulk density of the red soil layers was higher probably because the natural compact structure was maintained (Masto et al., 2015). The lower soil BD

Table 3. Percentages of different particle-size distributions of different weathering profiles of the four collapsing gullies.

| | Mass percentages of soil particle-size distribution (mm) | | | | | | | | | |
| | Gravel | Coarse sand | | Fine sand | | Silt | | | | Clay |
Soil layer code	2.0–1.0	1.0–0.5	0.5–0.25	0.25–0.15	0.15–0.05	0.05–0.02	0.02–0.01	0.01–0.005	0.005–0.002	<0.002
TC1	9.24±1.61b	7.13±0.10d	7.09±1.35b	3.97±0.64d	9.86±0.93c	6.55±1.67d	12.07±0.59a	5.16±0.58c	6.11±0.81b	32.81±1.46b
TC2	7.87±0.65b	6.55±0.12e	6.12±0.54c	6.10±0.07c	6.24±0.93d	16.67±1.04a	9.81±0.50b	6.18±1.07b	5.54±0.92c	28.91±0.62c
TC3	4.51±0.36c	4.91±0.24f	5.27±0.11d	6.72±0.85bc	10.55±1.14c	6.34±1.22d	9.74±1.16b	3.66±0.84d	7.26±0.21a	41.03±0.72a
TC4	3.05±0.55d	7.95±0.54c	9.78±1.08a	9.19±1.32a	17.66±1.57a	6.25±0.60d	10.97±0.96a	3.27±0.63d	5.69±0.55c	26.19±1.86d
TC5	5.34±0.71c	11.14±0.38b	11.75±0.78a	10.21±1.05a	13.68±1.45b	14.01±1.16b	9.44±0.17b	7.54±0.25a	6.64±0.79b	10.24±0.18e
TC6	19.84±2.28a	14.63±0.58a	11.95±1.23a	7.58±0.37b	16.46±1.04a	8.28±0.91c	8.48±0.98c	5.20±0.33c	3.71±0.13d	3.87±0.48f
GX1	8.99±0.37d	4.78±0.10d	4.43±0.29e	3.94±0.18e	12.77±0.34f	2.92±0.25e	5.49±0.78d	6.09±1.03e	13.92±1.65a	36.65±1.85a
GX2	8.12±0.31e	4.66±0.19d	4.41±0.05e	4.17±0.22e	13.62±0.31de	4.14±0.66d	7.92±1.27bc	7.00±1.10d	12.85±1.62a	33.10±1.80b
GX3	9.89±0.50c	5.65±0.21c	6.19±0.25d	5.32±0.41d	16.40±1.03c	9.24±0.33c	7.19±1.74c	8.50±0.65a	10.37±0.88b	21.25±1.14c
GX4	8.85±0.71d	5.68±0.30c	7.93±0.31b	8.68±0.53b	18.72±1.27b	8.80±0.45c	8.09±0.21b	7.65±0.48c	9.81±0.41bc	15.78±0.39d
GX5	9.71±1.30cd	5.03±0.25d	4.17±0.39e	4.91±0.42d	27.91±0.96a	11.14±0.54b	8.49±1.4b	6.68±1.43d	7.69±1.25d	14.29±0.55d
GX6	12.13±0.73b	7.90±0.19b	7.30±0.19c	8.69±0.40b	16.40±0.34c	12.44±0.52a	8.62±0.59b	8.24±0.53a	9.37±0.71c	8.90±0.42f
GX7	14.87±1.28a	8.87±0.14a	8.60±0.81ab	9.84±0.99a	14.60±0.72d	10.37±1.63bc	6.03±0.82d	8.83±0.17a	4.44±1.99e	13.55±1.39de
GX8	15.83±0.85a	8.80±0.07a	8.67±0.20a	8.09±0.62c	13.15±0.99ef	11.18±1.11ab	9.73±1.47a	7.68±0.31c	5.31±1.46e	11.55±1.11e
AX1	19.32±0.48c	7.55±0.42c	6.67±0.23c	3.86±0.18d	6.52±0.94d	5.04±0.95d	6.02±0.37d	3.63±0.47e	7.93±0.24c	33.47±1.39b
AX2	6.23±0.35e	5.34±0.16d	4.10±0.31d	2.90±0.23ef	4.42±0.33e	3.47±0.71e	4.01±0.19e	6.34±1.12c	11.53±1.90ab	51.66±1.54a
AX3	6.39±0.25e	5.66±0.21d	3.99±0.43d	3.21±0.13e	6.42±1.02d	4.19±0.97de	1.60±0.62f	5.64±1.35cd	9.61±0.69b	53.27±1.47a
AX4	8.65±0.74d	4.63±0.08e	3.31±0.16e	2.48±0.50f	12.22±1.02c	3.92±1.81e	8.27±1.17ab	11.65±0.56a	12.91±1.91a	31.96±0.55b
AX5	19.86±0.87bc	8.71±0.23b	6.08±0.29c	5.35±0.12c	14.30±1.81bc	8.62±0.48c	8.02±1.53b	8.35±0.37b	4.04±1.32d	16.68±1.10c
AX6	24.49±1.05a	10.01±0.42a	7.66±0.45b	6.44±1.02ab	15.82±1.44ab	10.71±0.50b	6.87±1.11cd	6.58±1.13c	4.27±0.07d	7.14±1.33d
AX7	19.15±0.35c	7.83±0.27c	7.04±0.57b	5.95±0.69b	15.96±0.78a	15.85±1.12a	8.00±0.74bc	8.00±0.48b	3.78±0.73d	8.45±0.31d
AX8	21.02±1.37b	10.93±0.43a	10.86±0.98a	7.94±1.76a	17.48±1.97a	8.73±1.08c	9.00±0.30a	5.01±0.27d	1.02±0.49e	8.00±1.25d
WH1	18.53±0.62f	5.67±0.12c	3.74±0.17c	2.30±0.39d	10.24±1.15a	9.33±1.30a	5.55±0.19d	4.59±0.62d	7.42±1.85d	32.62±1.30a
WH2	23.42±0.40d	5.78±0.09c	2.93±0.21de	2.29±0.05d	6.89±0.74c	7.34±0.56c	8.51±1.28a	3.70±0.55d	10.23±1.32c	28.92±2.22b
WH3	25.72±1.91b	5.92±0.29c	2.76±0.08e	1.97±0.05b	5.15±0.18d	5.74±0.53d	4.29±0.63e	8.72±0.93c	12.91±0.15b	26.83±1.82b
WH4	22.26±1.33de	6.39±0.21b	3.24±0.25d	2.06±0.10d	4.96±1.10d	5.45±1.25d	7.09±1.00bc	9.10±0.60c	16.07±1.60a	23.38±1.97c
WH5	24.53±0.62c	8.46±0.16a	4.29±0.27b	3.05±0.14c	5.67±1.34d	7.02±0.76c	4.04±0.94e	15.15±1.85a	10.23±1.03c	17.54±1.67d
WH6	27.73±0.23a	8.50±0.41a	5.00±0.49a	4.40±0.37b	3.06±0.38e	10.94±1.25a	6.98±1.34bc	12.39±0.65b	10.06±1.73c	10.93±1.38e
WH7	25.81±0.25b	8.54±0.05a	5.29±0.29a	5.57±0.24a	9.27±0.86ab	8.36±1.80ab	6.73±0.73c	14.46±1.25ab	5.56±0.38d	10.42±0.79e
WH8	25.16±0.82b	8.48±0.17a	5.42±0.08a	5.24±0.61a	8.43±0.49b	7.40±1.66bc	7.55±1.80ab	15.65±1.21a	10.91±0.57c	5.77±0.82f

Values with different letters are significantly different at the $P < 0.05$ level among the different soil layers of the same collapsing gully.

values of the sandy layer and detritus layer may be due to weak weathering and loose soil structure (Lan et al., 2013).

3.1.3 Total porosity (TP)

Unlike soil BD, the soil TP was comparatively high in the surface soil layer of GX and WH, but was the highest in the red soil layer of AX (Fig. 2c). From Table 2, it can be seen that the soil TP values were lower in the red soil layer, such as TC2 (44.11 %) and GX4 (46.02 %), which may be due to the weathering process of these soil layers, feldspar and mica in mineralized granites (Deng et al., 2016b).

3.1.4 Soil organic matter (SOM)

Soil organic matter (SOM) plays an important role in soil nutrient availability, and its increase may decrease the potential of soil erosion (Oliveira et al., 2015). As shown in Table 2, with the increase in depth, SOM contents in the soil layers of the four collapsing gullies showed a sharply decreasing trend ($P < 0.01$). The sandy soil layers and detritus layers showed relatively lower SOM contents than those in the red soil layers and surface layers (Fig. 2d). The AX1 had the highest SOM content (44.06 g kg^{-1}), followed by TC1 (23.37 g kg^{-1}), WH1 (15.17 g kg^{-1}) and AX2 (11.23 g kg^{-1}; Table 2), which is mainly due to the decomposition of surface litter in the ground surface. However, the sandy soil layer and the detritus layer are basically in the state of incomplete weathering, and there is no accumulation of SOM (Xia et al., 2015).

3.1.5 Cation exchange capacity (CEC)

Cationic exchange capacity (CEC) is a measure of the soil capacity to adsorb and release cations (Jordán et al., 2009; Khaledian et al., 2016; Muñoz-Rojas et al., 2016b). Similar to the SOM trend, CEC also decreased significantly from the topsoil layers to the subsoil layers in the four collapsing gullies. As shown in Table 2, the CEC values were the highest in the surface soil layer of the four gullies (1.29, 1.27, 1.21 and 1.33 g cm^{-3} for TC1, GX1, AX1 and WH1, respectively). The average CEC values followed the order of surface soil layer > red soil layer > sandy soil layer > detritus layer with significant difference ($P < 0.01$; Fig. 2e).

3.1.6 Free iron oxide (Fe$_d$)

Fe$_d$ is the secondary product formed by the weathering of the parent rock during soil formation. One Fe$_d$ state of the film surface is wrapped in the shape of clay minerals, and another state may be filled in the micropores of clay minerals (Cerdà et al., 2002; Lan et al., 2013). It is a unique and very important cementing material in weathered soil. As shown in Table 2, Fe$_d$ values were the lowest in the detritus layer of all the collapsing gullies (11.89, 9.41, 7.30 and 8.37 g kg^{-1} for TC, GX, AX and WH, respectively). The highest Fe$_d$ values of AX and WH were observed in the surface soil layer (31.03 and 28.40 g kg^{-1} for AX and WH), while those of TC and GX were observed in the red soil layer (27.37 and 26.59 g kg^{-1} for TC and GX). Overall, there are significant differences between surface soil layer, red soil layer, sandy

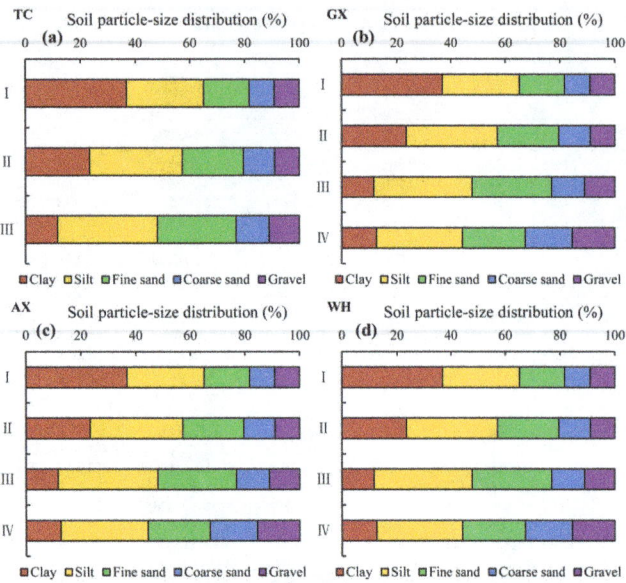

Figure 3. Averages of different particle-size distributions for different weathering profiles of the four collapsing gullies. **(a)** Tongcheng County, **(b)** Ganxian County, **(c)** Anxi County and **(d)** Wuhua County.

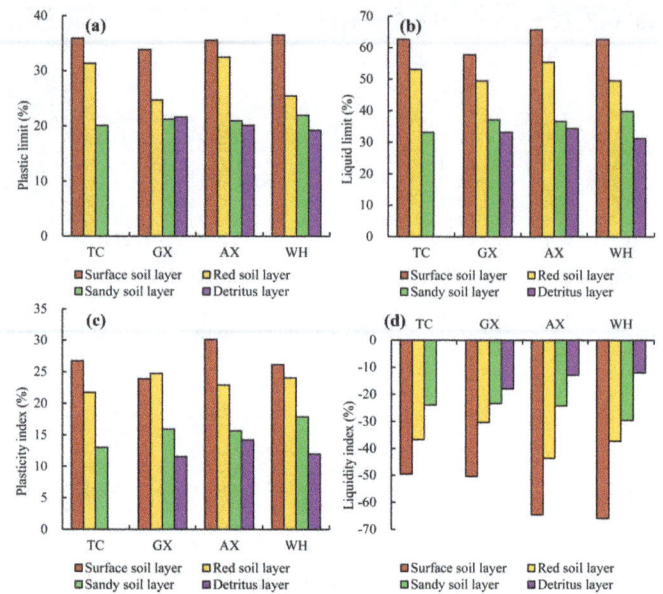

Figure 4. Averages of soil Atterberg limits for different weathering profiles of the four collapsing gullies. **(a)** Plastic limit, **(b)** liquid limit, **(c)** plasticity index and **(d)** liquidity index.

soil layer and detritus layer in different weathering profiles (Fig. 2f). These results show that the structural and mechanical properties are stronger in the surface soil layers and the red soil layers. However, when compared to the topsoil layers, the soil structure is loose and the cohesive strength is low in the sandy soil layer and detritus layer.

3.1.7 Particle-size distribution (PSD)

Soil particle-size distribution (PSD) is one of the most important physical attributes in soil systems (Hillel, 1980). PSD affects the movement and retention of water, solutes, heat, and air, and thus greatly affects soil properties (Arjmand Sajjadi et al., 2015). The highest clay contents were 41.03, 36.65, 53.27 and 32.62 % in TC, GX, AX and WH, respectively, and silt varied from 25.67 to 38.21 % in TC, 28.43 to 38.68 % in GX, 21.06 to 36.75 % in AX and 26.90 to 41.51 % in WH. The averages of particle-size distributions for different weathering profiles of the four collapsing gullies are shown in Fig. 3. The results indicated that the finer soil particles declined and the coarse soil particles increased from the surface layer to detritus layer. The surface layer of TC, GX and WH collapsing gullies had the greatest clay content of 32.81, 36.65 and 32.62 %, respectively, while the red soil layer of the AX collapsing gully showed the greatest clay content (45.63 %). This phenomenon can be attributed to the different weathering degree of granite: the grain size becomes coarser, the SiO_2 content and sand content increase, and the clay content decreases from the top to the bottom (Xu, 1996; Lin et al., 2015).

3.2 Soil Atterberg limits characteristics of weathering profiles of the collapsing gullies

All the measured soil plastic limit and liquid limit values varied significantly in the different soil layers. Table 4 lists the calculated values for the Atterberg limits, plasticity index and liquidity index. The average values for these properties are shown in Fig. 4 and the relationships of these values with soil depth are shown in Fig. 5.

3.2.1 Soil plastic limit and liquid limit

As shown in Table 4, soil plastic limit and liquid limit varied greatly from the top to the bottom of different soil layers. Specifically, the soil plastic limit ranged from 19.43 (TC6) to 35.93 % (TC1) with an average of 28.34 % in TC, 19.51 (GX6) to 33.82 % (GX1) with an average of 24.19 % in GX, 19.32 (AX7) to 36.03 % (AX2) with an average of 26.87 % in AX, and 18.91 (WH8) to 36.56 % (WH8) with an average of 23.98 % in WH. Consistent with the variation trend of plastic limit, the soil liquid limit was found to be highest in TC1 (62.68 %), GX1 (57.70 %), AX1 (65.71 %) and WH1 (62.70 %) in each weathering profile of the four collapsing gullies, and lowest in TC6 (30.91 %), GX6 (30.89 %), AX8 (32.48 %) and WH7 (30.77 %). The averages of soil plastic limit and liquid limit are shown in Table 4. The results indicated that, with declining weathering degree (from the surface layer to detritus layer), both of them decreased noticeably ($P < 0.01$; Figs. 5a; 7b). The surface layer of all the four collapsing gullies had the greatest soil Atterberg limits (35.93, 33.82, 35.58 and 36.56 % for the plastic limit, and

Table 4. Soil Atterberg limits of different weathering profiles of the four collapsing gullies.

Soil layer code	Plastic limit (%)	Liquid limit (%)	Plasticity index (%)	Liquidity index (%)
TC1	35.93 ± 0.69a	62.68 ± 1.32a	26.75 ± 2.01a	−49.55 ± 3.74d
TC2	31.73 ± 2.25b	53.09 ± 0.20bc	21.36 ± 2.05b	−47.08 ± 4.52d
TC3	30.51 ± 0.72b	56.03 ± 2.20b	25.52 ± 1.47a	−27.60 ± 1.59b
TC4	31.74 ± 0.56b	50.04 ± 0.23c	18.30 ± 0.33c	−35.54 ± 6.96c
TC5	20.73 ± 1.68c	35.31 ± 1.05d	14.58 ± 2.73d	−37.25 ± 6.96c
TC6	19.43 ± 2.07c	30.91 ± 0.25d	11.48 ± 1.82d	−10.57 ± 1.68a
GX1	33.82 ± 0.13a	57.70 ± 2.16a	23.88 ± 2.04ab	−50.36 ± 4.29e
GX2	27.04 ± 2.81b	52.91 ± 0.61b	25.87 ± 2.20a	−34.67 ± 2.94d
GX3	23.08 ± 0.45c	49.58 ± 0.96bc	26.50 ± 1.41a	−30.54 ± 1.62c
GX4	23.97 ± 2.39c	45.82 ± 3.61c	21.85 ± 1.22b	−25.80 ± 1.44bc
GX5	22.88 ± 1.98cd	43.32 ± 1.45c	20.44 ± 0.53b	−24.27 ± 0.63bc
GX6	19.51 ± 0.95d	30.89 ± 2.02e	11.38 ± 1.07d	−22.42 ± 2.10b
GX7	21.16 ± 1.53cd	34.25 ± 0.41d	13.09 ± 1.12c	−18.16 ± 1.57a
GX8	22.06 ± 0.59cd	32.15 ± 1.44de	10.09 ± 2.03d	−17.61 ± 3.56a
AX1	35.58 ± 1.70a	65.71 ± 0.02a	30.14 ± 1.72a	−64.57 ± 3.70d
AX2	36.03 ± 2.83a	60.67 ± 0.11ab	24.64 ± 2.72b	−52.16 ± 5.76c
AX3	35.42 ± 0.21a	57.01 ± 4.56b	21.59 ± 4.36bc	−52.00 ± 10.49c
AX4	25.84 ± 1.60b	48.34 ± 0.71c	22.49 ± 2.31bc	−26.59 ± 2.73b
AX5	22.34 ± 1.65bc	40.66 ± 0.12cd	18.32 ± 1.53c	−24.12 ± 2.00b
AX6	19.51 ± 0.44d	32.51 ± 1.18e	13.00 ± 0.74d	−24.27 ± 1.40b
AX7	19.32 ± 0.31d	36.26 ± 0.98d	16.94 ± 0.68cd	−13.35 ± 0.54a
AX8	20.95 ± 1.36c	32.48 ± 1.36e	11.53 ± 0.02e	−12.41 ± 0.01a
WH1	36.56 ± 0.99a	62.70 ± 1.04a	26.14 ± 0.05a	−65.91 ± 0.13e
WH2	26.01 ± 2.36b	52.20 ± 0.97b	26.19 ± 3.32a	−31.84 ± 4.03b
WH3	24.93 ± 0.17bc	46.86 ± 2.09c	21.93 ± 1.92b	−42.67 ± 3.74d
WH4	23.83 ± 0.10c	46.11 ± 0.86c	22.28 ± 0.96b	−38.60 ± 1.68bcd
WH5	22.25 ± 0.62c	39.11 ± 0.29d	16.87 ± 0.33c	−36.69 ± 0.70bc
WH6	19.74 ± 0.84d	34.22 ± 1.95e	14.48 ± 1.11cd	−13.38 ± 1.00a
WH7	19.56 ± 0.27d	30.77 ± 1.32f	11.21 ± 1.59d	−11.65 ± 1.63a
WH8	18.91 ± 1.44d	31.72 ± 0.48f	12.81 ± 1.93d	−12.24 ± 1.85a

62.68, 57.70, 65.71 and 62.70 % for the liquid limit, respectively). The plastic limit of the sandy soil layer and the detritus layer was significantly lower ($P < 0.01$) than that of the surface soil layer and the red soil layer, but with no significant difference between each other. As shown in Fig. 5, the soil Atterberg limits presented a nonlinear relationship with soil depth. Power function fitting showed that both the soil plastic limit and liquid limit had a remarkable negative correlation with the soil depth (Fig. 5a, $R^2 = 0.784$, $p < 0.001$ and Fig. 5b, $R^2 = 0.877$, $p < 0.0001$, respectively). Additionally, the soil plastic limit of the surface soil layer and the red soil layer ranged between 24.70 and 36.56 % with an average of 31.98 % and the liquid limit ranged between 49.43 and 65.71 % with an average of 57.02 %, which are higher compared with most types of soil (Reznik, 2016), but an opposite trend was observed in the sandy soil layer and the detritus layer. Our findings are in agreement with the previous studies by Zhuang et al. (2014) and Xia et al. (2015), who reported that the topsoil layers have a better ability to resist deformation than the subsoil layers. These results indicate that the change in water content has little influence on the sur-

face soil layer and the red soil layer, and the soil cannot be easily transformed into a liquid state by the rainfall erosion and runoff scouring. Conversely, the change in water content has a great influence on the sandy soil layer and the detritus layer, and with water content increasing, the soil can be changed from solid to liquid state.

3.2.2 Soil plasticity index and liquidity index

As shown in Table 4, there are considerable differences in soil plasticity index and liquidity index in the different weathering profiles of the four collapsing gullies. The soil plasticity index was highest in AX1 (30.14 %), followed by TC1 (26.75 %), GX3 (26.50 %) and WH2 (26.19 %), and it was also the highest in each soil layer. However, the plasticity index was lowest in the bottom soil layers (11.48, 10.09 and 11.53 % for TC6, GX8 and AX8, respectively) except for WH. Additionally, inconsistent with plasticity index, liquidity index was the lowest in the surface soil layer of each weathering profile (−49.55, −50.36, −64.57 and −65.91 % for TC1, GX1, AX1 and WH1, respectively). The highest liquidity indexes of TC, GX, AX and WH were −10.57 % in

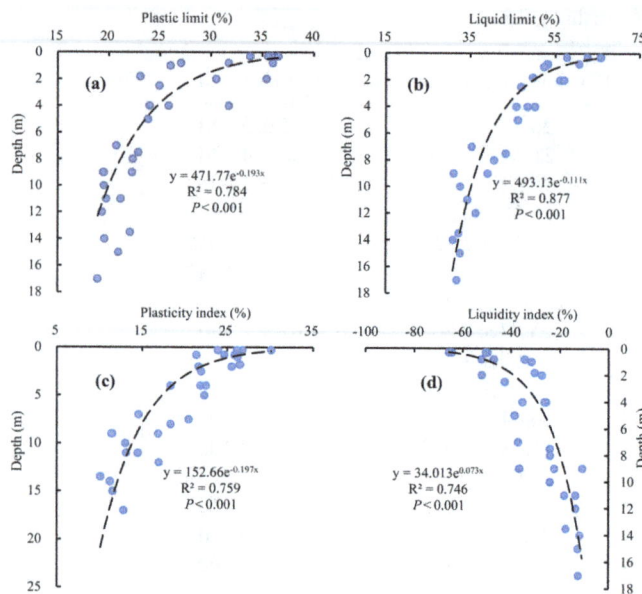

Figure 5. Relationship between soil Atterberg limits and soil depth. **(a)** Plastic limit, **(b)** liquid limit, **(c)** plasticity index and **(d)** liquidity index.

TC6, −17.61 % in GX8, −12.41 % in AX8 and −11.65 % in WH7, respectively. Figure 4a–d summarizes the statistics of soil plasticity index and liquidity index in all of the different weathering profiles. Significant differences were observed in all the measured plasticity and liquidity indexes between the surface soil layer, red soil layer, sandy soil layer and the detritus layer. The results indicated that the soil plasticity index decreased noticeably with the decline of weathering degree (from the surface layer to detritus layer), which is similar to the variation regularity of plastic limit and liquid limit.

The surface layer of the TC, AX and WH collapsing gullies had the greatest soil plasticity index (26.75, 30.14 and 26.14 %, respectively), but the greatest plasticity index (23.88 %) of the GX collapsing gully was found in the red soil layer. In contrast with the plasticity index, the liquidity index was significantly ($P < 0.01$) higher in the sandy soil layer and the detritus layer and was the lowest in the surface soil layer (−49.55, −50.36, −64.57 and −65.91 % for TC, GX, AX and WH, respectively; Fig. 4). Regression analyses were performed to determine the strength of relationships between the plasticity index, the liquidity index and soil depth (Fig. 5a–d). The nonlinear regression analyses showed that the plasticity index had a remarkable negative correlation with the soil depth (Fig. 5c, $R^2 = 0.759$, $P < 0.001$). However, there was a significant positive correlation between the soil liquidity index and the soil depth based on the power function fitting analysis (Fig. 5d, $R^2 = 0.746$, $P < 0.001$).

The differences in soil plasticity index and liquidity index between topsoil layers and subsoil layers may be related to the variation in the dynamics of the soil properties. As previously reported, changes in soil plasticity index and liquidity

index depend on soil properties such as clay and organic matter (Zhuang et al., 2014). The size of the plasticity index is directly related to the maximum possible bound water content of a certain mass of soil particles. However, the bound water content of soil is related to the size of soil particle, mineral composition, the composition and concentration of cation in the hydration membrane. Thus, the plasticity index is a comprehensive indicator for the reaction properties of clayey soil, which means the larger the index is, the higher the clay content will be (Husein et al., 1999). Our findings clearly demonstrated that the plasticity index of subsoil layers was significantly lower ($P < 0.01$) than that of topsoil layers in the different weathering profiles, implying that the content of fine particles in the soil gradually decreased with soil depth. Previous studies about soil texture classification are frequently based on soil plasticity index: a soil with a value between 10 and 17 % is defined as silty clay and that with a value greater than 17 % is classified as clay (Zentar et al., 2009; Marek et al., 2015). Based on this classification theory, most topsoil layers in the TC, GX, AX and WH collapsing gullies can be defined as clay, while the subsoil layers can be classified as silty clay, which is more susceptible to erosion.

However, the adsorption capacity of bound water varied under a different soil specific surface area and mineral composition. Therefore, given the same water content, for the soil with high viscosity, the water may be bound water, while for the soil with low viscosity, a considerable part of the water can be free water, which means that the soil state cannot be defined only by water content and we need another indicator, namely the liquidity index, to reflect the relationship between natural water content and Atterberg limits in the soil. The liquidity index is defined as the ratio of the difference between the natural moisture content and the plastic limit to the plastic limit (Sposito, 1989). When the natural moisture content is close to the plastic limit, the soil is hard; when it is close to the liquid limit, the soil is weak in cohesive strength. In engineering practice, the soil is in a hard state when the liquidity index is less than 0 (Zhuang et al., 2014). In our research, the liquidity indexes of all soils were less than 0, indicating that the soil of the different weathering profiles of the four collapsing gullies is hard in the natural state. Nevertheless, the subsoil layers of the collapsing gullies are more close to 0 than the topsoil layers in the liquidity index, indicating that the subsoil layers are weaker than the topsoil layers in cohesion strength.

3.2.3 Relationship between soil Atterberg limits and collapsing gully

In this study, the liquidity indexes of all soils were less than 0, indicating that the soils of the four collapsing gullies remain solid in natural state, with a high shear strength and strong resistance to water erosion, enabling the soil of granite weathering profile to maintain stability. From the soil Atterberg limits of all the soils, it can be seen that the plastic limit, liquid limit and plasticity index are higher in the surface soil layer and red soil layer, implying that the plastic state cannot be easily changed when the rain lasts a short time, such as moderate to light rain, which usually does not lead to the collapse and loss of the soils with high compaction and hardness. However, if the rainfall duration is long enough, the soil water content can reach a high level, leading to an increase in the soil self-weight, a decrease in the soil shear strength, and thus the collapse of the soils. The plastic limit, liquid limit and plasticity index of the sandy soil layer and detritus layer are significantly smaller than those of the surface soil layer and red soil layer, indicating that it is very easy for the soils in the sandy soil layer and detritus layer to reach the plastic limit in the case of short-term rainfall, and coupled with the looser soil and smaller soil shear strength, it is easy for them to collapse.

Because of the lower soil Atterberg limits of the collapsing gully in the subsoil layers, soil moisture absorption leads to the increase in water content after a long period of rain erosion and soil preferential flow. The sandy soil layer and detritus layer of the collapsing gully would be the first to reach or close to the plastic state in the same moisture conditions. Meanwhile, the shear strength of the two soil layers decreased rapidly, leading to the formation of the weak surface and then collapse or water erosion. The erosion is much more severe in the sandy soil layer and detritus layer than in the surface soil layer and red soil layer, resulting in the hollowing-out of the subsoil layers and the formation of a concave pit called "niche" in the engineering geology (Ding et al., 1995; Deng et al., 2016b). The formation and development of the niche is the preliminary stage of the formation of a collapsing gully. After niche formation, the surface soil layer and red soil layer lack support, giving rise to a total collapse by the soil self-weight. The occurrence of collapse forms the source of erosion, resulting in the formation of the collapsing gully.

In addition, as can be seen from Table 5, soil Atterberg limits of different weathering profiles of the Quaternary red clay are very different from those of granite soil, an Ultisol in southern China. The plastic limit, liquid limit, plasticity index and liquidity index of the Quaternary red clay show an upward trend first and then a downward trend. However, these values of the topsoil layers (A layer and B layer) are similar to those of the surface layer and red soil layer of granite, while the values of subsoil layers (C1 layer and C2 layer) are significantly higher than those of the sandy soil layers and detritus layers of granite. Therefore, under the condition of rainfall, even if the profile of the Quaternary red soil is exposed, the subsoil layers are not easy to be eroded. Purely because of these properties, the formation of a "niche" is difficult for the soil profile of the Quaternary red clay, and thus few collapsing gullies occurred in the Quaternary red clay. However, the stratigraphic characteristics of the soil Atterberg limits are particularly significant for granite soil, and the collapsing gully is most likely to occur on this parent material in the hilly region of southern China. There are many factors responsible for the occurrence of the collapsing gully (soil thickness, vegetation, climate, etc.), and soil Atterberg limits of different weathering profiles of granite soil may be just one of the necessary, rather than sufficient, conditions for the development of the collapsing gully. This will be further studied in the future work.

3.3　Effect of soil physicochemical properties on soil Atterberg limits

In this research, we examined the soil particle density (PD), bulk density (BD), total porosity (TP), soil organic matter (SOM), cation exchange capacity (CEC), free iron oxide (Fe_d) and particle-size distribution (PSD) in the different soil layers of the four collapsing gullies (TC, GX, AX and WH). The relationships between soil physicochemical properties and soil Atterberg limits are shown in Table 4.

3.3.1　Soil particle density (PD), bulk density (BD) and total porosity (TP)

Regression analyses were performed to determine the strength of relationships between Atterberg limits and soil particle density, bulk density and total porosity in the soils of the four collapsing gullies (TC, GX, AX and WH). Specifically, soil Atterberg limits had a very weak negative correlation with the soil BD ($R^2 = 0.044$, $P = 0.273$ for plastic limit; $R^2 = 0.021$, $P = 0.450$ for liquid limit) and PD ($R^2 = 0.023$, $P = 0.423$ for plastic limit; $R^2 = 0.002$, $P = 0.818$ for liquid limit), and a very weak positive correlation with the soil TP ($R^2 = 0.124$, $P = 0.057$ for plastic limit; $R^2 = 0.077$, $P = 0.139$ for liquid limit). Therefore, there was almost no significant correlation between soil Atterberg limits and PD, BD and TP in the soils of the four collapsing gullies.

3.3.2　Soil organic matter (SOM)

In Table 6, regression analyses showed that the soil organic matter had a significant and positive correlation with plastic limit ($R^2 = 0.816$, $P < 0.001$) and liquid limit ($R^2 = 0.785$, $P < 0.001$). This is probably because soil organic matter can promote organic colloid formation to affect the specific surface area, the water holding capacity of the soil particles and thus the soil liquid limit (Stanchi et al., 2012). With the increase in organic matter content, organic colloid also increased, implying that the greater the water holding capacity of the soil is, the greater the liquid limit will be. In this research, the soil Atterberg limits had a significant positive correlation with the organic matter. Similar results were also reported by Zhuang et al. (2014) and Husein et al. (1999), who both concluded that the plastic limit and the liquid limit of the soil increase with increasing organic content. According to the relationship between the Atterberg limits and the organic matter in the weathering profiles of granite soil, we

Earth Science: Current Studies

Table 5. Soil Atterberg limits of different weathering profiles of the Quaternary red clay in Xianning.

Soil layer	Plastic limit (%)	Liquid limit (%)	Plasticity index (%)	Liquidity index (%)
A	31.76 ± 0.11b	52.21 ± 0.15b	20.45 ± 0.26a	-71.10 ± 0.91c
B	35.37 ± 3.10a	56.16 ± 2.19a	20.79 ± 0.91a	-77.14 ± 3.39b
C1	28.80 ± 1.15bc	48.16 ± 2.78c	19.37 ± 1.64b	-54.23 ± 4.59a
C2	26.67 ± 0.61c	45.54 ± 0.38d	18.87 ± 0.24b	-51.91 ± 0.65a

Table 6. Regression and correlation analyses of soil Atterberg limits with soil physicochemical properties.

	Plastic limit			Liquid limit		
	Regression equations	R^2	P	Regression equations	R^2	P
Gravel content	$y = -5.083\ln(x) + 38.722$	0.258	0.004	$y = -8.323\ln(x) + 66.423$	0.219	0.009
Coarse sand content	$y = -8.895\ln(x) + 48.448$	0.208	0.011	$y = -21.66\ln(x) + 100.51$	0.362	<0.001
Fine sand content	$y = -4.772\ln(x) + 38.804$	0.155	0.031	$y = -9.633\ln(x) + 71.562$	0.178	0.020
Sand content	$y = -17.16\ln(x) + 90.809$	0.569	<0.001	$y = -32.52\ln(x) + 168.51$	0.644	<0.001
Silt content	$y = -19.2\ln(x) + 91.772$	0.314	0.001	$y = -28.59\ln(x) + 143.51$	0.213	0.010
Clay content	$y = 7.6773\ln(x) + 3.4506$	0.795	<0.001	$y = 14.915\ln(x) + 1.8834$	0.827	<0.001
BD	$y = -28.04\ln(x) + 34.789$	0.044	0.273	$y = -35.65\ln(x) + 56.651$	0.021	0.450
PD	$y = -49.17\ln(x) + 73.088$	0.023	0.423	$y = -27.35\ln(x) + 71.436$	0.002	0.818
TP	$y = 35.364\ln(x) - 110.82$	0.124	0.057	$y = 51.702\ln(x) - 154.49$	0.077	0.139
SOM	$y = 4.2553\ln(x) + 22.753$	0.816	<0.001	$y = 7.6856\ln(x) + 39.781$	0.785	<0.001
CEC	$y = 7.9009\ln(x) + 11.719$	0.657	<0.001	$y = 15.682\ln(x) + 17.359$	0.767	<0.001
Fe$_\mathrm{d}$	$y = 10.629\ln(x) - 4.226$	0.688	<0.001	$y = 21.885\ln(x) - 16.509$	0.837	<0.001

can conclude that the higher the content of organic matter is, the stronger the anti-erodibility of the soil will be. Thus, our research provides a theoretical basis for the prevention and control of collapsing gully by using green manure to improve soil organic matter in these areas.

3.3.3 Cation exchange capacity (CEC)

As shown in Table 6, there was a strong positive correlation between soil Atterberg limits and CEC ($R^2 = 0.657$, $P < 0.001$ for plastic limit; $R^2 = 0.767$, $P < 0.001$ for liquid limit). Similar results were reported by Cathy et al. (2008), who proposed that CEC can be used as an indicator for the mineral type and that it is highly correlated with plastic limit and liquid limit.

3.3.4 Free iron oxide (Fe$_\mathrm{d}$)

A positive significant correlation was observed between soil Atterberg limits and Fe$_\mathrm{d}$ ($R^2 = 0.688$, $P < 0.001$ for plastic limit; $R^2 = 0.837$, $P < 0.001$ for liquid limit; Table 6). This is consistent with the finding of Stanchi (2015), who reported that Atterberg limits were also affected by CEC. Therefore, Fe$_\mathrm{d}$ acts as an inorganic binding agent in structure formation and participates in reducing horizon vulnerability, as proposed by Sposito (1989).

3.3.5 Particle-size distribution (PSD)

Regression analyses were performed to determine the strength of relationships between soil Atterberg limits and the contents of gravel, coarse sand, fine sand, silt and clay in the soils of collapsing gullies (Table 6). The nonlinear regression analyses showed a strong positive correlation of the soil Atterberg limits with the clay content ($R^2 = 0.795$, $P < 0.001$ for plastic limit; $R^2 = 0.827$, $P < 0.001$ for liquid limit), a remarkable negative correlation with the content of sand ($R^2 = 0.569$, $P < 0.001$ for plastic limit; $R^2 = 0.644$, $P < 0.001$ for liquid limit) and a weak negative correlation with the silt content ($R^2 = 0.314$, $P = 0.001$ for plastic limit; $R^2 = 0.213$, $P = 0.010$ for liquid limit), gravel content ($R^2 = 0.258$, $P = 0.004$ for plastic limit; $R^2 = 0.219$, $P = 0.009$ for liquid limit), coarse sand content ($R^2 = 0.208$, $P = 0.011$ for plastic limit; $R^2 = 0.362$, $P < 0.001$ for liquid limit) and fine sand content ($R^2 = 0.155$, $P = 0.031$ for plastic limit; $R^2 = 0.178$, $P = 0.020$ for liquid limit). The significant negative correlation between soil Atterberg limits and sand may be attributed to porosity and specific surface area. When the sand content increases, the soil pores will increase and surface area will decrease, resulting in poor soil performance and facilitating water movement. Meanwhile, sandy soil is low in viscosity, loose and difficult to expand, leading to the slow rise of capillary water during water erosion. Therefore, the soil plastic limit and liquid limit will decrease with increasing sand content. Our results show that with declining weathering degree, the sand increased and the finer soil particles declined, which causes the decrease in soil Atterberg limits, and the subsoil layers are the first to be eroded (Zhuang et al., 2014).

Furthermore, there was a significant positive correlation between soil Atterberg limits and clay content, indicating that the clay content, despite its modest amount, plays a major role in determining the values of plastic limit and liquid limit. This also shows that, in the weathering profiles, the soil Atterberg limits increased with the increase in clay content, which is also reported by several other studies (Polidori, 2007; Baskan et al., 2009; Keller and Dexter, 2012). This result may be due to the effect of clay on soil plasticity in changing the arrangement of soil particles and cation exchange capacity. The connection form, the arrangement of soil particles and soil pore size will vary greatly with the clay content. Additionally, soil clay has a larger specific surface area, which will affect the soil water storage capacity. Therefore, the huge specific surface area enables the clay to have strong adsorption capacity, which affects the speed of water flow in the soil. Meanwhile, the mosaic of clay particles to the larger pores can also block the flow channels in the soil. All of these will affect the soil Atterberg limits, with the high clay content contributing to the directional arrangement of soil particles, leading to the increase in weakly bound water content, thereby increasing the plastic limit and liquid limit of the soil.

Overall, soil is a sphere of the earth system with a special structure and function. From the point of view of the earth system, soil science should not only study the soil material but also change towards the relationship between the soil and the earth system, which has a profound impact on the human living environment and global change research (Brevik et al., 2015; Keesstra et al., 2016). The results show that the relationship between soil Atterberg limits and the occurrence mechanism of collapsing gully, which can be used as a reference for the assessment of natural disasters occurring in the interaction between water and force in nature.

4   Conclusions

Based on the analyses of soil Atterberg limits, soil physicochemical properties, the influence factors on collapsing gully and the relationships between soil Atterberg limits and soil physicochemical properties of different weathering profiles of the four collapsing gullies in the hilly granitic region, the conclusions are summarized as follows.

Different weathering profiles exhibit a significant effect on soil Atterberg limits and soil physicochemical properties. The topsoil layers show the highest plastic limit, liquid limit, plasticity index, SOM, CEC and Fe_d; finer soil particles; and the lowest liquidity index, PD, and BD. As weathering degree decreases (from the surface layer to detritus layer), there is a sharp decrease in the plastic limit, liquid limit, plasticity index, SOM, CEC and Fe_d; a gradual increase in liquidity index; and a sharp increase in PD and BD first, followed by a slight decline. Additionally, the finer soil particles (silt and clay) decrease, and especially the clay contents decline noticeably, whereas the gravel and sand contents increase considerably. Therefore, the soils of subsoil layers very easily reach the soil Atterberg limits during rain, and coupled with the looser soil structure, it is easy for them to be eroded, resulting in the hollowing-out of these soil layers and the formation of a concave pit called a "niche" in engineering geology. After the niche formation, the topsoil layers lack support, leading to a total collapse in the soil by the soil self-weight and causing the formation of the collapsing gully. The regression analysis shows that soil Atterberg limits are significantly positively correlated with SOM, clay content, CEC and Fe_d; remarkably negatively correlated with sand content; and not obviously correlated with other properties. The results of this study demonstrate that soil Atterberg limits can be regarded as an informative indicator to reflect the weathering degree of different weathering profiles of the collapsing gully. Future research will include the relationship between soil Atterberg limits and soil mechanical properties.

Author contributions. Conceived and designed the experiments: Yusong Deng, Chongfa Cai and Jiazhou Chen. Performed the experiments: Yusong Deng and Dong Xia. Analyzed the data: Yusong Deng. Contributed reagents/materials/analysis tools: Yusong Deng, Dong Xia and Shuwen Ding. Wrote the paper: Yusong Deng, Chongfa Cai, Dong Xia, Shuwen Ding Jiazhou Chen and Tianwei Wang.

Competing interests. The authors declare that they have no conflict of interest.

Acknowledgements. Financial support for this research was provided by the National Natural Science Foundation of China (no. 41630858; 41601287 and 41571258) and National Science and technology basic work project (no. 2014 FY110200A16). We would like to thank several anonymous reviewers for their valuable comments on a previous version of the manuscript.

Edited by: M. Muñoz-Rojas

References

Anderson, J. M. and Ingram, J. S. I.: Tropical soil biology and fertility: a handbook of methods, Soil Sci., 157, 12–21, 1994.

Archer, J. R.: Soil consistency, in: Soil Physical Conditions and Crop Production, Ministry of Agriculture, Fisheries and Food, Tech. Bull., London: HMSO, 29, 289–297, 1975.

Arjmand Sajjadi, S. and Mahmoodabadi, M.: Aggregate breakdown and surface seal development influenced by rain intensity, slope gradient and soil particle size, Solid Earth, 6, 311–321, doi:10.5194/se-6-311-2015, 2015.

ASTMD 4318-10e1: Standard Test Methods for Liquid Limit, Plastic Limit, and Plasticity Index of Soils, ASTM International, West Conshohocken, PA, 2010.

Atterberg, A.: Die plastizität der tone, Internationale Mitteilungen für Bodenkunde, 1, 10–43, 1911.

Baskan, O., Erpul, G., and Dengiz, O.: Comparing the efficiency of ordinary Kriging and cokriging to estimate the Atterberg limits spatially using some soil physical properties, Clay Miner., 44, 181–193, 2009.

Baver, L. D.: The Atterberg consistency constants: Factors affecting their values and a new concept of their significance, J. Am. Soc. Agron., 22, 935–948, 1930.

Brevik, E. C., Cerdà, A., Mataix-Solera, J., Pereg, L., Quinton, J. N., Six, J., and Van Oost, K.: The interdisciplinary nature of SOIL, SOIL, 1, 117–129, doi:10.5194/soil-1-117-2015, 2015.

Campbell, D. J.: Liquid and plastic limits, in: Soil and environmental analysis-physical methods, edited by: Smith, K. A. and Mullins, C. E., Dekker Inc., New York, 349–375, 2001.

Casagrande, A.: Research on the Atterberg limits of soils, Public Roads, 13, 121–136, 1932.

Casagrande, A.: Notes on the design of the liquid limit device, Geáotechnique, 8, 84–91, 1958.

Cerdà, A.: The effect of season and parent material on water erosion on highly eroded soils in eastern Spain, J. Arid Environ., 52, 319–337, 2002.

Cerdà, A. and Doerr, S. H.: Soil wettability, runoff and erodibility of major dry-Mediterranean land use types on calcareous soils, Hydrol. Process., 21, 2325–2336, 2007.

Curtaz, F., Stanchi, S., D'Amico, M. E., Filippa, G., Zanini, E., and Freppaz, M.: Soil evolution after land-reshaping in mountains areas (Aosta Valley, NW Italy), Agr. Ecosyst. Environ., 199, 238–248, 2015.

Deng, Y. S., Ding, S. W., Liu, C. M., Xia, D., Zhang, X. M., and Lv, G. A.: Soil moisture characteristics of collapsing gully wall in granite area of southeastern Hubei, J. Soil Water Conserv., 29, 132–137, 2015 (in Chinese).

Deng, Y. S., Ding, S. W., Cai, C. F., and Lv, G. A.: Characteristic curves and model analysis of soil moisture in collapse mound profi les in southeast Hubei, Acta Pedologica Sinica, 53, 355–364, 2016a (in Chinese).

Deng, Y. S., Xia, D., Cai, C. F., and Ding, S. W.: Effects of land uses on soil physic-chemical properties and erodibility in collapsing-gully alluvial fan of Anxi County, China, J. Integr. Agr., 15, 1863–1873, 2016b.

Deng, Y. S., Cai, C. F., Xia, D., Ding, S. W, ang Chen, J. Z.: Fractal features of soil particle size distribution under different land-use patterns in the alluvial fans of collapsing gullies in the hilly granitic region of southern China, Plos one, 12, e0173555, doi:10.1371/journal.pone.0173555, 2017.

Ding, S. W., Cai, C. F., and Zhang, G. Y.: A study on gravitational erosion and the formation of collapsing gully in the granite area of Southeast Hubei, Journal of Nanchang College of Water Conservancy and Hydroelectric Power, S1, 50–54, 1995 (in Chinese).

Feng, M. H., Liao, C. Y., Li, S. X., and Lu, S. L.: Investigation on the present situation of collapsing gully in the south of China, People Yangtze River, 40, 66–68, 2009 (in Chinese).

Gee, G. W. and Bauder, J. W.: Particle size analysis, in: Methods of Soil Analysis, Part 1, Agronomy, edited by: Klute, A., Am. Soc. Agron. Inc., Madison, Wis, 9, 1986.

Hemmat, A., Aghilinategh, N., and Rezainejad, Y.: Long-term impacts of municipal solid waste compost, sewage sludge and farmyard manure application on organic carbon, bulk density and consistency limits of a calcareous soil in central Iran, Soil Till. Res., 108, 43–50, 2010.

Hillel, D.: Fundamentals of Soil Physics, Academic Press: New York, 1980.

Huang, B., Li, Z. W., Huang, J. Q., Liang, G., Nie, X. D., Wang, Y, Zhang, Y., and Zeng, G. M.: Adsorption characteristics of Cu and Zn onto various size fractions of aggregates from red paddy soil, J. Hazard. Mater., 264, 176–183, 2014.

Husein Malkawi, A. I., Alawneh, A. S., and Abu, O. T.: Effects of organic matter on the physical and the physicochemical properties of an illitic soil, Appl. Clay Sci., 14, 257–278, 1999.

Jiang, F. S., Huang, Y. H., Wang, M. K., Lin, J. S., Zhao, F., and Ge, H. L.: Effects of rainfall intensity and slope gradient on steep colluvial deposit erosion in southeast China, Soil Sci. Soc. Am. J., 78, 1741–1752, 2014.

Jong, E. D., Acton, D. F., and Stonehouse, H. B.: Estimating the Atterberg limits of southern Saskatchewan soils from texture and carbon contents, Can. J. Soil Sci., 70, 543–554, 1990.

Jordán, A., Zavala, L. M., Nava, A. L., and Alanís, N.: Occurrence and hydrological effects of water repellency in different soil and land use types in Mexican volcanic highlands, Catena, 79, 60–71, 2009.

Jordán, A., Ángel, J., Gordillo-Rivero, García-Moreno, J., Zavala, L. M., Granged, A. J. P., Gil, J., and Neto-Paixão, H. M.: Post-fire evolution of water repellency and aggregate stability in Mediterranean calcareous soils: a 6-year study, Catena, 118, 115–123, 2014.

Keesstra, S. D., Bouma, J., Wallinga, J., Tittonell, P., Smith, P., Cerdà, A., Montanarella, L., Quinton, J. N., Pachepsky, Y., van der Putten, W. H., Bardgett, R. D., Moolenaar, S., Mol, G., Jansen, B., and Fresco, L. O.: The significance of soils and soil science towards realization of the United Nations Sustainable Development Goals, SOIL, 2, 111–128, doi:10.5194/soil-2-111-2016, 2016.

Keller, T. and Dexter, A. R.: Plastic limits of agricultural soils as functions of soil texture and organic matter content, Soil Res., 50, 7–17, 2012.

Khaledian, Y., Kiani, F., Ebrahimi, S., Brevik, E. C., and Aitkenhead-Peterson, J.: Assessment and monitoring of soil degradation during land use change using multivariate analysis, Land Degrad. Dev., 128–141, doi:10.1002/ldr.2541, 2016.

Lan, H. X., Hu, R. L., Yue, Z. Q., Lee, C. F., and Wang, S. J.: Engineering and geological characteristics of granite weathering profiles in South China, J. Asian Earth Sci., 21, 353–364, 2003.

Lee, S. B., Chang, H. L., Jung, K. Y., Park, K. D., Lee, D., and Kim, P. J.: Changes of soil organic carbon and its fractions in relation to soil physical properties in a long-term fertilized paddy, Soil Till. Res., 104, 227–232, 2009.

Li, S. P.: Study on erosion law and control of slope disintegration in Guangdong province, Journal of Natural Disasters, 3, 68–74, 1992 (in Chinese).

Liang, Y., Ning, D. H., Pan, X. Z., Li, D. C., and Zhang, B.: Characteristics and treatment of collapsing gully in red soil region of southern China, Soil Water Conserv. China, 1, 31–34, 2009.

Lin, J. S., Huang, Y. H., Wang, M. K., Jiang, F. S., Zhang, X., and Ge, H.: Assessing the sources of sediment transported in gully systems using a fingerprinting approach: An example from South-east China, Catena, 129, 9–17, 2015.

Liu, X. L. and Zhang, D. L.: Distribution Characteristics and spatial variation of Benggang soil moistures: A case study of Liantanggang in Wuhua County, Guangdong, Tropical Geography, 35, 291–297, 2015 (in Chinese).

Luk, S. H., Yao, Q. Y., Gao, J. Q., Zhang, J. Q., He, Y. G., and Huang, S. M.: Environmental analysis of soil erosion in Guangdong province: a Deqing case study, Catena, 29, 97–113, 1997a.

Luk, S. H., Dicenzo, P. D., and Liu, X. Z.: Water and sediment yield from a small catchment in the hilly granitic region, South China, Catena, 29, 177–189, 1997b.

Marek, S. Ż., Williams, D. J., Song, Y.-F., and Wang, C.-C.: Smectite clay microstructural behaviour on the Atterberg limits transition, Colloid. Surface. A, 467, 89–96, 2015.

Masto, R. E., Sheik, S., Nehru, G., Selvi, V. A., George, J., and Ram, L. C.: Assessment of environmental soil quality around Sonepur Bazari mine of Raniganj coalfield, India, Solid Earth, 6, 811–821, doi:10.5194/se-6-811-2015, 2015.

McBride, R. A.: Soil consistency and lower plastic limits, in: Soil Sampling and Methods of Analysis, edited by: Carter, M. R. and Gregorich, E. G., 2nd Edn., CRC Press, 58, 761–769, 2008.

Mehra, O. P. and Jackson, M. L.: Iron oxide removal from soils and clays by a dithionite-citrate system buffered with sodium bicarbonate, Clay. Clay Miner., 7, 317–327, 1958.

Moreno-Ramón, H., Quizembe, S. J., and Ibáñez-Asensio, S.: Coffee husk mulch on soil erosion and runoff: experiences under rainfall simulation experiment, Solid Earth, 5, 851–862, doi:10.5194/se-5-851-2014, 2014.

Muñoz-Rojas, M., Erickson, T. E., Dixon, K. W., and Merritt, D. J.: Soil quality indicators to assess functionality of restored soils in degraded semiarid ecosystems, Restor. Ecol., 24, S43–S52, 2016a.

Muñoz-Rojas, M., Erickson, T. E., Martini, D., Dixon, K. W., and Merritt, D. J.: Soil physicochemical and microbiological indicators of short, medium and long term post-fire recovery in semiarid ecosystems, Ecol. Indic., 63, 14–22, 2016b.

Nelson, D. W. and Sommers, L. E.: Total carbon, organic carbon, and organic matter, Methods of Soil Analysis Part-chemical Methods, 1982.

Oliveira, S. P. D., Lacerda, N. B. D., Blum, S. C., Escobar, M. E. O., and Oliveira, T. S. D.: Organic carbon and nitrogen stocks in soils of northeastern Brazil converted to irrigated agriculture, Land Degrad. Dev., 26, 9–21, 2015.

Peng, F., Quangang, Y., Xue, X., Guo, J., and Wang, T.: Effects of rodent-induced land degradation on ecosystem carbon fluxes in an alpine meadow in the Qinghai-Tibet Plateau, China, Solid Earth, 6, 303–310, doi:10.5194/se-6-303-2015, 2015.

Perrin, A. S., Fujisaki, K., Petitjean, C., Sarrazin, M., Godet, M., Garric, B., and Brossard, M.: Conversion of forest to agriculture in Amazonia with the chop-and-mulch method: does it improve the soil carbon stock?, Agr. Ecosyst. Environ., 184, 101–114, 2014.

Polidori, E.: Relationship between the Atterberg limits and clay content, Soils Found., 47, 887–896, 2007.

Qiu, S. J.: The process and mechanism of red earth slope disintegration erosion, Bulletin of Soil and Water Conservation, 6, 31–40, 1994 (in Chinese).

Rashid, A. S. A., Kalatehjari, R., Noor, N. M., Yaacob, H., Moayedi, H., and Sing, L. K.: Relationship between liquidity index and stabilized strength of local subgrade materials in a tropical area, Measurement, 55, 231–237, 2014.

Reznik, Y. M.: Relationship between plastic limit values and fine fractions of soils, Geotechnical and Geological Engineering, 34, 403–410, 2016.

Rhoades, J. D.: Cation exchange capacity, in: Methods of Soil Analysis, Part 2, Chemical and Microbiological Properties, edited by: Page, A. L., 2nd Edn., Agronomy 9, ASA, SSSA, Madison, WI, USA, 149–157, 1982.

Seybold, C. A., Elrashidi, M. A., and Engel, R. J.: Linear regression models to estimate soil liquid limit and plasticity index from basic soil properties, Soil Sci., 173, 25–34, 2008.

Shahminan, D. N. I. A. A., Rashid, A. S. A., Bunawan, A. R., Yaacob, H., and Noor, N. M.: Relationship between strength and liquidity index of cement stabilized laterite for subgrade application, Int. J. Soil Sci., 9, 16–21, 2014.

Sharma, B. and Bora, P. K.: A study on correlation between liquid limit, plastic limit and consolidation properties of soils, Indian Geotechnical Journal, 45, 1–6, 2015.

Sheng, J. A. and Liao, A. Z.: Erosion control in south China, Catena, 29, 211–221, 1997.

S.I.S.S.: in: Angeli Milano, edited by: Angeli, F., Metodi di analisi fisica del suolo, Ministero per le Politiche Agricole, Rome (Italy), 1997.

Sposito, G.: The Chemistry of Soils, Oxford Univ. Press, New York, 1989.

Stanchi, S., Freppaz, M., and Zanini, E.: The influence of Alpine soil properties on shallow movement hazards, investigated through factor analysis, Nat. Hazards Earth Syst. Sci., 12, 1845–1854, doi:10.5194/nhess-12-1845-2012, 2012.

Stanchi, S., D'Amico, M., Zanini, E., and Freppaz, M.: Liquid and plastic limits of mountain soils as a function of the soil and horizon type, Catena, 135, 114–121, 2015.

Vacchiano, G., Stanchi, S., Ascoli, D., Marinari, G., Zanini, E., and Motta, R.: Soil-mediated effects of fire on Scots pine (Pinussylvestris L.) regeneration in a dry, inner-alpinevalley, Sci. Total Environ., 472, 778–788, 2014.

Wang, Y. H., Xie, X. D., and Wang, C. Y.: Formation mechanism of calamities due to Benggang processes of weathered granitic rocks, J. Mountain Sci., 6, 496–501, 2000.

Wroth, C. P. and Wood, D. M.: The correlation of index properties with some basic engineering properties of soils, Can. Geotech. J., 15, 137–145, 1978.

Xia, D., Deng, Y. S., Wang, S. L., Ding, S. W., and Cai, C. F.: Fractal features of soil particle-size distribution of different weathering profiles of the collapsing gullies in the hilly granitic region, South China, Nat. Hazards, 79, 455–478, 2015.

Xia, D., Ding, S. W., Long, L., Deng, Y. S., Wang, Q. X., Wang, S. L., and Cai, C. F.: Effects of collapsing gully erosion on soil qualities of farm fields in the hilly granitic region of south China, J. Integr. Agr., 15, 2873–2885, 2016.

Xu, J. X.: Benggang erosion: the influencing factors, Catena, 27, 249–263, 1996.

Yalcin, A.: The effects of clay on landslides: a case study, Appl. Clay Sci., 38, 77–85, 2007.

Zeng, Z. X.: Rock topography, Geological Publishing House, 1980 (in Chinese).

Zentar, R., Abriak, N. E., and Dubois, V.: Effects of salts and organic matter on Atterberg limits of dredged marine sediments, Appl. Clay Sci., 42, 391–397, 2009.

Zhang, S. and Tang, H. M. Experimental study of disintegration mechanism for unsaturated granite residual soil, Rock and Soil Mechanics, 6, 1668–1674, 2013 (in Chinese).

Zhang, X. J.: The practice and prospect of hill collapsing improving and development in southern China, China Water Resource, 4, 17–22, 2010 (in Chinese).

Zhang, X. M., Ding, S. W., and Cai, C.F.: Effects of drying and wetting on nonlinear decay of soil shear strength in slope disintegration erosion area, Transactions of the Chinese Society of Agricultural Engineering, 28, 241–245, 2012 (in Chinese).

Zhong, B. L., Peng, S. Y., Zhang, Q., Ma, H., and Gao, S. X.: Using an ecological economics approach to support the restoration of collapsing gullies in southern China, Land Use Policy, 32, 119–124, 2013.

Zhuang, Y. T., Huang, Y. H., Lin, J. S., Jiang, F. S., Zheng, Y., and Sun, S. X., Ding, Z. Q., and Yang, Y.: Study on liquid limit and plastic limit characteristics and factors of Benggang in red soil layer, Res. Soil Water Conserv., 21, 208–216, 2014 (in Chinese).

Zolfaghari, Z., Mosaddeghi, M. R., Ayoubi, S., and Kelishadi, H.: Soil Atterberg limits and consistency indices as influenced by land use and slope position in western Iran, J. Mt. Sci.-Engl., 12, 1471–1483, 2015.

7

Methods and uncertainty estimations of 3-D structural modelling in crystalline rocks

Raphael Schneeberger[1], Miguel de La Varga[2], Daniel Egli[1], Alfons Berger[1], Florian Kober[3], Florian Wellmann[2], and Marco Herwegh[1]

[1]Institute of Geological Sciences, University of Bern, Baltzerstrasse 1 + 3, 3012 Bern, Switzerland

[2]Graduate School AICES, RWTH Aachen University, Schinkelstrasse 2, 52062 Aachen, Germany

[3]Nagra, Hardstrasse 73, 5430 Wettingen, Switzerland

Correspondence to: Raphael Schneeberger (raphael.schneeberger@geo.unibe.ch)

Abstract. Exhumed basement rocks are often dissected by faults, the latter controlling physical parameters such as rock strength, porosity, or permeability. Knowledge on the three-dimensional (3-D) geometry of the fault pattern and its continuation with depth is therefore of paramount importance for applied geology projects (e.g. tunnelling, nuclear waste disposal) in crystalline bedrock. The central Aar massif (Central Switzerland) serves as a study area where we investigate the 3-D geometry of the Alpine fault pattern by means of both surface (fieldwork and remote sensing) and underground ground (mapping of the Grimsel Test Site) information. The fault zone pattern consists of planar steep major faults (kilometre scale) interconnected with secondary relay faults (hectometre scale). Starting with surface data, we present a workflow for structural 3-D modelling of the primary faults based on a comparison of three extrapolation approaches based on (a) field data, (b) Delaunay triangulation, and (c) a best-fitting moment of inertia analysis. The quality of these surface-data-based 3-D models is then tested with respect to the fit of the predictions with the underground appearance of faults. All three extrapolation approaches result in a close fit (> 10 %) when compared with underground rock laboratory mapping. Subsequently, we performed a statistical interpolation based on Bayesian inference in order to validate and further constrain the uncertainty of the extrapolation approaches. This comparison indicates that fieldwork at the surface is key for accurately constraining the geometry of the fault pattern and enabling a proper extrapolation of major faults towards depth. Considerable uncertainties, however, persist with respect to smaller-sized secondary structures because of their limited spatial extensions and unknown reoccurrence intervals.

1 Introduction

Geological information is inherently three-dimensional (3-D) in space but often represented in 2-D (Jones et al., 2009). With increasingly available computer power, 3-D modelling or geometrical visualizations have become widespread, as they can be performed on a desktop computer (e.g. Bistacchi et al., 2008; Caumon et al., 2009; Hassen et al., 2016; Sausse et al., 2010; Stephens et al., 2015). 3-D models widely serve as a basis for subsequent investigations, such as stress modelling or fluid flow modelling (e.g. Hassen et al., 2016; Stephens et al., 2015). Explicit structural modelling can further be subdivided into stochastic and deterministic methods. Deterministic approaches yield a single output for input parameters, analogous to drawing a map (e.g. Stephens et al., 2015), whereas as in stochastic approaches parameters are defined by a probabilistic density function with a component of randomness (e.g. Cherpeau and Caumon, 2015; González-Garcia and Jessell, 2016; Jørgensen et al., 2015; Koike et al., 2015).

When modelling a certain volume of Earth's intermediate deep subsurface (tens of metres to kilometres), as is often done for planning nuclear waste repositories, geothermal projects, or tunnelling work, 3-D structural modelling commonly starts from a known lithological and structural dataset from the Earth's surface or underground facilities such as tunnels or boreholes. Known information is then extrapolated towards the unknown. At the time of extrapolation, the validity cannot be proven unless additional information, such as geophysical, borehole, or excavation data, is integrated.

Previous studies report that this extrapolation represents a main uncertainty within 3-D structural modelling of known structures (e.g. Baumberger, 2015; Bistacchi et al., 2008). From environments with sparse data, for example, the topology of the fault network is known to be highly uncertain or prone to the existence of unknown faults (e.g. Cherpeau et al., 2012; Cherpeau and Caumon, 2015; Hollund et al., 2002).

More generally, for kilometre-scale models, uncertainties in accuracy related to input data (i.e. GPS location, dip–dip azimuth measurements) are small compared to the uncertainty related to the data interpolation between known locations or data extrapolation (Bond, 2015).

Uncertainties play an important role when considering decision-making based on information available from a 3-D model and have therefore been subject to extensive studies in the past (e.g. Bistacchi et al., 2008; Bond et al., 2007a; Lindsay et al., 2012; Tacher et al., 2006; Wellmann et al., 2010, 2014; Wellmann and Regenauer-Lieb, 2012; Yamamoto et al., 2014). Since models are a function of the data used, some of the approaches tend to analyse uncertainties in the input data before modelling (e.g. Bond et al., 2007b; Jones et al., 2009). Other approaches investigate the error propagation into the models, inferring the uncertainty after modelling (Jessell et al., 2010; Lindsay et al., 2012; Viard et al., 2011; Wellmann et al., 2010). Most of these published studies were performed within sedimentary environments where parameters such as stratigraphy, layer thickness, layer orientation, and structural setting are well constrained. Uncertainty estimation and its potential reduction are less well constrained for the structural modelling of basement rocks (e.g. Svensk Kärnbränslehantering AB, 2009), which are characterized by intrusive contacts and a complex arrangement of deformation structures.

In this study, we focus on deformed basement rocks and the extrapolation of faults to depth. We follow two main goals: (i) the application of an extrapolation workflow for three different techniques for projection of surface structures to depth considering associated projection uncertainties and (ii) the design and application of a probabilistic approach to compare different extrapolation techniques in order to validate the generated models.

We focus specifically on the combination of observations in outcrops at the surface with observations in an underground facility, allowing for an extrapolation modelling approach, and propose that it is possible to link these two types of observations in a probabilistic context by taking into account uncertainties in measurements and the exact tie between observed features at the surface and in the underground facility. We investigate a local case study in a relatively simple setting in crystalline rocks. The study area is characterized by well-exposed crystalline rocks of the Aar

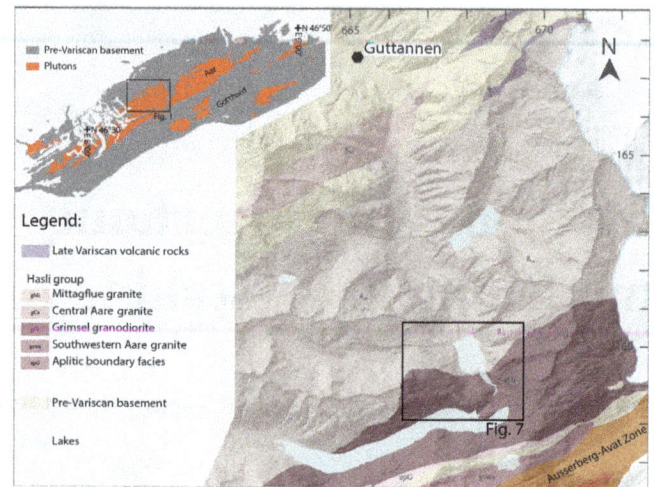

Figure 1. Geological map of the study area (modified after Berger et al., 2017). The coordinate system of the inset map is given in WGS84, whereas the coordinates of the local geological map are given in the Swiss coordinate system (CH1903).

massif in the central Swiss Alps (Fig. 1) and furthermore greatly benefits from subsurface information from the Grimsel Test Site (GTS) underground rock laboratory run by the Swiss Cooperative for Disposal of Radioactive Waste (Nagra).

This combination of good outcrop conditions at the surface and independent high-quality subsurface information allows for an extrapolation modelling approach and subsequent validation in a relatively simple and well-constrained high-topography crystalline setting.

2 Geological setting

The study site is located in the Haslital in the Central Alps (Switzerland; Fig. 1) within the Aar massif, an external crystalline massif in the Alps representing exhumed basement rocks of the former European continental margin and thus belonging to the palaeogeographic Helvetic domain of the Alps (e.g. Mercolli and Oberhänsli, 1988; Pfiffner, 2009; von Raumer et al., 2009).

Three different host rocks of magmatic origin occur in the study area: (i) Grimsel granodiorite (GrGr), (ii) Central Aare granite (CAGr), and (iii) meta-basic dykes (e.g. Abrecht, 1994; Keusen et al., 1989; Labhart, 1977; Stalder, 1964). The GrGr and the CAGr belong to the Haslital group, which is a Permian calc-alkaline magmatic differentiation suite (Berger et al., 2017; Schaltegger, 1990; Schaltegger and Corfu, 1992); the GrGr is the more primitive member. The two host rocks differ mainly in the relative amount of biotite, with ca. 11 vol% biotite in the GrGr compared to ca. 5 vol% biotite in the CAGr (Keusen et al., 1989). Intermingling structures observed in the field indicate a coeval viscous state

(Schneeberger et al., 2016). Furthermore, the concordant zircon and titanite U–Pb intrusion ages of both rock units are overlapping within error; the GrGr intrusion has a concordant titanite U–Pb intrusion age of 299 ± 2 Ma and the CAGr has an age of 299 ± 2 Ma (Schaltegger and Corfu, 1992). The granitoids intruded during late to post-Variscan extensional tectonics into a polymetamorphic pre-Variscan basement (Abrecht, 1994; Berger et al., 2017; Labhart, 1977; von Raumer et al., 2009; Schaltegger, 1990, 1994).

Meta-basic dykes, formerly called lamprophyres (Oberhänsli, 1986), intrude into the granitoid bedrock without altering the granitoid, indicating only slightly younger intrusion ages of former basic dykes with respect to the calc-alkaline granitoids.

The aforementioned rock types are subsequently overprinted by metamorphism and deformation related to Alpine orogeny. Peak metamorphic conditions reached $450 \pm 30\,°C$ and 6 ± 1 kbar (Challandes et al., 2008) at 22–20 Ma (Challandes et al., 2008; Rolland et al., 2009).

Several authors have described the deformation related to Alpine orogeny in the vicinity of the study area (e.g. Baumberger, 2015; Challandes et al., 2008; Choukroune and Gapais, 1983; Goncalves et al., 2012; Keusen et al., 1989; Marquer et al., 1985; Rolland et al., 2009; Steck, 1968; Wehrens et al., 2016, 2017). Ductile deformation is expressed by a pervasive foliation and localized high-strain zones (shear zones). The exact geometry of the 3-D shear zone network, which occurs at a variety of scales ranging from several kilometres down to millimetres, is complex (Choukroune and Gapais, 1983). It is, however, possible to extract a pattern of kilometre-long major shear zones interconnected by hectometre-long subordinate bridging structures. The major shear zones tend to be quasi-planar (Baumberger, 2015; Wehrens et al., 2017) and we therefore assume a considerably simplified shear zone pattern with quasi-planar to planar geometries of the major shear zones grouped according to their strike orientation (Fig. 2).

The kinematic framework of Alpine deformation is controversial. Several kinematic models have been proposed for the shear zone network genesis in the study area, including single-phase (Choukroune and Gapais, 1983) and multistage evolution models (Herwegh et al., 2017; Rolland et al., 2009; Steck, 1968; Wehrens et al., 2016, 2017). This study aims to reconstruct the present day 3-D geometry. Although the kinematic evolution is beyond the scope of this study, the generated models have been validated for kinematic inconsistency with respect to the known tectonic framework (e.g. models with unrealistic dip values have been removed; dip $< 60°$ or north verging). The different orientations of the structures are therefore used without kinematic implications. The major orientation of structures within the area are NE–SW (group A), E–W (group B), and NW–SE (group C) trending (Schneeberger et al., 2016; Wehrens et al., 2017).

The pervasive foliation and the highly localized shear zones form mechanical anisotropies, which favour subse-

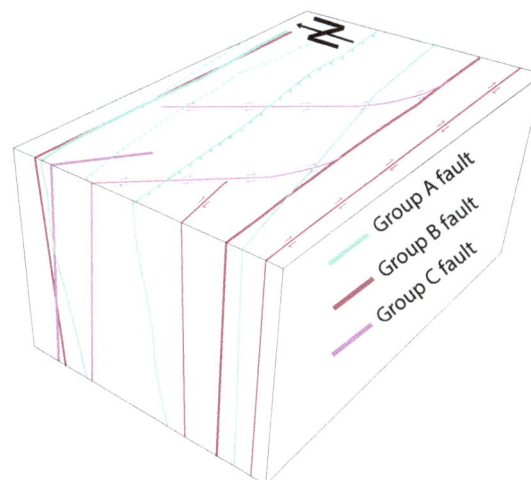

Figure 2. Schematic bloc diagram showing geometrical relationships between faults of different orientation groups (modified after Wehrens et al., 2017).

quent brittle localization (Belgrano et al., 2016; Kralik et al., 1992). Deformation in the brittle regime is expressed by fracturing and cataclasis, often resulting in fault gouges (Bense et al., 2014; Wehrens et al., 2016). The spatial distribution of fractures and their reactivation in the form of fault gouge development is heterogeneous (Bossart and Mazurek, 1991; Mazurek, 2000).

Although the shear zones experienced a severe ductile deformation history, most of them were reactivated in a brittle manner during the exhumation history (Wehrens et al., 2017). Subsequently, we therefore use the term fault as a summary term for high T ductile shear zones, low T ductile shear zones, and their reactivation by brittle shearing leading to cohesive (protocataclasite, cataclasite) or non-cohesive (fault gouge) fault rocks.

Present day seismic activity (Pfiffner and Deichmann, 2014) indicates ongoing recent tectonic activity in the deep subsurface of the Aar massif.

Glaciation and glacial retreat contributed to the latest history of the area (Wirsig et al., 2016). Basal erosion and the latest young (17.7 ka, Wirsig et al., 2016) retreat ages produced excellent outcrop conditions, as most outcrops are glacially polished and above the treeline, exposing bare bedrock.

Owing to deglaciation, exfoliation jointing occurred (Ziegler et al., 2013). Given the restricted near-surface occurrence of these exfoliation joints and their small dimensions, we exclude these deformation features from further consideration in this study.

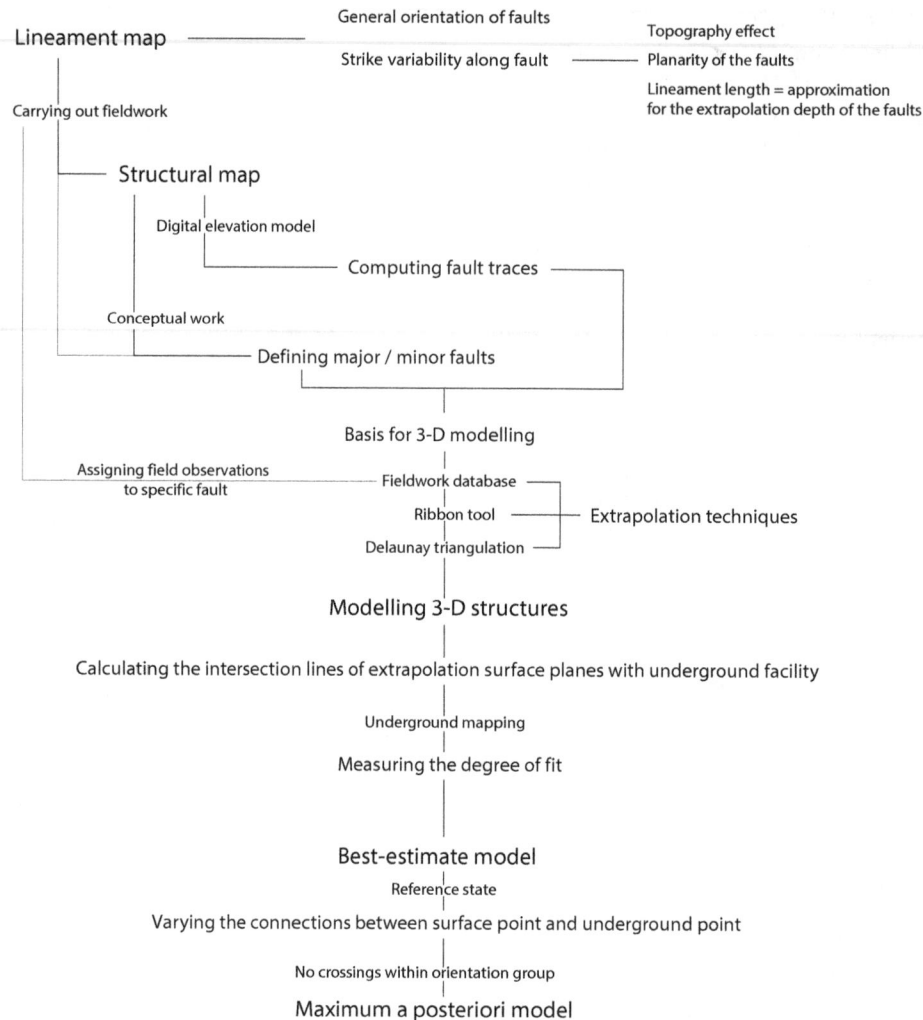

Figure 3. Employed modelling workflow to generate a 3-D structural model of the area based on a surface lineament map. As a major step, the workflow also considers the uncertainty related to the connection between mapped faults at the surface and underground.

3 Methods

3.1 Extrapolation workflow

In order to represent the 3-D geometry of faults, we developed a workflow based on a combination of remote sensing and fieldwork (Fig. 3).

As a first step, we generated a lineament map using remote sensing data. We use the term lineament as defined by Gabrielsen and Braathen (2014) and O'Leary et al. (1976): a lineament is a mappable linear or curvilinear feature identified by remote sensing, possibly representing the intersection between a planar to subplanar structural anisotropy and the Earth's surface. Lineament mapping followed the methodology presented by Baumberger (2015). Aerial photographs (swisstopo) and a digital elevation model (DEM; swisstopo) with resolutions of 0.5 and 2 m, respectively, served as a basis.

Using the DEM, hillshade images (i.e. greyscale relief images) with distinct illumination angles (0–360° illumination azimuth with 45° steps constant at a 30° altitude angle) were calculated, resulting in eight hillshade images and illuminating different parts of the investigation area. On a pixel-based map, the possible strike angle of a line depends on the number of pixels of the raster matrix in which the line is enclosed (Heilbronner and Barrett, 2014). Our approach requires an angular resolution < 10°, and thus a minimum length of 10 pixels for a specific lineament was necessary to fulfil this criterion. Hence, shorter lineaments (< 5 m) were discarded. Lineaments were manually digitized and are composed of a minimum of two endpoints and potentially several points in between.

The strike of lineaments was defined as the angle measured clockwise from north. Two different approaches to analyse the strike of lineaments were compared: (i) single strike values from endpoint to endpoint and (ii) strike values for individual segments between a lineament's nodes. In both approaches, a weight is added to the strike proportional to the length of the lineament.

In addition to the aforementioned remote sensing approach, conventional structural surface mapping over an area of $13\,km^2$ was performed. Spatially restricted outcrop observations at the surface were extrapolated along strike using the lineament map; thus combining fieldwork and remote sensing allowed us to obtain a structural surface map. Ductile deformation was mapped by differentiating pervasive background strain and localized high-strain zones (shear zones). At the surface, mapping of brittle deformation focused on the occurrence of fault gouges. In addition, mapping in the GTS underground facility was performed similarly to surface mapping on a decametre scale and in more detail regarding brittle structures (Schneeberger et al., 2016).

Structural modelling was performed using Move™ software (Midland Valley) on two distinct scales: a local scale (decametre) for the GTS and a regional scale (kilometre) for the entire study area. Underground 3-D structural modelling was performed on the basis of underground mapping and drill core data, which resulted in fault traces and orientations. This information provided the basis for the 3-D reconstruction of fault planes. Regional 3-D structural modelling was performed following published workflows using the surface fault map as a basis (e.g. Baumberger, 2015; Bistacchi et al., 2008; Kaufmann and Martin, 2009; Zanchi et al., 2009). Surface faults were extrapolated to depth by assigning a dip value to individual surface traces, where a trace is the intersection between the Earth's surface and a fault. Three different extrapolation approaches were applied: (i) extrapolation along measured dip and dip azimuth (fieldwork-based approach). Data from outcrops were considered within an orthogonal distance of $< 20\,m$ to inferred fault traces and a strike differing less than $20°$ compared to the fault's mean strike as defined by remote sensing. The fault's mean strike was calculated via linear regression through all points defining its trace. (ii) Delaunay triangulation is a meshing algorithm that produces a triangulation for several points such that for a given point cloud, no point of the point cloud is inside the circumcircle of any triangle connecting three points of the point cloud (Delaunay, 1934). It results in a 3-D surface interpolating the selected points. Based on this surface, the entire fault trace can be extrapolated. Noise can arise because of rugosities of the fault planes, uncertainties in tracing the fault intersection at the surface, and too-low vertical variations in topography. In the case of near planar faults, the noise is reduced in the case of high variations in altitude between valleys and mountain peaks and by preferring projections of long fault traces over those of short fault segments. (iii) The ribbon tool is a Move™ internal interpolation algorithm based on a three-points approach in which three points form a triangle and the orientation is averaged over a defined number of triangles (Midland Valley). The maximum dip orientation of each average triangle is represented as a

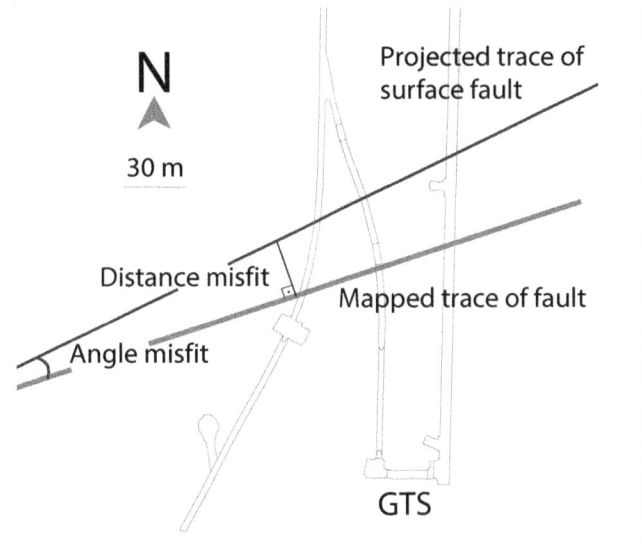

Figure 4. Schematic drawing of hypothetical example for validation of 3-D models based on angular and distance misfit (map view). The contours of the GTS are shown in grey with a mapped fault trace and a fault trace resulting from projection of the fault plane from the surface.

stick at the location of the starting point. The combination of all sticks along a trace results in a plane for the given trace. More details on the method used by the ribbon tool are given in Fernandez (2005) and Baumberger (2015).

For each approach, the surface fault trace was extrapolated to depth using the obtained specific orientation. Subsequently, the intersection line between the extrapolated plane and a horizontal plane at GTS elevation (approx. 1730 m a.s.l.) was calculated. Then, the resulting intersection lines were compared with the underground structural map in order to find the "best-fitting" underground structure to the obtained intersection line. The degree of fit between the intersection line at the surface and the trace of the underground structure was estimated using the orthogonal distance (distance misfit) starting from the intersection with the main gallery and the angular difference (angle misfit) between the two linear features (Fig. 4). Only structures within the same orientation group (groups A, B, C) were compared.

Furthermore, the degree of fit was compared between the different extrapolation approaches and thus for every surface fault. Considering all approaches, a best-fitting underground fault was assigned based on the aforementioned criteria. This assignment served as basis for the following structural modelling step in which every surface fault was linearly interpolated with the assigned best-fitting underground fault, yielding a "best-estimate" model.

3.2 Bayesian inference

For a better description of the system taking into consideration the inherent uncertainty in the extrapolation methods above, we performed a Bayesian inference on the basis of a GTS parallel cross section. Bayes' theorem,

$$p(\theta|y) = \frac{p(\theta)p(y|\theta)}{p(y)},$$

provides a formal way to update probability distributions for model parameters θ when new data y are obtained. The final goal is to obtain the posterior distribution $p(\theta|y)$ of the parameters θ given the observations y. This distribution is proportional to the distribution of prior parameters $p(\theta)$ and likelihood functions $p(y|\theta)$, which determine how likely these parameters are given specific observations y. The term $p(y)$ is a normalization constant commonly referred to as evidence or marginal likelihood (see for example MacKay, 2003, for more details).

In this study scenario, we assign a parameter to each surface fault at the tunnel level. We represented the uncertainty about the exact value with a Gaussian distribution and a constant standard deviation of 40 m in the horizontal axis (Fig. 5) based on the dip uncertainty of 10° based on the dip variation in multiple orientation measurements along a single fault (Fig. 5). As a mean value, we assign the best-estimate model from the previous interpolation. Interpolated planes were grouped according to their orientation into three separate groups (identified by A, B, C in the following; Fig. 2).

Within each orientation group, we expect faults to be mostly parallel with limited intersections based on field observations. To capture this idea, we assigned a penalty factor that reduces the log-likelihood of a parameter set for an increasing number of intersections (by 0.05 per intersection, to be precise). The number of intersections per iteration was calculated using the Bentley–Ottmann algorithm (Supplement; Balaban, 1995; Bentley and Ottmann, 1979).

The described Bayesian inference cannot be performed directly due to the complexity of multiple parameters in several groups and the non-linearities due to the fault intersections. We therefore apply a computational sampling method based on an adaptive Metropolis MCMC approach (Haario et al., 2001) implemented in the probabilistic programming package PyMC 2 (Patil et al., 2010) and previously successfully used in a geological context (de la Varga and Wellmann, 2016).

Final posteriors were discretized to match the locations of measured faults in the underground tunnel by a simple nearest location classifier (Fig. 5). Therefore, the final result of the inference is a discretized distribution of each of the parameters. In order to compare the 3-D models obtained by the three extrapolations approaches, we then use the maximum a posteriori value, i.e. the highest probability value of the posterior distributions.

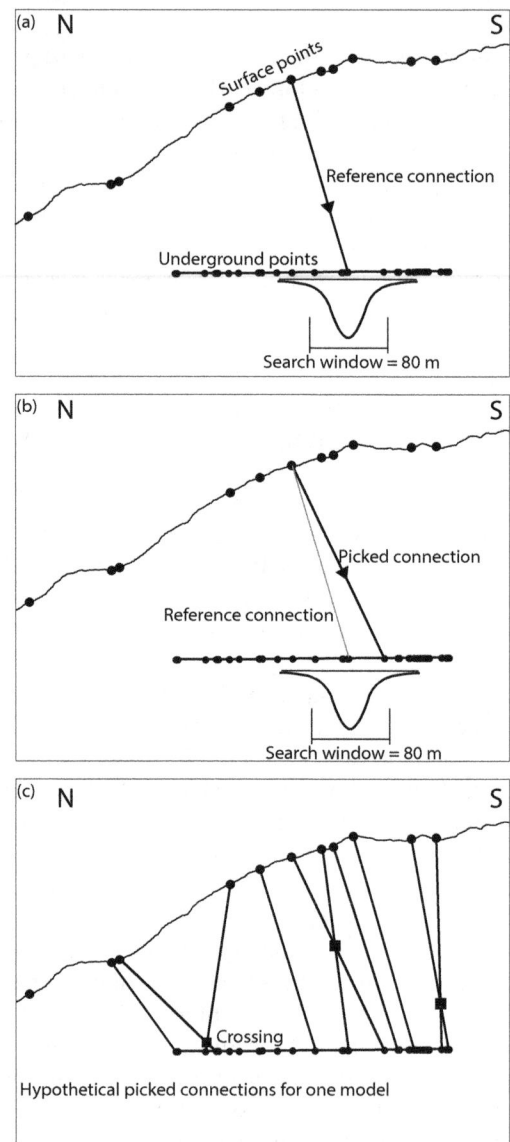

Figure 5. Schematic cross section illustrating the statistical modelling methodology for one example. **(a)** Reference state is defined by aforementioned workflow. A search window of 80 m is assigned to the reference underground point. **(b)** The code picks a possible underground point within the 80 m search window based on a normal Gaussian distribution and calculates the connection. **(c)** For every surface point one underground point is picked and connected by a projection line (hypothetical fault plane). By connecting each surface point with one underground point, a connection pattern is formed. Ten thousand runs were performed, each connecting different surface and underground points. The 10 000 connection patterns are compared with each other and evaluated by a penalty criterion. This penalty criterion addresses the number of crossings and rate solutions with a lowest and highest number of fault plane crossings, yielding a probability for connecting a specific surface point to a certain underground point.

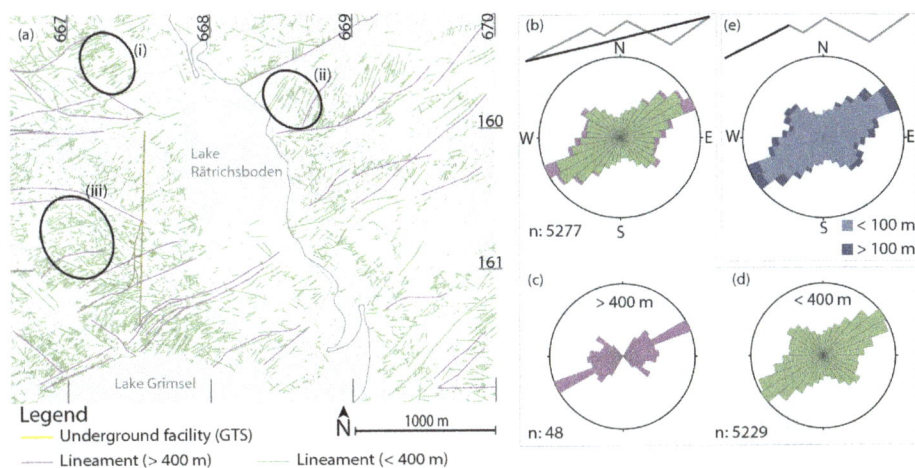

Figure 6. (a) Lineament map of the study area with underground rock laboratory. Topography contours are based on swissALTI3D (reproduced by permission from swisstopo; BA17063). **(b)–(d)** Length-weighted rose diagrams showing the endpoint-to-endpoint strike of all lineaments **(b)**, lineaments longer than 400 m **(c)**, and lineaments shorter than 400 m **(d)**. **(e)** Length-weighted rose diagram showing the orientation of each segment of all lineaments.

4 Results

4.1 Lineament map

In total, 5277 lineaments with a spatially heterogeneous distribution and lengths ranging from 5 to 1941 m were mapped (Fig. 6a). Lineaments are generally more concentrated along topographic highs and lows. Within certain areas (areas i and ii in Fig. 6a), the lineament's strike tends to be parallel to the dip azimuth of the slopes, yielding uniform orientations. In contrast, in domains with relatively low topographic variations, a variety of strike orientations become discernible (area iii in Fig. 6a).

Looking at the bulk data, lineaments show a major NE–SW and a minor NW–SE trend (Fig. 6b). Long lineaments (> 400 m) are mainly oriented NE–SW (Fig. 6c), whereas short lineaments show a considerable variation in strike (Fig. 6d). The two methods for estimating lineament orientations, endpoint to endpoint (Fig. 6b) and individual segments (Fig. 6e), yield similar results.

4.2 Field observations and data

Data obtained from fieldwork combined with a compilation of several published maps (Baumberger, 2015; Keusen et al., 1989; Vouillomaz, 2009; Wehrens et al., 2017; Wicki, 2011) yielded a surface fault map (Fig. 7, see also Schneeberger et al., 2016). Based on their orientation, we discriminated different groups of faults (Fig. 7): group A are mainly steep SE-dipping faults. Their average orientation (dip azimuth to dip) is 149/74. Group A faults mostly show steeply plunging stretching lineations resulting from ductile shearing. Group A can be correlated with faults formed during the Handegg phase (22–17 Ma) as defined by Wehrens et

al. (2017), while group B and group C would correspond to faults formed during the Oberaar phase (14–12 Ma; Wehrens et al., 2017). Group B are mainly steep S-dipping (mean orientation: 178/72) faults. Lastly, group C are SW-dipping faults coeval with group B with an average orientation of 196/72. Group C faults are subparallel to meta-basic dykes and often co-occur spatially with the latter. Groups B and C mostly show oblique to horizontal stretching lineations. For multiple orientation measurements along individual faults, the standard deviation of the mean dip azimuth was below 15° and the mean dip below 10°. Generally, the GrGr-dominated southern area shows an increased number of faults (Figs. 7 and 8). Detailed underground mapping resulted in a lithological (Fig. 8a) and a structural map of the GTS (Fig. 8b).

Meta-basic dykes occur as three distinct swarms, two located within the CAGr domain (Fig. 8a). The northern two swarms strike NW–SE, whereas the southern swarm strikes E–W; however, it is less clearly marked. Numerous dykes are overprinted by an Alpine foliation, which is sometimes oblique to the dyke boundary. Furthermore, dykes are often overprinted by localized ductile and brittle deformation expressed by shear zones and fault gouges.

Faults occur along three NE–SW trending swarms, two E–W trending swarms, and two NW–SE trending swarms, leading to a heterogeneous strain distribution along the underground facility (Fig. 8b).

The NE–SW trending swarms correspond to group A faults with an average spacing of ca. 16 m. In total, 31 group A faults were mapped underground. They can be further subdivided into 17 moderately to steeply dipping faults (between 45 and 75°) and 14 sub-vertically dipping ones (> 80°).

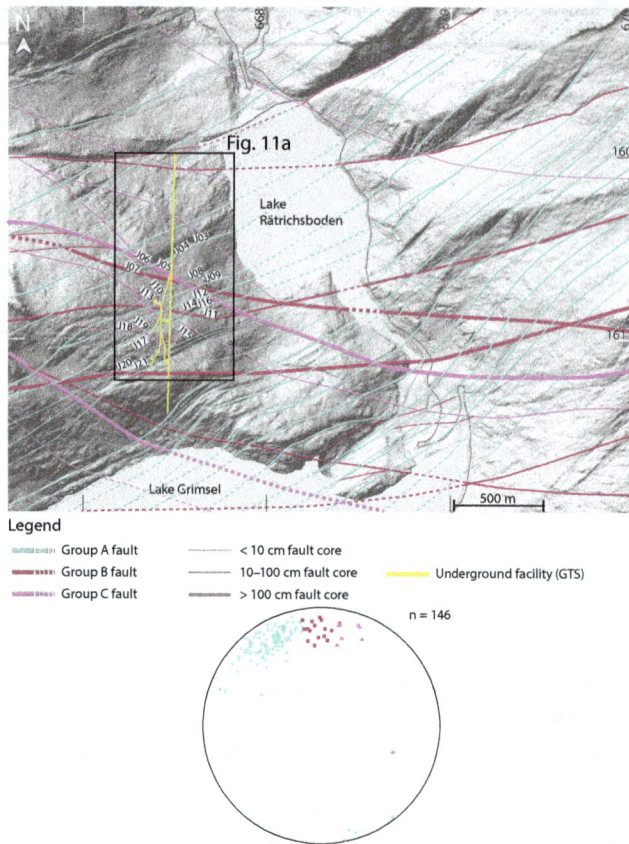

Figure 7. Surface fault map with faults grouped by strike orientation (group A, B, C). Hillshade image underlying the map is based on swissALTI3D (reproduced by permission from swisstopo; BA17063). Fault exposure lines are dashed over uncertain areas and labelled in cases for which a connection to GTS exists. Lower hemisphere equal area projection with plane poles grouped according to strike. The map is based on the Swiss coordinate system.

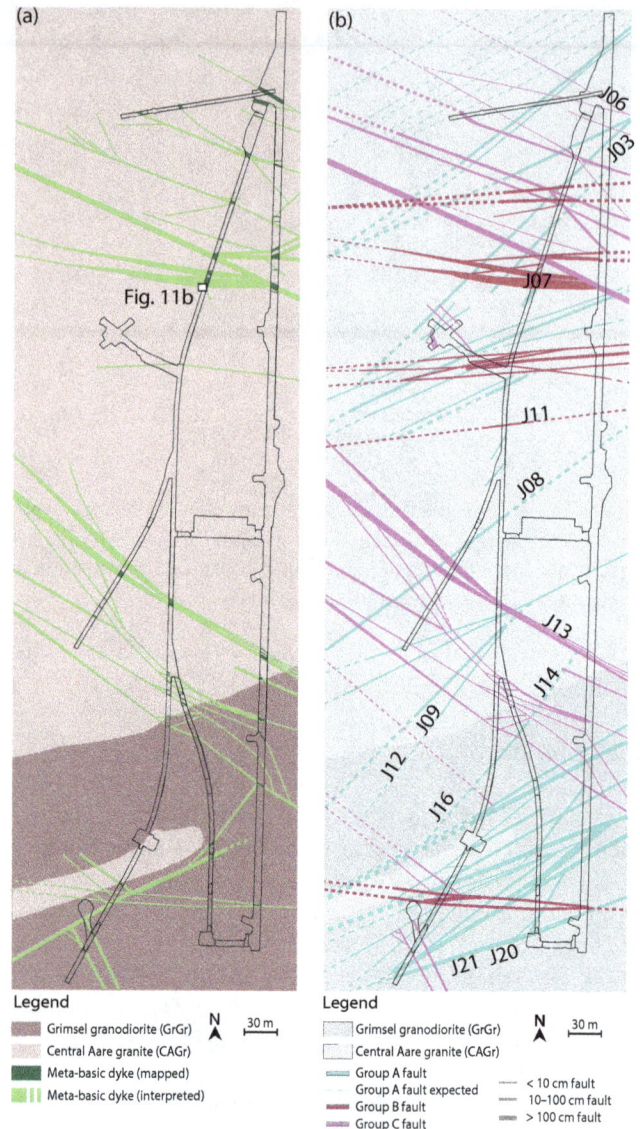

Figure 8. (a) Petrographic underground map. **(b)** Structural mapping (1 : 1000) of the underground rock laboratory (GTS) with faults grouped according to their strike. Indicated labels correspond to surface fault labelling and represent the maximum a posteriori interpolation.

The E–W trending swarms correspond to faults with orientations that are similar to group B. In total, 12 of these E–W striking faults were mapped.

The NW–SE trending fault swarms are localized mainly along dykes (Fig. 8) and represent group C structures. In total, 25 NW–SE striking faults occur within the GTS.

Faults in the CAGr (northern part) seem to preferentially localize along pre-existing anisotropies, i.e. high-temperature brittle fractures (biotite coating) or meta-basic dykes, and thus form discrete faults (centimetre sized) with marked contacts to the host rock. In contrast, faults in the GrGr-dominated southern part form strain gradients over larger distances (metres). This observation is in agreement with the findings of Wehrens et al. (2017).

4.3   3-D structural modelling

The GTS model size is $600 \times 250 \times 100$ m, whereas the regional model size was 4×3 km with a projection depth reach-

ing the underground facility for all faults. The projection depth was defined arbitrarily but no larger than half of the fault trace's length. All 3-D models are provided in the Supplement.

4.3.1   GTS model

The combination of the above-presented underground map with measured surface orientation data resulted in a 3-D geometric visualization of meta-basic dykes and faults mapped underground. Swarms of meta-basic dykes tend to join towards less numerous dykes with depth. Based on geometrical considerations, we infer the occurrence of three major dykes from which all others either fan out or form relay structures between the major dykes. Based on the field observation that the major faults and relay structures dip steeply sub-vertically towards the south, we discriminated 8 major group A faults and 23 relay structures. Major group A faults occur within each NE–SW trending swarm discriminated on the map view. Group B deformation structures can be further subdivided into six major and seven relay faults. Group C deformation structures can be subdivided into 6 major and 32 relay deformation structures, some of which are very short (14 m).

4.3.2 Regional model

The surface fault map (Fig. 7) served as a basis for the generation of the three different kilometre-scale 3-D models (see above). All three modelling approaches yielded the 3-D geometrical visualization of the surface fault pattern. They all share the same fault traces at the model surface. As mentioned above, projection specific dip values were used for each of the models. However, not all surface faults were extrapolated with each approach. Of the 21 possible surface faults, 10 were extrapolated with the fieldwork-based approach, 11 using the Delaunay triangulation, and 13 with the ribbon tool method. Missing projections can be due to a lack of outcrop description or the absence of sufficient topographic relief for remote-sensing-based approaches.

By combining all three approaches, at least 1 (but up to 3) degrees of fit with underground faults were calculated for each surface fault. Based on the different degrees of fit, a best-fitting underground structure was assigned to each surface fault. By linearly interpolating the two traces, we obtained a model which we called best-estimate model. In total, 11 group A faults reach the GTS. From the total 11, 7 have a dip < 80°, which would correspond to the major structures defined in the above-presented GTS-scale model, whereas the 4 steeper faults correspond to relay structures. Moreover, two group B and eight group C faults connect the surface with the GTS. The combination of all faults yields an average spacing of 25.4 m, and faults appear to converge with depth.

4.3.3 Bayesian inference

For each model that is obtained when each surface point (intersection between surface fault and 2-D section along the GTS) is interpolated with a specific underground point, the number of intersections was calculated and the likelihood of the model compiled based on the number of intersections. In total 10 000 models were calculated and for each a probability for a certain interpolation of a specific surface point with an underground point was obtained (Fig. 9). For certain surface points, a clear maximum a posteriori value was

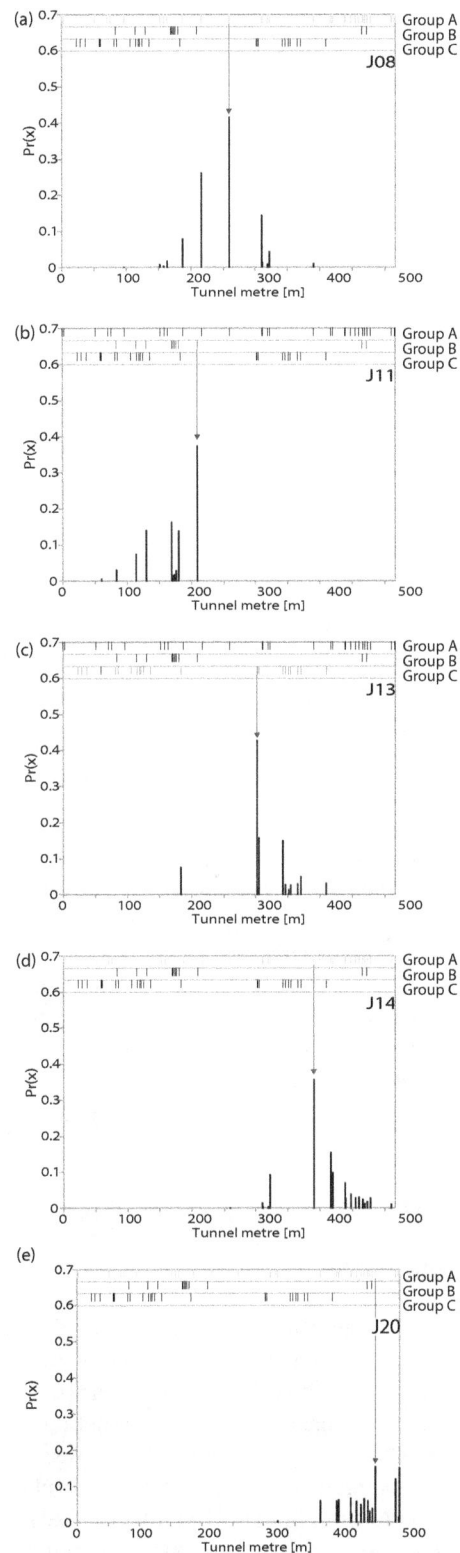

Figure 9. Probability distributions of five selected examples: panels **(a)** to **(d)** show the highest probabilities achieved, whereas panel **(e)** shows an example without clear maximum probability. On top, the positions of the underground deformations zones are indicated and grouped according to their strike. Additionally, the maximum a posteriori interpolation is highlighted with an arrow.

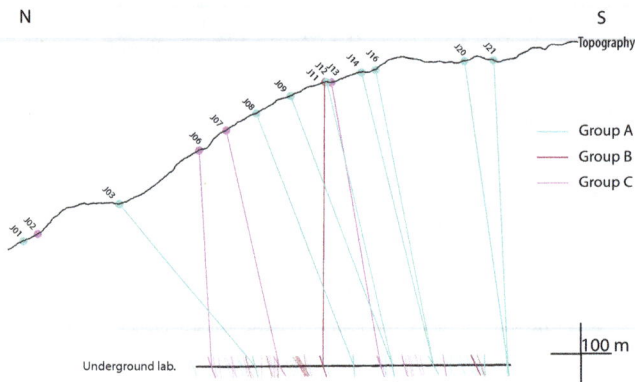

Figure 10. Cross section showing maximum a posteriori connections between surface and underground faults. Faults are grouped and coloured according to their strike. Underground faults are represented by short ticks; the less transparent ones have a connection to the surface.

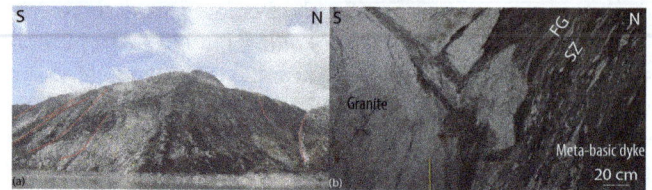

Figure 11. (a) Mountainside with incisions and exfoliation joints. **(b)** Detailed picture of underground outcrop showing outcrop conditions and key structural features: a ductile shear zone (SZ) and a fault gouge (FG).

found (Fig. 9a–d); however, for other surface points no underground point could be assigned unambiguously (Fig. 9e).

Based on the maximum a posteriori value, a 3-D structural model was obtained by linearly interpolating each surface point to the underground point with the maximum a posteriori value. We call this model the "maximum a posteriori" model (Fig. 10). Note that the maximum a posteriori model only adds information to the initial model through consideration of a likelihood, i.e. the assumption that crossing faults at a large scale are unlikely. Note that the smaller-scaled relay structures are not considered in this approach.

This maximum a posteriori interpolation model served as a basis for comparing different employed extrapolation approaches. The comparison did not yield a clear "best" extrapolation approach; however, it seems that fieldwork-based approach results in the most accurate extrapolation.

5 Discussion

5.1 Lineament map

A comparison of the remote-sensing-based lineament map and field data showed that in intact granitic rocks, purely ductile shear zones without later brittle overprinting are not detected by remote sensing. Brittle deformation generating fractures, cataclasites, or even fault gouges responsible for mechanical weakening is necessary to form morphologically detectable structures (Fig. 11a; Baumberger, 2015). Moreover, the orientations of the slopes play an important role, as faults striking in the down-dip directions of slopes are prone to the most effective erosion processes driven by gravity. Different orientations observed on the lineament map (Fig. 6a, areas i and ii) for the eastern and western flank of the Hasli valley are interpreted to result from such preferential erosion. In contrast, the surface area (iii) in Fig. 6a is nearly horizontal, thus reflecting a homogeneously eroded pattern of intersection for lineaments. The dependence on erosion for the formation of morphological incisions leads to the observed heterogeneous lineament density distribution as ridges and valleys show higher lineament densities.

Endpoint-to-endpoint strike and the strikes of individual segments of lineaments are very similar (Fig. 6b and e), indicating only small variation in the strike of the lineaments themselves. Therefore, underlying structures should be quasilinear to linear in 2-D and planar in 3-D. We also observe that the longest lineaments are NE–SW striking and that the variability shown in Fig. 6e is mostly due to varying strike orientations of very short lineaments (< 20 m). In addition to the NE-SW striking maximum, few long lineaments strike NW–SE. Both major orientations are similar to those reported from field observations (Figs. 7 and 8) and correlate with previous studies (Rolland et al., 2009; Steck, 1968; Wehrens et al., 2017), indicating that lineament maps are suitable to obtain the general trend of steep faults in well-exposed crystalline terrain. Much care is needed, however, when further interpreting lineament maps, as the geologic meaning of the lineament is ambiguous and lineament maps are strongly operator dependent (e.g. Scheiber et al., 2015).

5.2 Field observations and data

Differences between the surface map and the underground map are relatively small. The spacing of faults at the surface is lower, but general orientations are comparable (Figs. 7 and 8) and the two mappings are thus discussed jointly.

Faults commonly show little variation in orientation along strike as evidenced by consistent orientations of the dip and dip azimuth of multiple outcrop descriptions along the same fault. In conjunction with the small variability in strike for lineaments, this is clear evidence for the planarity of large-scale faults. At the surface, the 2-D length of faults is between 229 and 5591 m (mean 2199 ± 1603 m). Therefore, extrapolation of surface faults to depths similar to the depths of the underground faults, which have an overburden between 420 and 520 m, is well in the projection depth range assuming a circular shape for the plane as a minimum estimation for their lateral extent.

Localization processes seem to differ between the two host rocks (CAGr and GrGr; Wehrens et al., 2017). The higher amount of biotite in the GrGr could influence the rock's rheology towards more ductile behaviour. In contrast, the relatively higher amount of quartz and K-feldspars renders the CAGr more brittle than GrGr at similar pressure P and temperature T conditions and thus enforces brittle fracturing and possible subsequent ductile shear zone widening, as observed in other crystalline rocks (Guermani and Pennacchioni, 1998; Mancktelow and Pennacchioni, 2005; Wehrens et al., 2016, 2017). Hence any mechanical anisotropy, such as along pre-existing structures in the form of magmatic shear zones, meta-basic dykes, or aplitic dykes, served in the CAGr as sites for strain localization when suitably oriented with respect to the stress field.

5.3 3-D structural modelling

Our 3-D structural models were generated as a contribution to a project monitoring several parameters, such as microseismicity and in situ stress conditions, on the kilometre scale (large-scale monitoring; Nagra). Therefore, 3-D structural models were required mostly for visualization purposes. A deterministic explicit modelling workflow was required, as often is used in applied projects. It is, however, clear that for model updating, an implicit modelling approach would result in faster data handling. The deterministic approach was chosen because we attempted to obtain a geometrically satisfying product within the simplest geological setting possible. Furthermore, we were interested in the actual geometry of the faults dissecting granitoid rock bodies. Lastly, the uncommonly well-constrained setting of our study site (high-resolution underground data) was used to test and potentially validate extrapolation techniques for common application. Therefore, the underground data were only integrated as validation and not as a constraint during interpolation.

5.3.1 Three different approaches to obtain extrapolation 3-D structural models (kilometre-scale models)

Uncertainty related to the assignment of specific dip values to lineament traces (Baumberger, 2015; Bistacchi et al., 2008) led to the comparison of three different approaches. Validation attempts by comparison with underground mapping are purely geometrical and were based on two criteria, namely angle and distance misfit (Fig. 4). All three extrapolation approaches yielded similar results and no significant differences were observed. Moreover, in order to allow for a thorough comparison between the different extrapolation approaches solely based on the angle and distance misfit, the underground faults would need to be homogeneously distributed, which is not the case (Fig. 8).

The validation procedure could be refined using fault core thickness. However, fault thickness varies substantially along

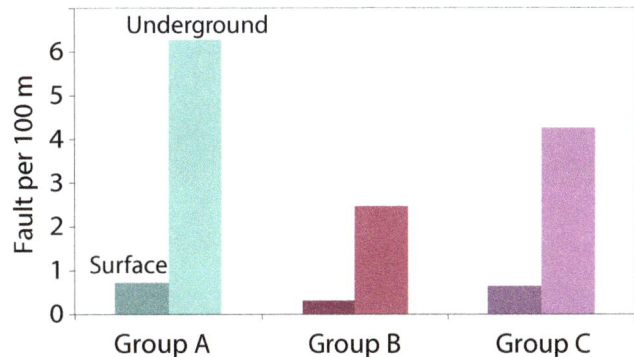

Figure 12. Histogram showing the number of faults grouped per strike at the Earth's surface and underground (GTS).

strike (e.g. Torabi and Berg, 2011) and thus is not a clear distinction criterion.

In addition to the average dip, the maximum and minimum dips could be used, which would yield a projection cone similar to the uncertainty visualization suggested by Baumberger (2015). Applying this approach to a restricted area, such as the underground rock laboratory investigated in this study, resulted in total coverage and no possible distinction between different faults. However, for a final representation of the uncertainty related to the dip value on a regional scale (kilometre scale), the approach of visualizing projection cones would suit.

5.3.2 GTS (decametre-scale model) compared with kilometre-scale "best-estimate" model

As a result of differences in outcrop conditions, the number of observed faults is significantly higher in the underground laboratory compared to the surface (Fig. 12). Underground, nearly 100 % of polished outcrop is accessible along the tunnel walls (Fig. 11b), whereas at the surface faults are often covered with vegetation, even in relatively vegetation-poor domains.

Furthermore, we observe convergence of surface faults with depth in our best-estimate model, which could be a modelling artefact. The N–S extent of surface area is larger compared to the GTS area, leaving a northern and a southern surface part underneath for which no underground data exist (Fig. 10). Hence faults in these domains are forced by the model set-up to be connected to the underground, leading to artificial fault orientations in these two cases (N and S rims of Fig. 13a). For that reason, only faults in the central part of the best-estimate model will be further considered.

5.3.3 "Maximum a posteriori" model

A comparison of numerous models obtained from Bayesian inference was performed by calculating the number of intersections. The fewer the intersections, the more probable the

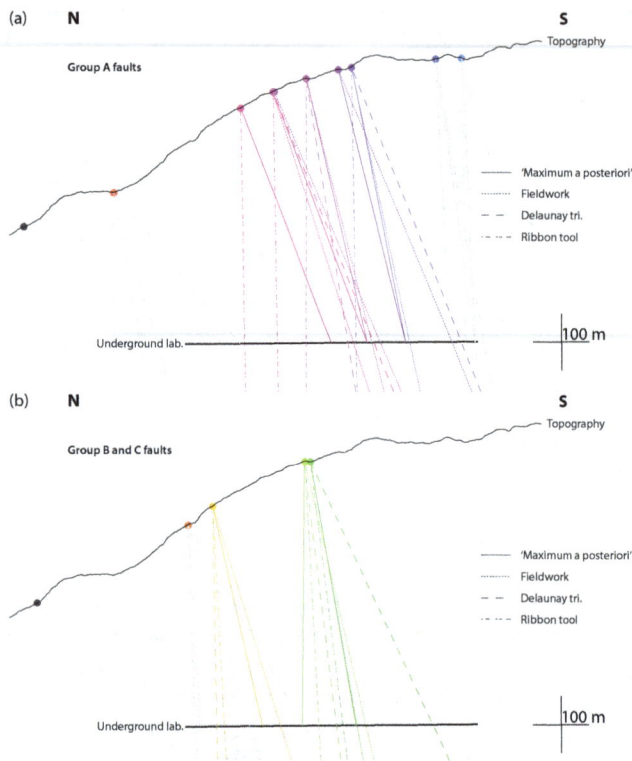

Figure 13. Comparison of maximum a posteriori interpolation with three extrapolation approaches used to assign dip to fault exposure line. Figure subdivided into **(a)** group A (NE–SW), **(b)** group B (E–W), and group C (NW–SE). Group B and group C are displayed jointly as group B, which contains only two faults.

model was considered. Assuming no intersections within a large-scale fault set is simplistic, but from field observations it seems plausible (Fig. 7) as a first approach for faults belonging to a specific orientation group (group A, B, C). The maximum a posteriori model is based on a N–S cross section along the GTS, and this orientation implies that faults would only cross if their dip varies strongly. Such a strong variation in the dip value is improbable based on measured dip values (Fig. 7). Therefore, this assumption seems feasible.

This simplistic representation of nature enabled us to obtain a probability for all possible interpolations between a specific surface point and all underground points of the corresponding orientation group. As previously mentioned, the margins of the interpolation space show boundary artefacts, and thus the following surface points at the model margin were not further considered: J03, J06, J20, and J21 (Fig. 8). As expected, probability densities are skewed towards the area of lesser fault density (Fig. 9). At this point it is important to remember that probability density is given as area, and therefore we cannot directly compare the discretized posteriors since they are a function of the distance between nearby faults.

We compared the initial three extrapolation techniques based on the maximum a posteriori model (Fig. 13). When comparing group A faults, fieldwork-based extrapolation closely fit the maximum a posteriori interpolation, which indicates either that the fieldwork-based model yields the best results or validates the Bayesian inference approach depending on whether the reference state is the statistical interpolation or the measured field data. Generally, dips of the maximum a posteriori models are slightly steeper than measured dips during fieldwork (Fig. 14a). However, the dip differences between the fieldwork-based extrapolation 3-D structural model and the maximum a posteriori interpolation model are small (Fig. 14b). Also, the dip differences between the ribbon-tool-based 3-D structural model and the maximum a posteriori interpolation model are small, but dips obtained via the ribbon tool are systematically steeper, which does not correspond to the measured dips (Fig. 14a). The extrapolation 3-D structural model obtained via Delaunay triangulation is less close to the maximum a posteriori interpolation model and obtained dips vary substantially.

The comparison for the group B and C faults is less clear. Fieldwork-based and ribbon tool extrapolations are close to the maximum a posteriori model (Fig. 13). Therefore, we conclude that fieldwork is still necessary for 3-D structural modelling in crystalline environments and that the ribbon tool (Move™) offers numerous options to tune the obtained plane; however, this tuning requires a profound conceptual background model.

5.3.4 Possible model refinements

Presented surface models include only major faults (Fig. 15). However, for further applications, such as groundwater flow modelling or slip tendency analysis, not only major faults are of interest but also their relay structures. Based on the orientation information gained from the regional kilometre-scale models and on the intersection pattern observed during lineament mapping, it is possible to infer a near surface 3-D model not only with the major fault but also the relay structure. Furthermore, the increased level of detail in the GTS model (decametre scale) forms a similar model in the underground. The unknown space between the two models would require probabilistic modelling with several key parameters, for example fault spacing, fault orientations, apertures, or cross-cutting relationships.

6   Conclusions

The exceptional opportunity for a surface and underground data comparison over 3-D structural modelling approaches led us to the following conclusions:

Lineament maps enable the identification of major faults but are highly sensitive to preferential erosion.

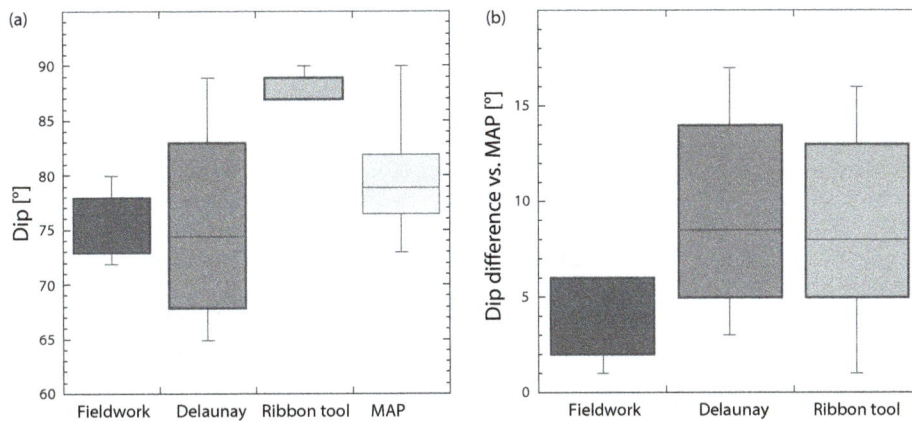

Figure 14. (a) Box plot showing dip value for different extrapolation approaches and for maximum a posteriori (MAP) interpolation. **(b)** Box plots for dip comparison between different extrapolation approaches and the maximum a posteriori (MAP) interpolation.

Figure 15. Representation in 3-D of the maximum a posteriori model of fault geometry with three different angles of view. N is indicated by the black triangle. The black tunnel is 717 m long.

Structural surface mapping allowed for a discrimination of three orientation groups of faults.

Comparison based on geometrical criteria (distance and angle misfit) of the three approaches to extrapolate to depth surface traces yielded comparable results for all extrapolation approaches.

Interpolation of surface data with underground data based on a Bayesian inference problem showed that the fieldwork-based approach is the most accurate extrapolation technique. However, this could also validate the interpolation approach.

We conclude, similarly to Zanchi et al. (2009), that for 3-D structural modelling a high-topography area within crystalline bedrock, classical fieldwork as an information source and as a basis for a conceptual background model on which interpolations or extrapolations performed within 3-D structural modelling can be examined for validity. In terms of general fault networks, our approach can be applied to (i) pervasive regional fault or fracture patterns. Currently it will fail in the case of (ii) discrete large-scale faults (e.g. strike-slip faults) consisting of one fault core and an associated damage zone. In such cases, more elaborate probabilistic models have to be generated in future, including 3-D variations in terms of spacing and orientation of secondary faults and splay faults. (i) Even with limited or missing underground information, our approach can be used to predict a surface-based 2-D model including a probability evaluation (e.g. variable dip

angles) with depth. If available, this evaluation can be tested with individual depth points such as drill core information. Additionally, an expansion towards 3-D would require probability attributes for dip azimuths.

Competing interests. The authors declare that they have no conflict of interest.

Acknowledgements. This study was funded by the LASMO project run by Nagra, RWM, and SURAO. We thank the Nagra staff in the underground rock laboratory for the excellent working environment. We thank the reviewers, Clare Bond, Gautier Laurent, and an anonymous reviewer, whose comments greatly improved the paper.

Edited by: Bernhard Grasemann

References

Abrecht, J.: Geologic units of the Aar massif and their pre-Alpine rock associations: a critical review, Schweizerische Mineral. und Petrogr. Mitteilungen, 74, 5–27, https://doi.org/10.5169/seals-56328, 1994.

Balaban, I. J.: An optimal algorithm for finding segments inter-
sections, in: Proceedings of the eleventh annual symposium on
Computational geometry – SCG '95, pp. 211–219, ACM Press,
New York, New York, USA, 1995.

Baumberger, R.: Quantification of lineaments: Link between internal 3D structure and surface evolution of the Hasli valley (Aar
massif, Central Alps, Switzerland), PhD thesis, University of
Bern, Switzerland, 2015.

Belgrano, T. M., Herwegh, M., and Berger, A.: Inherited structural controls on fault geometry, architecture and hydrothermal
activity?: an example from Grimsel Pass, Switzerland, Swiss
J. Geosci., 109, 345–364, https://doi.org/10.1007/s00015-016-
0212-9, 2016.

Bense, F. A., Wemmer, K., Löbens, S., and Siegesmund, S.: Fault
gouge analyses: K-Ar illite dating, clay mineralogy and tectonic
significance – a study from the Sierras Pampeanas, Argentina,
Int. J. Earth Sci., 103, 189–218, https://doi.org/10.1007/s00531-
013-0956-7, 2014.

Bentley, J. L. and Ottmann, T. A.: Algorithms for Reporting and
Counting Geometric Intersections, IEEE Trans. Comput., C-28,
643–47, https://doi.org/10.1109/TC.1979.1675432, 1979.

Berger, A., Mercolli, I., Herwegh, M., and Gnos, E.: Explanatory
notes accompanying the geological map of the Aar massif and
the Tavetsch- and Gotthard Nappes, Geological Special Map 129,
Swisstopo, Wabern, Switzerland, 2017.

Bistacchi, A., Massironi, M., Dal Piaz, G. V., Dal Piaz, G., Mo-
nopoli, B., Schiavo, A., and Toffolon, G.: 3D fold and fault
reconstruction with an uncertainty model: An example from
an Alpine tunnel case study, Comput. Geosci., 34, 351–372,
https://doi.org/10.1016/j.cageo.2007.04.002, 2008.

Bond, C. E.: Uncertainty in structural interpretation:
Lessons to be learnt, J. Struct. Geol., 74, 185–200,
https://doi.org/10.1016/j.jsg.2015.03.003, 2015.

Bond, C. E., Gibbs, A. D., Shipton, Z. K., and Jones, S.:
What do you think this is? "Conceptual uncertainty"
In geoscience interpretation, GSA Today, 17, 4–10,
https://doi.org/10.1130/GSAT01711A.1, 2007a.

Bond, C. E., Shipton, Z. K., Jones, R. R., Butler, R. W. H., and
Gibbs, A. D.: Knowledge Transfer in a digital world: Field data
acquisition, uncertainty and data management, Geosphere, 3,
568–576, https://doi.org/10.1130/GES00094.1, 2007b.

Bossart, P. and Mazurek, M.: Structural geology and water flow-
paths in the migration shear-zone, Nagra technical report NTB
91-12, Wettingen, Switzerland, 1991.

Caumon, G., Collon-Drouaillet, P., Le Carlier de Veslud,
C., Viseur, S., and Sausse, J.: Surface-Based 3D Modeling of Geological Structures, Math. Geosci., 41, 927–945,
https://doi.org/10.1007/s11004-009-9244-2, 2009.

Challandes, N., Marquer, D., and Villa, I. M.: P-T-t modelling, fluid
circulation, and 39 Ar–40 Ar and Rb-Sr mica ages in the Aar
Massif shear zones (Swiss Alps), Swiss J. Geosci., 101, 269–
288, https://doi.org/10.1007/s00015-008-1260-6, 2008.

Cherpeau, N. and Caumon, G.: Stochastic structural modelling in sparse data situations, Pet. Geosci., 21, 233–247,
https://doi.org/10.1144/petgeo2013-030, 2015.

Cherpeau, N., Caumon, G., Caers, J., and Lévy, B.: Method
for Stochastic Inverse Modeling of Fault Geometry and Connectivity Using Flow Data, Math. Geosci., 44, 147–168,
https://doi.org/10.1007/s11004-012-9389-2, 2012.

Choukroune, P. and Gapais, D.: Strain pattern in the Aar Granite

(Central Alps): orthogneiss developed by bulk inhomogeneous
flattening, J. Struct. Geol., 5, 411–418, 1983.

Delaunay, B.: Sur la sphère vide. A la mémoire de Georges Voronoi,
Bull. l'Académie des Sci. l'URSS. Cl. des Sci. mathématiques
Nat., 6, 793–800, 1934.

de la Varga, M. and Wellmann, J. F.: Structural geologic modeling
as an inference problem: A Bayesian perspective, Interpretation,
4, SM1–SM16, 2016.

Fernández, O.: Obtaining a best fitting plane through
3D georeferenced data, J. Struct. Geol., 27, 855–858,
https://doi.org/10.1016/j.jsg.2004.12.004, 2005.

Gabrielsen, R. H. and Braathen, A.: Models of fracture lineaments – Joint swarms, fracture corridors and faults in crystalline
rocks, and their genetic relations, Tectonophysics, 628, 26–44,
https://doi.org/10.1016/j.tecto.2014.04.022, 2014.

Goncalves, P., Oliot, E., Marquer, D., and Connolly, J. A. D.: Role
of chemical processes on shear zone formation: an example
from the Grimsel metagranodiorite (Aar massif, Central Alps), J.
Metamorph. Geol., 30, 703–722, https://doi.org/10.1111/j.1525-
1314.2012.00991.x, 2012.

González-Garcia, J. and Jessell, M.: A 3D geological model
for the Ruiz-Tolima Volcanic Massif (Colombia): Assessment of geological uncertainty using a stochastic approach
based on Bézier curve design, Tectonophysics, 687, 139–157,
https://doi.org/10.1016/j.tecto.2016.09.011, 2016.

Guermani, A. and Pennacchioni, G.: Brittle precursors of plastic
deformation in a granite: an example from the Mont Blanc massif
(Helvetic, western Alps), J. Struct. Geol., 20, 135–148, 1998.

Haario, H., Saksman, E., and Tamminen, J.: An adaptive metropolis
algorithm, Bernoulli, 7, 223–242, 2001.

Hassen, I., Gibson, H., Hamzaoui-Azaza, F., Negro, F., Rachid,
K., and Bouhlila, R.: 3D geological modeling of the Kasser-
ine Aquifer System, Central Tunisia: New insights into
aquifer-geometry and interconnections for a better assessment of groundwater resources, J. Hydrol., 539, 223–236,
https://doi.org/10.1016/j.jhydrol.2016.05.034, 2016.

Heilbronner, R. and Barrett, S.: Image Analysis in Earth Sciences,
Springer, Berlin Heidelberg, 2014.

Herwegh, M., Berger, A., Baumberger, R., Wehrens, P.,
and Kissling, E.: Large-Scale Crustal-Block-Extrusion
During Late Alpine Collision, Sci. Rep., 7, 413,
https://doi.org/10.1038/s41598-017-00440-0, 2017.

Hollund, K., Mostad, P., Fredrik Nielsen, B., Holden, L., Gjerde, J.,
Consursi, M. J., McCann, A. J., Townsend, C., and Sverdrup, E.:
Havana – a fault modeling tool, Nor. Pet. Soc. Spec. Publ., 11,
157–171, 2002.

Jessell, M. W., Aillères, L., and de Kemp, E. A.: Towards an integrated inversion of geoscientific data:
What price of geology?, Tectonophysics, 490, 294–306,
https://doi.org/10.1016/j.tecto.2010.05.020, 2010.

Jones, R. R., McCaffrey, K. J. W., Clegg, P., Wilson, R. W., Holli-
man, N. S., Holdsworth, R. E., Imber, J., and Waggott, S.: Integration of regional to outcrop digital data: 3D visualisation
of multi-scale geological models, Comput. Geosci., 35, 4–18,
https://doi.org/10.1016/j.cageo.2007.09.007, 2009.

Jørgensen, F., Høyer, A., Sandersen, P. B. E., and He, X.:
Combining 3D geological modelling techniques to address
variations in geology, data type and density – An example from Southern Denmark, Comput. Geosci., 81, 53–63,
https://doi.org/10.1016/j.cageo.2015.04.010, 2015.

Kaufmann, O. and Martin, T.: 3D geological modelling from boreholes, cross-sections and geological maps, application over former natural gas storages in coal mines, Comput. Geosci., 35, 70–82, https://doi.org/10.1016/S0098-3004(08)00227-6, 2009.

Keusen, H. R., Ganguin, J., Schuler, P., and Buletti, M.: Felslabor Grimsel: Geologie, Nagra technical report NTB 87-14, Baden, Switzerland, 1989.

Koike, K., Kubo, T., Liu, C., Masoud, A., Amano, K., Kurihara, A., Matsuoka, T., and Lanyon, B.: 3D geostatistical modeling of fracture system in a granitic massif to characterize hydraulic properties and fracture distribution, Tectonophysics, 660, 1–16, https://doi.org/10.1016/j.tecto.2015.06.008, 2015.

Kralik, M., Clauer, N., Holnsteiner, R., Huemer, H., and Kappel, F.: Recurrent fault activity in the Grimsel Test Site (GTS, Switzerland): revealed by Rb-Sr, K-Ar and tritium isotope techniques, J. Geol. Soc. London., 149, 293–301, 1992.

Labhart, T.: Aarmassiv und Gotthardmassiv, Gebruder Borntraeger Verlagsbuchhandlung, Berlin and Stuttgart, 173 p., 1977.

Lindsay, M. D., Aillères, L., Jessell, M. W., de Kemp, E. A., and Betts, P. G.: Locating and quantifying geological uncertainty in three-dimensional models: Analysis of the Gippsland Basin, southeastern Australia, Tectonophysics, 546–547, 10–27, https://doi.org/10.1016/j.tecto.2012.04.007, 2012.

MacKay, D. J. C.: Information Theory, Inference, and Learning Algorithms, 4th edition, Cambridge University Press, Cambridge, UK, 2003.

Mancktelow, N. S. and Pennacchioni, G.: The control of precursor brittle fracture and fluid–rock interaction on the development of single and paired ductile shear zones, J. Struct. Geol., 27, 645–661, https://doi.org/10.1016/j.jsg.2004.12.001, 2005.

Marquer, D., Gapais, D., and Capdevila, R.: Chemical-changes and mylonitization of a granodiorite within low-grade metamorphism (Aar massif, central Alps), Bull. Mineral., 108, 209–221, 1985.

Mazurek, M.: Geological and hydraulic properties of water-conducting features in crystalline rocks, in: Hydrogeology of crystalline rocks, vol. 34, edited by: Stober, I. and Bucher, K., pp. 3–26, Springer, the Netherlands, 2000.

Mercolli, I. and Oberhänsli, R.: Variscan tectonic evolution in the Central Alps: a working hypothesis, Schweizerische Mineral. und Petrogr. Mitteilungen, 68, 491–500, https://doi.org/10.5169/seals-52084, 1988.

Oberhänsli, R.: Geochemistry of meta-lamprophyres from the Central Swiss Alps, Schweizerische Mineral. und Petrogr. Mitteilungen, 66, 315–342, 1986.

O'Leary, D. W., Friedman, J. D., and Pohn, H. A.: Lineament, linear, lineation?: Some proposed new standards for old terms, Geol. Soc. Am. Bull., 87, 1463–1469, https://doi.org/10.1130/0016-7606(1976)87<1463:LLLSPN>2.0.CO;2, 1976.

Patil, A., Huard, D., and Fonnesbeck, C. J.: PyMC: Bayesian stochastic modelling in Python, J. Stat. Softw., 35, 1–81, 2010.

Pfiffner, O. A.: Geologie der Alpen, Haupt Verlag, Bern, Stuttgart, Wien, 2009.

Pfiffner, O. A. and Deichmann, N.: Seismotektonik der Zentralschweiz, Wettingen, Switzerland, 2014.

Rolland, Y., Cox, S. F., and Corsini, M.: Constraining deformation stages in brittle–ductile shear zones from combined field mapping and 40Ar/39Ar dating: The structural evolution of the Grim-

sel Pass area (Aar Massif, Swiss Alps), J. Struct. Geol., 31, 1377–1394, https://doi.org/10.1016/j.jsg.2009.08.003, 2009.

Sausse, J., Dezayes, C., Dorbath, L., Genter, A., and Place, J.: 3D model of fracture zones at Soultz-sous-Forêts based on geological data, image logs, induced microseismicity and vertical seismic profiles, Comptes Rendus-Geosci., 342, 531–545, https://doi.org/10.1016/j.crte.2010.01.011, 2010.

Schaltegger, U.: The Central Aar Granite: Highly differentiated calc-alkaline magmatism in the Aar massif (Central Alps, Switzerland), Eur. J. Mineral., 2, 245–259, 1990.

Schaltegger, U.: Unravelling the pre-Mesozoic history of Aar and Gotthard massifs (Central Alps) by isotopic dating: a review, Schweizerische Mineral. und Petrogr. Mitteilungen, 74, 41–51, https://doi.org/10.5169/seals-56330, 1994.

Schaltegger, U. and Corfu, F.: The age and source of late Hercynian magmatism in the central Alps: evidence from precise U-Pb ages and initial Hf isotopes, Contrib. Mineral. Petrol., 111, 329–344, 1992.

Scheiber, T., Fredin, O., Viola, G., Jarna, A., Gasser, D., and Łapińska-Viola, R.: Manual extraction of bedrock lineaments from high-resolution LiDAR data: methodological bias and human perception, GFF, 5897, 1–11, https://doi.org/10.1080/11035897.2015.1085434, 2015.

Schneeberger, R., Berger, A., Herwegh, M., Eugster, A., Kober, F., Spillmann, T., and Blechschmidt, I.: GTS Phase VI – LASMO: Geology and structures of the GTS and Grimsel region, Nagra Arbeitsbericht NAB 16-27, Wettingen, Switzerland, 2016.

Stalder, H. A.: Petrographische und mineralogische Untersuchungen im Grimselgebiet (Mittleres Aarmassiv), PhD thesis, University of Bern, Switzerland, 1964.

Steck, A.: Die alpidischen Strukturen in den Zentralen Aaregranite des westlichen Aarmassivs, Eclogae Geol. Helv., 61, 19–48, https://doi.org/10.5169/seals-163584, 1968.

Stephens, M. B., Follin, S., Petersson, J., Isaksson, H., Juhlin, C., and Simeonov, A.: Review of the deterministic modelling of deformation zones and fracture domains at the site proposed for a spent nuclear fuel repository, Sweden, and consequences of structural anisotropy, Tectonophysics, 653, 68–94, https://doi.org/10.1016/j.tecto.2015.03.027, 2015.

Svensk Kärnbränslehantering AB: Site description of Laxemar a completion of the site investigation phase, available at: http://www.skb.se/upload/publications/pdf/TR-09-01del1w (last access: 26 September 2017), 2009.

Tacher, L., Pomian-Srzednicki, I., and Parriaux, A.: Geological uncertainties associated with 3-D subsurface models, Comput. Geosci., 32, 212–221, https://doi.org/10.1016/j.cageo.2005.06.010, 2006.

Torabi, A. and Berg, S. S.: Scaling of fault attributes: A review, Mar. Pet. Geol., 28, 1444–1460, https://doi.org/10.1016/j.marpetgeo.2011.04.003, 2011.

Viard, T., Caumon, G., and Lévy, B.: Adjacent versus coincident representations of geospatial uncertainty: Which promote better decisions?, Comput. Geosci., 37, 511–520, https://doi.org/10.1016/j.cageo.2010.08.004, 2011.

Vouillomaz, J.: Strain localization along shear zones in the Juchlistock area, Master thesis, University of Bern, Switzerland, 2009.

von Raumer, J. F., Bussy, F., and Stampfli, G. M.: The Variscan evolution in the External massifs of the Alps and place in their

Variscan framework, Comptes Rendus Geosci., 341, 239–252, https://doi.org/10.1016/j.crte.2008.11.007, 2009.

Wehrens, P., Berger, A., Peters, M., Spillmann, T., and Herwegh, M.: Deformation at the frictional-viscous transition: Evidence for cycles of fluid-assisted embrittlement and ductile deformation in the granitoid crust, Tectonophysics, 693, 66–84, https://doi.org/10.1016/j.tecto.2016.10.022, 2016.

Wehrens, P., Baumberger, R., Berger, A., and Herwegh, M.: How is strain localized in a mid-crustal basement section? Spatial distribution of deformation in the Aar massif (Switzerland), J. Struct. Geol., 94, 47–67, https://doi.org/10.1016/j.jsg.2016.11.004, 2017.

Wellmann, J. F. and Regenauer-Lieb, K.: Uncertainties have a meaning: Information entropy as a quality measure for 3-D geological models, Tectonophysics, 526–529, 207–216, https://doi.org/10.1016/j.tecto.2011.05.001, 2012.

Wellmann, J. F., Horowitz, F. G., Schill, E., and Regenauer-Lieb, K.: Towards incorporating uncertainty of structural data in 3D geological inversion, Tectonophysics, 490, 141–151, https://doi.org/10.1016/j.tecto.2010.04.022, 2010.

Wellmann, J. F., Lindsay, M., Poh, J., and Jessell, M.: Validating 3-D Structural Models with Geological Knowledge for Improved Uncertainty Evaluations, Energy Procedia, 59, 374–381, https://doi.org/10.1016/j.egypro.2014.10.391, 2014.

Wicki, T.: 3D-shear zone pattern in the Grimsel area: ductile to brittle deformation in granitic rocks, Master thesis, University of Bern, Switzerland, 2011.

Wirsig, C., Zasadni, J., Ivy-Ochs, S., Christl, M., Kober, F., and Schlüchter, C.: A deglaciation model of the Oberhasli, Switzerland, J. Quat. Sci., 31, 46–59, https://doi.org/10.1002/jqs.2831, 2016.

Yamamoto, J. K., Koike, K., Kikuda, A. T., da Cruz Campanha, G. A., and Endlen, A.: Post-processing for uncertainty reduction in computed 3D geological models, Tectonophysics, 633, 232–245, https://doi.org/10.1016/j.tecto.2014.07.013, 2014.

Zanchi, A., Francesca, S., Stefano, Z., Simone, S., and Graziano, G.: 3D reconstruction of complex geological bodies: Examples from the Alps, Comput. Geosci., 35, 49–69, https://doi.org/10.1016/j.cageo.2007.09.003, 2009.

Ziegler, M., Loew, S., and Moore, J. R.: Distribution and inferred age of exfoliation joints in the Aar Granite of the central Swiss Alps and relationship to Quaternary landscape evolution, Geomorphology, 201, 344–362, https://doi.org/10.1016/j.geomorph.2013.07.010, 2013.

8

Interpretation of zircon coronae textures from metapelitic granulites of the Ivrea–Verbano Zone, Northern Italy: two-stage decomposition of Fe–Ti oxides

Elizaveta Kovaleva[1,2], **Håkon O. Austrheim**[3], **and Urs S. Klötzli**[2]

[1]Department of Geology, University of the Free State, Bloemfontein, 9300, 205 Nelson Mandela Drive, Free State, South Africa

[2]Department of Lithospheric Research, Faculty of Earth Sciences, Geography and Astronomy, University of Vienna, Althanstrasse 14, Vienna, 1090, Austria

[3]Section of Physics of Geological processes, Department of Geoscience, University of Oslo, Oslo, 0316, Norway

Correspondence to: Elizaveta Kovaleva (kovalevae@ufs.ac.za)

Abstract. In this study, we report the occurrence of zircon coronae textures in metapelitic granulites of the Ivrea–Verbano Zone. Unusual zircon textures are spatially associated with Fe–Ti oxides and occur as (1) vermicular-shaped aggregates 50–200 μm long and 5–20 μm thick and as (2) zircon coronae and fine-grained chains, hundreds of micrometers long and ≤ 1 μm thick, spatially associated with the larger zircon grains. Formation of such textures is a result of zircon precipitation during cooling after peak metamorphic conditions, which involved: (1) decomposition of Zr-rich ilmenite to Zr-bearing rutile, and formation of the vermicular-shaped zircon during retrograde metamorphism and hydration; and (2) recrystallization of Zr-bearing rutile to Zr-depleted rutile intergrown with quartz, and precipitation of the submicron-thick zircon coronae during further exhumation and cooling. We also observed hat-shaped grains that are composed of preexisting zircon overgrown by zircon coronae during stage (2). Formation of vermicular zircon (1) preceded ductile and brittle deformation of the host rock, as vermicular zircon is found both plastically and cataclastically deformed. Formation of thin zircon coronae (2) was coeval with, or immediately after, brittle deformation as coronae are found to fill fractures in the host rock. The latter is evidence of local, fluid-aided mobility of Zr. This study demonstrates that metamorphic zircon can nucleate and grow as a result of hydration reactions and mineral breakdown during cooling after granulite-facies metamorphism. Zircon coronae textures indicate metamorphic reactions in the host rock and establish the direction of the reaction front.

1 Introduction

1.1 Growth of metamorphic zircon

Growth of zircon during metamorphism is described in a variety of lithologies and can occur at different metamorphic grades, from high- (e.g., Fraser et al., 1997, 2004; Degeling et al., 2001; Möller et al., 2003; Wu et al., 2006; Harley et al., 2007; Zhao et al., 2015) to low-temperature metamorphism, including low-temperature hydrothermal reactions (e.g., Dempster et al., 2004, 2008; Rasmussen, 2005; Hay and Dempster, 2009; Hay et al., 2010; Kohn et al., 2015). Growth and new precipitation of zircon has been traced under temperatures as low as 250 °C (Rasmussen, 2005). Metamorphic zircon can (a) precipitate from fluid or melt (e.g., Rasmussen, 2005; Kohn et al., 2015), (b) can result from breakdown of Zr-bearing phases (major or accessory) (e.g., Davidson and van Breemen, 1988; Fraser et al., 1997, 2004; Degeling et al., 2001), or (c) can exsolve from Zr-bearing phases (e.g., Bingen et al., 2001; Tomkins et al., 2007).

a. New metamorphic zircon precipitates during exhumation and cooling of the host rock (Rasmussen, 2005; Kohn et al., 2015) as a result of partial or complete dissolution of preexisting zircon (e.g., Dempster et al., 2008) in partial melt or metamorphic fluid during high-temperature metamorphism (Ewing et al., 2014). Solubility of zircon in most natural fluids is very low (Tromans, 2006); thus, zircon dissolution models mostly describe interactions with the melt (e.g., Harrison and Watson, 1983; Watson and Harrison, 1983). However, under high (prograde) temperatures and in fluids of favorable composition, zircon can be dissolved without melt involvement (e.g., Rubatto et al., 2008; Hay and Dempster, 2009; Hay et al., 2010; Ewing et al., 2014; Kohn et al., 2015).

b. Zircon growth and overgrowth formation during cooling stage and/or retrograde metamorphism may also result from metamorphic reactions and breakdown of other Zr-bearing minerals (Fraser et al., 1997, 2004; Degeling et al., 2001; Möller et al., 2002, 2003; Tomkins et al., 2007). Fraser et al. (1997) and Möller et al. (2002) suggested that the source of newly precipitated zircon is Zr-bearing rock-forming phases (e.g., garnet), which experience breakdown and release Zr. The released Zr is not compatible with the breakdown product (e.g., with cordierite) and thus has to form a separate Zr phase, which could be zircon (Degeling et al., 2001; Möller et al., 2003). Zircon precipitation from other phases may also be facilitated by fluid. For example, Fraser et al. (2004) documented zircon rims precipitated during cooling stage from the hydrous fluid phase, which originated locally as a result of chlorite breakdown. The reactions with zircon precipitation in metamorphic rocks may be more efficient in the zones available for fluid infiltration, like fractures and shear zones (Bingen et al., 2001).

c. Zircon exsolution has been observed in nature (e.g., Ewing et al., 2013; Pape et al., 2016) and has been demonstrated experimentally with Zr-rich rutile (Tomkins et al., 2007). Resulting zircon appears as thin exsolution lamellae or as small individual euhedral grains within rutile. Similarly, the metapelites from the Ivrea–Verbano Zone (IVZ) reveal thin zircon needles in rutile and chains of fine zircon grains framing rutile (Ewing et al., 2013; Pape et al., 2016).

In this contribution, we investigate unusual zircon textures, such as coronae found in dehydrated metapelitic granulites of the IVZ. We start with a review of the process of zircon precipitation from various Zr-bearing phases, followed by an overview of known examples of zircon coronae. After a short geological background of the unit, we describe the sampled outcrop as well as the sample itself macroscopically. Then a short exposition of applied methods and microscopic description of the studied sample are presented, followed by a detailed depiction of observed zircon microstructures and textures. For the sake of completion, we also include micro-

probe data of the studied sample. In the discussion, we suggest mineral reactions that could result in the formation of observed zircon coronae textures and then discuss the implications of our findings.

1.2 Zr-bearing phases potentially associated with zircon precipitation

Zircon dissolution and growth during metamorphism are not independent processes but must be coupled with the breakdown–growth of other phases and/or with various mineral–fluid reactions in the host rock (e.g., Tomkins et al., 2007; Austrheim et al., 2008). Metamorphic precipitation of zircon could be a result of the breakdown of and/or exsolution from various Zr-bearing phases (Davidson and van Breemen, 1988) such as garnet, amphibole, (clino)pyroxene and ilmenite (e.g., Fraser et al., 1997; Degeling et al., 2001; Möller et al., 2003; Söderlund et al., 2004; Harley et al., 2007; Kelsey et al., 2008; Morisset and Scoates, 2008), hemo-ilmenite (Morisset et al., 2005), baddeleyite (Bingen et al., 2001; Söderlund et al., 2004), rutile (Harley et al., 2007; Tomkins et al., 2007; Morisset and Scoates, 2008; Kelsey and Powell, 2011; Ewing et al., 2013, 2014; Pape et al., 2016), epidote, titanite (Kohn et al., 2015), chlorite (Fraser et al., 2004), and biotite (Vavra et al., 1996). Zircon coronae have been reported around Martian baddeleyite as a result of shock metamorphism (Moser et al., 2013). In mafic metamorphic rocks, precipitation of zircon is commonly associated with the Fe–Ti oxides (Bingen et al., 2001; Ewing et al., 2013, 2014) due to similar chemical properties of Zr and Ti.

Zirconium and titanium both belong to group 4 in the periodic table, have close chemical properties and are usually regarded as relatively immobile trace elements (e.g., Mohamed and Hassanen, 1996). In the group of incompatible cations, Zr and Ti belong to high field strength (HFS) elements, which are smaller and are highly charged compared with large ion lithophile (LIL) elements. The chemical similarities result in a positive correlation between Zr and Ti for most rock suites and in their ability to replace each other in oxides (e.g., Morisset et al., 2005). The fact that Zr oxides and Ti oxides are spatially related in many rocks confirms chemical similarities between Zr and Ti.

Thus, rutile and ilmenite are the main minerals interpreted to influence the Zr mass balance in metabasites in the absence of other Zr phases (e.g., Ferry and Watson, 2007; Tomkins et al., 2007; Morisset and Scoates, 2008; Ewing et al., 2013). In the absence of zircon, rutile can be the main phase holding Zr and Hf in the absence of zircon (Ewing et al., 2014). Zirconium is a common component of rutile, in which its content can reach 10 000 ppm (e.g., Ewing et al., 2013); thus, Zr distributions generally reflect the formation and decomposition of rutile. The temperature dependence of Zr solubility in rutile can have a fundamental impact on the zircon growth rate (Kohn et al., 2015) and controls zircon stability (Kelsey and Powell, 2011). The zirconium-in-rutile thermometer for the rutile–quartz–zircon system was calibrated by a number of authors (e.g., Watson et al., 2006; Ferry and Watson, 2007; Tomkins et al., 2007; Lucassen et al., 2010; Ewing et al., 2013), who have shown a large temperature- and pressure-dependent solubility of Zr in rutile. Zircon growth is frequently associated with the oxide transition from Zr-rich rutile to ilmenite during late-stage exhumation and cooling under a large variety of $P-T$ conditions (Ewing et al., 2013). Magmatic and metamorphic ilmenite can also contain significant amounts of Zr (Bingen et al., 2001; Morisset et al., 2005; Charlier et al., 2007), up to more than 500 ppm (e.g., Morisset and Scoates, 2008). Consistently, many authors describe zircon precipitation on ilmenite (e.g., Bingen et al., 2001; Austrheim et al., 2008; Morisset and Scoates, 2008) (see below).

1.3 Occurrences of fine-grain zircon and zircon corona textures

In igneous rocks zircon usually forms euhedral elongated single crystals that are shaped by a combination of prismatic and pyramidal faces, whereas metamorphic zircon is characterized by roundish or irregular shapes (Corfu et al., 2003). Rarely, zircon has unusual saccharoidal or needle-shaped morphology or forms coronae (Corfu et al., 2003 and references therein). Mineral–fluid interactions, decomposition of Zr-bearing minerals, and exsolution from Zr-bearing accessory and rock-forming minerals can result in such unusual zircon textures (e.g., Corfu et al., 2003 and references therein; Dempster et al., 2004, 2008; Rasmussen, 2005), even at low metamorphic grades (e.g., Dempster et al., 2008).

In natural samples, there are several documented examples of zircon coronae textures from igneous and metaigneous rocks (e.g., Bingen et al., 2001; Söderlund et al., 2004; Austrheim et al., 2008), as well as from metapelites of the IVZ (Pape et al., 2016). Such textures are found in rocks of different metamorphic grades, ranging from prehnite-pumpellyite to eclogite facies (Austrheim et al., 2008). It has been suggested that coronae textures may evolve in magmatic rocks as a result of slow cooling (Morisset et al., 2005) and in metamorphic rocks due to mineral–fluid reactions or exsolution with fluid-aided diffusion along grain boundaries during progressive metamorphism (e.g., Bingen et al., 2001).

One of the first descriptions of zircon coronae in mafic metaigneous rocks was done by Söderlund et al. (2004). The authors attributed formation of secondary fine-grained zircon to the breakdown of baddeleyite in the presence of silica (saccharoidal zircon) and to consumption of minerals that have trace amounts of Zr, such as ilmenite (coronitic zircon). Both of these textural types of secondary zircon precipitated under prograde heating. Bingen et al. (2001) reported hat-shaped zircon grains and coronae around ilmenite in granulites and amphibolites. Charlier et al. (2007) and Austrheim et al. (2008) reported fine-grained zircon chains around, but at a distance from ilmenite and rutile grains in metagabbros.

These authors suggested that zircon chains had grown around primary Fe–Ti oxides and, therefore, trace the former grain boundaries. Fine-grained zircon was reported to frame some rutile grains in the metapelitic septae from the IVZ (Ewing et al., 2013; Pape et al., 2016). Morisset and Scoates (2008) reported 1–100 µm thick zircon coronae around ilmenite in mafic plutonic rocks. They consider it to be a result of Zr diffusion from ilmenite during slow cooling, aided by hydrothermal fluid.

In this study, we report the two textural types of zircon coronae, characterized by various thickness and aspect ratio, occurring within Fe–Ti oxides in granulitic metapelites. We present evidence that these textures formed as a product of the breakdown of Fe–Ti oxides, which helps to understand the initial mineral paragenesis of the host rock and reveals former reaction fronts.

2 Geological background and sampled locality

The IVZ in the Southern Alps (northern Italy) consists of a NE–SW trending, steeply dipping sequence of metasedimentary and metaigneous basic rocks, ultrabasic mantle tectonites and a large underplated mafic igneous complex (Fig. 1a) (e.g., Brodie and Rutter, 1987; Brodie et al., 1992; Rutter et al., 2007). The sequence predominantly consists of metasedimentary rocks in the SE and metabasic rocks and strongly depleted metapelites in the NW. Metamorphic grade increases progressively from amphibolite facies in the SE to granulite facies in the NW. The IVZ is generally accepted as a section through the lower continental crust that experienced regional metamorphism during the uppermost Paleozoic and was tectonically overturned and uplifted. The IVZ is delimited by the Insubric Line in the NW and the Pogallo Line in the SE (Brodie and Rutter, 1987; Barboza et al., 1999; Rutter et al., 2007; Quick et al., 2009).

The sampled outcrop near the village Cuzzago (Val d'Ossola) shows massive, non-foliated granulite-facies metasediments, known as stronalites. Stronalite is defined as granulite-facies metapelite, consisting of garnet, sillimanite and biotite with leucocratic patches and veins, composed of quartz, plagioclase and K-feldspar (Bea and Montero, 1999) or as granoblastic graphite–sillimanite–garnet gneiss, one of the components of the IVZ septa (Barboza et al., 1999). Local foliation and/or compositional layering of stronalites is moderately folded (e.g., Kovaleva et al., 2014, their Fig. 1C). Stronalites are broken by orthogonal sets of fractures and crosscut by a contrasting layer of darker gneiss (45°59′46.46″ N, 8°21′38.65″ E, sampled rock; Fig. 1b), which is macroscopically massive to weakly foliated, broken by abundant faults normal to foliation. The foliation of the layer strikes NW (310°, angle 77°) and the lineation plunges to the NE (34° towards 038°). No obvious kinematic indicators were observed in the host stronalites or in the sampled dark gneiss. However, detailed structural in-

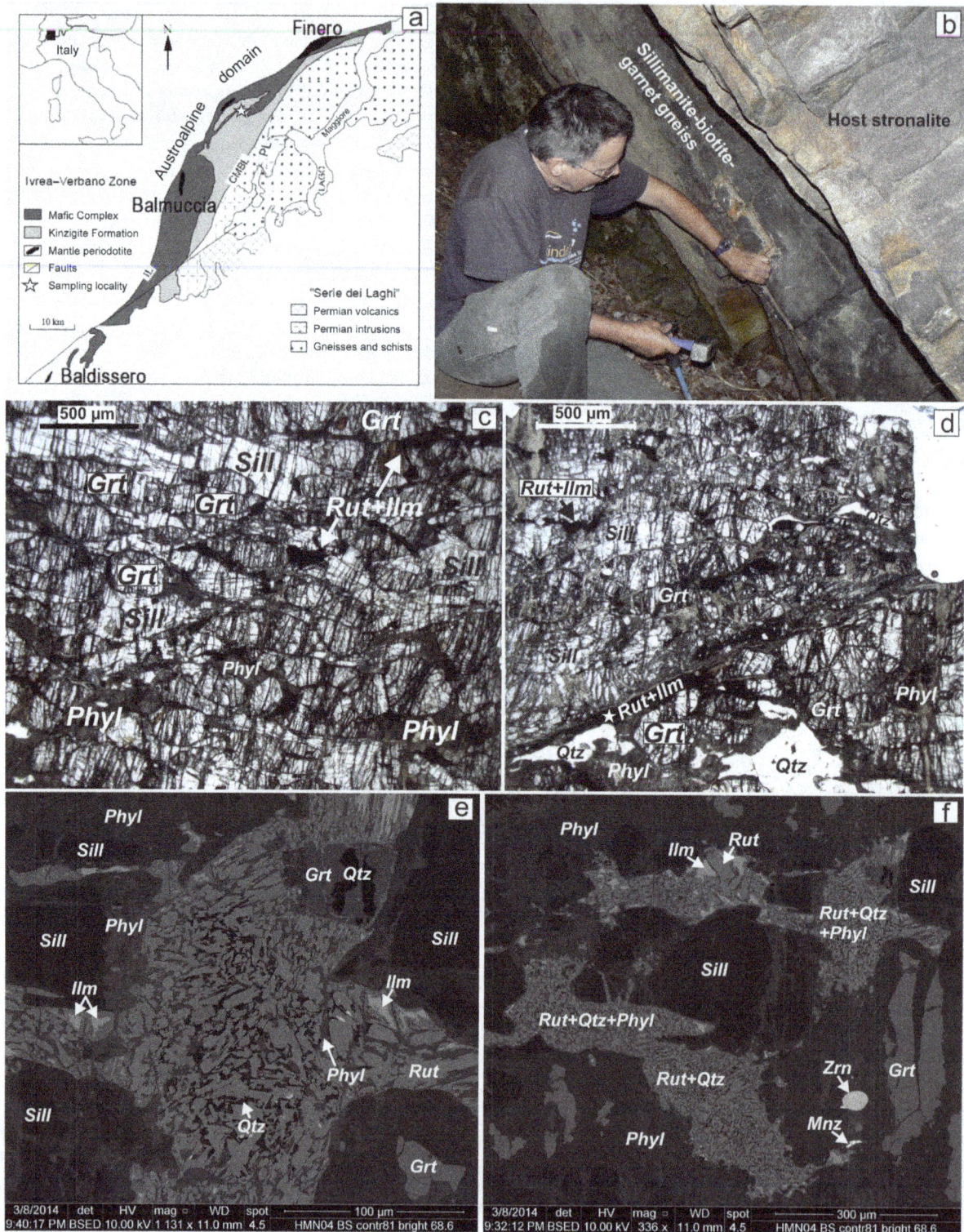

Figure 1. (a) Geological map of the Ivrea–Verbano Zone after Zanetti et al. (1999) with the sampling location indicated by a star. **(b)** Field photograph of the sampled outcrop with the dyke-shaped body of the sillimanite–biotite–garnet gneiss, interpreted as restitic, hosted by mylonitized and fractured stronalite. **(c–d)** Plain-polarized light photomicrographs. Two generations of veins are visible: black veins and pockets (Fe–Ti oxides) and brown-grey material (mixture of fine-grained phyllosilicates and K-feldspar). White star in panel **(d)** indicates the position of Fig. 4a. **(e–f)** BSE images with mineral paragenesis. Note in panel **(e)** that the rutile aggregate contains ilmenite cores (bright grey) and forms intergrowths with two different phases: phyllosilicates from the reaction rim (grey shade, slightly darker than rutile) and quartz (the darkest phase), all indicated by arrows. Sill: sillimanite, Grt: garnet, Qtz: quartz, Ilm: ilmenite, Rut: rutile, Rut + Qtz: rutile–quartz intergrowths, Phyl: fine-grained mixture of phyllosilicates, Zrn: zircon detrital grain, and Mnz: monazite.

vestigations of the shear zones in the neighboring Val Strona revealed numerous structures that provide consistent evidence of sinistral shear (Siegesmund et al., 2008, and references therein).

The sampled gneiss consists of sillimanite–biotite–garnet intergrowths (Fig. 1c–d). Such restitic mineral assemblage in granulitic metapelites is interpreted to form due to partial melting and separation of leucosome (e.g., Barboza et al., 1999; Luvizotto and Zack, 2009; Ewing et al., 2013; Pape et al., 2016). Thus, the sampled gneiss is a restite, resulting from migmatization at peak granulite-facies conditions (e.g., Ewing et al., 2013). Partial melting is also responsible for apparent layering of host stronalites (Bea and Montero, 1999; Siegesmund et al., 2008), which are composed of alternating leucocratic and melanocratic layers.

The investigated sample came from the northwestern part of the IVZ, were metapelites and metagabbro were re-equilibrated under granulite-facies conditions prevailing during crustal attenuation–extension and contemporaneous magmatic underplating between 315 and 270 Ma (Rutter et al., 2007; Quick et al., 2009; Sinigoi et al., 2011; Klötzli et al., 2014). $P - T$ estimates in the neighboring valleys, Val Strona and Val d'Ossola, indicated granulite-facies $P - T$ metamorphic conditions in the metapelites at a maximum of $750 \pm 50 \,°C$ and $0.6 \pm 0.1 \,GPa$ (Sills, 1984). Zr-in-rutile temperatures of up to $850–930 \,°C$ were obtained for granulite-facies metapelites from Val d'Ossola (Luvizotto and Zack, 2009).

Ubiquitous faulting of restitic material and both faulting and folding of host stronalite is due to intensive deformation taking place in granulitic metapelites during a long-time span after peak metamorphism (Siegesmund et al., 2008). According to Siegesmund et al. (2008) brittle and ductile deformation acted simultaneously during formation of shear zones, and their close interactions resulted in complex deformation microstructures.

3   Sample preparation and analytical methods

Zircon textures have been examined in situ using polished thin sections that were mechanically prepared with a final polish using $0.25 \,\mu m$ diamond paste. Zircon grains were identified by backscattered-electron (BSE) imaging and were additionally characterized by cathodoluminescence (CL) imaging for the internal growth features using an FEI Inspect S scanning electron microscope equipped with a Gatan MonoCL system (Faculty of Earth Sciences, Geography and Astronomy, University of Vienna, Austria). Imaging conditions were at 10 kV accelerating voltage, CL image resolution of 1500×1500 to 2500×2500 pixels using a dwell time of $80.0–150.0 \,ms$ and a probe current of $4.5–5.0 \,nA$. Qualitative chemical compositions of host phases to zircon were made using an energy-dispersive X-ray (EDX) spectrometer. Orientation contrast images of zircon grains (e.g.,

Fig. 2) were taken using a forescatter electron (FSE) detector on a chemically polished sample surface. The FSE detector is mounted on the electron backscatter detector (EBSD) tube of a FEI Quanta 3D field emission gun (FEG) instrument (Faculty of Earth Sciences, Geography and Astronomy, University of Vienna, Austria), which is equipped with a Schottky field emission electron source. Electron beam conditions were 15 kV accelerating voltage and 2.5–4 nA probe current using the analytic mode. Stage settings were at 70° tilt and 14–16 mm working distance. Full quantitative chemical compositions of host minerals (Tables 1–2) were determined by Cameca SX 100 electron microprobe equipped with four wavelength-dispersive spectrometers (WDSs) and an EDX system for high quality of quantitative chemical analyses (Faculty of Earth Sciences, Geography and Astronomy, University of Vienna, Austria). Operating conditions were 15 kV accelerating voltage and 100 nA probe current. The detection limits in parts per million (ppm) for each microprobe analysis point are presented in Table S1 in the Supplement.

4   Results

4.1   Microscopic description

The generally restitic mineralogy of the sample is composed of garnet, biotite and sillimanite with minor amounts of cordierite, ilmenite, rutile, K-feldspar and quartz (Fig. 1c–f). The primary mineralogy indicates prograde–peak mineral paragenesis, which consists of biotite, sillimanite and garnet. The foliation is formed by a fabric of elongated garnet and sillimanite crystals 0.5–1 mm in length that compose 80–90 % of the sample (Fig. 1c–d). The stretching lineation is formed by elongated biotite crystals. Biotite contains numerous micrometer-sized apatite needles and is mostly replaced by chlorite. Primary metamorphic fabric is crosscut by several generations of veins and/or fractures (Fig. 2c–f), which were formed during cataclastic deformation and shear zone development (e.g., Siegesmund et al., 2008). Fractures are filled with post-peak and late hydration mineral assemblages. The earlier generation of veins is mostly composed of Fe–Ti oxides and their intergrowths with quartz (Fig. 1c–d, black material). Fe–Ti oxides form aggregates with lobate boundaries with the primary minerals (garnet and sillimanite) (Fig. 1e–f). The network of veins of the later generation cross-cuts the veins of the earlier generation or follows their contacts. These later veins are more abundant than earlier ones and are composed of fine-grained phyllosilicates, such as chlorite, muscovite and/or phengite, and may also contain K-feldspar patches in the vein cores (Fig. 1c–d, grey-brown material; Table 1). Abundance of phyllosilicates indicates post-metamorphic hydration reactions. Large (up to 2 mm in length and 0.3 mm thick) elongate quartz aggregates generally follow the vein distribution (Fig. 1d, bottom part).

Table 1. Results of the microprobe analyses of the rock-forming silicates; n.d.: not detected. Grt: garnet, Bt: biotite, Chl: chlorite, Phyl: phyllosilicate(s), Pheng: phengite, and Mus: muscovite.

Mineral	SiO$_2$	TiO$_2$	Al$_2$O$_3$	MgO	CaO	MnO	FeO	BaO	Na$_2$O	K$_2$O
Grt core	38.90	0.02	22.22	9.33	1.51	0.31	28.60	n.d.	n.d.	n.d.
Grt core	38.14	0.01	21.69	6.19	1.46	0.43	32.92	n.d.	n.d.	n.d.
Grt core	38.16	0.01	21.70	6.37	1.46	0.49	32.51	n.d.	0.01	n.d.
Grt core	39.01	n.d.	22.15	9.19	1.51	0.36	28.52	n.d.	0.02	n.d.
Grt core	39.10	0.02	22.09	9.26	1.50	0.32	28.80	n.d.	0.02	n.d.
Grt core	38.92	0.03	22.11	9.13	1.46	0.30	28.84	n.d.	0.03	n.d.
Grt core	38.83	0.01	22.00	8.34	1.41	0.35	29.92	n.d.	0.03	n.d.
Grt core	37.80	0.02	22.03	6.75	1.49	0.55	31.69	n.d.	0.01	n.d.
Grt rim	38.81	0.01	22.12	8.58	1.54	0.34	29.44	n.d.	0.01	n.d.
Grt rim	38.81	0.03	22.06	8.28	1.53	0.34	30.31	n.d.	0.03	n.d.
Grt rim	38.95	0.01	21.82	7.73	1.52	0.40	30.77	n.d.	n.d.	n.d.
Grt rim	38.35	0.01	21.95	7.77	1.51	0.42	30.35	n.d.	0.02	n.d.
Grt rim	38.61	0.01	21.59	5.67	1.46	0.52	33.25	n.d.	0.01	n.d.
Grt rim	38.39	0.07	21.78	5.96	1.46	0.53	32.82	n.d.	n.d.	n.d.
Grt rim	37.93	0.21	21.49	5.65	1.52	0.55	32.93	n.d.	0.02	n.d.
Grt rim	38.30	0.07	21.86	6.82	1.45	0.43	31.69	n.d.	n.d.	n.d.
Grt rim	38.34	0.04	21.75	7.15	1.48	0.43	31.37	n.d.	0.01	n.d.
Grt rim	37.35	0.01	21.12	3.85	1.50	0.87	35.51	n.d.	n.d.	n.d.
Grt rim	37.38	0.03	21.12	3.86	1.45	0.81	35.22	n.d.	n.d.	n.d.
Grt rim	37.65	n.d.	21.30	3.98	1.43	0.78	35.34	n.d.	n.d.	n.d.
Bt core	37.15	6.27	15.51	15.79	0.03	0.03	10.75	n.d.	0.11	9.83
Bt core	37.15	1.32	16.53	18.79	0.13	0.03	11.36	n.d.	0.10	9.16
Bt rim	34.81	1.74	17.17	14.41	0.04	n.d.	17.50	n.d.	0.12	7.83
Bt rim	35.22	2.82	16.59	13.55	0.02	0.04	18.27	n.d.	0.10	8.15
Bt rim	35.61	1.46	16.14	14.38	0.16	0.02	17.94	n.d.	0.09	7.38
Bt rim	34.75	1.27	17.16	13.37	0.03	0.04	19.60	n.d.	0.08	7.27
Chl over Bt	29.75	0.03	18.37	16.40	0.10	0.05	22.91	n.d.	n.d.	0.14
Chl over Bt	27.72	0.01	19.47	15.12	0.08	0.06	25.34	n.d.	n.d.	0.05
Chl over Bt	27.29	0.21	19.76	15.84	0.03	0.07	24.21	n.d.	n.d.	0.03
Chl over Bt	27.64	0.28	20.07	16.34	0.04	0.04	23.41	0.04	n.d.	0.04
Chl over Bt	28.09	0.40	19.07	15.97	0.06	0.06	24.07	0.04	n.d.	0.20
Chl new	27.14	0.02	21.48	14.10	0.03	0.09	24.90	n.d.	n.d.	0.07
Chl new	24.84	0.03	22.94	12.98	0.05	0.06	26.44	n.d.	n.d.	0.06
Chl new	28.47	0.02	18.16	15.53	0.06	0.05	24.92	n.d.	0.01	0.06
Chl new	29.00	0.08	18.92	15.73	0.09	0.06	24.07	n.d.	0.02	0.21
Chl new	29.44	0.06	19.27	15.65	0.07	0.06	24.44	n.d.	0.02	0.26
Chl new	28.70	0.02	18.38	15.42	0.08	0.07	25.45	0.05	0.04	0.09
Chl new	29.30	0.04	17.69	15.27	0.08	0.06	25.39	0.02	0.01	0.10
Chl new	31.95	4.44	17.50	11.51	0.04	0.05	21.81	0.04	0.04	4.64
Phyl matrix	44.34	n.d.	31.75	3.84	0.13	n.d.	6.06	0.28	0.33	7.70
Phyl matrix	49.04	0.01	32.24	1.68	0.04	n.d.	1.63	0.17	0.18	10.58
Phyl matrix	45.92	0.01	32.36	2.78	0.12	n.d.	4.12	0.47	0.32	8.63
Phyl matrix	43.30	0.18	32.40	4.12	0.15	0.05	7.16	n.d.	0.33	7.12
Phyl matrix	47.74	0.12	33.88	1.98	0.17	0.01	3.05	n.d.	0.42	8.67
Phyl matrix	44.89	0.21	32.84	3.28	0.10	n.d.	4.87	n.d.	0.37	8.39
Phyl matrix	44.91	0.02	32.38	3.38	0.17	0.02	5.17	n.d.	0.38	7.81
Phyl matrix	38.78	0.05	28.79	6.94	0.10	0.02	12.02	0.19	0.27	4.91
Phyl matrix	45.22	0.03	32.53	3.29	0.09	0.03	4.49	0.73	0.37	8.28
Pheng/Mus	47.45	0.07	35.68	1.31	0.09	n.d.	1.65	0.32	0.40	9.53
Pheng/Mus	48.41	0.17	35.13	1.34	0.14	n.d.	1.46	0.37	0.43	9.19
Pheng/Mus	46.62	0.12	35.34	1.43	0.10	0.01	1.94	0.36	0.45	9.07

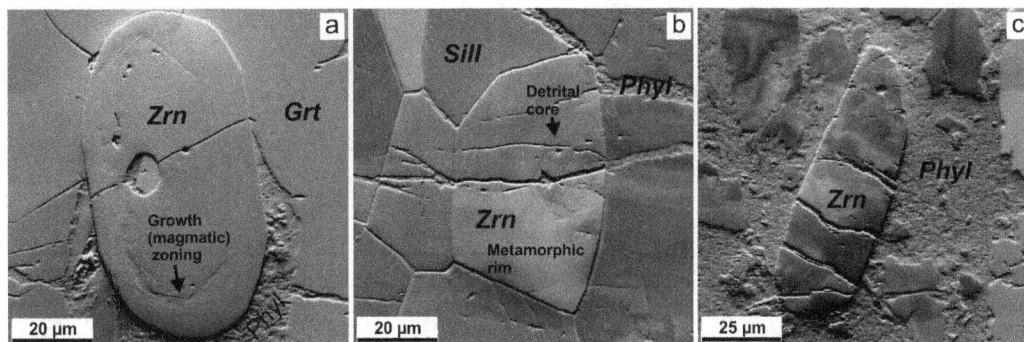

Figure 2. Orientation contrast images of detrital zircon in the sampled gneiss: **(a)** zircon grain hosted by garnet; note the concentric growth zoning. **(b)** Zircon grain hosted by sillimanite, note the small detrital core (right hand side) and wide metamorphic rim. **(c)** Zircon grain hosted by a fine-grained matrix that fills the veins; note intensive change in orientation contrast, especially conspicuous in the upper part of the grain. Orientation contrast image indicates the crystal-plastic deformation of the zircon grain and surrounding mineral fragments (garnet and sillimanite). Fracture surfaces appear to be dissolved. Mineral abbreviations as in Fig. 1.

Table 2. Results of the microprobe analyses of the Fe–Ti oxides, n.d.: not detected. Ilm: ilmenite and Rut: rutile.

Mineral	Ta$_2$O$_5$	SiO$_2$	TiO$_2$	Al$_2$O$_3$	Cr$_2$O$_3$	Nb$_2$O$_3$	MgO	MnO	FeO	NiO	Total
Ilm	0.04	0.03	52.82	0.02	n.d.	0.11	0.05	0.85	45.18	n.d.	99.15
Ilm	0.03	0.14	53.07	0.12	0.02	0.12	0.07	0.85	44.30	0.02	98.73
Ilm	n.d.	0.15	53.72	0.12	n.d.	0.04	0.03	0.89	43.56	n.d.	98.52
Rut	n.d.	0.60	98.31	0.22	0.08	0.25	0.04	n.d.	0.44	n.d.	99.92
Rut	n.d.	0.80	98.46	0.26	0.08	0.40	0.01	n.d.	0.40	n.d.	100.42
Rut	n.d.	2.10	93.19	1.42	0.06	0.22	0.60	n.d.	1.46	n.d.	99.03

Veins and fractures form a conjugated orthogonal network, stretching in at least two directions in a 2-D section.

Accessory minerals are zircon and monazite (e.g., Figs. 1f, 2). Where hosted by garnet, zircon forms roundish elongated crystals with aspect ratios from 1 : 1 to 1 : 3 and lengths from 30 to 100 μm (Fig. 2a). Where forming intergrowths with sillimanite, zircon reveals well-developed faces and forms triple junctions with the adjacent sillimanite grains (Fig. 2b), reflecting equilibration growth with sillimanite during prograde and peak metamorphism. Where hosted by fine-grained material that fills fractures, zircon crystals are elongated, with an aspect ratio from 1 : 2 to 1 : 3; these grains are fractured and fragmented. The fragments have irregular dissolved boundaries and show evidence of crystal-plastic deformation (Fig. 2c). Vermicular- and hat-shaped zircon aggregates and zircon coronae are spatially associated with each other and occur within ilmenite–rutile–quartz or rutile–quartz clusters and/or intergrowths (Figs. 3–5), which fill transgranular fractures and pockets in the gneiss (e.g., Fig. 4a).

4.2   Zircon microstructures and textures

Zircon textures reported in this study are coronae, by which we mean thin envelopes or shells in 3-D. Accordingly, in the 2-D plane of a sample they have thread- or worm-like shapes (depending on the thickness and aspect ratio). Zircon coronae in our sample occur as two main textural types. The first type is referred to as vermicular-shaped (coarser-grained) aggregates, which have a thickness ≥ 5 μm and an aspect ratio of 1 : 4 to 1 : 20. The second type is referred to as coronae (finer-grained) zircon aggregates, which have a thickness ≤ 1 μm and an aspect ratio of approximately 1 : 100 (e.g., Figs. 3–4). There is also a third (subordinate) coronae type: hat-shaped aggregates that are the result of zircon coronae overgrowth preexisting (probably detrital) grains.

4.2.1   Vermicular textures

This textural type occurs as lamellae-like intergrowths with rutile. Some vermicular-shaped zircon grains are hosted by thin rutile-quartz intergrowths, in which rutile forms < 1 μm thin and 1–3 μm long needles (Fig. 3b, matrix). Such needle shapes are evidence of rapid rutile recrystallization and re-equilibration during the metamorphic evolution.

Figure 3a shows a zircon aggregate composed of three large vermicular-shaped grains (indicated by V, enlarged in Fig. 3b, c, e). These vermicular grains are 5 to 15 μm thick and 20 to 50 μm long and have diffuse or auroral-light (Corfu et al., 2003) CL zoning (Fig. 3b, c, e). Vermicular-shaped grains have curved (Fig. 3b) or ragged (Fig. 3e) boundaries, a crescent-like shape (Fig. 3c, e) and are commonly broken

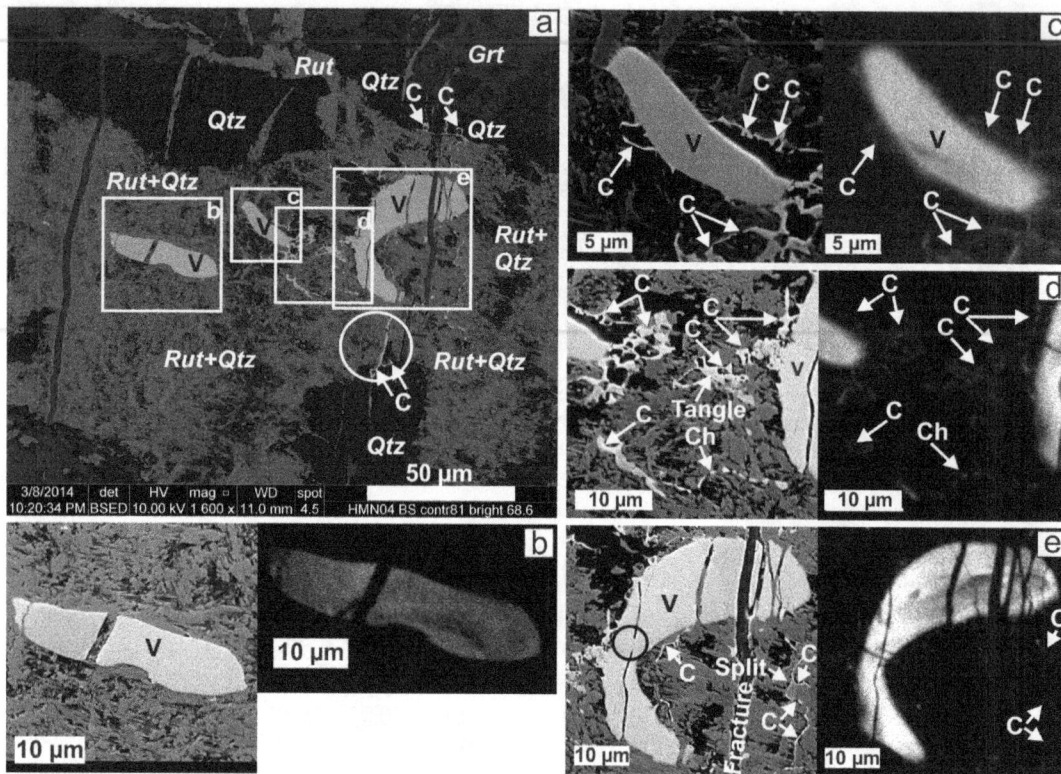

Figure 3. (a) BSE image of the vermicular-shaped zircon aggregates ("V") and zircon coronae ("C"). Arrows indicate zircon coronae that trace the quartz–rutile boundary or fill the cavities in quartz; the circle highlights a corona that fills the fracture. Mineral abbreviations as in Fig. 1. **(c–e)** Enlarged BSE (left) and CL (right) images of the areas indicated in panel **(a)**. V highlights the vermicular-shaped zircon grains and C points to zircon coronae (the difference between V and C is in thickness). "Ch" in panel **(d)** points to the chain of submicron-sized zircon grains, and "Tangle" points to the tangled occurrence of coronae. Arrows in panel **(e)** indicate the directions of the reaction fronts, and "Split" points to the branching of zircon coronae. The circle in panel **(e)** highlights a partially healed fracture in vermicular zircon.

with transgranular fractures (Fig. 3b, e). Some of these fractures are traced in the host rock and filled with fine-grained phyllosilicates (Fig. 3e), which suggests that vermicular zircon predates the cataclasis and hydration with phyllosilicate growth. Furthermore, some fractures in vermicular grains (Fig. 3e) are partially healed by low-CL zircon material. This indicates that some precipitation of zircon has occurred after fracturing.

Another example of vermicular-shaped zircon aggregate is presented in Fig. 4. This texture is found in a vein of the early generation filled by the rutile–quartz intergrowths and elongate aggregates of quartz (Figs. 1d, 4a). A large zircon aggregate has a W shape and consists of two major fragments (Fig. 4b). The thickness of the vermicular zircon varies from 5 to 20 µm, and the total length is about 200 µm. The W-shaped vermicular aggregate shows diffuse CL zoning (Fig. 4d). The lower part of this aggregate used to extend to the right (Fig. 4c) and connect with the smaller vermicular grain at the right-hand side from the W-shaped grain (Fig. 4a). This 50 µm long extension was removed by subsequent polishing. The lower right tip of the aggregate drops below the surface plane of the thin section. The CL image

and the EDX map of Zr distribution reveal the blurred trace around the lower right tip (Fig. 4d–e, grey arrows). This indicates that the zircon aggregate continues deeper into the sample at a shallow angle, and its signal is documented by CL and EDX from a few micrometers below the surface. As such, the aggregate represents an envelope in 3-D. The W-shaped zircon grain is plastically deformed in its central part, which is indicated by an orientation contrast image (Fig. 4f). Rotation of the lattice reaches 7° with respect to the undeformed lattice (Kovaleva et al., 2016), indicating that this vermicular zircon grain predated shearing and ductile deformation.

Vermicular aggregates presented in Figs. 3 and 4 are associated with coronae textures, unlike aggregates in Fig. 5a. These aggregates are fractured and hosted by a rutile–phyllosilicate aggregate, which fills the pocket between sillimanite and garnet (V in Fig. 5a).

4.2.2 Coronae textures

The matrix around some vermicular grains (Figs. 3c–e, 4c) contains abundant and continuous thin zircon coronae (C in Figs. 3a, c–e, 4a–e, 5) and fine-grained zircon chains (Ch in

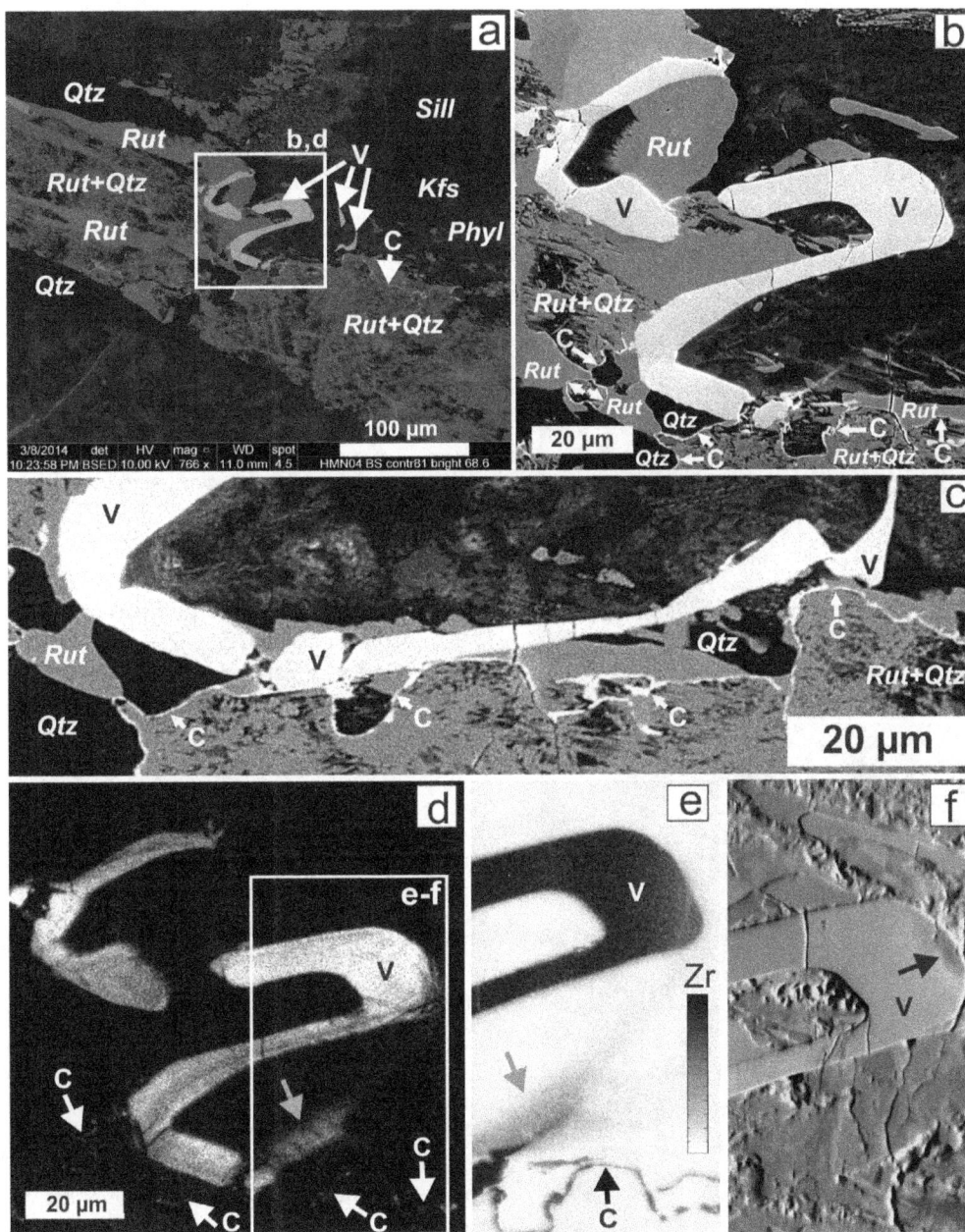

Figure 4. (a) BSE image of the zircon aggregate, which forms intergrowth with rutile and quartz. White rectangle highlights the area enlarged in panels **(b)** and **(d)**. **(b)** Enlarged BSE image of the area indicated in panel **(a)**. Arrows point out the direction of reaction fronts. **(c)** Enlarged area of the lower part of panel **(b)**. The middle part of vermicular texture ("V") is present (it was subsequently polished away and thus absent in panels **(a–b)**. Arrows point out the direction of the reaction front. **(d)** Enlarged CL image of the area indicated in panel **(a)**. White arrows as in panel **(b)**; grey arrow points to the wedged zircon texture that continues below the surface. **(e)** Qualitative EDX intensity map for Zr of the area indicated in panel **(d)**. Black arrows point out the direction of reaction front; grey arrow as in panel **(d)**. **(f)** Orientation contrast image of the area indicated in panel **(d)**. Arrow points to plastically deformed tip of vermicular zircon grain. V in panels **(a–f)** highlights vermicular zircon grains and C highlights zircon coronae. Mineral abbreviations as in Fig. 1; Kfs: K-feldspar.

Figs. 3d, 4e). Coronae are ≤ 1 µm thick and are up to 200 µm long. They can be observed anywhere around and within rutile–quartz–phyllosilicates, ilmenite–rutile and ilmenite–rutile–quartz aggregates (Figs. 3–5). Coronae are distributed rather randomly and are commonly found in the presence of larger zircon grains, which can be vermicular-shaped aggregates (e.g., Figs. 3–4), detrital grains (e.g., Fig. 5a, b, d) or peak metamorphic grains (e.g., Fig. 5c). At the same time, not all rutile–quartz intergrowths (Fig. 1e–f, 5a) and not all

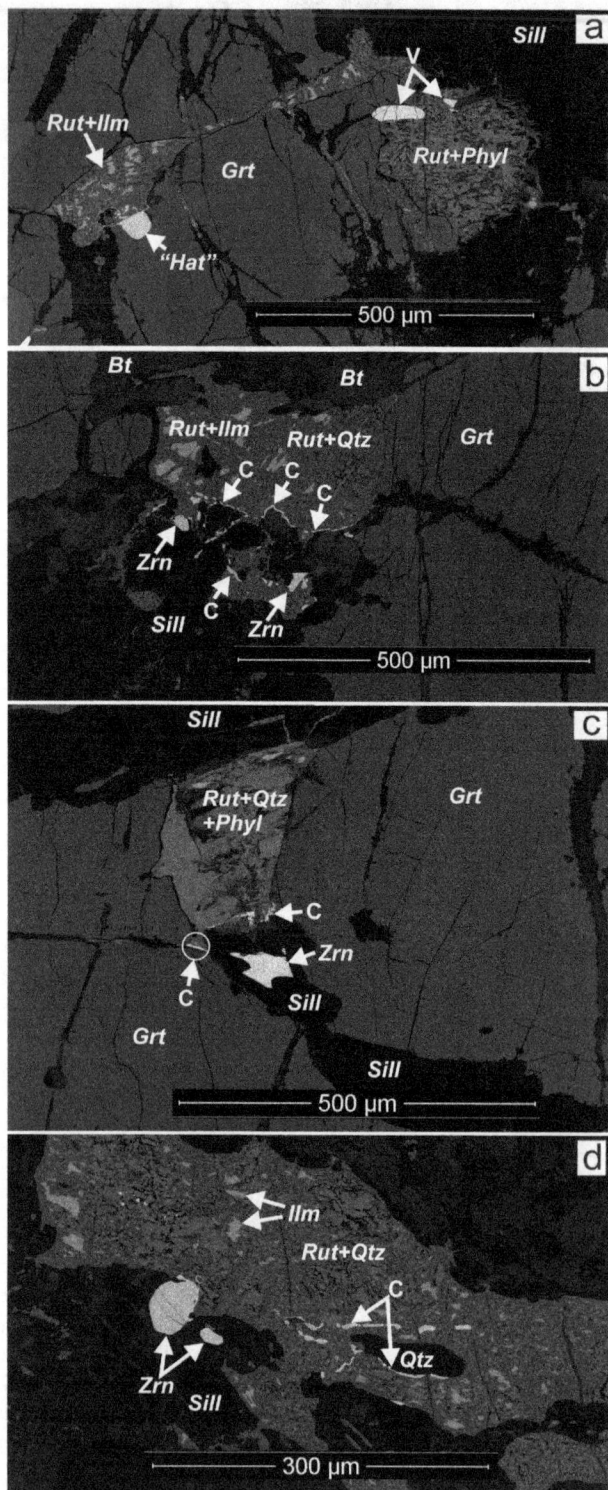

Figure 5. BSE images of mineral reactions that contain detrital zircon grains and associated zircon coronae. Mineral abbreviations are as in Fig. 1: Bt: biotite; Hat: hat-shaped zircon aggregate. "V" highlights the vermicular zircon grains, "C" points to the zircon coronae, and Zrn is the preexisting detrital and metamorphic zircon grains. Arrows in panel (b) show the direction of the mineral reaction front.

vermicular-shaped grains (V in Fig. 5a) are associated with zircon coronae.

Both zircon coronae and fine-grained chains have distinguishable CL responses (Figs. 3c–e, 4d). Coronae occur as continuous threads that form splits (Figs. 3e, 4e) or isolated tangles (Fig. 3d). An especially dense network of zircon coronae is presented in Fig. 3d. Coronae are commonly attached to the larger vermicular zircon grains (Figs. 3c–e, 4b–c). Some coronae follow the phase boundaries between rutile–quartz aggregates and quartz (Figs. 3a, 4c, 5d) or rutile–quartz aggregates and sillimanite (C in Fig. 5b–c). Some thin zircon coronae extend outside of rutile–quartz intergrowths and fill fractures in quartz and garnet (Figs. 3a and 5c accordingly, circles), which suggests zircon coeval-to post-fracturing precipitation.

4.2.3 "Hat" textures

The rutile–ilmenite intergrowths, adjacent to a rectangular zircon grain, fill a fracture in garnet adjacent to a rectangular zircon grain (Fig. 5a). Zircon coronae trace the boundary between garnet and rutile–ilmenite intergrowths and are connected to the rectangular zircon grain, so that the latter acquires a hat-like shape (Hat in Fig. 5a, after Bingen et al., 2001). Another example of similarly formed zircon aggregate does not have such a well-developed hat shape. It is a roundish zircon grain (Zrn in Fig. 5d), which is spatially associated with coronae and has short coronae outgrowths pointed towards the rutile–quartz aggregate.

4.3 Microprobe data

Mineral electron microprobe data are presented in Tables 1 and 2. These data are used to determine the temperature of metamorphism using the garnet–biotite thermometer and to support the suggestion of possible mineral–fluid reactions. The X_{Fe} of garnet is systematically lower in the cores than in the rims and in the smaller fragments. The same applies to the biotite (Table 1). Garnet rims are also systematically enriched in Mn, compared to the cores. Compositions of Fe–Ti oxides (Table 2) demonstrate that rutile is much higher in SiO_2 content than ilmenite. Rutile is also slightly enriched in such elements as Al, Cr and Nb, but lower in Mn (Table 2) compared to ilmenite.

5 Discussion

5.1 Mineral reactions

Mineral textures in the studied sample provide important information about the reactions that could have enhanced the growth of zircon coronae during metamorphism. We suggest the following reaction sequence: (1) formation of peak metamorphic phases and partial melting of the metapelites; (2) initial resorption of peak metamorphic phases and crystallization of interstitial ilmenite with lobate boundaries in fractures; (3) retrograde metamorphism, further resorption and fracturing of the high-temperature phases, hydration reactions with formation of phyllosilicates and decomposition of ilmenite to rutile; and (4) further cooling and recrystallization of rutile.

Temperature estimations were done using the garnet–biotite thermometer using microprobe data (Table 1). Various calibrations of this geothermometer (Thompson, 1976; Holdaway and Lee, 1977; Ferry and Spear, 1978; Hodges and Spear, 1982; Perchuk and Lavrent'eva, 1983; Bhattacharya et al., 1992) gave temperatures for garnet–biotite preserved cores of 570–700 °C, for inner rims 800–860 °C, and for outer rims 820–1090 °C. Estimations were done for pressures of 0.7 and 1.0 GPa. Pressure variations did not have any significant effect on the resulting temperatures. It is, however, possible, that garnet and biotite rims were affected by diffusion from the host environment during retrograde metamorphism. Mineral textures (Figs. 1e–f, 2c, 4a, 5) and microprobe analyses (Table 1) indicate that the initial granulite-facies garnet, biotite and sillimanite were intensely altered and resorbed. The fragments of garnet and sillimanite have ragged edges and are plastically deformed, dissolved and altered. Mn and Fe, enriched in the rims of large garnet grains and in small garnet fragments, suggest garnet resorption and hydration during retrograde metamorphism (e.g., Tuccillo et al., 1990) after peak metamorphism. Thus, garnet rims do not indicate peak metamorphic temperatures; therefore, the rim temperatures are erroneous. More likely, peak metamorphic temperatures in this IVZ section were between 700 and 860 °C (temperatures obtained for the cores and inner rims), in agreement with previous estimations (Sills, 1984; lowermost estimations of Luvizotto and Zack, 2009).

Pockets and fractures in garnet and sillimanite are filled with Fe–Ti oxide aggregates with lobate boundaries. Thus, ilmenite probably crystallized from the partial melt and/or fluid after the formation of peak metamorphic phases and after their resorption was initiated. Occasionally, veins with Fe–Ti oxides are associated with the quartz aggregates; thus, quartz possibly formed in the early generation of veins together with ilmenite. Ubiquitous fracturing of the rock (e.g., Fig. 1c–d) and crystal-plastic deformation of zircon (Figs. 2c, 4f) indicate the extreme conditions of post-peak metamorphism deformation (e.g., Siegesmund et al., 2008).

Further retrograde (greenschist facies) metamorphism led to hydration reactions and formation of veins filled with phyllosilicates and K-feldspar. The following features are regarded as evidence of intensive mineral–fluid reactions in the dry restitic granulite-facies rock (e.g., Rajesh et al., 2013): reaction rims around fragments of granulite-facies minerals (e.g., Fig. 1e), fine-grained phyllosilicate mixture that fills fractures (e.g., Fig. 1c–d, f), quartz veins (Figs. 1d, 4a), alteration of biotite with chloritization and exsolution of apatite needles. Water-rich fluids could have been sourced from the decomposing biotite (e.g., Pape et al., 2016). Rare il-

menite cores are surrounded by rutile rims (Figs. 1e–f, 5a–b, d). Thus, post-peak, trace-element-rich ilmenite was partially or entirely decomposed to rutile, which resulted in the migration of excess Fe into the matrix and into the garnet and biotite rims. Fe from ilmenite and Mg diffusing out of garnet and biotite rims are needed to compensate for the formation of the large volume of Mg–Fe phyllosilicates in the second generation of veins (Figs. 1c–d, 2c; Table 1). K from biotite and Al from sillimanite would allow and/or favor the growth of K-feldspar in the veins (e.g., Fig. 4a). Excess Hf and Zr from breaking down ilmenite are responsible for the formation of zircon intergrowths with rutile (thick vermicular-shaped grains, e.g., Figs. 3b, c, e, 4a–c). Newly formed rutile is enriched in trace elements, possibly due to the decreased volume of Fe–Ti oxides (e.g., Austrheim et al., 2008). This rutile is also enriched in silica (Table 2) that requires sourcing SiO_2 from the environment and may indicate solid solution of SiO_2 in rutile (Taylor-Jones and Powell, 2015), which would play a role in a further reaction. Excess Si, possibly derived from the fragmentation and dissolution of garnet and sillimanite, would form quartz veins and react with Zr to form zircon coronae. As for the apatite formation, P, F, Cl and OH could be derived either from decomposed biotite or were delivered by the water-rich fluid as components of a water brine from, for example, dissolution of monazite. However, the occurrence of apatite needles inside altered biotite grains points to genetic relationships between these two minerals.

Further cooling caused nonequilibrium recrystallization of rutile. Ewing et al. (2013) described partial replacement of rutile by other phases, characteristic for all granulitic metapelites from the IVZ. In our sample, we observe recrystallization of rutile with the formation of fine rutile–quartz intergrowths and thin zircon coronae around them. This took place during later stages of the rock evolution, when the temperature decreased and caused the Zr solubility in rutile to decrease. Therefore, we suggest the following stylized reaction (R1):

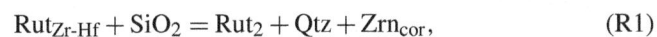

$$Rut_{Zr-Hf} + SiO_2 = Rut_2 + Qtz + Zrn_{cor}, \qquad (R1)$$

in which Rut_{Zr-Hf} ($= SiO_2$-, Zr- and Hf-bearing rutile) resulted from decomposition of ilmenite and Rut_2 ($= SiO_2$-, Zr- and Hf- depleted rutile) forms intergrowth with quartz (Qtz) and zircon coronae (Zrn_{cor}). Quartz was exsolved from SiO_2-rich rutile and formed thin intergrowths with the newly crystallized, trace-element-depleted rutile (e.g., Figs. 3, 4c). Zircon coronae produced in this way are very thin, suggesting slow reaction rates and a limited solubility of Zr in rutile at a low temperature (Tomkins et al., 2007; Ewing et al., 2013, 2014; Pape et al., 2016).

5.2 Zircon textures

Zircon grains hosted by garnet, sillimanite and fine-grained phyllosilicate matrix (Fig. 2a–c accordingly) represent detrital grains, enclosed within the main mineral phases during metamorphism. Zircon in garnet shows euhedral shapes and concentric growth zoning (Fig. 2a), indicating capture of detrital grains by metamorphic garnet. Zircon enclosed in sillimanite has detrital cores that are overgrown by metamorphic rims, which are in growth equilibrium with sillimanite (Figs. 2b, 5c). After the peak metamorphic conditions these detrital zircon grains seem to have been mostly inert and are therefore well-preserved. Zircon grains hosted by the fine-grained phyllosilicate matrix in hydration veins are the most deformed and fractured and show dissolved and/or corroded surfaces (Fig. 2c). These latter grains were probably exposed to the post-peak metamorphic fluids. The dissolved material from their surfaces might have been transported with a fluid and serve as a source for the zircon coronae precipitation. We suggest, however, that this was not the main source of Zr for coronae zircon and that coronae mainly precipitated from Fe–Ti oxides.

Vermicular-shaped aggregates of zircon (or thick coronae) and thin coronae (Figs. 3–5) have a different origin than the detrital grains (Fig. 2). It has been shown that zircon can grow from other Zr-bearing phases as a result of mineral reactions and as a mineral response to the changing conditions (Bingen et al., 2001; Söderlund et al., 2004; Austrheim et al., 2008; Ewing et al., 2013, 2014; Kohn et al., 2015; Pape et al., 2016). Metamorphic (coronae) zircon in granulite-facies rocks may not be a product of peak metamorphism, but precipitate during the retrograde evolution (Tomkins et al., 2007). Zircon coronae textures are the evidence of zircon formation due to breakdown of Zr-bearing Fe–Ti oxides (e.g., Davidson and van Breemen, 1988; Fraser et al., 2004; Degeling et al., 2001). However, taking into account complexity of the textures and the fact that they were formed in more than one stage, we do not entirely exclude the possibility of exsolution of zircon from, for example, Zr-bearing rutile or exsolution from ilmenite before its breakdown (e.g., Bingen et al., 2001; Tomkins et al., 2007; Ewing et al., 2013; Pape et al., 2016). The possibility of the two-stage exsolution of Zr from Fe–Ti oxides was suggested by Ewing et al., (2013). Our textural observations are consistent with this idea. The sketch in Fig. 6 shows stages (1) and (2) of zircon coronae formation:

1. After the peak metamorphic conditions, ilmenite was the main host phase for Zr, together with the primary detrital zircon (Bingen et al., 2001). At the initial cooling stage it partially decomposed to rutile (Ewing et al., 2013). The expelled Zr was not entirely incorporated into the growing rutile and precipitated as new zircon (Fig. 6). Formation of zircon vermicular aggregates preceded brittle and ductile deformation of the rock. Vermicular grains in 3-D volume represent curved envelope-type aggregates (Fig. 4b–e), thus resembling coronae in shape (e.g., Bingen et al., 2001). However, they are thicker than what was previously observed

for zircon. Therefore, we interpret vermicular grains as evolved coronae. The thickness of these coronae should be controlled by reaction and cooling rates (Kohn et al., 2015). At comparatively high temperatures and slow reaction rates, zircon coronae grew thick, and formed lamellae-like intergrowths with the newly forming rutile (e.g., Fig. 4a–b). Formation of similar exsolution lamellae was described for many metamorphic minerals (e.g., Zhang and Liou, 2000).

2. In contrast with the thick coronae, formation of thin zircon coronae during Reaction (R1) occurred at lower temperatures, simultaneous with or soon after fracturing, as some of these coronae fill fractures (Figs. 3a, e, 5c). Fracture filling also indicates local Zr mobility, aided by fluid. At lower temperatures rutile recrystallizes and progressively incorporates less Zr (Ewing et al., 2013) than the high-temperature rutile, according to Zr-in-rutile thermometer models (Watson et al., 2006; Ferry and Watson, 2007). Thus, the excess Zr in the cooling system should be hosted by other Zr-bearing phases, most commonly by zircon (e.g., Pape et al., 2016). Crystallization of thin ($\leq 1\,\mu m$) zircon coronae and thin needle-shaped 1–$3\,\mu m$ long rutile grains indicates rapid cooling resulting in non-equilibrium recrystallization of Zr-bearing rutile, when Zr and Hf were expelled from the host grain (Ewing et al., 2013). This occurred after the initial cooling during the exhumation stage (Ewing et al., 2014; Kohn et al., 2015). The rutile grains that did not recrystallize usually occur in intimate contact with the rutile–quartz intergrowths and are separated from them by zircon coronae (e.g., Fig. 4c). Thus, it is possible to indicate the direction of the recrystallization front. Earlier zircon grains serve as the nucleation spots for thin zircon coronae, which is similar to the low-temperature textures described by Rasmussen (2005).

Zircon coronae in our sample are different from those described in Bingen et al. (2001), Charlier et al. (2007), and Morisset and Scoates (2008), who only observed coronae at the boundary of the (former) ilmenite grains. The described textures are also different from the coronae reported by Austrheim et al. (2008) and Pape et al. (2016), in which zircon forms continuous chains or closed contours of small grains. However, zircon coronae in all cases (described in earlier literature and here) represent 3-D shells and/or envelopes around the reacting grains (Bingen et al., 2001). Textures, indicating reaction fronts of rutile recrystallization, have not been found by Pape et al. (2016), even though these authors searched these features. In contrast, in our sample we observe former reaction fronts formed by tangled and split zircon coronae within recrystallized rutile aggregates. Split coronae may show different reaction fronts converging to one point (Figs. 3e, 4e). The reaction fronts moved from rutile–quartz intergrowths towards unreacted rutile, forming rutile–quartz

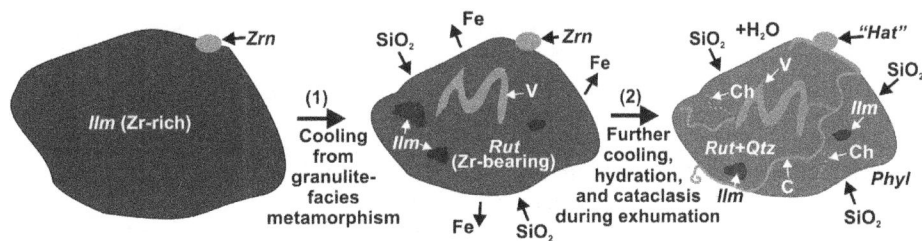

Figure 6. Sketch of the formation stages of zircon coronae. Post-peak metamorphic Zr-rich ilmenite fills the pocket between peak metamorphic minerals and has lobate boundaries. During the initial retrograde cooling (1) it decomposes to Zr- and SiO_2-bearing rutile and vermicular-shaped zircon aggregates. The system loses Fe and requires SiO_2 from the surrounding phases; the volume of Fe–Ti oxide decreases. Further cooling, hydration and cataclasis during exhumation (2) results in recrystallization of Zr-bearing rutile to rutile–quartz intergrowths with precipitation of thin zircon coronae. At this point the reaction requires SiO_2 and aqueous fluid from the surroundings. Abbreviations as in Figs. 1, 3 and 5.

embayments in the latter, rimmed by zircon coronae (e.g., Fig. 4c, arrows show the directions of the reaction front). The chains of small zircon grains are effectively the same as zircon coronae and are similar to those described in Austrheim et al. (2008). The hat-shaped zircon grains are formed by coronae that are connected to the larger zircon grains (Fig. 5a) and thus represent aggregates formed by different zircon generations.

Not all rutile aggregates in our sample are associated with zircon coronae. Similarly, the diversity in appearance of rutile grains from the same sample was described by Pape et al. (2016) for IVZ metapelites. This can be due to (a) thin section cut that does not reveal associated coronae or (b) only local recrystallization of rutile (e.g., due to locally elevated strain or inhomogeneous distribution of fluid), so that the rest of the rutile still contains a significant amount of Zr. In the case of (b), Zr-in-rutile thermobarometry can be applied to the Zr-enriched rutile to estimate the temperature of ilmenite decomposition and coeval formation of vermicular zircon (e.g., Ewing et al., 2013).

6 Conclusions and implications

In our study, we demonstrate that zircon coronae can form within and around Fe–Ti oxides in metapelites during cooling and hydration after peak granulite-facies metamorphism. Zircon formed as a result of breakdown (exsolution) of ilmenite and rutile. Formation of zircon coronae occurred in two distinct stages and resulted in (1) thick (5–20 µm) vermicular-shaped grains presumably formed during breakdown of Zr-bearing ilmenite to Zr-bearing rutile, and in (2) thin (≤ 1 µm) corona aggregates and submicron-grain chains formed due to low-temperature recrystallization of Zr-bearing rutile (Fig. 6). Two zircon-forming episodes were separated in time and represent two evolution stages of the sampled rock and could therefore be connected with the evolution of the Ivrea–Verbano Zone on a larger scale.

We report a new textural relationship between zircon and host rutile grains, as only exsolution needles of zircon in rutile and small zircon grains framing rutile were described in metapelites before (e.g., Ewing et al., 2013; Pape et al., 2016). We describe zircon coronae in metasedimentary rocks, in contrast with the previous authors, who reported similar textures in metaigneous rocks (e.g., Bingen et al., 2001; Austrheim et al., 2008 and references therein).

The detailed study of zircon corona textures can have a significant influence on the trace element balance calculations for the bulk rock, provides a tool for the reconstruction of metamorphic mineral–fluid reactions and helps derive the direction of rutile recrystallization reaction fronts. Moreover, precipitated zircon can potentially be used in geochronology for in situ dating of metamorphic evolution stages and may yield the isotopic age of metamorphic reactions (e.g., Charlier et al., 2007; Ewing et al., 2013). The trace elements in zircon can be measured to fingerprint different fluid infiltration–recrystallization events. They can be used in thermobarometry for estimating the $P - T$ conditions of the ilmenite breakdown and formation of Zr-bearing rutile and the $P - T$ conditions of the Zr exsolution from rutile. For the latter, Zr-in-rutile, Ti-in-zircon and Si-in-rutile thermometers can be applied (e.g., Ewing et al., 2013; Pape et al., 2016).

Author contributions. EK and UK were responsible for sampling. EK performed laboratory work, SEM and EMPA analysis, data reduction and analysis, and drafted the paper. HA and UK conceptualized the study, oversaw the progression of the work and advised on interpretation.

Competing interests. The authors declare that they have no conflict of interest.

Acknowledgements. This study was funded by the University of Vienna (doctoral school "DOGMA", project IK 052). The authors acknowledge access to the laboratory for scanning electron microscopy and focused ion beam applications, Faculty of Earth Sciences, Geography and Astronomy at the University of Vienna (Austria), and specifically Gerlinde Habler, who acquired the orientation contrast images presented in this study. The authors are grateful to Rainer Abart, Claudia Beybel, Franz Biedermann, Sigrid Hrabe, Hugh Rice and all colleagues of the FOR741 research group for fruitful discussions; to the Geologische Bundesanstalt (GBA) of Austria and Christian Auer for access to the SEM; to the Department of Geology at the University of the Free State for the support in writing this paper. Comments of Nigel Kelly, Fernando Corfu and Roberto Weinberg helped to improve the text greatly.

Edited by: Roberto Weinberg

References

Austrheim, H., Putnis, C. V., Engvik, A. K., and Putnis, A.: Zircon coronae around Fe–Ti oxides: a physical reference frame for metamorphic and metasomatic reactions, Contrib. Mineral. Petr., 156, 517–527, 2008.

Barboza, S. A., Bergantz, G. W., and Brown, M.: Regional granulite facies metamorphism in the Ivrea zone: is the Mafic Complex the smoking gun or a red herring?, Geology, 27, 447–450, 1999.

Bea, F. and Montero, P.: Behavior of accessory phases and redistribution of Zr, REE, Y, Th, and U during metamorphism and partial melting of metapelites in the lower crust: An example from the Kinzigite Formation of Ivrea-Verbano, NW Italy, Geochim. Cosmochim. Acta, 63, 1133–1153, 1999.

Bhattacharya, A., Mohanty, L., Maji, A., Sen, S. K., and Raith, M.: Non-ideal mixing in the phlogopite-annite binary: constraints from experimental data on Mg-Fe partitioning and a reformulation of the biotite-garnet geothermometer, Contrib. Mineral. Petrol., 111, 87–93, 1992.

Bingen, B., Austrheim, H., and Whitehouse, M.: Ilmenite as a source for zirconium during high-grade metamorphism? Textural evidence from the Caledonides of western Norway and implications for zircon geochronology, J. Petrol., 42, 355–375, 2001.

Brodie, K. H. and Rutter, E. H.: Deep crustal extensional faulting in the Ivrea zone of northern Italy, Tectonophysics, 140, 193–212, 1987.

Brodie, K. H., Rutter, E. H., and Evans, P.: On the structure of the Ivrea-Verbano Zone (northern Italy) and its implications for present-day lower continental crust geometry, Terra Nova, 4, 34–39, 1992.

Charlier, B., Skår, Ø., Korneliussen, A., Duchesne, J.-C., and Auwera, J. V.: Ilmenite composition in the Tellnes Fe–Ti deposit, SW Norway: fractional crystallization, postcumulus evolution and ilmenite–zircon relation, Contrib. Mineral. Petr., 154, 119–134, 2007.

Corfu, F., Hanchar, J. M., Hoskin, P. W. O., and Kinny, P.: Atlas of zircon textures, in: Zircon, Reviews in Mineralogy and Geochemistry, 53, edited by: Hanchar, J. M. and Hoskin, P. W. O., Mineralogical Society of America, Washington, DC, USA, 468–500, 2003.

Davidson, A. and van Breemen, O.: Baddeleyite-zircon relationships in coronitic metagabbro, Grenville Province, Ontario: im-

plications for geochronology, Contrib. Mineral. Petr., 100, 291–299, 1988.

Degeling, H., Eggins, S., and Ellis, D. J.: Zr budgets for metamorphic reactions, and the formation of zircon from garnet breakdown, Min. Mag., 65, 749–758, 2001.

Dempster, T. J., Hay, D. C., and Bluck, B. J.: Zircon growth in slate, Geology, 32, 221–224, 2004.

Dempster, T. J., Hay, D. C., Gordon, S. H., and Kelly, N. M.: Microzircon: origin and evolution during metamorphism, J. Metamorph. Geol., 26, 499–507, 2008.

Ewing, T. A., Hermann, J., and Rubatto, D.: The robustness of the Zr-in-rutile and Ti-in-zircon thermometers during high-temperature metamorphism (Ivrea-Verbano Zone, northern Italy), Contrib. Mineral. Petr., 165, 757–779, 2013.

Ewing, T. A., Rubatto, D., and Hermann, J.: Hafnium isotopes and Zr/Hf of rutile and zircon from lower crustal metapelites (Ivrea-Verbano Zone, Italy): Implications for chemical differentiation of the crust, Earth. Planet. Sc. Lett., 389, 106–118, 2014.

Ferry, J. M. and Spear, F. S.: Experimental calibration of the partitioning of Fe and Mg between biotite and garnet, Contrib. Mineral. Petrol., 66, 113–117, 1978.

Ferry, J. M. and Watson, E. B.: New thermodynamic models and revised calibrations for the Ti-in-zircon and Zr-in-rutile thermometers, Contrib. Mineral. Petr., 154, 429–437, 2007.

Fraser, G., Ellis, D., and Eggins, S.: Zirconium abundance in granulite-facies minerals, with implications for zircon geochronology in high-grade rocks, Geology, 25, 607–610, 1997.

Fraser, G. L., Pattison, D. R. M., and Heaman, L. M.: Age of the Ballachulish and Glencoe Igneous Complexes (Scottish Highlands), and paragenesis of zircon, monazite and baddeleyite in the Ballachulish Aureole, J. Geol. Soc. London, 161, 447–462, 2004.

Harley, S. L., Kelly, N. M., and Möller, A.: Zircon behaviour and the thermal histories of mountain chains, Elements, 3, 25–30, 2007.

Harrison, M. T. and Watson, E. B.: Kinetics of zircon dissolution and zirconium diffusion in granitic melts of variable water content, Contrib. Mineral. Petr., 84, 66–72, 1983.

Hay, D. C. and Dempster, T. J.: Zircon behavior during low-temperature metamorphism, J. Petrol., 50, 571–589, 2009.

Hay, D. C., Dempster, T. J., Lee, M. R., and Brown, D. J.: Anatomy of a low temperature zircon outgrowth, Contrib. Mineral. Petr., 159, 81–92, 2010.

Hodges, K. V. and Spear, F. S.: Geothermometry, geobarometry and the Al_2SiO_5 triple point at Mt. Moosilauke, New Hampshire, Am. Mineral., 67, 1118–1134, 1982.

Holdaway, M. J. and Lee, S. M.: Fe-Mg cordierite stability in high grade pelitic rocks based on experimental, theoretical and natural observations, Contrib. Mineral. Petr., 63, 175–198, 1977.

Kelsey, D. E. and Powell, R.: Progress in linking accessory mineral growth and breakdown to major mineral evolution in metamorphic rocks: a thermodynamic approach in the Na_2O-CaO-K_2O-FeO-MgO-Al_2O_3-SiO_2-H_2O-TiO_2-ZrO_2 system, J. Metamorph. Geol., 29, 151–166, 2011.

Kelsey, D. E., Clark, C., and Hand, M.: Thermobarometric modelling of zircon and monazite growth in melt-bearing systems: examples using model metapelitic and metapsammitic granulites, J. Metamorph. Geol., 26, 199–212, 2008.

Klötzli, U. S., Sinigoi, S., Quick, J. E., Demarchi, G., Tassinari, C. C. G., Sato, K., and Günes, Z.: Duration of igneous activity in the Sesia Magmatic System and implications for high-temperature metamorphism in the Ivrea–Verbano deep crust, Lithos, 206–207, 19–33, 2014.

Kohn, M. J., Corrie, S. L., and Markley, C.: The fall and rise of metamorphic zircon, Am. Mineral., 100, 897–908, 2015.

Kovaleva, E., Klötzli, U., Habler, G., and Libowitzky, E.: Finite lattice distortion patterns in plastically deformed zircon grains, Solid Earth, 5, 1099–1122, https://doi.org/10.5194/se-5-1099-2014, 2014.

Kovaleva, E., Klötzli, U., and Habler. G.: On the geometric relationship between deformation microstructures in zircon and the kinematic framework of the shear zone, Lithos, 262, 192–212, 2016.

Lucassen, F., Dulski, P., Abart, R., Franz, G., Rhede, D., and Romer, R. L.: Redistribution of HFSE elements during rutile replacement by titanite, Contrib. Mineral. Petr., 160, 279–295, 2010.

Luvizotto, G. L. and Zack, T.: Nb and Zr behavior in rutile during high-grade metamorphism and retrogression: an example from the Ivrea-Verbano Zone, Chem. Geol., 261, 303–317, 2009.

Mohamed, F. H. and Hassanen, M. A.: Geochemical evolution of arc-related mafic plutonism in the Umm Naggat district, Eastern Desert of Egypt, J. Afr. Earth. Sci., 22. 269–283, 1996.

Möller, A., O'Brien, P. J., Kennedy, A., and Kröner, A.: Polyphase zircon in ultrahigh-temperature granulites (Rogaland, SW Norway): constraints for Pb diffusion in zircon, J. Metamorph. Geol., 20, 727–740, 2002.

Möller, A., O'Brien, P. J., Kennedy, A., and Kröner, A.: Linking growth episodes of zircon and metamorphic textures to zircon chemistry: an example from the ultrahigh-temperature granulites of Rogaland (SW Norway), in: Geochronology: Linking the isotopic record with petrology and textures, edited by: Vance, D., Müller, W., and Villa, I. M., J. Geol. Soc. London Sp. Publ., 220, 65–81, London, UK, 2003.

Morisset, C. E. and Scoates, J. S.: Origin of zircon rims around ilmenite in mafic plutonic rocks of proterozoic anorthosite suites, Can. Mineral., 46, 289–304, 2008.

Morisset, C. E., Scoates, J. S., and Weis, D.: Exsolution origin for zircon rims around hemo-ilmenite in magmatic Fe–Ti oxide deposits, Geohimica et Cosmochimica Acta, 15th Annual V. M. Goldschmidt Conference, 21–25 May 2005, Moscow, Idaho, USA, A16, 2005.

Moser, D. E., Chamberlain, K. R., Tait, K. T., Schmitt, A. K., Darling, J. R., Barker, I. R., and Hyde, B. C.: Solving the Martian meteorite age conundrum using micro-baddeleyite and launch-generated zircon, Nature, 499, 454–458, 2013.

Pape, J., Mezger, K., and Robyr, M.: A systematic evaluation of the Zr-in-rutile thermometer in ultra-high temperature (UHT) rocks, Contrib. Mineral. Petr., 171, 44, https://doi.org/10.1007/s00410-016-1254-8, 2016.

Perchuk, L. L. and Lavrent'eva, I. V.: Experimental investigation of exchange equilibria in the system cordierite-garnet-biotite, in: Kinetics and equilibrium in mineral reactions, edited by: Saxena, S. K., Springer, New York, USA, 199–239, 1983.

Quick, J. E., Sinigoi, S., Peressini, G., Demarchi, G., Wooden, J. L., and Sbisà, A.: Magmatic plumbing of a large Permian caldera exposed to a depth of 25 km, Geology, 37, 603–606, 2009.

Rajesh, H. M., Belyanin, G. A., Safonov, O. G., Kovaleva, E. I., Golunova, M. A., and Van Reenen, D. D.: Fluid-induced dehydration of the paleoarchean Sand River biotite–hornblende gneiss, Cen-tral Zone, Limpopo Complex, South Africa, J. Petrol., 54, 41–74, 2013.

Rasmussen, B.: Zircon growth in very low grade metasedimentary rocks: evidence for zirconium mobility at ∼ 250 °C, Contrib. Mineral. Petr., 150, 146–155, 2005.

Rubatto, D., Müntener, O., Barnhoorn, A., and Gregory, C.: Dissolution-reprecipitation of zircon at low-temperature, high-pressure conditions (Lanzo Massif, Italy), Am. Mineral., 93, 1519–1529, 2008.

Rutter, E. H., Brodie, K. H., James, T., and Burlini, L.: Large-scale folding in the upper part of the Ivrea-Verbano zone, NW Italy, J. Struct. Geol., 29, 1–17, 2007.

Siegesmund, S., Layer, P., Dunkl, I., Vollbrecht, A., Steenken, A., Wemmer, K., and Ahrendt, H.: Exhumation and deformation history of the lower crustal section of the Valstrona di Omegna in the Ivrea Zone, Southern Alps, Geol. Soc. London. Sp. Publ., 298, 45–68, 2008.

Sills, J. D.: Granulite facies metamorphism in the Ivrea zone, NW Italy, Schweiz, Miner. Petrog., 64, 169–191, 1984.

Sinigoi, S., Quick, J. E., Demarchi, G., and Klötzli, U.: The role of crustal fertility in the generation of large silicic magmatic systems triggered by intrusion of mantle magma in the deep crust, Contrib. Mineral. Petr., 162, 691–707, 2011.

Söderlund, P., Söderlund, U., Möller, C., Gorbatschev, R., and Rodhe, A.: Petrology and ion microprobe U-Pb chronology applied to a metabasic intrusion in southern Sweden: A study on zircon formation during metamorphism and deformation, Tectonics, 23, TC5005, https://doi.org/10.1029/2003TC001498, 2004.

Taylor-Jones, K. and Powell, R.: Interpreting zirconium-in-rutile thermometric results, J. Metamorph. Geol., 33, 115–122, 2015.

Thompson, A. B.: Mineral reactions in pelitic rocks. II. Calculation of some $P - T - X$(Fe-Mg) phase relations, Am. J. Sci., 276, 425–454, 1976.

Tomkins, H. S., Powell, R., and Ellis, D. J.: The pressure dependence of the zirconium-in-rutile thermometer, J. Metamorph. Geol., 25, 703–713, 2007.

Tromans, D.: Solubility of crystalline and metamict zircon: A thermodynamic analysis, J. Nucl. Mater., 357, 221–233, 2006.

Tuccillo, M. E., Essene, E. J., and van der Pluijm, B. A.: Growth and retrograde zoning in garnets from high-grade metapelites: Implications for pressure-temperature paths, Geology, 18, 830–842, 1990.

Vavra, G., Gebauer, D., Schmid, R., and Compston, W.: Multiple zircon growth and recrystallization during polyphase Late Carboniferous to Triassic metamorphism in granulites of the Ivrea zone (Southern Alps): An ion microprobe (SHRIMP) study, Contrib. Mineral. Petr., 122, 337–358, 1996.

Watson, E. B. and Harrison, M. T.: Zircon saturation revisited: temperature and composition effects in a variety of crustal magma types, Earth Planet. Sc. Lett., 64, 295–304, 1983.

Watson, E. B., Wark, D. A., and Thomas, J. B.: Crystallization thermometers for zircon and rutile, Contrib. Mineral. Petr., 151, 413–433, 2006.

Wu, Y.-B., Zheng, Y.-F., Zhao, Z.-F., Gong, B., Liu, X., and Wu, F.-Y.: U–Pb, Hf and O isotope evidence for two episodes of fluid-assisted zircon growth in marble-hosted eclogites from the Dabie orogeny, Geochim. Cosmochim. Acta, 70, 3743–3761, 2006.

Zanetti, A., Mazzucchelli, M., Rivalenti, G., and Vannucci, R.: The Finero phlogopite-peridotite massif: an example of subduction-related metasomatism, Contrib. Mineral. Petr., 134, 107–122, 1999.

Zhang, R. Y. and Liou, J. G.: Exsolution minerals from ultrahigh-pressure rocks, in: Ultra-high pressure metamorphism and geodynamics in collision-type orogenic belts, edited by: Ernst, W. G. and Liou, J. G., Bellwether Publisher for Geological Society of America, Columbia, 216–228, 2000.

Zhao, L., Li, T., Peng, P., Guo, J., Wang, W., Wang, H., Santosh, M., and Zhai, M.: Anatomy of zircon growth in high pressure granulites: SIMS U–Pb geochronology and Lu–Hf isotopes from the Jiaobei Terrane, eastern North China Craton, Gondwana Res., 28, 1373–1390, 2015.

Strain field evolution at the ductile-to-brittle transition

Thomas Chauve[1], Maurine Montagnat[1], Cedric Lachaud[1], David Georges[1], and Pierre Vacher[2]

[1]Université Grenoble Alpes, CNRS, IRD, G-INP, IGE, 38041 Grenoble, France

[2]Laboratoire SYMME, Université de Savoie Mont Blanc, BP 80439, 74944 Annecy le Vieux CEDEX, France

Correspondence to: M. Montagnat (maurine.montagnat@univ-grenoble-alpes.fr)

Abstract. This paper presents, for the first time, the evolution of the local heterogeneous strain field around intra-granular cracking in polycrystalline ice, at the onset of tertiary creep. Owing to the high homologous temperature conditions and relatively low compressive stress applied, stress concentration at the crack tips is relaxed by plastic mechanisms associated with dynamic recrystallization. Strain field evolution followed by digital image correlation (DIC) directly shows the redistribution of strain during crack opening, but also the redistribution driven by crack tip plasticity mechanisms and recrystallization. Associated local changes in microstructure induce modifications of the local stress field evidenced by crack closure during deformation. At the ductile-to-brittle transition in ice, micro-cracking and dynamic recrystallization mechanisms can co-exist and interact, the later being efficient to relax stress concentration at the crack tips.

1 Introduction

The evaluation and the characterization of strain heterogeneities is of primary importance in material sciences at various scales of observation. Plastic strain localization in metals plays a crucial role on the propagation of fracture and on the response to fatigue conditions, and Portevin-Le-Chatelier is a strong example of plastic strain heterogeneities development during mechanical tests in some metal alloys (see Antolovich and Armstrong, 2014, for a review). Similarly, strain heterogeneities and localization are known to strongly influence the rheological behavior of the Earth lithosphere, in particular to explain post-seismic deformation (Tommasi et al., 2009; Vauchez et al., 2012).

In the context of ice sheet flow, successive layers of ice with slightly different viscosity can experience different strain history as a result of strain localization initiated by bedrock topography (Paterson, 1994; Durand et al., 2004, 2007). Strain localization can induce flow disturbances that can mix the climatic signal and counteract the search for the oldest ice (Dahl-Jensen et al., 2013; Fischer et al., 2013). These flow disturbances can form as folding that is observed at large scale from ice-penetrating radar surveys now able to highlight deep stratigraphy (MacGregor et al., 2015; Panton and Karlsson, 2015; Bons et al., 2016), but also at smaller scales from microstructure observations (Jansen et al., 2016).

During ductile deformation of ice in natural or laboratory conditions (at high homologous temperature of $\sim 0.97\ T_m$, low strain rate of $\sim 10^{-7} \mathrm{s}^{-1}$ and low stress of 0.5–1 MPa), plastic deformation is mainly accommodated by the glide of basal dislocations (Duval et al., 1983). The resulting strongly anisotropic viscoplastic behavior of the single crystal leads to the development of strong strain heterogeneities during deformation of polycrystalline ice (Duval et al., 1983).

Strain heterogeneities evaluated during transient creep of ice were shown to reach local values higher than 10 times the macroscopic strain, and to settle into bands whose dimensions are higher than the grain size. Strain localization bands may follow grain boundaries, but they also cross entire grains, and there is no statistical link between the crystallographic orientation and the amount of local strain (Grennerat et al., 2012). These first measurements of strain localization during laboratory experiments were restricted to transient (or primary) creep conditions, in ductile conditions ($\sigma < 0.5$ MPa and $T > 0.97\ T_m$) and prior to any microstructure modification due to dynamic recrystallization.

More generally, creep of isotropic polycrystalline ice is characterized by a three-stage behavior, with a strong decrease in strain rate during primary creep, down to a minimum reached at about 1 % strain, also called secondary creep, immediately followed by a increase in strain rate to reach tertiary creep at about 10 % strain (see Jacka and Maccagnan, 1984; Duval et al., 1983, for instance).

At the onset of tertiary creep, for experiment performed at low strain rate ($< 10^{-7}\,\mathrm{s}^{-1}$) or low stress ($< 0.5\,\mathrm{MPa}$), dynamic recrystallization mechanisms occur increasingly to relax the kinematic hardening and enable for further ductile deformation to occur (Duval et al., 1983). Dynamic recrystallization leads to strong modification in microstructure and texture (Duval, 1979; Jacka and Maccagnan, 1984; Montagnat et al., 2015) through various mechanisms such as nucleation of new grains or polygonization associated with sub-grain boundaries and bulging, recently characterized by cryo-EBSD (electron backscatter diffraction; Chauve et al., 2017). While Piazolo et al. (2015) showed that sub-grain boundary formation such as kink bands could be correlated with heterogeneities of local stress (simulated with a full-field crystal plasticity code, CraFT), Chauve et al. (2015) were able to directly associate nucleation mechanisms (polygonization, bulging) with local modification of the strain field estimated in situ from digital image correlation (DIC) measurements.

During experiments performed at higher imposed stress (typically above 0.9 MPa), the increase in strain rate after secondary creep can also be associated with the occurrence of micro-cracking without a total collapse of the sample (Schulson et al., 1984; Batto and Schulson, 1993; Schulson and Duval, 2009). The local stress field is therefore relaxed by cracks opening at or close to grain boundaries, and depending on the boundary conditions, crack propagation can occur at various rates. This mechanical response is typical of a ductile-to-brittle transition (Schulson and Buck, 1995; Schulson and Duval, 2009).

In this domain, most of the studies performed so far, some of which are mentioned here, have focused on macroscopic parameters (deformation and creep curves, evaluation of the effect of temperature and grain size on strength) and optical observations of the full sample to characterize the nature of the cracks (Batto and Schulson, 1993; Iliescu and Schulson, 2004). From these observations, a theoretical framework was elaborated based on the assumption of the formation of wing cracks at the tip of initial cracks to relax the local stress field (Renshaw and Schulson, 2001). In particular, the conditions required to form these secondary cracks were shown to control the ductile-to-brittle transition under compression. More recently, Snyder et al. (2016) showed that this model was able to take into account the effect of a pre-strain, including recrystallization mechanisms, on the increase in ductile-to-brittle transition strain rate for ice.

At the ductile-to-brittle transition, mixture of creep by dislocations and cracking will occur, and it is related to the ability of the material to relax the stress accumulated at the tip of the initial cracks. For instance, Batto and Schulson (1993) showed that a small amount of creep relaxation at the crack tip could be enough to postponed the transition to brittle behavior (in time or in strain-rate level). The mechanism of relaxation of the stress produced by a crack opening in mode I through rapid multiplication of dislocations at the crack tip was pioneered by Rice and Thomson (1974) and has been reviewed by Argon (2001) for metallic materials. More recently, Martínez-Pañeda and Niordson (2016) were able to simulate the complexity of the effect of strain gradient plasticity on the level of stress at the crack tip and on crack-tip blunting. Crack-tip-initiated plasticity is a crucial mechanism to explain a ductile-like behavior at the ductile-to-brittle transition.

In the present work we use the DIC technique, already well proven on ice, to evaluate the strain field evolution during a creep experiment on ice polycrystal performed at the ductile-to-brittle transition. After a brief presentation of the experimental setup (Sect. 2), Sect. 3 will explore stress conditions during which strain-rate increase with tertiary creep results from local cracking. We will see that plasticity is strongly active at the crack tips as evidenced by the occurrence of dynamic recrystallization mechanisms. These mechanisms, by modifying the microstructure, indeed play a crucial role to reduce and redistribute the local stress concentration that appears at the crack tips during the ductile-to-brittle transition.

2 Experimental setup

Unconfined uniaxial creep tests have been carried out on polycrystalline columnar ice samples of type S2 (Ple and Meyssonnier, 1997). Parallelepipedic samples ($\sim 90 \times 90 \times 15$ mm) were built and the column axes were positioned perpendicularly to the larger surface, and to the compression axis (Fig. 1). By doing so, the samples provide a "2-D–1/2" microstructure, from which surface characterization satisfactorily reflects volume behavior. Sample microstructure and texture were measured using an automatic ice texture analyzer (AITA; Wilson et al., 2003; Peternell et al., 2011), which is an optical technique measuring the c axis (or optical axis) orientation (azimuth θ and colatitude ϕ) with a spatial resolution from 50 to 5 µm, and an angular resolution of about 3°. Although large areas can be analyzed (up to 120×120 mm), this technique requires the preparation of thin sections of ice (~ 0.3 mm thick), and is then destructive. By taking advantage of the columnar microstructure, we were able to compare pre- and post-deformation microstructures by carefully extracting thin layers of ice before and after the test (Fig. 1). Details of the procedure for sample preparation can be found in Grennerat et al. (2012) and Chauve et al. (2015).

During the experiment, DIC analyses were performed over the full surface of the samples by following the procedure adapted to ice by Grennerat et al. (2012). DIC provides in

Table 1. Characteristics of the DIC measurements.

Camera	DIC spatial resolution	DIC strain resolution		
		$\sigma_{\varepsilon_{xx}}$	$\sigma_{\varepsilon_{yy}}$	$\sigma_{\varepsilon_{xy}}$
Phase One 80 Mpx	$0.19\,\mathrm{mm\,pix}^{-1}$	4.10^{-3}	3.10^{-3}	4.10^{-3}

situ measurements of the displacement and therefore strain field on the sample surface, from the correlation of surface images of a grey-level speckle that follows the sample deformation. By taking advantage of the 2-D–1/2 configuration, we assumed the surface strain field to be as representative as possible of the volume deformation. This configuration makes it possible to compare the microstructures measured by AITA (before and after the test) to the strain field evaluated by DIC (Fig. 1).

The spatial resolution strongly depends on the quality of the speckle, the illumination and the sensitivity of the camera used. In the following experiments, we used a Phase One 80 Mpx camera, the speckle was made of shoe polish that offers good cohesion with the ice surface and good illumination was obtained thanks to two neon lamps. From this, we ended up with a spatial resolution of $0.19\,\mathrm{mm\,pix}^{-1}$, and a strain resolution between 3.10^{-3} and 4.10^{-3} for the different strain components (see Table 2).

Displacement and total strain data were extracted using the 7-D software from Vacher et al. (1999). This DIC method provides a set of displacement vectors over a given grid, defined for the DIC calculation as a function of the speckle and picture qualities (Vacher et al., 1999). From the displacement field components, the total strain components are extracted by using Green–Lagrange expression. Please note that the elastic and plastic components can not be separated, and strain field refers to the total strain field. In the case of ice, elasticity is very low and nearly isotropic, and can be neglected (Schulson and Duval, 2009). In-plane components of strain are therefore provided (ε_{xx}, ε_{yy} and ε_{xy}), from which an equivalent strain ($\varepsilon_{\mathrm{eq}} = \sqrt{\frac{2}{3}\left(\varepsilon_{xx}^2 + \varepsilon_{yy}^2 + 2\varepsilon_{xy}^2\right)}$) and principal strain components are calculated. The later will be plotted along their principal directions in the following figures.

Discontinuities such as cracks produce displacements whose translation in terms of strain is not direct but could be estimated as shown by Nguyen et al. (2011). In the present study, we simply use the direction of the principal strain components calculated around a crack to interpret the direction of the crack opening (or closing), since the displacement produced is small enough to be followed by the speckle on each side of the crack.

Since all surfaces except the loaded ones remained free (unconfined tests), a limited amount of out-of-plane shear cannot be excluded. The effect of a deformation going out of the plane xOy was estimated in previous analyses performed by Grennerat et al. (2012) and shown to remain low, within

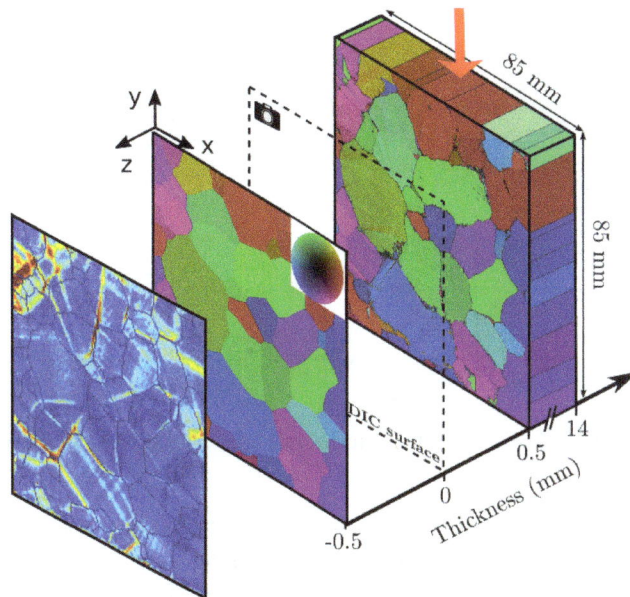

Figure 1. Scheme of the experimental setup showing the shape of the sample with the direction of imposed stress (red arrow). Position 0 corresponds to the sample surface (during the test) on top of which the speckle is marked. The microstructure analyzed by AITA prior to deformation is located at about $-0.5\,\mathrm{mm}$ and the one after deformation is at about $0.5\,\mathrm{mm}$ from the sample surface ($0.5\,\mathrm{mm}$ corresponds to the ice thickness needed to make the thin section). The strain-field image, measured at position 0, is added in the front plan for illustration.

the limit of the small macroscopic deformations reached in the present study (less than 5.5 %). In order to reduce the noise and this out-of-plane strain effect on the evaluation of the strain evolution during the experiment, we calculated the strain field during short increments of macroscopic deformation of 0.1 to 0.5 %. Additionally, observation of the incremental strain field enables individualizing consecutive events that would be hidden in a strain field calculation integrating the whole experiment duration.

Table 2 summarizes the experimental conditions of the test used as an illustration in this paper, and Fig. 2 provides the creep curves. The minimum strain rate is reached at about 0.5 % of compressive macro-strain, slightly before the standard 1 % value. This can be attributed to a microstructure effect since our 2-D–1/2 samples contain only few grains and do not form good representative volume elements. In the following, a negative sign will be given to the compressive strain, at the macroscopic and local scale.

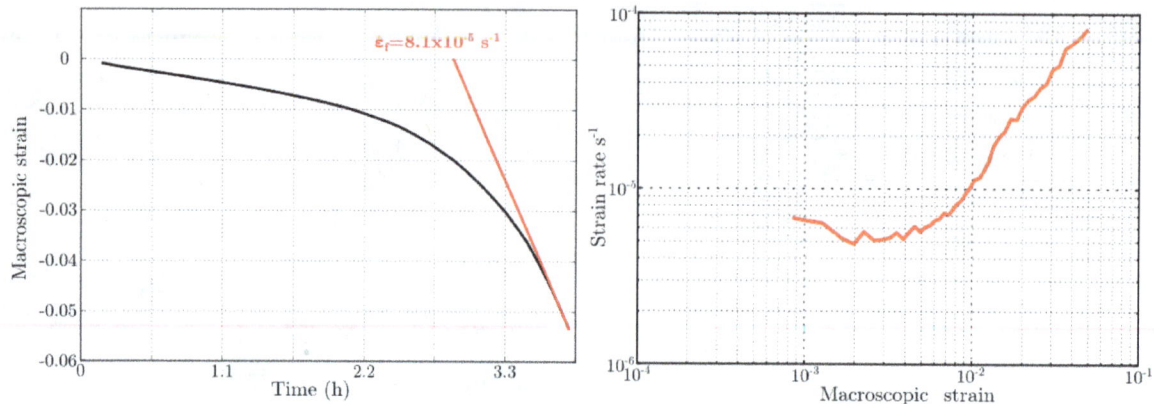

Figure 2. Evolution of the macroscopic strain and strain rate measured by DIC. Values less than 10^{-3} macro-strain were not calculated.

Table 2. Experimental conditions at the ductile-to-brittle transition for the illustrative test presented here. The minimum creep rate is reached at about -0.5% strain.

Stress	Temperature	Strain rate (s^{-1})	
(MPa)	(°C)	mini ($\varepsilon_{yy} = -0.5\%$)	end ($\varepsilon_{yy} = -5.5\%$)
1.0	-7	5.0×10^{-6}	8.1×10^{-5}

3 Strain field evolution at the ductile-to-brittle transition

The macroscopic strain curve reveals an increase in strain rate after -0.5% of ε_{yy} (vertical) macro-strain (Fig. 2). At -0.5% of macro-strain the minimum strain rate is $5.0 \times 10^{-6}\,s^{-1}$ and at the end of the experiment the strain rate reaches $8.1 \times 10^{-5}\,s^{-1}$, evidencing an acceleration at the onset of tertiary creep captured here.

The initial microstructure of the sample, the ending one and an optical observation of the full sample at the end of the test are shown in Fig. 3. Thanks to the transparency of ice, cracks and de-cohesion features can be observed with natural light. They appear as grey and black areas in Fig. 3. Both features were clearly distinguished and analyzed by Weiss and Schulson (2000). From the c axis orientation color scale, one can see that the initial texture is not isotropic. On top of the expected columnar grain-shape effect, we therefore expect, as observed (Fig. 2), a macroscopic mechanical response different from the one of an isotropic granular sample.

The global strain field measured prior to any visible crack opening on the speckle, at -0.5% of macro-strain (at the minimum creep rate), is represented in Fig. 4 via the equivalent strain ε_{eq} at two different spatial resolutions in order to illustrate the structure of strain heterogeneities. Similarly to what was already observed by Grennerat et al. (2012), the deformation is organized into bands crossing most of the sample. The main orientation of the bands is about 20 to 30° E from the compression direction. Local equivalent strain am-

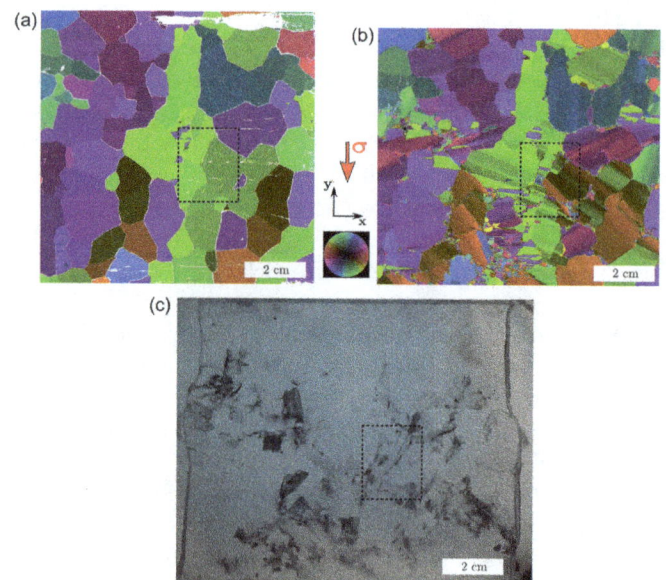

Figure 3. Microstructure (color-coded c axis orientation, from AITA analysis) before deformation **(a)** and after -5.5% of compressive creep at $-7\,°C$ under $1\,MPa$ **(b)**. **(c)** Raw picture of the sample taken in natural light at the end of the compressive test. Black areas result from light diffusion by cracks and de-cohesion features. The dashed black rectangle shows the area studied in detail in the paper.

plitude in the deformation bands can reach more than 10%, for a ε_{yy} macro-strain of about -0.5%.

In the following, focus will be given to a small area located within the dashed black rectangle of Fig. 3. The initial microstructure and orientations of the grains in this area, the final microstructure where cracks, sub-grain boundaries and small nucleated grains appear and a picture of the speckle from the surface of the sample where crack locations are visible (arrows 1 to 4) are shown in Fig. 5. Very small grains visible inside the cracks are artifacts from the thin sectioning process (shaving produces small chips that fill the crack in-

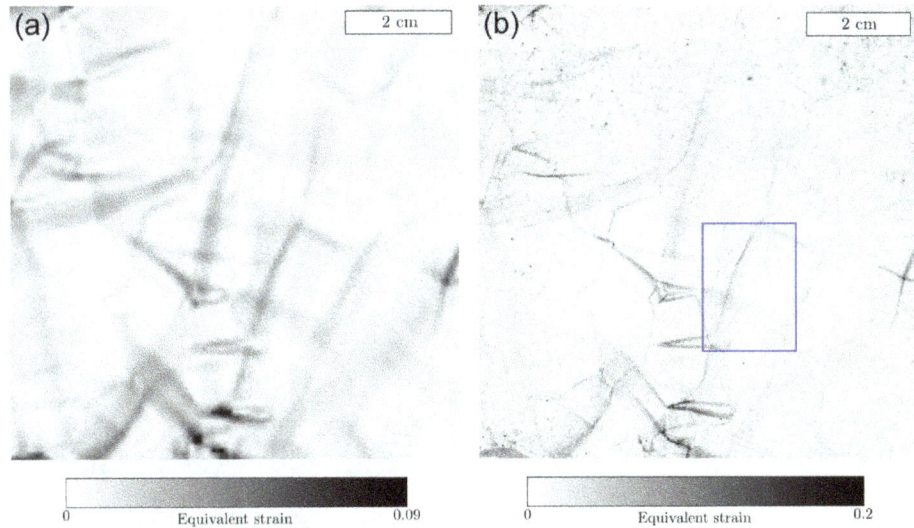

Figure 4. Map of equivalent strain (ε_{eq}) after -0.47% of compressive deformation. The blue rectangle shows the area studied in detail in the paper. **(a)** Spatial resolution of $0.76\,\mathrm{mm\,pix}^{-1}$. **(b)** Spatial resolution of $0.19\,\mathrm{mm\,pix}^{-1}$.

terior), but new grains from dynamic recrystallization mechanisms (DRX) can be distinguished away from the crack interior. See for instance the new blue area in between the bottom of crack 2 and top of crack 3. Also the dark blue grain at the bottom of crack 3, and the pink–purple ones surrounding crack 4. None of these new grains were pre-existing in the initial microstructure, therefore illustrating DRX nucleation. Grain boundaries surrounding new small grains can appear more irregular than in reality because of intrinsic limitation of the AITA observation based on thin sections (about 0.3 mm thick). A lot of sub-grain boundaries similar to the tilt and kink bands characterized in Chauve et al. (2015, 2017) are visible after deformation. A tilt band is composed of basal edge dislocations and can accommodate a large misorientation, as observed here. A kink band is composed of two nearby tilt bands that accommodate opposite misorientations. For instance, a highly misoriented tilt band is visible as a sharp transition between orange and light brown close to cracks 2 and 3 (ellipse). Between cracks 2 and 1, the two close color transitions (from green to blue to green, and from brown to green to brown) illustrate misorientations resulting from kink bands.

Intra-granular cracks (cracks 1, 3 and 4) and cracks along grain boundaries (crack 2) are observed. Observed intra-granular cracks do not always cross the entire grain, such as crack 1, which seems interrupted in the middle of the grain (Fig. 5). This final microstructure very likely results from strong strain heterogeneities at grain boundaries and within grain interiors.

In the following, we track the history of formation of the four cracks labeled in Fig. 3 by analyzing the strain field evolution through the principal strain components, such as in Chauve et al. (2015). The principal strain component rep-

resentation enables us to distinguish among the components of the local strain.

Within this area of interest, the deformation before the apparition of any crack is localized in two main bands, one crossing the full area at about 10 to 20° E from the vertical (compression) direction, and another one in the bottom part of the area, nearly perpendicular to the first one (Fig. 6, left). Both bands are following some boundaries and crossing some grains. The principal strain components are typical of a local pure shear configuration.

The accumulated strain field just before (at $\varepsilon_{yy} = -1.35\%$), and just after the opening of cracks 1, 2 and 3 (at $\varepsilon_{yy} = -1.46\%$) is shown in Fig. 6. As an illustration, the location of cracks 2 and 3 is shown by a blue ellipse, and the small dark dots within this ellipse highlight an apparent high equivalent strain due the speckle modifications from cracking. Cracks 2 and 3 occurred at the side of the main deformation bands, but not on these bands. Crack 2 appears to be the closest to the grain boundary, although grain boundaries cannot be positioned precisely enough on top of the image after deformation (Grennerat et al., 2012).

As mentioned in Sect. 2, maps of accumulated strain include some noise, and likely some out-of-plane component of strain. In order to provide a more precise description of the relationship between strain field and crack opening, the following analyses will be performed based on strain-field increments measured during adapted macro-strain increments. This procedure strongly reduces the effect of accumulated noise and out-of-plane strain is negligible during small strain increments.

As a departure point, the strain field during increment of macro-strain between -1.34 and -1.35%, just before any visible crack event, is shown in Fig. 7, by both ε_{eq} (grey

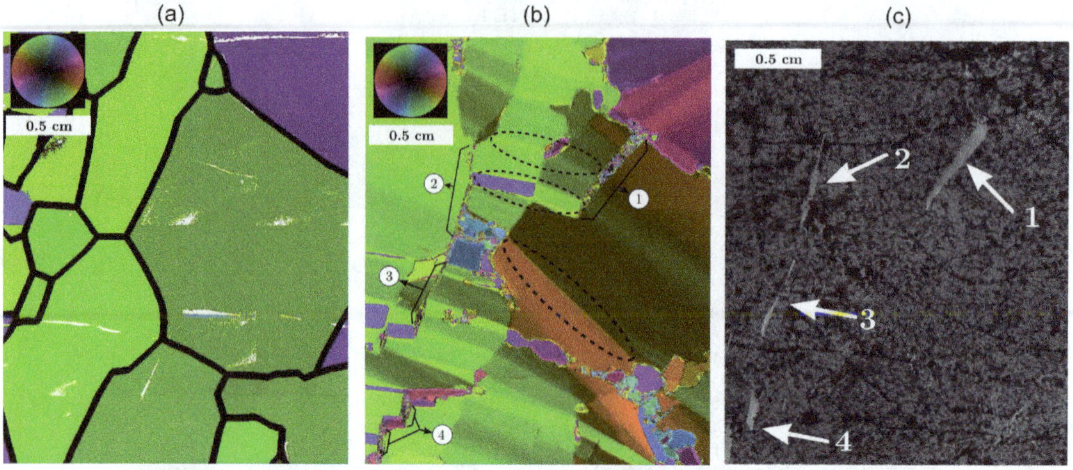

Figure 5. Studied area extracted from the sample of Fig. 3. **(a)** Initial microstructure measured by AITA. The black lines show the superimposed and artificially enlarged grain boundaries. The irregular white lines and small white areas are measurement artifacts. **(b)** Microstructure measured by AITA after deformation. Crack locations are highlighted by black full lines and labeled. Areas where recrystallization by subgrain rotation took place are shown by black dashed ellipses. **(c)** Raw picture of the surface speckle at the end of the test where cracks 1 to 4 can be seen by speckle discontinuities.

color scale) and the projections of the principal strain components (arrows). During this strain increment, strain field is very similar to the one accumulated during the entire experiment before cracking events (Fig. 6, left). Blue ellipses were drawn on the position of future opening of cracks 1, 2 and 3. Local strain within the ellipses is low compared to the one accommodated by the two main bands.

The strain field measured during increments at later steps of the experiment is presented on the different parts of Fig. 8. Cracks 1, 2 and 3 are first observed to open between -1.35 and -1.46% of macro-strain (Fig. 8a). Cracks can be visualized on the speckle as discontinuities, and the DIC calculation provides an apparent strain characterized by a pure extension which provides the main direction of the crack opening. Cracks 1 to 3 could have opened mainly in mode I since no shear component was measured by DIC in the area of crack formation just before the opening (Fig. 7), within the limit of resolution of our observations. During this strain increment, associated with the crack opening, strain is still localized in the nearly horizontal band. The nearly vertical deformation band, which was accommodating a lot of deformation before the cracks start to open, is not active anymore (compare Figs. 6 and 8).

During the next increment (Fig. 8b), cracks continue opening, as illustrated by the tension component evaluated by DIC around the crack sides. From the final microstructure picture (Fig. 6), we see that the top right part of crack 1 is connected to a grain boundary but the bottom part of this crack remains inside the grain, while no clear strain localization can be observed at this position (within our limit of accuracy).

About 30 min later, between macro-strain of -2.40 and -2.59% (Fig. 8c), the strain field evaluation (together with

speckle observation) tends to show that cracks 1 and 2 are still expanding when crack 3 remains stable, since DIC calculation shows no more tensile strain components in the area of crack 3.

During this increment, crack 4 appears at the bottom left and two new deformation bands appear at the lower tip of crack 2, and at the bottom tip of crack 3, both being parallel to the main transverse deformation band observed from the beginning of this sequence. These new lines of strain localization end up joining each other and the initial transverse deformation band during the macro-strain interval between -3.12 and -3.37% (Fig. 8d). By looking at the final microstructure (Fig. 5), these new deformation bands appear to be localized in an area where new grains recrystallized. Since we follow the strain field evolution during the test, we are able to verify that the new grains formed after the apparition of the cracks as in Chauve et al. (2015).

At the same time, crack 3 closes, as evidenced by the thinning of the corresponding white zone in the speckle image. Strain in the crack 3 area turns into a pure compressive component (blue arrows, Fig. 8d), which is likely to be responsible for this crack closure. Similarly, during the last increment of deformation (between -5.05 and -5.50% of macro-strain), pure compressive principal strain components are calculated in most of the observed crack discontinuities (Fig. 9). Together with the visual observation of crack evolution on the speckle images, these observations reveal a crack closure mechanism.

During this last increment, strain field is also characterized by several new bands of strain localization in the area (Fig. 9). By observing the final microstructure, we can attribute this strain localization to the formation of high-angle

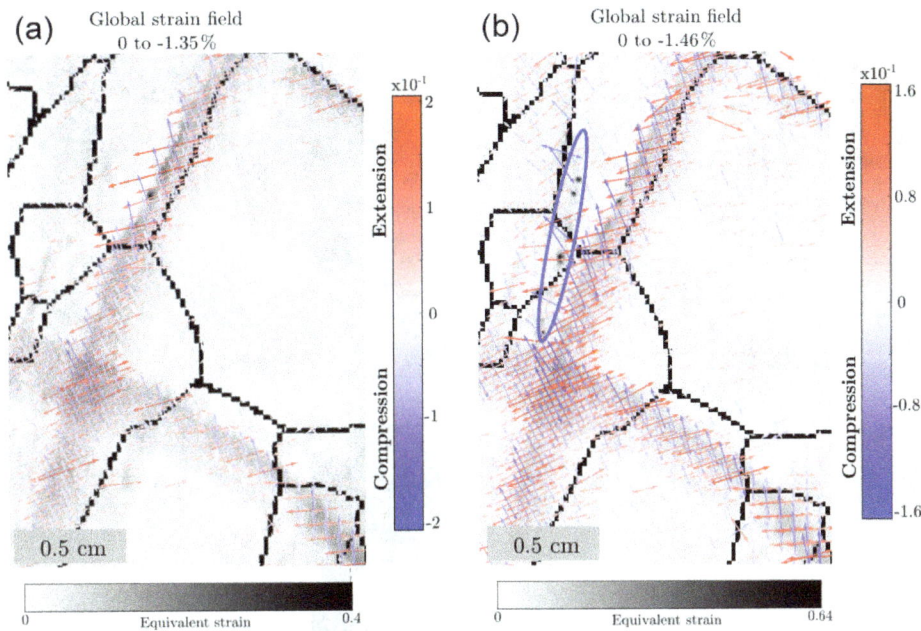

Figure 6. Evaluation of the total accumulated strain field in the focused area from Fig. 3 represented by ε_{eq} and by the principal components. **(a)** Total strain accumulated after $-1.35\,\%$ of macro-strain. **(b)** Total strain accumulated after $-1.46\,\%$ of macro-strain, just after the crack formation. The blue ellipse shows the area of formation of crack 2.

sub-grain boundaries and kink bands. Their likely locations are shown by dashed black ellipses in Figs. 5 and 9 to facilitate the observation. In particular, the two kink bands marked by the top black dashed ellipses seem to be localized at the tips of cracks 2 and 1. Please note that crack 1 bottom tip localized in the grain interior strongly coincides with the edge of a high-angle sub-grain boundary.

To summarize, by measuring the strain field evolution during the onset of tertiary creep, at the ductile-to-brittle transition, we were able to follow crack formation close to grain boundaries and within grain interiors, as well as their consequences on the local strain field. Some cracks appear at the side of high strain localization bands, where stress must have concentrated in "hard" zones for deformation. Following the crack opening, we observe a strong redistribution of the local strain, with the disappearance of one of the major localization band. Additionally, we show that stress concentration at the crack tips can be efficiently relaxed by dynamic recrystallization mechanisms (nucleation and sub-grain boundary formation), and that the stress redistribution induced by crack opening and microstructure changes due to DRX mechanisms can lead to the closure of cracks during the test. The occurrence of dynamic recrystallization mechanisms is here strongly enhanced by the high homologous temperature conditions of the experiment.

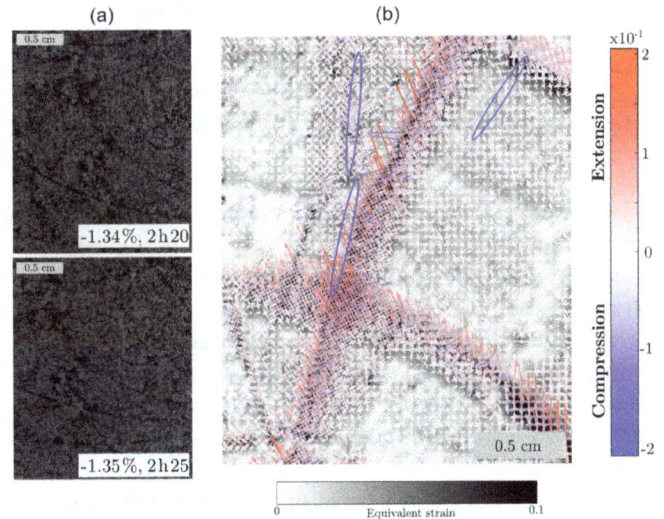

Figure 7. Strain field increment during the 5 min before the apparition of cracks between -1.34 and $-1.35\,\%$ of macroscopic strain. **(a)** Pictures of the speckled surface used for the DIC. **(b)** Principal component of the strain field superimposed on the equivalent strain field (ε_{eq}).

4 Discussion – mechanisms to relax local stress concentration

During compressive tests on an isotropic material, the maximum shear stress occurs at $45°\,$E from the compression direction (Tresca criterion). For material with plastic

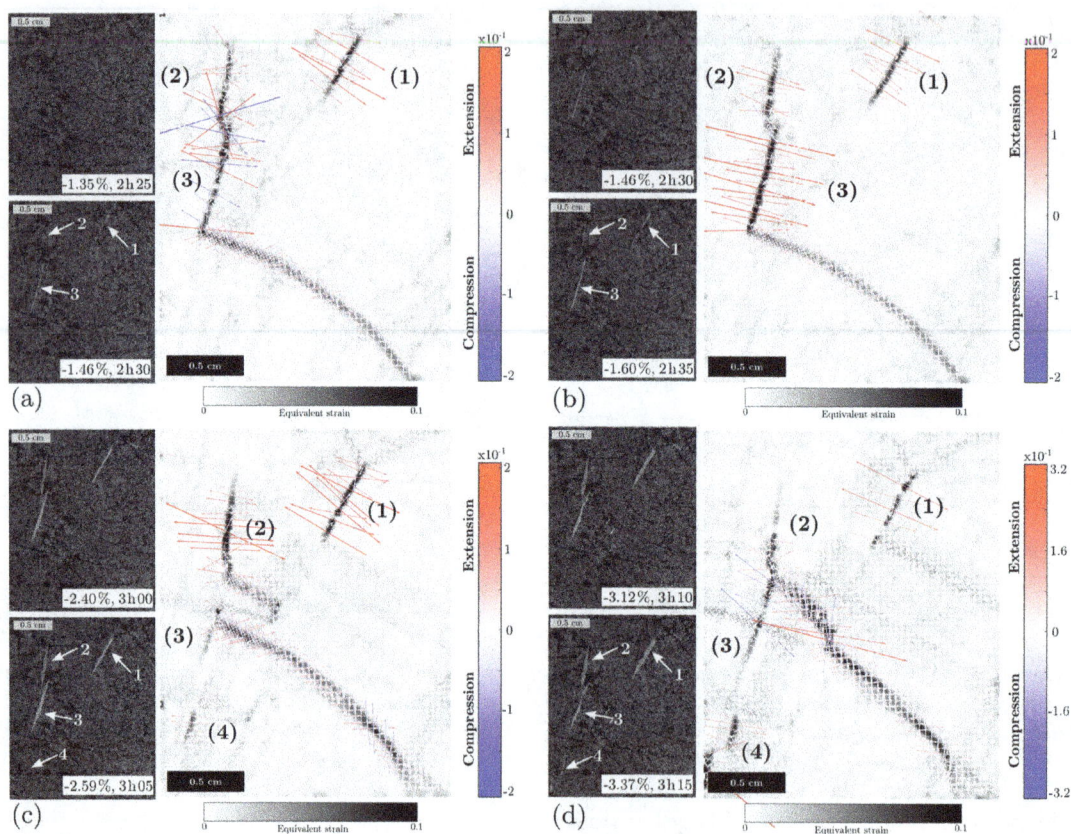

Figure 8. Four steps of 5 min strain field increment during crack opening. **(a, c)** Pictures of the speckled surface used for the DIC. **(b, d)** Principal component of the strain field superimposed on the equivalent strain field. **(a)** Increment between -1.35 and -1.46%. **(b)** Increment between -1.46 and -1.60%. **(c)** Increment between -2.40 and -2.59%. **(d)** Increment between -3.12 and -3.37%.

anisotropy such as ice, a redistribution of stress is expected to occur which depends on the orientation relationship between grains. Such a redistribution has been simulated by full-field crystal plasticity approaches by Lebensohn et al. (2004) and Grennerat et al. (2012) for instance. Although the stress field is not experimentally accessible so far, these modeling results were validated by a comparison between predicted and measured strain field magnitudes and heterogeneities (Grennerat et al., 2012).

At the onset of tertiary creep in laboratory deformed ice, strain rate increases thanks to accommodating processes. As summarized by Schulson and Duval (2009), depending on the deformation conditions (temperature, imposed stress or imposed strain rate), accommodation can take place through dynamic recrystallization or micro-cracking. To our knowledge, no direct observations exist of the effect of micro-cracking on the redistribution of strain and therefore on local stress relaxation. The results presented here fill this gap by exploring the ductile-to-brittle transition where micro-cracking and plasticity can coexist. Common features with previous observations made by Grennerat et al. (2012) and Chauve et al. (2015) are the strong strain heterogeneities, with local strains more than 10 to 20 times as great as the

macroscopic strain. Although influenced by the boundary conditions, grain interactions tend to deviate the strain concentration from the main 45° E directions. While the work of Grennerat et al. (2012) remained in the primary creep regime, and mostly concentrated on sample-scale field characterizations, Chauve et al. (2015) went a step further and found that DRX mechanisms explained the interplay between local changes in microstructure and strain-field evolution. In particular, strain was shown to re-localize close to the newly formed grain boundaries and sub-grain boundaries. This has also been observed in the present study.

Compared with previous works, conditions imposed during the experiment presented here induced local cracking at the onset of tertiary creep (which occurs before 1 % of macro-strain for the sample studied very likely because of the influence of a non-isotropic texture and of a columnar microstructure). Most of the local cracks observed were intra-granular. Cracks appeared in areas near strain localization bands, but not within these bands, as evidenced by Fig. 6 and by comparing Figs. 7 and 8a (cracks 1 and 2). These observations highlight the fact that local stresses can be concentrated at the side of high strained region. This can result from strain incompatibilities between regions of different orientations,

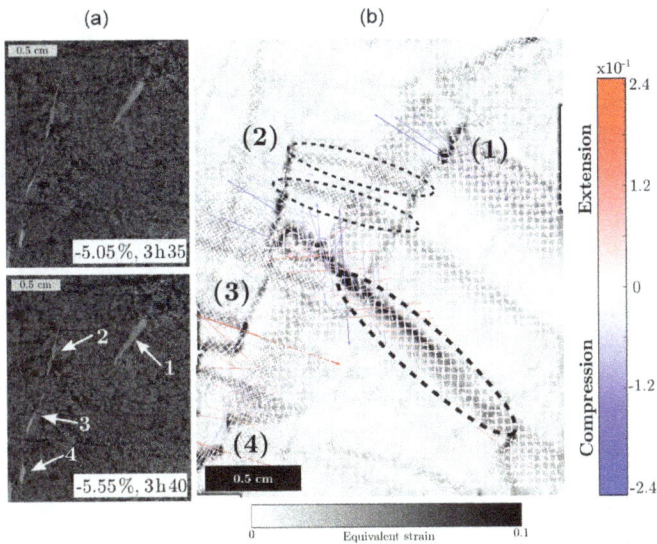

Figure 9. Increment of deformation during the last 5 min of the test (between −5.05 and −5.55 %). Kink band formation (within dashed black ellipses) at the crack tips and crack closure are observed. (a) Pictures of the speckled surface used for DIC. (b) Principal components of the strain field superimposed on the equivalent strain field.

with regions with locally low Schmid factors (relative to the local stress tensor) behaving as solid inclusions in composite materials. The likely impact of low local Schmid factors might be strengthened by the strong viscoplastic anisotropy of ice that renders some orientations strongly unfavorable for basal dislocation slip.

Crack formation relaxes these high local stresses, and meanwhile, stress concentration is translated at the crack tips. Previous studies on columnar ice performed at higher strain rate ($\dot{\varepsilon} = 4 \times 10^{-3}\,\mathrm{s}^{-1}$) but similar temperature ($T = -10\,°\mathrm{C}$; Batto and Schulson, 1993; Iliescu and Schulson, 2004) evidenced the typical mechanism of wing-crack formation at the crack tips. Wing cracks appear as the result of tensile stress concentration at the crack tips and can lead to the overall failure of the sample by propagating through it, or by connecting to other cracks. Recently, a similar mechanism of wing cracks propagation has been characterized by DIC in a soft rock by Nguyen et al. (2011), and they were able to quantify the different fracture modes (opening, closing and shearing) thanks to local strain measurements.

As the experiment presented here is performed in conditions equivalent to a lower strain rate (although through imposed load conditions) compared to Batto and Schulson (1993), the stress concentration at the crack tips is not relaxed by the formation of wing cracks but by plasticity mechanisms in the creep zone at the tip. Dislocations are therefore expected to nucleate and propagate at the crack tips as shown by Rice and Thomson (1974). Recently, Argon (2001) showed that both nucleation of dislocations at the crack tip, and the

mobility of the nucleated dislocations come into play to induce the stress relaxation responsible for a crack arrest. Considering the high-temperature conditions of our experiments, the dislocation multiplication leads to dynamic recrystallization mechanisms to occur in the creep zone near crack tips. Indeed, nucleation of new grains is observed very close to the crack tips of cracks 2, 3 and 4 (Fig. 5) and dislocation substructures as sub-grains are formed, for instance around crack tips of cracks 1 and 2 (Fig. 9). These observations reveal that plasticity-driven recrystallization mechanisms are efficient to relax the local tensile stresses initiated at the crack tips.

Local stresses associated with grain interactions during deformation of ice was indeed shown to be strongly heterogeneous, and to be responsible for the initiation of sub-grain boundaries at the end of primary creep (Piazolo et al., 2015). Observation of crack initiation near grain boundaries and within grain interior is another evidence of such local stress concentration.

By following the strain field evolution all along the tests, we observe the closure of some parts of the cracks, in areas where nucleation and sub-grain boundary formation were the most active. Crack closure is evidenced by the representation of principal strains which directions evolve from a tension component to a compressive component that ensures the recovering of continuity (Fig. 8d). In order to obtain a local closure of cracks, the stress field components should be drastically modified, and possibly turn to a compressive component. The new microstructure formed by recrystallization mechanisms must therefore drive a redistribution of the local stress field to enable such a modification, still compatible with the macroscopic stress conditions.

Ductile fracture occurring at elevated temperature in metals can be related to void propagation, growth and coalescence. Recently, Shang et al. (2017) showed that DRX mechanisms induced a softening that reduces the local stress concentration, which serves as the driving force for this void-induced ductile fracture. Similar observations of a ductile-to-brittle transition in Olivine driven by plasticity mechanisms was thoroughly studied by Druiventak et al. (2011). In samples deformed at 20, 300 and 600 °C they observed microcracking at grain boundaries and in the grain interiors, but also arrays of dislocations related to crystal plasticity. Similarly to our observations, at the highest temperature, plasticity took place in the form of strongly misoriented undulatory extinctions (associated with various types of dislocations), deformation lamellae and 3-D dislocation cells inducing strong modifications of the microstructure. Our results therefore present some interest beyond the ice community. Similar procedures could very interestingly be applied to a wide range of materials in order to estimate the role of the level of plastic anisotropy on strain localization and on the efficiency of plasticity-driven recrystallization mechanisms to relax the local stress field at the crack tips.

On top of the mechanical meaning of these observations, we highlight the fact that, since we were able to follow local crack closure during the test, care must be taken when performing experimental tests in conditions close to the ductile-to-brittle transition (typically at strain rates above $10^{-6}\,s^{-1}$, or compressive stress above 0.9 MPa), and at high temperature. Micro-cracking and DRX mechanisms can influence the local stress relaxation, and therefore the mechanical response, without leaving any track in the final microstructure.

5 Concluding remarks

The present work reveals, for the first time in ice, the evolution of the heterogeneous strain field during the onset of tertiary creep, in conditions where local cracking occurs to relax the local stress field. This observation was made possible by taking advantage of samples with 2-D–1/2 microstructures from which surface observations reflect bulk behavior.

While strain field localizes into bands with a length larger than the grain dimensions, cracks appear to relax stress concentration at the side of the strain localization bands, where deformation by dislocation glide must have been impeded by low local Schmid factor conditions.

Relaxation of the local stress field by crack opening results in a local redistribution of the strain field, as evidenced by the abrupt weakening of some deformation bands after cracking. At the crack tips, where stress concentrates, plasticity-driven dynamic recrystallization mechanisms are observed as new small grains and high-angle sub-grain boundaries in the final microstructure. The new formed boundaries also appear visible on strain field patterns during the test, as new strain concentration areas. While induced by local stress concentration at the crack tip, recrystallization mechanisms in turn generate a stress field redistribution as a result of microstructure modifications. This redistribution is indirectly evidenced by the modification of the measured strain field in the area, but also by the original observation of local crack closure, likely associated with a measured local compressive stress in place of the initial tensile stress responsible for the observed mode I crack opening. To conclude, the main results show that micro-cracking and dynamic recrystallization mechanisms both resulting from a strongly heterogeneous stress field can coexist locally and that these mechanisms are efficient to relax local stresses at the ductile-to-brittle transition. Hence one should be careful when working at the frontiers of this transition since recrystallization can hide local cracking in the final microstructures.

Author contributions. TC, DG and CL performed the laboratory experiments. DG and TC provided the data treatment. MM and TC analyzed the data and wrote the paper. PV provided some support for the DIC analyses and interpretation.

Competing interests. The authors declare that they have no conflict of interest.

Special issue statement. This article is part of the special issue "Analysis of deformation microstructures and mechanisms on all scales". It is a result of the EGU General Assembly 2016, Vienna, Austria, 17–22 April 2016.

Acknowledgements. Financial support by the French "Agence Nationale de la Recherche" is acknowledged (project DREAM, ANR-13-BS09-0001-01). This work benefited from support from the INSIS and INSU institutes of CNRS. It was supported by a grant from Labex OSUG@2020 (ANR10 LABEX56) and from INP-Grenoble and UJF within the "Grenoble Innovation Recherche AGIR" proposal. Support from the imagery center IRIS of Grenoble-INP is acknowledged. Maurine Montagnat benefited from a visitor research fellowship from WSL (Switzerland) in 2016–2017.

Edited by: Ilka Weikusat

References

Antolovich, S. D. and Armstrong, R. W.: Plastic strain localization in metals: origins and consequences, Prog. Mater. Sci, 59, 1–160, https://doi.org/10.1016/j.pmatsci.2013.06.001, 2014.

Argon, A.: Mechanics and physics of brittle to ductile transitions in fracture, J. Eng. Mater.-T. ASME, 123, 1–11, 2001.

Batto, R. A. and Schulson, E. M.: On the ductile-to-brittle transition in ice under compression, Acta Metall. Mater., 41, 2219–2225, https://doi.org/10.1016/0956-7151(93)90391-5, 1993.

Bons, P. D., Jansen, D., Mundel, F., Bauer, C. C., Binder, T., Eisen, O., Jessell, M. W., Llorens, M.-G., Steinbach, F., Steinhage, D., and Weikusat, I.: Converging flow and anisotropy cause large-scale folding in Greenland's ice sheet, Nat. Commun., 7, 1–6, https://doi.org/10.1038/ncomms11427, 2016.

Chauve, T., Montagnat, M., and Vacher, P.: Strain field evolution during dynamic recrystallization nucleation; A case study on ice, Acta Mater., 101, 116–124, https://doi.org/10.1016/j.actamat.2015.08.033, 2015.

Chauve, T., Montagnat, M., Barou, F., Hidas, K., Tommasi, A., and Mainprice, D.: Investigation of nucleation processes during dynamic recrystallization of ice using cryo-EBSD, Philos. T. R. Soc. A, 375, 1–20, https://doi.org/10.1098/rsta.2015.0345, 2017.

Dahl-Jensen, D., Albert, M. R., Aldahan, A., et al.: Eemian interglacial reconstructed from a Greenland folded ice core, Nature, 493, 489–494, 2013.

Druiventak, A., Trepmann, C. A., Renner, J., and Hanke, K.: Low-temperature plasticity of olivine during high stress deformation of peridotite at lithospheric conditions – An experimental study, Earth Planet. Sc. Lett., 311, 199–211, 2011.

Durand, G., Graner, F., and Weiss, J.: Deformation of grain boundaries in polar ice, Europhys. Lett., 67, 1038, https://doi.org/10.1209/epl/i2004-10139-0, 2004.

Durand, G., Gillet-Chaulet, F., Svensson, A., Gagliardini, O., Kipfstuhl, S., Meyssonnier, J., Parrenin, F., Duval, P., and Dahl-Jensen, D.: Change in ice rheology during climate variations – implications for ice flow modelling and dating of the EPICA Dome C core, Clim. Past, 3, 155–167, https://doi.org/10.5194/cp-3-155-2007, 2007.

Duval, P.: Creep and recrystallization of polycrystalline ice, Bull. Mineral, 102, 80–85, 1979.

Duval, P., Ashby, M., and Anderman, I.: Rate controlling processes in the creep of polycrystalline ice, J. Phys. Chem., 87, 4066–4074, 1983.

Fischer, H., Severinghaus, J., Brook, E., Wolff, E., Albert, M., Alemany, O., Arthern, R., Bentley, C., Blankenship, D., Chappellaz, J., Creyts, T., Dahl-Jensen, D., Dinn, M., Frezzotti, M., Fujita, S., Gallee, H., Hindmarsh, R., Hudspeth, D., Jugie, G., Kawamura, K., Lipenkov, V., Miller, H., Mulvaney, R., Parrenin, F., Pattyn, F., Ritz, C., Schwander, J., Steinhage, D., van Ommen, T., and Wilhelms, F.: Where to find 1.5 million yr old ice for the IPICS "Oldest-Ice" ice core, Clim. Past, 9, 2489–2505, https://doi.org/10.5194/cp-9-2489-2013, 2013.

Grennerat, F., Montagnat, M., Castelnau, O., Vacher, P., Moulinec, H., Suquet, P., and Duval, P.: Experimental characterization of the intragranular strain field in columnar ice during transient creep, Acta Mater., 60, 3655–3666, 2012.

Iliescu, D. and Schulson, E. M.: The brittle compressive failure of fresh-water columnar ice loaded biaxially, Acta Mater., 52, 5723–5735, https://doi.org/10.1016/j.actamat.2004.07.027, 2004.

Jacka, T. H. and Maccagnan, M.: Ice crystallographic and strain rate changes with strain in compression and extension, Cold Reg. Sci. Technol., 8, 269–286, 1984.

Jansen, D., Llorens, M.-G., Westhoff, J., Steinbach, F., Kipfstuhl, S., Bons, P. D., Griera, A., and Weikusat, I.: Small-scale disturbances in the stratigraphy of the NEEM ice core: observations and numerical model simulations, The Cryosphere, 10, 359–370, https://doi.org/10.5194/tc-10-359-2016, 2016.

Lebensohn, R. A., Liu, Y., and Ponte-Castañeda, P.: On the accuracy of the self-consistent approximation for polycrystals: comparison with full-field numerical simulations, Acta Mater., 52, 5347–5361, 2004.

MacGregor, J. A., Fahnestock, M. A., Catania, G. A., Paden, J. D., Prasad Gogineni, S., Young, S. K., Rybarski, S. C., Mabrey, A. N., Wagman, B. M., and Morlighem, M.: Radiostratigraphy and age structure of the Greenland Ice Sheet, J. Geophys. Res.-Earth, 120, 212–241, 2015.

Martínez-Pañeda, E. and Niordson, C. F.: On fracture in finite strain gradient plasticity, Int. J. Plasticity, 80, 154–167, https://doi.org/10.1016/j.ijplas.2015.09.009, 2016.

Montagnat, M., Chauve, T., Barou, F., Tommasi, A., Beausir, B., and Fressengeas, C.: Analysis of dynamic recrystallization of ice from EBSD orientation mapping, Front. Earth Sci., 3, 1–13, https://doi.org/10.3389/feart.2015.00081, 2015.

Nguyen, T. L., Hall, S. A., Vacher, P., and Viggiani, G.: Fracture mechanisms in soft rock: Identification and quantification of evolving displacement discontinuities by extended digital image correlation, Tectonophysics, 503, 117–128, https://doi.org/10.1016/j.tecto.2010.09.024, 2011.

Panton, C. and Karlsson, N. B.: Automated mapping of near bed radio-echo layer disruptions in the Greenland Ice Sheet, Earth Planet. Sc. Lett., 432, 323–331, 2015.

Paterson, W. S. B.: The physics of glaciers, Pergamon, Oxford, 1994.

Peternell, M., Russell-Head, D., and Wilson, C.: A technique for recording polycrystalline structure and orientation during in situ deformation cycles of rock analogues using an automated fabric analyser, J. Microsc., 242, 181–188, 2011.

Piazolo, S., Montagnat, M., Grennerat, F., Moulinec, H., and Wheeler, J.: Effect of local stress heterogeneities on dislocation fields: Examples from transient creep in polycrystalline ice, Acta Mater., 90, 303–309, https://doi.org/10.1016/j.actamat.2015.02.046, 2015.

Ple, O. and Meyssonnier, J.: Preparation and Preliminary Study of Structure-Controlled S2 Columnar Ice, J. Phys. Chem. B, 101, 6118–6122, https://doi.org/10.1021/jp963256t, 1997.

Renshaw, C. E. and Schulson, E. M.: Universal behaviour in compressive failure of brittle materials, Nature, 412, 897–900, https://doi.org/10.1038/35091045, 2001.

Rice, J. R. and Thomson, R.: Ductile versus brittle behaviour of crystals, Philos. Mag., 29, 73–97, 1974.

Schulson, E. and Buck, S.: The ductile-to-brittle transition and ductile failure envelopes of orthotropic ice under biaxial compression, Acta Metall. Mater., 43, 3661–3668, 1995.

Schulson, E. M. and Duval, P.: Creep and Fracture of Ice, Cambridge University Press, https://doi.org/10.1017/CBO9780511581397, 2009.

Schulson, E. M., Lim, P. N., and Lee, R. W.: A brittle to ductile transition in ice under tension, Philos. Mag. A, 49, 353–363, https://doi.org/10.1080/01418618408233279, 1984.

Shang, X., Cui, Z., and Fu, M. W.: Dynamic recrystallization based ductile fracture modeling in hot working of metallic materials, Int. J. Plasticity, 95, 105–122, https://doi.org/10.1016/j.ijplas.2017.04.002, 2017.

Snyder, S. A., Schulson, E. M., and Renshaw, C. E.: Effects of prestrain on the ductile-to-brittle transition of ice, Acta Mater,, 108, 110–127, https://doi.org/10.1016/j.actamat.2016.01.062, 2016.

Tommasi, A., Knoll, M., Vauchez, A., Signorelli, J., Thoraval, C., and Logé, R.: Structural reactivation in plate tectonics controlled by olivine crystal anisotropy, Nat. Geosci., 2, 423–427, https://doi.org/10.1038/ngeo528, 2009.

Vacher, P., Dumoulin, S., Morestin, F., and Mguil-Touchal, S.: Bidimensional strain measurement using digital images, P. I. Mech. Eng. C-J. Mec., 213, 811–817, 1999.

Vauchez, A., Tommasi, A., and Mainprice, D.: Faults (shear zones) in the Earth's mantle, Tectonophysics, 558, 1–27, 2012.

Weiss, J. and Schulson, E. M.: Grain-boundary sliding and crack nucleation in ice, Philos. Mag. A, 80, 279–300, https://doi.org/10.1080/01418610008212053, 2000.

Wilson, C., Russell-Head, D., and Sim, H.: The application of an automated fabric analyzer system to the textural evolution of folded ice layers in shear zones, Ann. Glaciol., 37, 7–17, 2003.

10

Land use change affects biogenic silica pool distribution in a subtropical soil toposequence

Dácil Unzué-Belmonte[1,*], **Yolanda Ameijeiras-Mariño**[2], **Sophie Opfergelt**[2], **Jean-Thomas Cornelis**[3], **Lúcia Barão**[4], **Jean Minella**[5], **Patrick Meire**[1], **and Eric Struyf**[1]

[1]EcosystemManagement Research Group, Department of Biology, University of Antwerp, Universiteitsplein 1C, 2610 Wilrijk, Belgium
[2]Earth and Life Institute, Environmental Sciences, Université catholique de Louvain, Croix du Sud 2 bte L7.05.10, 1348 Louvain-la-Neuve, Belgium
[3]Department Biosystem Engineering (BIOSE), Gembloux Agro-Bio Tech (GxABT), University of Liège (ULg), Avenue Maréchal Juin, 27, 5030 Gembloux, Belgium
[4]ICAAM, Instituto de Ciências Agrárias e Ambientais Mediterrânicas, University of Évora, Apartado 94, 7002-554 Évora, Portugal
[5]Universidade Federal de Santa Maria (UFSM), Department of Soil Science, 1000 Avenue Roraima, Camobi, CEP 97105-900 Santa Maria, RS, Brazil
[*] *Invited contribution by Dácil Unzué-Belmonte, recipient of the EGU Soil System Sciences Outstanding Student Poster Award 2014.*

Correspondence to: Dácil Unzué-Belmonte (dacil.unzuebelmonte@uantwerpen.be)

Abstract. Land use change (deforestation) has several negative consequences for the soil system. It is known to increase erosion rates, which affect the distribution of elements in soils. In this context, the crucial nutrient Si has received little attention, especially in a tropical context. Therefore, we studied the effect of land conversion and erosion intensity on the biogenic silica pools in a subtropical soil in the south of Brazil. Biogenic silica (BSi) was determined using a novel alkaline continuous extraction where Si / Al ratios of the fractions extracted are used to distinguish BSi and other soluble fractions: Si / Al > 5 for the biogenic AlkExSi (alkaline-extractable Si) and Si / Al < 5 for the non-biogenic AlkExSi. Our study shows that deforestation can rapidly (< 50 years) deplete the biogenic AlkExSi pool in soils depending on the slope of the study site (10–53 %), with faster depletion in steeper sites. We show that higher erosion in steeper sites implies increased accumulation of biogenic Si in deposition zones near the bottom of the slope, where rapid burial can cause removal of BSi from biologically active zones. Our study highlights the interaction of erosion strength and land use for BSi redistribution and depletion in a soil toposequence, with implications for basin-scale Si cycling.

1 Introduction

The terrestrial Si cycle has received increased attention in the past two decades. Multiple studies show its complexity, with a strong interaction among primary lithology and weathering, biotic Si uptake, the formation of secondary pedogenic phases and environmental controls such as precipitation, temperature and hydrology (Struyf and Conley, 2012). Lithology controls the primary source of Si through the weathering of silicate minerals of the bedrock (Drever, 1994). This process provides Si to the soil solution in the form of monosilicic acid (H_4SiO_4), also referred to as dissolved silicon (DSi). This DSi is taken up by plants and is resupplied to the soil in the form of relatively soluble (compared to crystalline silicates) biogenic silicates (BSi) upon plant die-off, usually in the form of phytoliths (plant silica bodies) (Piperno, 2006). Biogenic silica is one of the most soluble forms of Si in soils (e.g., Van Cappellen, 2003), although some pedogenic compounds have similar reactivities (Sauer et al., 2006; Sommer et al., 2006; Vandevenne et al., 2015a). During soil formation, the DSi released to the soil solution through the dissolution of lithogenic and biogenic

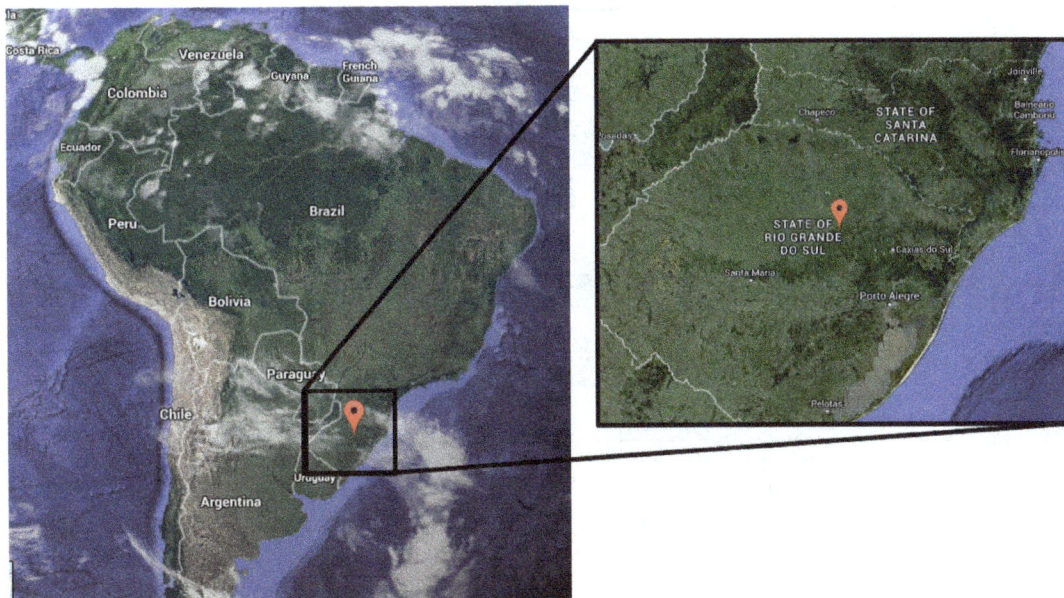

Figure 1. Location of study site.

silicates contributes to the neoformation of pedogenic silicates, i.e., secondary phyllosilicates (Sommer et al., 2006). The biogenic control on the DSi availability in soil increases with weathering degree. Soil mineralogy, strongly governed by geological and climatic conditions, therefore plays a key role in the DSi transfer from soil to plants (Cornelis and Delvaux, 2016). The complex interactions described above, which act to control the Si cycle in terrestrial ecosystems, are often referred to as the "ecosystem Si filter" (Struyf and Conley, 2012), and ultimately determine an important part of the Si fluxes towards rivers.

Land use change is a particularly interesting global change driver to address in this context. Dissolution of soil BSi increases immediately after deforestation (Conley et al., 2008), increasing DSi fluxes out of the soil and the ecosystem. However, in the long term, Struyf et al. (2010) showed a decrease in overall DSi fluxes from cultivated land. The conversion from forest to croplands decreases the soil biogenic Si stock, the most important contributor to the easily available Si pool for plants. The decrease in soil biogenic Si stock has been related to two important factors. The first factor is the harvesting of crops (Guntzer et al., 2012; Meunier et al., 1999; Vandevenne et al., 2012). Harvest prevents the return of plant phytoliths to the soil, depleting the phytolith pool. The resulting decrease in DSi availability also reduces the formation of non-biogenic secondary Si fractions (Barão et al., 2014). A thorough analysis separating both biogenic and non-biogenic fractions is crucial in this regard, since traditional extraction procedures to quantify biogenic Si may also dissolve non-biogenic Si fractions. The second factor affecting BSi losses is erosion. In cultivated catchments, preferential BSi mobilization is associated with erosion during strong rainfall

events (Clymans et al., 2015). During such events, biogenic Si can represent up to 40 % of the easily soluble Si inputs to rivers (Smis et al., 2011). Clymans et al. (2015) found that Si mobilization did not depend on tillage technique or crop type but solely on soil loss rate due to erosion.

While it is now accepted that cultivation can cause significant changes in soil Si pools and Si fluxes in temperate climates (Keller et al., 2012), the effect of cultivation on (sub)tropical soil Si pools or on soils of volcanic origin is poorly known. Only specific ecosystems, such as rice fields, have been studied (Guntzer et al., 2012) in this regard. Yet, the increasing demand for firewood, timber, pasture and food crops is causing an increase in land conversion to croplands, implying ongoing rapid land degradation in tropical and subtropical forests (Von Braun, 2007; Hall et al., 1993). The aim of our study was to investigate the interactive effects of land use change and terrain slope (as a proxy for erosion) on the distribution of the BSi pool in a subtropical soil system derived from a basaltic parent material. For this purpose, we studied terrestrial Si pools in a natural forest and cultivated land, in gently and steeply sloped locations, applying a recently developed alkaline extraction technique that permits the biogenic and non-biogenic phases to be distinguished.

2 Methods

2.1 Study area

The study area is situated near Arvorezinha, in the south of Brazil (28°56′ S, 52°6′ W) (Fig. 1). Four sites with identical climatic conditions (warm temperate, fully humid with warm summer, Cfb (Kottek et al., 2006), were selected. An-

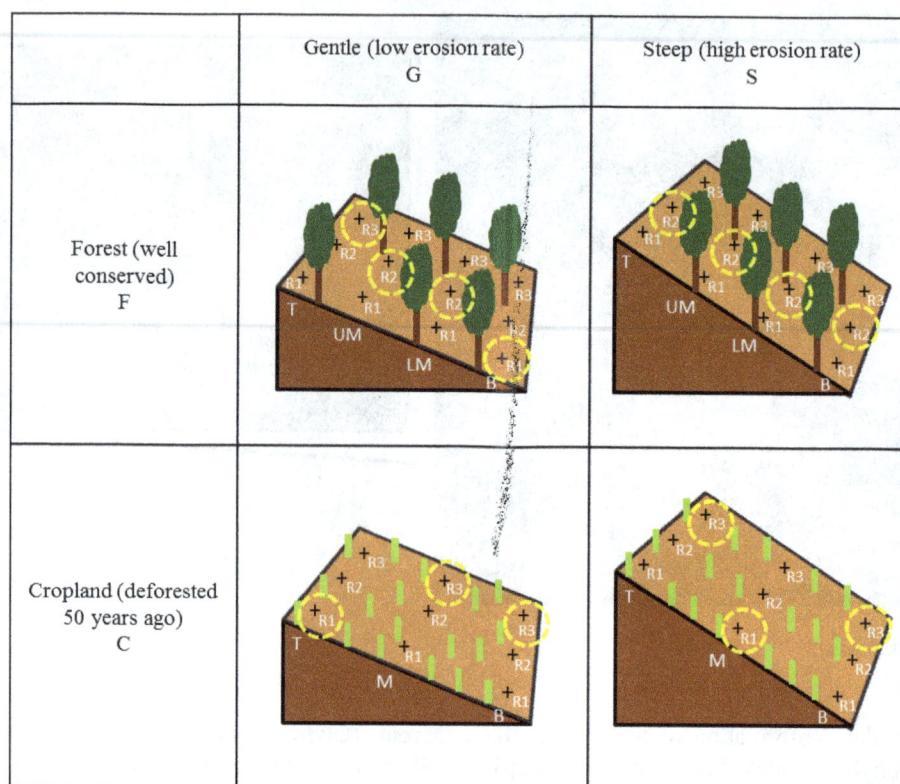

Figure 2. Diagram of the studied sites and the abbreviations used in the text ordered by ecosystem (F: forest; C: cropland), slope (G: gentle; S: steep), position (T: top; UM: upper middle; LM: lower middle; M: middle; B: bottom) and replicate (R1: replicate 1; R2: replicate 2; R3: replicate 3). Plus signs represent sampling points and yellow circles the selected pits.

nual mean temperature is between 14 and 18 °C and annual mean precipitation between 1700 and 1800 mm (Minella et al., 2014). The four sampling sites also have the same parent material (rhyodacite). The mineralogy was similar in all sites with sanidine and quartz as the main minerals (Ameijeiras-Mariño, 2017). Soil type corresponded to an Acrisol in three of the sites and a Leptosol (IUSS Working Group WRB, 2015) in the fourth (steep slope of the cropland), with pH values between 4.7 and 5.9. They represent two land uses, a well-conserved forest and a cropland, and two slope steepnesses (a steep and a gentle slope), resulting in four different factor combinations (see Fig. 2).

The forest site consists of a semi-deciduous forest with *Araucaria angustifolia*, *Luehea divaricata*, *Nectandra grandiflora* and *Campomanesia guaviroba* as the dominant species. Within the same forest area, two adjacent sites with different slopes were chosen: a gentle slope (maximum 10°) and a steeper slope (maximum 18°). On the gentle slope, some scattered small patches of yerba mate crop (*Ilex paraguariensis*) were recently planted (< 3 years ago), occupying less than 10 % of the study site. All studied sampling locations were separated at least 5 m from these mate patches.

The cropland sites were located in two geographically separated areas, 1.4 km apart. Deforestation occurred around

50 years ago and they have since experienced the same historical agricultural practices. Intensive soil tillage occurred from the time of deforestation to 2003, when a cover cropping and a minimum tillage practice was introduced (Minella et al., 2014). The actual soil tillage is traditional, based on topsoil mixing and making ridges and furrows. Crops in the gently sloping cropland (maximum 7°) rotate between soybean (*Glycine max*) in summer and black oat (*Avena sativa*) in winter. Some cattle occasionally graze during the vegetative stage, and after the oat is harvested the biomass is left to produce mulch (cover) to soybean seeding based on the no-till system. The cropland of the steep slope (maximum 18°) rotates between tobacco (*Nicotiana* sp.) or maize (*Zea mays*) in summer and fallow or black oat in winter. The winter crop on this slope is also left behind to produce cover for the next crop.

2.2 Soil sampling

Bulk soil samples ($n = 297$) were collected during the summer of 2014. In the forest sites, four positions along the slope (from top to bottom) were selected. In the croplands, due to time constraints during the field campaign, only three positions along the slope (from top to bottom) were selected. Three replicate soil pits were dug per position and soil samples were collected every 10 cm (from top to 50 cm deep) and every 20 cm (from 50 to 110 cm deep) (Fig. 2). Deeper depths were sampled every 50 cm until 200 cm deep or until the saprolite was reached. At each depth, 10 cm of soil (around 2 kg) was collected. At larger sample intervals, the 10 cm sample was collected in the middle of the depth interval. Soil samples were mixed, dried ($\sim 40\,^\circ$C), gently crushed and sieved (2 mm) prior to analysis.

Kopecky ring samples were also collected at each sampled depth. Samples were weighted before and after drying at 105 $^\circ$C in order to calculate bulk densities.

2.3 Analysis

One pit per position was selected as a representative pit due to the impossibility of carrying out the novel alkaline extraction analyses on such a high number of samples (297), resulting in a total of 81 samples. The selection avoided pits containing large inclusions (visually) or pits shallower than the other two replicas. The abbreviations and selected pits are shown in Fig. 2.

2.3.1 Physicochemical analyses

A portion of the bulk samples was crushed and a subsample was heated at 105 and 1000 $^\circ$C to obtain the dry weight and the loss on ignition. The total element content was obtained through borate fusion (Chao and Sanzolone, 1992) of another subsample of the crushed sample; 100 mg was fluxed at 1000 $^\circ$C for 5 min in a graphite crucible with 0.4 g of lithium tetraborate and 1.6 g of lithium metaborate, then cooled and dissolved in 100 mL of 2 M HNO_3 under magnetic agitation at 90–100 $^\circ$C. Elemental contents were determined by inductively coupled plasma–atomic emission spectrometry (ICP-AES); the total reserve of bases (TRB = [Ca] + [Mg] + [K] + [Na]) was calculated afterwards. TRB is commonly used as a weathering index as it estimates the content of weatherable minerals (Herbillon, 1986).

Particle size distribution was executed with a Beckman Coulter device (LSTM-13320) to quantify the sand (2 mm–50 µm), silt (50–2 µm) and clay (<2 µm) fractions.

The mineralogy of sand and silt fractions was determined by powder X-ray diffraction (XRD, Cu Ka, D8). Clay fraction mineralogy was assessed by XRD after K^+ and Mg^{2+} saturation, ethylene glycol solvation and thermal treatments at 300 and 550 $^\circ$C (Robert and Tessier, 1974).

2.3.2 Alkaline continuous extraction

All samples from selected pits ($n = 81$), together with some additional depths from other pits, were analyzed for biogenic and non-biogenic Si content, resulting in a total of 145 bulk soil samples (84 on the forest sites and 61 on the croplands). Samples were analyzed in a continuous flow analyzer (Skalar, Breda, the Netherlands), using a continuous alkaline

extraction recently adapted for soils by Barão et al. (2014). The extraction in 180 mL of 0.5 M NaOH, at 85 $^\circ$C runs for half an hour. Dissolved Si and dissolved aluminum (Al) are measured continuously (with the spectrophotometric molybdate blue method and the lumogallion fluorescence method, respectively), obtaining two dissolution curves which are fitted with first-order Eq. (1).

$$\text{Si}_t \ (\text{mg g}^{-1}) = \left(\sum_{i=1}^{n} \text{AlkExSi}_i \times \left(1 - e^{-k_i \times t} \right) \right) + b \times t,$$

$$\text{Al}_t \ (\text{mg g}^{-1}) = \left(\sum_{i=1}^{n} \frac{\text{AlkExSi}_i}{\text{Si}/\text{Al}_i} \times \left(1 - e^{-k_i \times t} \right) \right) + \frac{b \times t}{\text{Si}/\text{Al}_{\min}}, \quad (1)$$

where Si_t and Al_t are the concentrations of Si and Al, respectively, at any given time. The equations consist of two parts: the mineral fraction, which has a linear dissolution behavior (DeMaster, 1981; Koning et al., 2002), and the fractions exhibiting nonlinear dissolving behavior. For the mineral fraction, the model renders a linear dissolution rate (b) and the Si / Al ratio (Si / Al$_{\min}$) of that linear fraction. Nonlinearly dissolving fractions are characterized by the total amount of Si (alkaline extractable Si (AlkExSi), mg g^{-1} dry weight of initial sample mass), the Si / Al ratio (concentration of Si over concentration of Al) of that fraction and its dissolution rate (k, min^{-1}). Assuming the same Si and Al release rate from the same compound and relating the Si and Al concentration equations through the Si / Al ratio, with the three parameters estimated (AlkExSi, k and Si / Al ratio) the different fractions dissolving nonlinearly are distinguished. The same model is fitted with one, two or three first-order equations (summation to n in the formula) and the solution showing least error (F test) from the three fits is kept. For the nonlinear fractions, the Si / Al ratio of the fraction is used to determine its origin. Barão et al. (2014) recognized the following fractions: fractions showing a Si / Al ratio > 5 were considered as indicative of a biogenic fraction, as the concentration of Al in phytoliths is low (Bartoli, 1985; Piperno, 2006). A fraction showing a Si / Al ratio < 5 was considered as representative of non-biogenic or pedogenic Si fractions (clay minerals, oxides and organo-Al complexes). We opted to discard fractions that represent less than 0.1 mg Si g^{-1}, as they are smaller than or equal to the detection limit of the method (Barão et al., 2015). Fractions with $k < 0.1$ were also discarded, as they represent near linearly dissolving fractions.

2.3.3 Post-data treatments

AlkExSi pools or stocks every 10 cm depth (kg Si m^{-2}) for selected pits were calculated according to Eq. (2).

$$\text{AlkExSi stock (kg Si m}^{-2}) = \frac{[\text{AlkExSi}] \times \text{BD} \times h}{100}, \quad (2)$$

where [AlkExSi] is the concentration (mg g^{-1}) obtained in the alkaline continuous extraction, BD is the bulk density

$(\mathrm{g\,cm^{-3}})$ of that sample, h is the thickness of the depth interval of the sample (cm) and 100 is a conversion factor from $\mathrm{mg\,cm^{-2}}$ to $\mathrm{kg\,m^{-2}}$. This calculation takes into account the bulk density of each sample, correcting the amount of AlkExSi per gram of dried soil according to the water content at that specific soil depth. It also calculates the amount of AlkExSi in relation to the thickness of the interval collected (10 cm). For larger intervals, where only 10 cm was collected at mid-interval depth, values of the non-sampled depths were linearly interpolated between two known values. The result is given in kilograms per square meter, in our case at 10 cm deep intervals.

In order to estimate the total biogenic and non-biogenic AlkExSi pools per pit, the sum of all 10 cm depth biogenic and non-biogenic AlkExSi pools of each pit was made.

Once having the biogenic and the non-biogenic AlkExSi pools per pit, averages between the three (for the croplands) or four (for the forests) selected pits were made in order to assign average biogenic and non-biogenic AlkExSi pool values to the slope and to be able to compare AlkExSi pools between different sites. Then, comparisons between the different study sites were made. In order to compare the biogenic and non-biogenic AlkExSi pools from the forests with the croplands, two different methods were considered, taking into consideration that the number of positions along the slope in the forest sites is higher than in the cropland sites (four and three, respectively): Average 1, using all available measurements for the forest (the four positions along the slope) and cropland sites, and Average 2, using a pre-calculated average between upper and lower middle position measurements in the forest sites.

To study the accumulation of biogenic and non-biogenic AlkExSi pools at the bottom of the slope we have calculated the accumulation (AC) using the pool in the bottom compared to the summed pools along the slope for the forests (Eq. 3) and the croplands (Eq. 4). The closer the AC value is to 100 %, the higher the accumulation results.

$$AC_{\mathrm{Forest}} = \qquad\qquad\qquad (3)$$
$$\frac{AlkExSi_{\mathrm{bottom}}}{AlkExSi_{\mathrm{top}} + AlkExSi_{\mathrm{upper\ middle}} + AlkExSi_{\mathrm{lower\ middle}} + AlkExSi_{\mathrm{bottom}}} \times 100,$$

$$AC_{\mathrm{Cropland}} = \qquad\qquad\qquad (4)$$
$$\frac{AlkExSi_{\mathrm{St-bottom}}}{AlkExSi_{\mathrm{top}} + AlkExSi_{\mathrm{middle}} + AlkExSi_{\mathrm{bottom}}} \times 100.$$

Statistical differences between biogenic AlkExSi pool averages for top pits, middle slope pits, bottom pits and differences between biogenic AlkExSi pool averages from the top pit and the bottom pit within the same slope were tested pair by pair for significance at the 5 % level confidence using a Student t test assuming unequal variances.

3 Results

3.1 Soil physicochemical characteristics

Results from total element content, particle size, bulk density and TRB values for selected pits are shown in Tables S2–S4 in the Supplement. The XRD mineralogical analysis of the bedrock (rhyodacitic volcanic rocks) reveals that sanidine (feldspar group) is the most abundant mineral (45–55 %), followed by very fine-grained quartz ($\sim 38\,\%$) embedded in a matrix of hematite, goethite and clays ($\sim 8\,\%$) (Ameijeiras-Mariño, 2017). Bulk densities of selected pits ranged from 0.7 to $1.54\,\mathrm{mg\,cm^{-3}}$.

3.2 AlkExSi concentrations

AlkExSi values ($\mathrm{mg\,g^{-1}}$ dried soil) with the corresponding k values and Si / Al ratio per fraction are presented in Table S1. In order to distinguish fractions according to the Si / Al ratio, the thresholds applied by Barão et al. (2014) were used: fractions showing Si / Al ratios above 5 were considered to be biogenic, and fractions showing Si / Al ratios below 5 were considered to be non-biogenic fractions.

Figure 3 shows the concentrations of biogenic (Si / Al > 5) and non-biogenic AlkExSi fractions (Si / Al < 5) within the soil profiles of selected pits. Overall, the highest concentrations of biogenic AlkExSi appear in the top of the profiles or near the surface and decrease with depth. Biogenic AlkExSi is also more abundant at the bottom positions of the slopes. On the other hand, non-biogenic AlkExSi fractions are generally absent in the top soil layers and increase in concentration with depth.

3.3 AlkExSi pools

The biogenic and non-biogenic AlkExSi pools of selected pits at 10 cm intervals are presented in Table S5.

Figure 4 shows the biogenic and non-biogenic AlkExSi pools as a soil profile cut from the top to the bottom of the slope, for the four study sites.

The averages of biogenic and non-biogenic AlkExSi pools per position, land use and slope are shown in Table 1. As mentioned, other averaged AlkExSi pools were calculated when comparing forest to cropland ("Average 2" in Table 1). The pre-calculated average between the upper-middle and lower-middle position was used in the calculation for "Average 2" (Table 1) (i.e., values used for the gentle slope "Average 2" calculation were 16.7 (top), 16.1 (middle) and 6.79 $\mathrm{kg\,m^{-2}}$ (bottom)).

While the gentle and the steep slope of the forest showed near-equal biogenic AlkExSi pools (+10 % for the steep slope), non-biogenic AlkExSi pool might be higher on the steep slope (+81 % for the steep slope).

In the cropland, results were slightly different. Both AlkExSi pools were higher on the gently sloped cropland

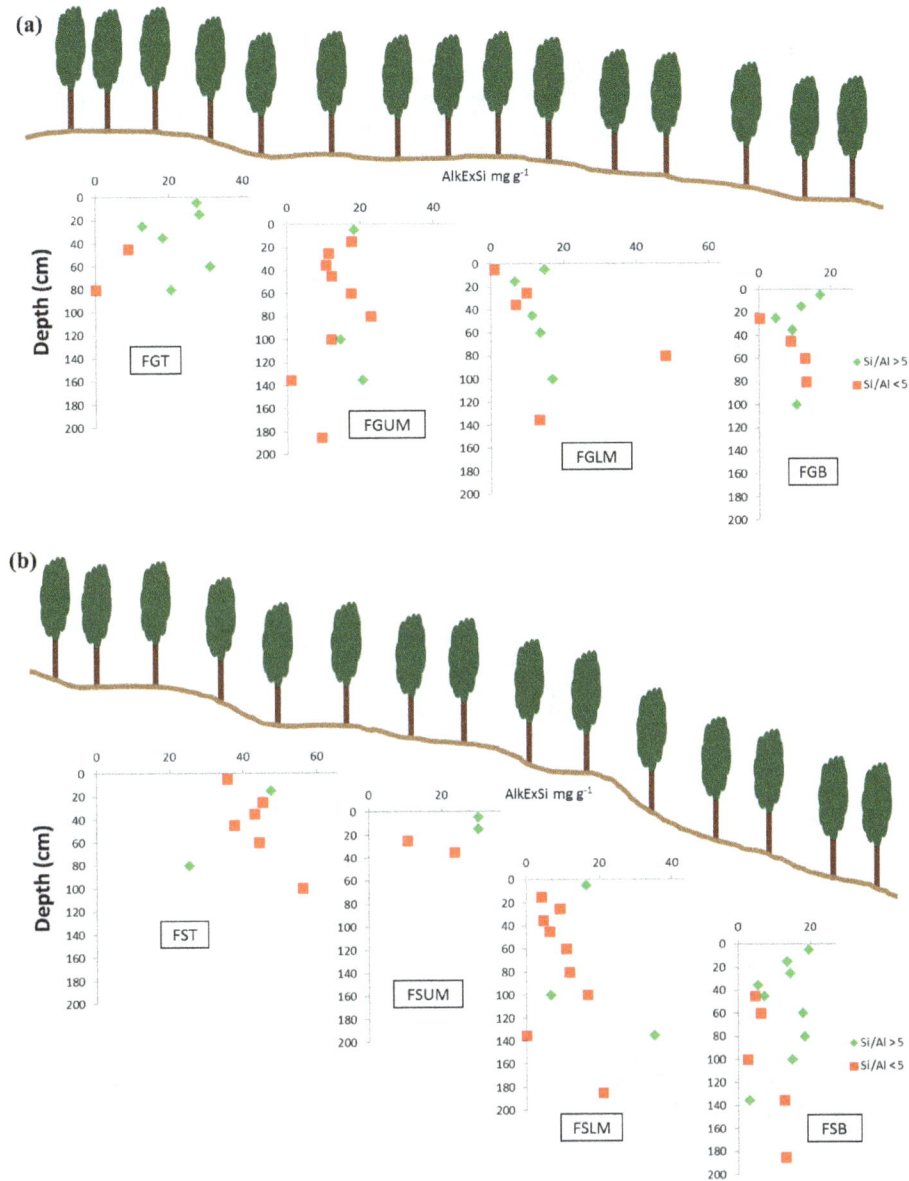

Figure 3.

(+35 % for the biogenic AlkExSi pool and +85 % for the non-biogenic AlkExSi pool).

When comparing gently sloped forest and cropland (using "Average 2" for forests), there was only a small difference for biogenic AlkExSi pool (−12 % for the cropland), but non-biogenic AlkExSi might be higher in the cropland (+ 57 % for the cropland).

On the steep slopes, it was clear that both AlkExSi pools were much lower in the cropland compared to the forest (−53 % for the biogenic AlkExSi pool and −90 % for the non-biogenic AlkExSi pool).

The sum of the AlkExSi pools of selected pits per land use and slope is shown in the Table 1 ("Total (sum)"). The accumulation of the biogenic and non-biogenic AlkExSi pools

at the bottom position of each slope is also shown in Table 2. Both steep slopes clearly showed higher accumulation of both pools at the lowest position than the gentle slopes, with the exception of the non-biogenic AlkExSi pool in the steep slope of the cropland.

Pairs showing significant differences are represented with the same letter in Table 1.

4 Discussion

One of the most striking observations in our study is the interaction between slope and land use effect. On the steep slope, there is a decrease in AlkExSi pools from forest to cropland.

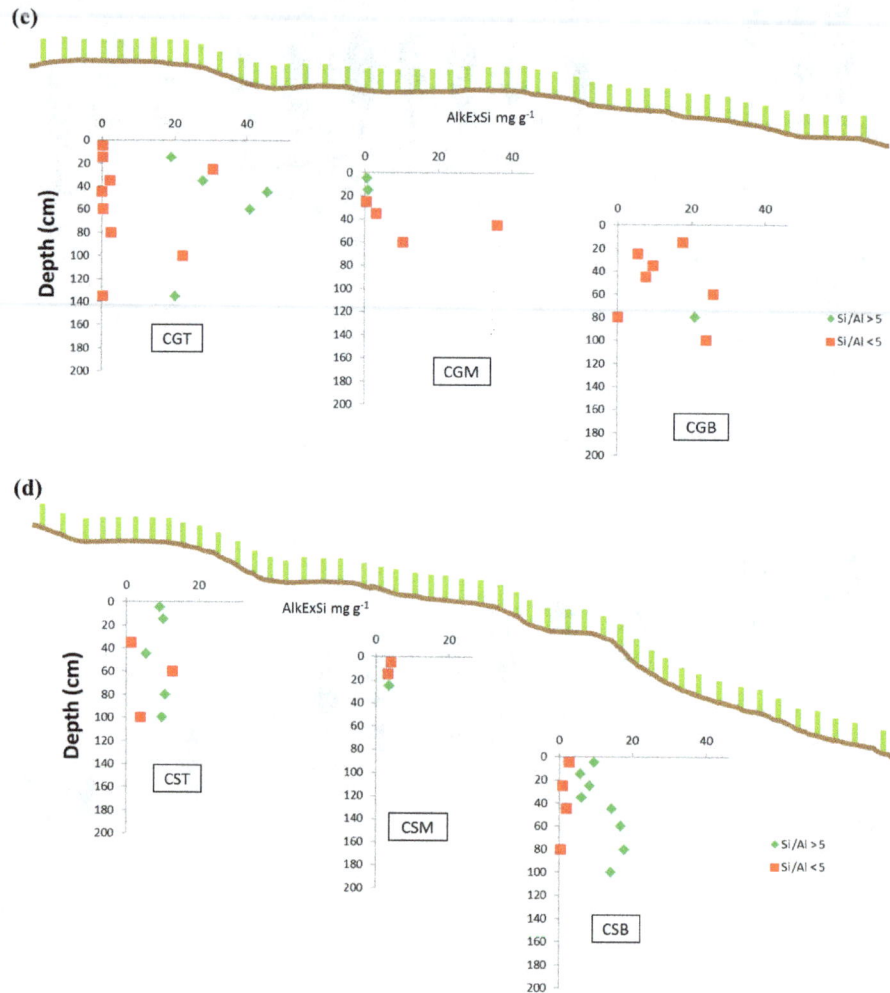

Figure 3. Biogenic and non-biogenic AlkExSi concentrations (mg g^{-1} dried soil) from selected pits of the sites studied: **(a)** gentle slope of the forest, **(b)** steep slope of the forest, **(c)** gentle slope of the cropland and **(d)** steep slope of the cropland. Graphs from left to right: top, upper middle, lower middle (or middle) and bottom pit.

In contrast, the gentle slopes had similar biogenic AlkExSi pools. It is also clear that there is redistribution of biogenic AlkExSi towards the bottom positions of the slope on steeply sloped croplands and forests.

4.1 Redistribution of AlkExSi concentrations in depth and along the toposequence

In general, the distribution of biogenic AlkExSi shows the same pattern within each pit: the concentration decreases with depth and highest concentrations are found at the bottom of the slope (with the exception of the gentle slope of the cropland). This agrees with earlier observations on the distribution of BSi along a toposequence in several soil catenas from temperate areas (Saccone et al., 2007). The distribution of non-biogenic AlkExSi shows a complementary pattern. Non-biogenic AlkExSi fractions are rarely present at the top of the profiles but higher concentrations are found

in deeper layers. Similar patterns were reported in a study carried out in arkosic sediment soils in California (Kendrick and Graham, 2004) and for temperate Luvisols in Belgium and Sweden (Barão et al., 2014; Vandevenne et al., 2015a). Upon leaching of DSi after BSi dissolution, the DSi infiltrates and reacts to form, for example, secondary clays. It can also be adsorbed onto oxides. The rate of adsorption of DSi by oxides is determined by water infiltration rate, pH, water residence time and weathering intensity (Cornelis et al., 2011; Jones and Handreck, 1963). A large amount of oxides in soil (see "Mineralogy" in Table S4), high DSi supply, strong water infiltration rates and high pH may result in larger concentrations of Si absorbed by oxides. Our studied sites satisfy these conditions with the exception of the pH (4.7–5.9). Uehara and Gillman (1981) suggested that weathered soil systems can result in a desilicated soil enriched in Fe and Al oxides, with pH close to neutral values. Similar

Table 1. Biogenic and non-biogenic AlkExSi pools (kg Si m^{-2}), of the selected pits, for the two ecosystems (forest, cropland), for the different slopes (gentle, steep), along different positions along the slope (top, upper middle, lower middle and bottom). Total (sum), Average 1 (averaged pool between all selected pits) and Average 2 (for the forest sites: averaged pool between top, pre-calculated average between the upper-middle and the lower-middle pit (i.e., for the biogenic AlkExSi pool of FG: 16.1 kg Si m^{-2}) and bottom pits) of biogenic and non-biogenic AlkExSi pools per site. Accumulation of the biogenic and non-biogenic AlkExSi pools (see Eqs. 3 and 4).

| | Forest | | | | Cropland | | | |
| | Gentle | | Steep | | Gentle | | Steep | |
	Biogenic	Non-biogenic	Biogenic	Non-biogenic	Biogenic	Non-biogenic	Biogenic	Non-biogenic
Top	17.6[a]	1.32	9.06	31.0	30.7[b]	15.0	7.50[ab]	3.25
Upper middle	19.3**[ab]	22.1	6.98**[cd]	3.98	0.21[ad]	8.50	0.43[bc]	0.93
Lower middle	12.9	10.3	25.8	25.7				
Bottom	6.79[ab]	7.63	24.8[ac]	20.3	5.38[cd]	16.0	15.8[bd]	0.80
Total (sum)	56.6	41.3	66.6	81.0	36.3	39.5	23.8	4.97
Average 1	14.2 ± 5	10.3 ± 7.6	16.6 ± 8.7	20.3 ± 10	12.1 ± 13	13.2 ± 3.3	7.92 ± 6.3	1.66 ± 1
Average 2	13.5 ± 5	8.4 ± 6.1	16.7 ± 6	22.1 ± 6.7				
Accumulation	12 %*	18 %	37 %	25 %	15 %*	41 %	67 %*	16 %

[a, b, c, d] Averages by row showing the same letter are statistically different ($p < 0.005$). * Difference between the top and the bottom pit averages from that slope is statistically significant ($p < 0.005$). ** Statistical comparison between middle position between forest sites and cropland sites were calculated taking the two middle pits (upper and lower middle) for the forests.

processes might occur in our soils, although they are not desilicated, but do show a high weathering intensity.

Biogenic Si concentrations from Vandevenne et al. (2015a) in temperate Luvisols were 1 order of magnitude lower than in our study. The high silica content of the rhyodacite bedrock in our study sites, together with high weathering rates characteristic of tropical and subtropical soils (Drever, 1994), supplies a large amount of DSi to the soil. In addition, weathering stimulated by plants is particularly strong in the tropics (Blecker et al., 2006; Kelly et al., 1998); turnover rates of nutrients are also higher in tropical and subtropical ecosystems than in temperate regions (Alexandre et al., 1997; Derry et al., 2005), due to high water availability and temperature. Meunier et al. (2010) showed that the DSi supply from the dissolution of basalts was 1.8 times higher than the DSi produced from the dissolution of the litter in a Leptosol of La Réunion (Indian Ocean).

4.2 Effects of erosion and land use change on the biogenic AlkExSi pool along the toposequence

For cropland, it is well documented that the harvest of crops exports large amounts of BSi from the system. This generates BSi-depleted systems in the long term (e.g., Vandevenne et al. 2015b). Results from Clymans et al. (2011) in long-term croplands from Sweden showed a BSi pool reduction of 10 % compared to a forested system.

Guntzer et al. (2012) showed the importance of crop rotation in the turnover and accumulation of phytoliths in soil. The accumulation of phytoliths is also influenced by the geochemical stability of phytoliths (Song et al., 2012). However,

the crops rotating in both fields are different and have different Si demands. Maize and black oat are known to have high Si content, while tobacco and soy do not (Currie and Perry, 2007; Piperno, 2006). The turnover between maize/tobacco and fallow/black oat on the steep slope might be an explanation for the smaller biogenic AlkExSi pool at this site. Moreover, the higher erosion rate increases the biogenic AlkExSi deposition at the bottom of the steeply sloped cropland. In fact, the TRB in this slope was higher than at any of the other sites (the lower the TRB, the more weathered the soil is, or vice versa – the higher the TRB, the closer the soil is to the composition of the bedrock), suggesting that all weathered material has been already eroded and the saprolite is closer to the surface.

It is interesting to note that a redistribution of biogenic AlkExSi occurs along the slope (Fig. 4). A higher slope degree, and thus higher erosion rate, provokes the loss of material through water erosion and tillage (Govers et al., 1996), transporting material downslope and resulting in an accumulation of the biogenic AlkExSi pool at the bottom of the slope. In the gently sloped sample site, biogenic AlkExSi is more stable at the higher positions of the slope, while in the steep slope it accumulates at the bottom.

The biogenic AlkExSi pool in the gentle slope of the forest was ~ 14 kg Si m^{-2}. A high rate of phytolith production in this forest, corresponding to a high Si demand from trees and efficient internal recycling, can maintain the BSi stock of the soil system. Ferrasols in Congolese equatorial forests had a phytolith pool 5 times smaller than the present results (2.66 kg m^{-2} in Alexandre et al., 1997) and the amorphous

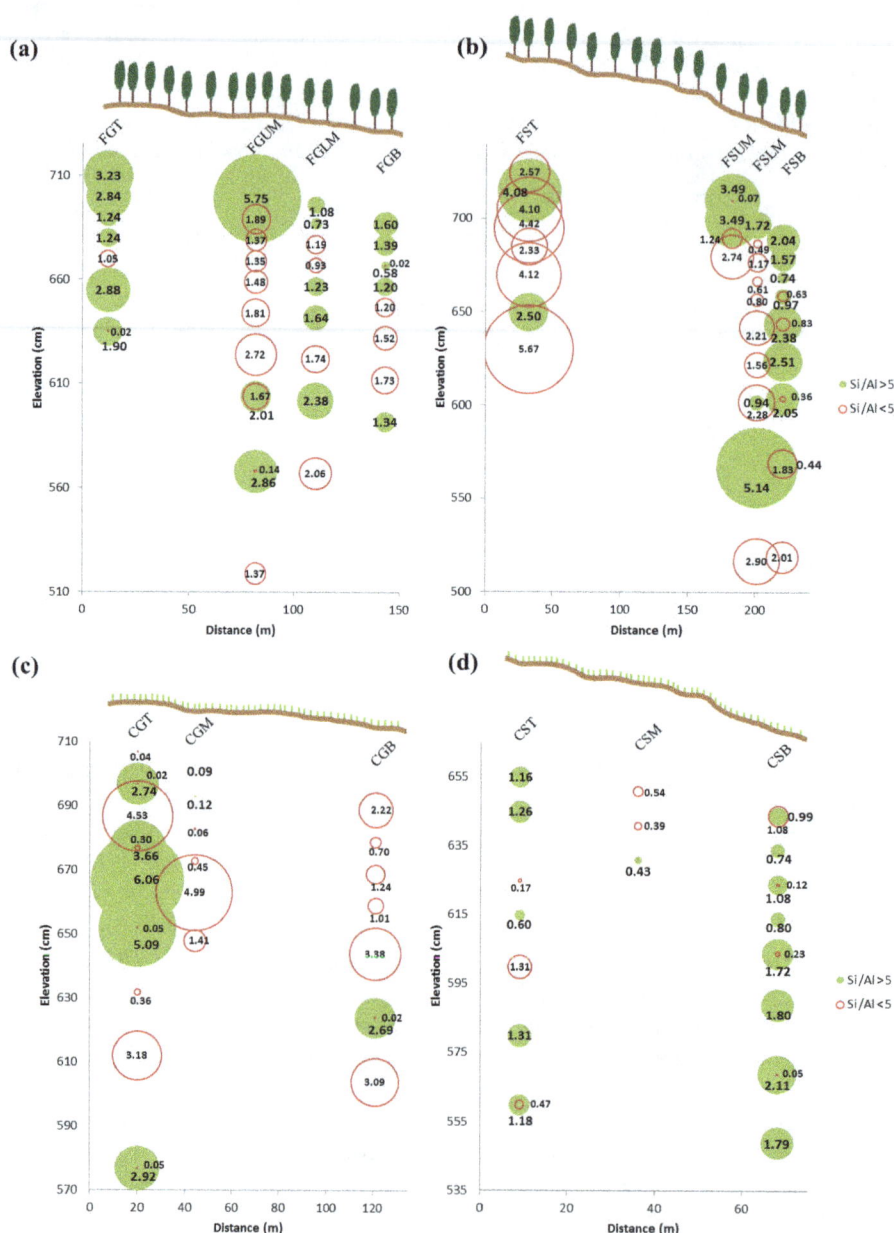

Figure 4. Biogenic and non-biogenic AlkExSi pools (kg m^{-2}) in the studied sites: **(a)** gentle slope of the forest (FG), **(b)** steep slope of the forest (FS), **(c)** gentle slope of the cropland (CG) and **(d)** steep slope of the cropland (CS). Green bubbles represent biogenic AlkExSi pools. Red empty bubbles represent non-biogenic AlkExSi pools. Labels show values of the pools (kg m^{-2}). Note that the x scales are different.

silica pool from temperate forests in Sweden was close to half our observations (6.7 kg m^{-2} in Clymans et al., 2011).

The biogenic AlkExSi pool was not enriched at the lowest position of the gently sloped forest. This suggests that the physical erosion at this site is low. In the steeply sloped forest, higher erosion rate apparently did provoke the physical loss of biogenic AlkExSi, potentially decreasing the amount of Si recycled by the vegetation. BSi is consequently transported to the bottom of the slope before it can dissolve and be recycled by plants, resulting in an accumulation of BSi at the bottom of the slope (AC of 37 %). However, this ap-

parent effect is not statically confirmed probably due to the strong variability of biogenic AlkExSi pools within the top and the bottom pits in the steep-sloped forest. Larger biogenic AlkExSi pools are also found at the lower-middle position, which suggests that the accumulation of eroded material also occurs at the lower-middle slope. Both (lower-middle and bottom) pits together accumulate the 76 % of the total biogenic AlkExSi pool of the slope. These deposition zones could serve as a location for permanent BSi storage.

The average biogenic AlkExSi pool size followed the sequence FS > FG > CG > CS. Overall, cropland gentle and steep slopes had 10 and 53 % lower biogenic AlkExSi pool, respectively, compared to well-conserved forest. This loss of biogenic AlkExSi has previously been described in other studies. Vandevenne et al. (2015b) showed similar results for temperate Belgian Luvisols, where croplands showed a decrease in total biogenic AlkExSi of 35 % compared to the temperate forest. Results from Clymans et al. (2011) support the same pattern, showing smaller AlkExSi pools in cultivated systems in Sweden. Our results are apparently in contrast with results from Struyf et al. (2010), who showed a large reduction in DSi export after deforestation in croplands deforested > 250 years ago. Nevertheless, the absence of a larger decrease in the gently sloped cropland may indicate that deforestation occurred too recently to see such a decrease, only triggered by harvest. Opfergelt et al. (2010) found phytoliths from the previous forested system in croplands of Cameroon deforested in the early 1950s. However, top and bottom positions do not differ statistically between the cropland and its forest counterpart for any of the slopes. The difference relies only on the mid-positions, where erosion is higher (Doetterl et al., 2015), highlighting the importance of erosion as an added factor, as a consequence of the agricultural tillage (Govers et al., 1996).

A depletion of > 50 % is seen at the steep slope of the cropland compared to its forested counterpart. Although it has been shown that an increase in erosion rate occurs after the conversion from forests to croplands (Vanacker et al., 2014) and this may affect both croplands, Montgomery and Brandon (2002) described how the erosion rate depends directly on the slope and stressed the importance of landslides. The consequence is an increase in the accumulation of biogenic AlkExSi pool at the bottom of the steep slope of the cropland (AC of 67 %).

4.3 Importance of scales and methods

The present study clearly shows how deforestation may have a strong impact on the silica cycling in subtropical soils under steep slopes, and potentially also on gentler slopes in the long term. The croplands in earlier studies, e.g., Vandevenne et al. (2015b), had usually been cultivated for more than 200 years, and BSi depletion was explained as a result of long-term cultivation. However, the croplands in the present study were deforested 50 years ago, highlighting how fast the biogenic AlkExSi pool can be depleted from the soil system when physical erosion is high.

Our results confirm the importance of using a continuous extraction to determine BSi pools in soils (Barão et al., 2014). The non-biogenic AlkExSi fractions would have been determined as BSi if conventional alkaline extractions, applying only analyses during the linear phase of the extraction, had been used (i.e., adaptations of the method from DeMaster, 1981). We acknowledge that some difficulties still remain when applying the method we have used. The dissolution in NaOH does not show a true reactivity within soils: the non-

biogenic AlkExSi fractions probably have lower solubility in soils (Ronchi et al., 2015) or water (Unzué-Belmonte et al., 2016) than BSi. Using the Si / Al ratio thresholds described for temperate soils to determine the character of the fractions in a different soil introduces some concerns. Without physical extraction we cannot verify that fractions showing specific ratios (below 5) correspond to the same pedogenic compounds as those found in temperate soils. The method is also unable to distinguish, among the Si / Al > 5 fractions, between phytoliths and opal-A/CT. Under a silica-saturated system, silica can precipitate in amorphous structures called opal-A, that in further transformations could be transformed into opal-CT and finally microquartz (Chadwick et al., 1987; Drees et al., 1989). Opal deposits were identified at more than 1 m deep layers in temperate pastures (Vandevenne et al., 2015b) and the tropics (Alexandre et al., 1997). Moreover, results from (Saccone et al., 2007) showed that the amounts of easily soluble silica were larger in deeper horizons, agreeing with the possibility of having opal-A at deeper layers in our systems. Despite some concerns, the method used allowed us to identify a new non-biogenic AlkExSi pool which might have been affected by land use and erosion as well.

4.4 Effects of erosion and land use change on the non-biogenic AlkExSi pool along the toposequence

The averaged total pool of non-biogenic AlkExSi followed the sequence FS > CG > FG > CS. A study in Belgian Luvisols under long-term cropland management (Vandevenne et al., 2015a) showed a larger non-biogenic AlkExSi pool in the croplands relative to a forested site. The authors explained the result by the fact that the high Si demand from the crops increases the weathering rate of the mineral phases, transforming low-solubility compounds into high-solubility ones (with the caveat that solubility is determined in NaOH). A combination of a relatively short time period since deforestation and the increased demand for Si by the crops compared to forest species could thus explain the larger non-biogenic AlkExSi pool in gently sloping cropland, compared to forests.

However, the non-biogenic AlkExSi pool of the steeply sloping cropland is almost non-existent. As with the biogenic AlkExSi pool, the high Si demand by crops together with the higher erosion rate results in a complete depletion of the non-biogenic AlkExSi pool in the steeply sloped cropland.

The steeply sloped forest showed a larger non-biogenic AlkExSi pool, mainly accumulated at top and bottom positions (Fig. 4). It is clear that the continuous long-term biogenic AlkExSi deposition at bottom positions (apparent also at the lower-middle position) triggers the formation of new non-biogenic AlkExSi phases that correspond with lower TRB values. Weathering degree has previously been correlated to the amount of pedogenic silica accumulation in sedimentary soils (Kendrick and Graham, 2004). Further, clay minerals and Si adsorbed onto oxides were reported by Delvaux et al. (1989) and Opfergelt et al. (2009), respectively, to be largest at most weathered sites in a study carried out in volcanic soils from Cameroon.

4.5 Implications

We show how slope and land use change have strong interacting effects on the distribution of the AlkExSi pool in a subtropical soil. In general, our study agrees well with earlier findings in temperate climates: landscape cultivation diminishes soil BSi stocks. Even though deforestation occurred only 50 years ago, the biogenic AlkExSi pool in the steeply sloped cropland was only 50 % of the pool in steeply sloped forests. In contrast, on the gentle slopes, no similar depletion was observed. This highlights the importance of erosion strength for the rate of depletion. To our knowledge, almost no studies have included slope as a potential factor (Ibrahim and Lal, 2014). It could therefore also be relevant to include erosion rates in studies of BSi in temperate ecosystems.

The presence of phytoliths from the past in soils helps to reconstruct former vegetation (Kirchholtes et al., 2015; Rovner, 1971). Here, we consider the biogenic Si pool as a single biogeochemical pool that is able to supply readily available DSi for plants. Although the presence of two Si pools within the plant is well documented (Fraysse et al., 2009; Watteau and Villemin, 2001) and different pools may show different solubilities, the higher solubility of phytoliths in soils compared to non-biological solid Si phases has been confirmed by several studies (Fraysse et al., 2006; Lindsay, 1979; Ronchi et al., 2015; Sommer et al., 2013). Moreover, Alexandre et al. (1997) described how 92 % of the BSi in top soil is rapidly recycled, while only 8 % seems to be permanently stored due to a lower turnover.

The silicon and carbon cycles are closely related through the production of phytoliths. A recent study showed a positive relation between soil organic carbon (SOC) and amorphous silica content along a toposequence and along the depth profile (Ibrahim and Lal, 2014). However, a comparison between their results and ours is not possible due to the different methods used to extract the silica fractions. The assumed tight relationship between both elements together with the SOC depletion (reported at 45 %) after 11–50 years of conversion from forest to cropland (Wei et al., 2014) hints at similar mechanisms behind both observations. Some studies have indicated that silica could act as a "carbon protector" through phytolith formation: carbon is occluded within the phytoliths and remains stored until they dissolve (Song et al., 2014). Although there are different opinions regarding this topic (Santos and Alexandre, 2017) some have suggested that atmospheric carbon sequestration could be enhanced through phytolith production and subsequent burial (Li et al., 2013; Parr et al., 2010; Song et al., 2016).

Our study highlights the accumulation of biogenic AlkExSi at deposition zones in croplands. Very little is known on the potential Si sink associated with such deposition zones, as little research has actually focused on Si biogeochemistry in these zones. Deposition of BSi here could be an important sink for Si in the long term. As shown earlier in tidal marshes (Struyf et al., 2007), rapid accumulation of BSi can prevent its complete dissolution, resulting in long-term burial and removal from the global biogeochemical Si cycle.

Competing interests. The authors declare that they have no conflict of interest.

Acknowledgements. We thank BELSPO for funding project SOGLO (The soil system under global change, P7/24), all the members of the SOGLO project and the University of Santa Maria for their help during field work in Brazil. Dácil Unzué-Belmonte also thanks the Soil System Sciences Division of the European Geoscience Union (EGU) for awarding her the best Outstanding Student Poster Award at the 2014 EGU Assembly.

Edited by: Miriam Muñoz-Rojas

References

Alexandre, A., Meunier, J.-D., Colin, F., and Koud, J.-M.: Plant impact on the biogeochemical cycle of silicon and related weathering processes, Geochim. Cosmochim. Ac., 61, 677–682, https://doi.org/10.1016/S0016-7037(97)00001-X, 1997.

Ameijeiras-Mariño, Y.: Changes in soil chemical weathering intensity after forest conversion to cropland: importance of slope geomorphology, in: The response of soil chemical weathering to physical erosion: integrating the impact of forest conversion, edited by: Université catholique de Louvain (UCL), PhD thesis, Belgium, 2017.

Barão, L., Clymans, W., Vandevenne, F., Meire, P., Conley, D. J., and Struyf, E.: Pedogenic and biogenic alkaline-extracted silicon distributions along a temperate land-use gradient, Eur. J. Soil Sci., 65, 693–705, https://doi.org/10.1111/ejss.12161, 2014.

Barão, L., Vandevenne, F., Clymans, W., Frings, P., Ragueneau, O., Meire, P., Conley, D. J., and Struyf, E.: Alkaline-extractable silicon from land to ocean: A challenge for biogenic silicon determination, Limnol. Oceanogr. Methods, 13, 329–344, https://doi.org/10.1002/lom3.10028, 2015.

Bartoli, F.: Crystallochemistry and surface properties of biogenic opal, J. Soil Sci., 36, 335–350, https://doi.org/10.1111/j.1365-2389.1985.tb00340.x, 1985.

Blecker, S. W., Mcculley, R. L., Chadwick, O. A., and Kelly, E. F.: Biologic cycling of silica across a grassland bioclimosequence, Global Biogeochem. Cy., 20, 1–11, https://doi.org/10.1029/2006GB002690, 2006.

Campforts, B., Van de Broek, M., Trigalet, S., Schoonejans, J., Robinet, J., Unzué-Belmonte, D., Ameijeiras-Mariño, Y., Vermeire, M. L., and Minella, J.: Soil texture data Brazilian SOGLO field sites, available at: https://doi.org/10.17632/pcbpbx5x7n.1 (last access: 27 June 2017), 2016.

Chadwick, O. A., Hendricks, D. M., and Nettleton, W. D.: Silica in Duric Soils: II. Mineralogy1, Soil Sci. Soc. Am. J., 51, 982–985, https://doi.org/10.2136/sssaj1987.03615995005100040029x, 1987.

Chao, T. T. and Sanzolone, R. F.: Decomposition techniques, J. Geochemical Explor., 44, 65–106, 1992.

Clymans, W., Struyf, E., Govers, G., Vandevenne, F., and Conley, D. J.: Anthropogenic impact on amorphous silica pools in temperate soils, Biogeosciences, 8, 2281–2293, https://doi.org/10.5194/bg-8-2281-2011, 2011.

Clymans, W., Struyf, E., Van den Putte, A., Langhans, C., Wang, Z., and Govers, G.: Amorphous silica mobilization by inter-rill erosion: Insights from rainfall experiments, Earth Surf. Proc. Land., 40, 1171–1181, https://doi.org/10.1002/esp.3707, 2015.

Conley, D. J., Likens, G. E., Buso, D. C., Saccone, L., Bailey, S. W., and Johnson, C. E.: Deforestation causes increased dissolved silicate losses in the Hubbard Brook Experimental Forest, Glob. Chang. Biol., 14, 2548–2554, https://doi.org/10.1111/j.1365-2486.2008.01667.x, 2008.

Cornelis, J. and Delvaux, B.: Soil processes drive the biological silicon feedback loop, Funct. Ecol., 1, 1298–1310, https://doi.org/10.1111/1365-2435.12704, 2016.

Cornelis, J.-T., Delvaux, B., Georg, R. B., Lucas, Y., Ranger, J., and Opfergelt, S.: Tracing the origin of dissolved silicon transferred from various soil-plant systems towards rivers: a review, Biogeosciences, 8, 89–112, https://doi.org/10.5194/bg-8-89-2011, 2011.

Currie, H. A. and Perry, C. C.: Silica in Plants: Biological, Biochemical and Chemical Studies, Ann. Bot., 100, 1383–1389, https://doi.org/10.1093/aob/mcm247, 2007.

Delvaux, B., Herbillon, A. J., and Vielvoye, L.: Characterization of a weathering sequence of soils derived from volcanic ash in Cameroon. Taxonomic, mineralogical and agronomic implications, Geoderma, 45, 375–388, https://doi.org/10.1016/0016-7061(89)90017-7, 1989.

DeMaster, D. J.: The supply and accumulation of silica in the marine environment, Geochim. Cosmochim. Ac., 45, 1715–1732, https://doi.org/10.1016/0016-7037(81)90006-5, 1981.

Derry, L. A., Kurtz, A. C., Ziegler, K., and Chadwick, O. A.: Biological control of terrestrial silica cycling and export fluxes to watersheds, Nature, 433, 728–31, https://doi.org/10.1038/nature03299, 2005.

Doetterl, S., Cornelis, J.-T., Six, J., Bodé, S., Opfergelt, S., Boeckx, P., and Van Oost, K.: Soil redistribution and weathering controlling the fate of geochemical and physical carbon stabilization mechanisms in soils of an eroding landscape, Biogeosciences, 12, 1357–1371, https://doi.org/10.5194/bg-12-1357-2015, 2015.

Drees, L. R., Wilding, L. P., Smeck, N. E., and Senkayi, A. L.: Silica in soils: quartz and disorders polymorphs, in Minerals in soil environments, 914–974, Soil Science Society of America, Madison, 1989.

Drever, J. I.: The effect of land plants on weathering rates of silicate minerals, Geochim. Cosmochim. Ac., 58, 2325–2332, https://doi.org/10.1016/0016-7037(94)90013-2, 1994.

Fraysse, F., Pokrovsky, O. S., Schott, J., and Meunier, J.: Surface properties, solubility and dissolution kinetics of bamboo phytoliths, 70, 1939–1951, https://doi.org/10.1016/j.gca.2005.12.025, 2006.

Fraysse, F., Pokrovsky, O. S., Schott, J., and Meunier, J.-D.: Surface chemistry and reactivity of plant phy-

toliths in aqueous solutions, Chem. Geol., 258, 197–206, https://doi.org/10.1016/j.chemgeo.2008.10.003, 2009.

Govers, G., Quine, T. A., Desmet, P. J. J., and Walling, D. E.: The relative contribution of soil tillage and overland flow erosion to soil redistribution on agricultural land, Earth Surf. Proc. Land., 21, 929–946, https://doi.org/10.1002/(SICI)1096-9837(199610)21:10<929::AID-ESP631>3.0.CO;2-C, 1996.

Guntzer, F., Keller, C., Poulton, P. R., McGrath, S. P., and Meunier, J. D.: Long-term removal of wheat straw decreases soil amorphous silica at Broadbalk, Rothamsted, Plant Soil, 352, 173–184, https://doi.org/10.1007/s11104-011-0987-4, 2012.

Hall, T. B., Rosillo-Calle, F., Williams, R. H., and Woods, J.: Biomass for energy: supply prospects, in: Renewable Energy: Sources for Fuels and Electricity, edited by: Hall, R. H., Kelly, T. B., Reddy, H., Williams, A. K. N., 593–651, Island Press, Washington DC, 1993.

Herbillon, A. J.: Chemical estimation of weatherable minerals present in the diagnostic horizons of low activity clay soils, in: Proceedings of the 8th International Soil Classification Workshop: Classification, Characterization and Utilization of Oxisols, Part 1, edited by: EMBRAPA, 39–48, Rio de Janeiro, 1986.

Ibrahim, M. A. and Lal, R.: Catena Soil carbon and silicon pools across an un-drained toposequence in central Ohio, Catena, 120, 57–63, https://doi.org/10.1016/j.catena.2014.04.006, 2014.

IUSS Working Group WRB: World Reference Base for Soil Resources 2014, update 2015, International soil classification system for naming soils and creating legends for soil maps, World Soil Resources Reports No. 106, FAO, Rome, 2015.

Jones, L. H. P. and Handreck, K. A.: Effects of Iron and Aluminium Oxides on Silica in Solution in Soils, Nature, 198, 852–853, https://doi.org/10.1038/198852a0, 1963.

Keller, C., Guntzer, F., Barboni, D., Labreuche, J., and Meunier, J. D.: Impact of agriculture on the Si biogeochemical cycle: Input from phytolith studies, C. R. Geosci., 344, 739–746, https://doi.org/10.1016/j.crte.2012.10.004, 2012.

Kelly, E. F., Chadwick, O. A., and Hilinski, T. E.: The effect of plants on mineral weathering, Biogeochemistry, 42, 21–53, https://doi.org/10.1007/978-94-017-2691-7_2, 1998.

Kendrick, K. J. and Graham, R. C.: Pedogenic silica accumulation in chronosequence soils, southern California, Soil Sci. Soc. Am. J., 68, 1295–1303, 2004.

Kirchholtes, R. P. J., van Mourik, J. M., and Johnson, B. R.: Phytoliths as indicators of plant community change: A case study of the reconstruction of the historical extent of the oak savanna in the Willamette Valley Oregon, USA, Catena, 132, 89–96, https://doi.org/10.1016/j.catena.2014.11.004, 2015.

Koning, E., Epping, E., and Van Raaphorst, W.: Determining Biogenic Silica in Marine Samples by Tracking Silicate and Aluminium Concentrations in Alkaline Leaching Solutions, Aquat. Geochem., 8, 37–67, https://doi.org/10.1023/A:1020318610178, 2002.

Kottek, M., Grieser, J., Beck, C., Rudolf, B., and Rubel, F.: World Map of the Köppen-Geiger climate classification updated, Meteorol. Z., 15, 259–263, https://doi.org/10.1127/0941-2948/2006/0130, 2006.

Li, Z., Song, Z., Parr, J. F., and Wang, H.: Occluded C in rice phytoliths: implications to biogeochemical carbon sequestration, Plant Soil, 370, 615–623, https://doi.org/10.1007/s11104-013-1661-9, 2013.

Lindsay, W. L.: Chemical equilibria in soils, Wiley, New York, available at: http://soils.ifas.ufl.edu/lqma/SEED/CWR6252/Handout/Chemicalequilibira.pdf (last access: 27 June 2017), 1979.

Meunier, J., Colin, F., and Alarcon, C.: Biogenic silica storage in soils, Geology, 27, 835–838, https://doi.org/10.1130/0091-7613, 1999.

Meunier, J. D., Kirman, S., Strasberg, D., Nicolini, E., Delcher, E., and Keller, C.: The output and bio-cycling of Si in a tropical rain forest developed on young basalt flows (La Reunion Island), Geoderma, 159, 431–439, https://doi.org/10.1016/j.geoderma.2010.09.010, 2010.

Minella, J. P. G., Walling, D. E., and Merten, G. H.: Establishing a sediment budget for a small agricultural catchment in southern Brazil, to support the development of effective sediment management strategies, J. Hydrol., 519, 2189–2201, https://doi.org/10.1016/j.jhydrol.2014.10.013, 2014.

Montgomery, D. R. and Brandon, M. T.: Topographic controls on erosion rates in tectonically active mountain ranges, Earth Planet. Sci. Lett., 201, 481–489, https://doi.org/10.1016/S0012-821X(02)00725-2, 2002.

Opfergelt, S., de Bournonville, G., Cardinal, D., André, L., Delstanche, S., and Delvaux, B.: Impact of soil weathering degree on silicon isotopic fractionation during adsorption onto iron oxides in basaltic ash soils, Cameroon, Geochim. Cosmochim. Ac., 73, 7226–7240, https://doi.org/10.1016/j.gca.2009.09.003, 2009.

Opfergelt, S., Cardinal, D., André, L., Delvigne, C., Bremond, L., and Delvaux, B.: Variations of $\delta 30Si$ and Ge / Si with weathering and biogenic input in tropical basaltic ash soils under monoculture, Geochim. Cosmochim. Ac., 74, 225–240, https://doi.org/10.1016/j.gca.2009.09.025, 2010.

Parr, J., Sullivan, L., Chen, B., Ye, G., and Zheng, W.: Carbon bio-sequestration within the phytoliths of economic bamboo species, Glob. Chang. Biol., 16, 2661–2667, https://doi.org/10.1111/j.1365-2486.2009.02118.x, 2010.

Piperno, D. R.: Phytoliths: A Comprehensive Guide for Archaeologists and Paleoecologists, Altamira Press, San Diego, 2006.

Robert, M. and Tessier, D.: Méthode de préparation des argiles des sols pour des études minéralogiques, Ann. Agron., 25, 859–882, 1974.

Ronchi, B., Barão, L., Clymans, W., Vandevenne, F., Batelaan, O., Govers, G., Struyf, E., and Dassargues, A.: Factors controlling Si export from soils: A soil column approach, Catena, 133, 85–96, https://doi.org/10.1016/j.catena.2015.05.007, 2015.

Rovner, I.: Potential of opal phytoliths for use in paleoecological reconstruction, Quat. Res., 1, 343–359, https://doi.org/10.1016/0033-5894(71)90070-6, 1971.

Saccone, L., Conley, D. J., Koning, E., Sauer, D., Sommer, M., Kaczorek, D., Blecker, S. W., and Kelly, E. F.: Assessing the extraction and quantification of amorphous silica in soils of forest and grassland ecosystems, Eur. J. Soil Sci., 58, 1446–1459, https://doi.org/10.1111/j.1365-2389.2007.00949.x, 2007.

Santos, G. M. and Alexandre, A.: Earth-Science Reviews The phytolith carbon sequestration concept: Fact or fiction? A comment on "Occurrence, turnover and carbon sequestration potential of phytoliths in terrestrial ecosystems by Song et al. https://doi.org/10.1016/j.earscirev.2016.04.007", Earth Sci. Rev., 164, 251–255, https://doi.org/10.1016/j.earscirev.2016.11.005, 2017.

Sauer, D., Saccone, L., Conley, D. J., Herrmann, L., and Sommer, M.: Review of methodologies for extracting plant-available and amorphous Si from soils and aquatic sediments, Biogeochemistry, 80, 89–108, https://doi.org/10.1007/s10533-005-5879-3, 2006.

Smis, A., Van Damme, S., Struyf, E., Clymans, W., Van Wesemael, B., Frot, E., Vandevenne, F., Van Hoestenberghe, T., Govers, G., and Meire, P.: A trade-off between dissolved and amorphous silica transport during peak flow events (Scheldt river basin, Belgium): Impacts of precipitation intensity on terrestrial Si dynamics in strongly cultivated catchments, Biogeochemistry, 106, 475–487, https://doi.org/10.1007/s10533-010-9527-1, 2011.

Sommer, M., Kaczorek, D., Kuzyakov, Y., and Breuer, J.: Silicon pools and fluxes in soils and landscapes – a review, J. Plant Nutr. Soil Sci., 169, 310–329, https://doi.org/10.1002/jpln.200521981, 2006.

Sommer, M., Jochheim, H., Höhn, A., Breuer, J., Zagorski, Z., Busse, J., Barkusky, D., Meier, K., Puppe, D., Wanner, M., and Kaczorek, D.: Si cycling in a forest biogeosystem – the importance of transient state biogenic Si pools, Biogeosciences, 10, 4991–5007, https://doi.org/10.5194/bg-10-4991-2013, 2013.

Song, Z., Wang, H., Strong, P. J., Li, Z., and Jiang, P.: Plant impact on the coupled terrestrial biogeochemical cycles of silicon and carbon: Implications for biogeochemical carbon sequestration, Earth-Sci. Rev., 115, 319–331, https://doi.org/10.1016/j.earscirev.2012.09.006, 2012.

Song, Z., Müller, K., and Wang, H.: Biogeochemical silicon cycle and carbon sequestration in agricultural ecosystems, Earth-Sci. Rev., 139, 268–278, https://doi.org/10.1016/j.earscirev.2014.09.009, 2014.

Song, Z., McGrouther, K., and Wang, H.: Occurrence, turnover and carbon sequestration potential of phytoliths in terrestrial ecosystems, Earth-Sci. Rev., 158, 19–30, https://doi.org/10.1016/j.earscirev.2016.04.007, 2016.

Struyf, E. and Conley, D. J.: Emerging understanding of the ecosystem silica filter, Biogeochemistry, 107, 9–18, https://doi.org/10.1007/s10533-011-9590-2, 2012.

Struyf, E., Temmerman, S., and Meire, P.: Dynamics of biogenic Si in freshwater tidal marshes: Si regeneration and retention in marsh sediments (Scheldt estuary), Biogeochemistry, 82, 41–53, https://doi.org/10.1007/s10533-006-9051-5, 2007.

Struyf, E., Smis, A., Van Damme, S., Garnier, J., Govers, G., Van Wesemael, B., Conley, D. J., Batelaan, O., Frot, E., Clymans, W., Vandevenne, F., Lancelot, C., Goos, P., and Meire, P.: Historical land use change has lowered terrestrial silica mobilization, Nat. Commun., 1, 129, https://doi.org/10.1038/ncomms1128, 2010.

Uehara, G. and Gillman, G.: The Mineralogy, Chemistry, and Physics of Tropical Soils With Variable Charge Clays, West View Press, Boulder, CO, USA, 1981.

Unzué-Belmonte, D., Struyf, E., Clymans, W., Tischer, A., Potthast, K., Bremer, M., Meire, P., and Schaller, J.: Fire enhances solubility of biogenic silica, Sci. Total Environ., 572, 1289–1296, https://doi.org/10.1016/j.scitotenv.2015.12.085, 2016.

Unzué-Belmonte, D., Ameijeiras-Mariño, Y., Trigalet, S., Schoonejans, J., Campforts, B., Van de Broek, M., Robinet, J., and Minella, J.: Alkaline Extractable Silica – SOGLO Project, available at: https://doi.org/10.17632/r996jnwhtg.1, last access: 27 June 2017.

Vanacker, V., Bellin, N., Molina, A., and Kubik, P. W.: Erosion regulation as a function of human disturbances to vegetation cover: a conceptual model, Landsc. Ecol, 29, 293–309, https://doi.org/10.1007/s10980-013-9956-z, 2014.

Van Cappellen, P.: Biomineralization and global biogeochemical cycles, Rev. Mineral. Geochemistry, 54, 357–381, https://doi.org/10.2113/0540357, 2003.

Vandevenne, F., Struyf, E., Clymans, W., and Meire, P.: Agricultural silica harvest: have humans created a new loop in the global silica cycle?, Front. Ecol. Environ., 10, 243–248, https://doi.org/10.1890/110046, 2012.

Vandevenne, F. I., Barão, L., Ronchi, B., Govers, G., Meire, P., Kelly, E. F., and Struyf, E.: Silicon pools in human impacted soils of temperate zones, Global Biogeochem. Cy., 29, 1439–1450, https://doi.org/10.1002/2014GB005049, 2015a.

Vandevenne, F. I., Delvaux, C., Hughes, H. J., André, L., Ronchi, B., Clymans, W., Barão, L., Cornelis, J.-T., Govers, G., Meire, P., and Struyf, E.: Landscape cultivation alters $\delta 30Si$ signature in terrestrial ecosystems, Sci. Rep., 5, 7732, https://doi.org/10.1038/srep07732, 2015b.

Von Braun, J.: The world food situation: new driving forces and required actions, Food Policy Reports 18, International Food Policy Research Institute, Washington DC, 2007.

Watteau, F. and Villemin, G.: Ultrastructural study of the biogeochemical cycle of silicon in the soil and litter of a temperate forest, Eur. J. Soil Sci., 52, 385–396, https://doi.org/10.1046/j.1365-2389.2001.00391.x, 2001.

Wei, X., Shao, M., Gale, W., and Li, L.: Global pattern of soil carbon losses due to the conversion of forests to agricultural land, Sci. Rep., 1, 6–11, https://doi.org/10.1038/srep04062, 2014.

11

First magmatism in the New England Batholith, Australia: forearc and arc–back-arc components in the Bakers Creek Suite gabbros

Seann J. McKibbin[1,2,a], **Bill Landenberger**[1], and **C. Mark Fanning**[2]

[1]School of Environmental and Life Sciences, University of Newcastle, University Drive, Callaghan, 2308, Australia
[2]Research School of Earth Sciences, Australian National University, Bldg. 61, Mills Road, Canberra, 0200, Australia
[a]now at: Analytical, Environmental and Geo-Chemistry, Vrije Universiteit Brussel, Pleinlaan 2, Brussels 1050, Belgium

Correspondence to: Seann J. McKibbin (seann.mckibbin@gmail.com, seann.mckibbin@vub.ac.be)

Abstract. The New England Orogen, eastern Australia, was established as an outboard extension of the Lachlan Orogen through the migration of magmatism into forearc basin and accretionary prism sediments. Widespread S-type granitic rocks of the Hillgrove and Bundarra supersuites represent the first pulse of magmatism, followed by I- and A-types typical of circum-Pacific extensional accretionary orogens. Associated with the former are a number of small tholeiite–gabbroic to intermediate bodies of the Bakers Creek Suite, which sample the heat source for production of granitic magmas and are potential tectonic markers indicating why magmatism moved into the forearc and accretionary complexes rather than rifting the old Lachlan Orogen arc. The Bakers Creek Suite gabbros capture an early (~ 305 Ma) forearc basalt-like component with low Th / Nb and with high Y / Zr and Ba / La, recording melting in the mantle wedge with little involvement of a slab flux and indicating forearc rifting. Subsequently, arc–back-arc like gabbroic magmas (305–304 Ma) were emplaced, followed by compositionally diverse magmatism leading up to the main S-type granitic intrusion (~ 290 Ma). This trend in magmatic evolution implicates forearc and other mantle wedge melts in the heating and melting of fertile accretion complex sediments and relatively long (~ 10 Myr) timescales for such melting.

1 Introduction

The New England Orogen (NEO) is the youngest and easternmost component in the Tasmanides accretionary orogenic system and of the Australian continental craton (e.g. Cawood et al., 2011; Fig. 1a). The NEO has similarities to its older neighbour the Lachlan Orogen, such as west-dipping subduction (e.g. Leitch, 1974, 1975), a general tectonic regime switching between crustal thinning and thickening (Collins, 2002; Brown, 2003), and granitic magmatism spanning a compositional range between peraluminous and metaluminous end members (S-type and I-type for sedimentary and igneous sources respectively: Hensel et al., 1985; Chappell and White, 2001; Collins and Richards, 2008). However, the NEO represents eastward migration of magmatic activity into the Devonian–Carboniferous forearc basin and accretionary prism sediments on the margins of the Lachlan Orogen (Jenkins et al., 2002). These sediments, derived from "calc-alkaline" arc rocks, inherited juvenile isotopic characters that were passed on to their derivative granitic melts by rapid subduction cycling (Kemp et al., 2009).

In the Southern NEO, termination of the Carboniferous magmatic arc and replacement by widespread and relatively disorganised magmatism (Collins et al., 1993; Caprarelli and Leitch, 1998; Jenkins et al., 2002) culminated in the first phase of construction of the New England Batholith (Shaw and Flood, 1981), with emplacement of the contrasting Bundarra and Hillgrove S-type granitoid supersuites at ~ 290 Ma (Rosenbaum et al., 2012). They differ from each other with the former being a voluminous, compositionally homogenous belt, while the latter is variably foliated and generally more mafic in composition (Shaw and Flood, 1981), and it is associated with high-temperature low-pressure (HTLP) metamorphic complexes (Farrell, 1988; Dirks et al., 1992) as well as small, mafic to intermediate intrusive bodies referred to as the Bakers Creek Suite (Jenkins et al., 2002). Following

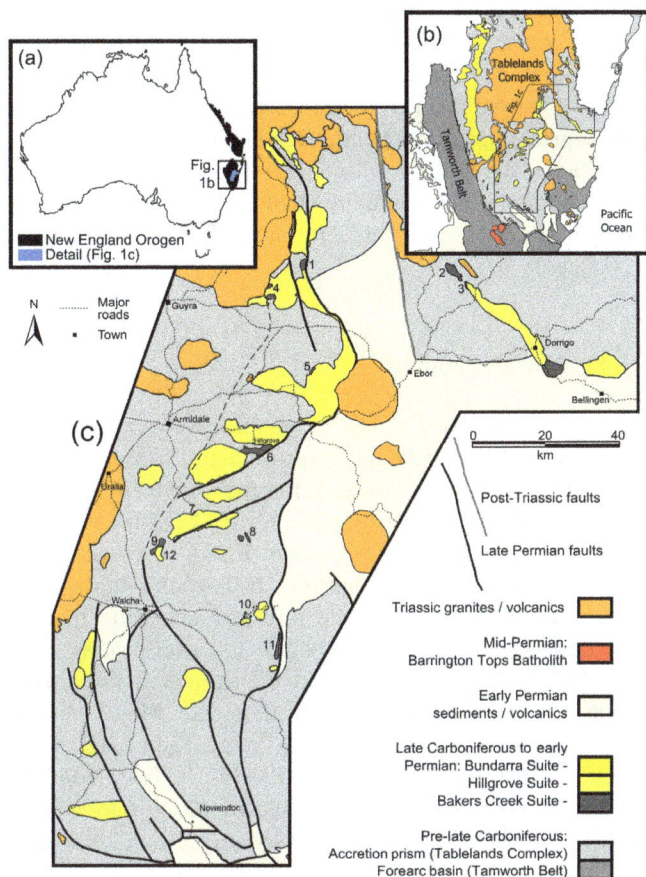

Figure 1. (a) New England Orogen **(b)** and field area (orange; c) on the Australian continent. **(b)** Southern New England Orogen outlining Tamworth Belt, Tablelands Complex, and study area (Fig. 1c). **(c)** Hillgrove and Bakers Creek Suite plutons in the Tablelands Complex; overlying Tertiary basalt omitted for clarity. Sampled Bakers Creek plutons: (1) Mornington; (2) Big Bull; (3) Charon Creek; (4) Days Creek; (5) Camperdown; (6) Bakers Creek; (7) Barney House; (8) Cheyenne; (9) Woodburn; (10) Moona Plains; (11) Apsley River; (12) East Lake (Hillgrove Suite).

and Stern, 2006), and early arc tholeiite (EAT; Todd et al., 2012) affinities. Each of these has distinctive trace element compositions that can potentially be recognised in palaeo-arc systems (Dilek and Furnes, 2014; Pearce, 2014). We present here a study of the geochemistry of the Bakers Creek Suite with emphasis on samples from uncontaminated, mafic plutons, and U–Pb chronology of these earliest magmatic rocks in the New England Batholith. Furthermore, we identify forearc and back-arc components and address the tectonic setting and mechanisms by which magmatism began in this section of an ancient extensional accretionary orogen.

2   Regional geology

The Southern NEO is built upon a metasedimentary base comprising the Tablelands Complex (an old accretionary prism) and the Tamworth Belt (a forearc basin), separated by the Peel–Manning Fault System (Leitch, 1974; Korsch, 1977; Glen and Roberts, 2012; Li et al., 2015). Both are related to a poorly exposed Devonian–Late Carboniferous magmatic arc on the margins of the Lachlan Orogen (Leitch, 1975). In the Tablelands Complex (Fig. 1b), high temperature and low pressure metamorphism overprints the accretion–subduction sequences (Wongwibinda and Tia Complexes; Farrell, 1988; Hand, 1988; Dirks et al., 1992; Phillips et al., 2008; Craven et al., 2012). Subsequently, intrusion of the Hillgrove Suite biotite granites and granodiorites (± garnet, hornblende) took place, forming a discontinuous belt of scattered plutons (Flood and Shaw, 1977; Shaw and Flood, 1981). Spatially associated with the Hillgrove granitoids are the small plutons of the Bakers Creek Suite, a diverse group of mafic to intermediate bodies ranging from two-pyroxene (± olivine) gabbros and related cumulate rocks through hornblende–biotite diorites to mafic hornblende-bearing granodiorites (Jenkins et al., 2002). The Hillgrove and Bakers Creek mafic plutons have been exhumed from depth as a result of early Permian rifting and subsequent thrusting during the Hunter-Bowen Orogeny (Fig. 1b; Landenberger et al., 1995; Li et al., 2014; Shaanan et al., 2015). Also present are the voluminous and more strongly peraluminous S-type granites of the Bundarra Suite, lying in a continuous north-trending belt to the west of the Hillgrove Suite (Flood and Shaw, 1977; Shaw and Flood, 1981). In contrast to the Hillgrove Suite, the Bundarra Suite granites are generally non-foliated, have no mafic plutons associated with them, and are not associated with metamorphic complexes, despite generally contemporaneous intrusion (Rosenbaum et al., 2012).

Mafic, primitive members of the Bakers Creek Suite include the small Barney House and Big Bull gabbros, while larger plutons such as the Days Creek gabbro and Apsley River Complex exhibit more complex characteristics of differentiation (e.g. samples BHC2, CC26A, G39, and GK2 respectively from Jenkins et al., 2002). Sampling was undertaken with a focus on mafic plutons such as the Bar-

minor magmatic activity of other types (e.g. I-type intrusions dated by Roberts et al., 1995; Donchak et al., 2007; Cross et al., 2009; Phillips et al., 2011) and a temporal magmatic gap associated with orogeny (Rosenbaum et al., 2012), the New England Batholith was overwhelmed by voluminous I-type magmatism from ~ 265 Ma (Li et al., 2012).

The mafic mantle-derived plutons of the Bakers Creek Suite, while small and variably evolved, ultimately record the conditions of mantle partial melting and subduction zone contributions to the first magmatism in the New England Batholith. New advances in the understanding of the geochemistry of arc-related magmas have established roles for the various mafic magmas emplaced during subduction zone initiation and migration. These include basalts with forearc (FAB; Reagan et al., 2010; Meffre et al., 2012; Ribeiro et al., 2013), back-arc (BAB; Langmuir et al., 2006; Pearce

Figure 2. Photos of the Days Creek Gabbro showing **(a)** massive gabbro enclosing domain of gabbro pegmatite. Camera lens has a diameter of 5 cm; **(b)** finer-grained dolerite pluton margin hosting felsic veins. Hammer length 30 cm.

ney House, Big Bull, and Days Creek gabbros. The Barney House and Big Bull gabbros are small (scale of tens to hundreds of metres) and consist of finely crystalline gabbro, often hosting plagioclase phenocrysts, in contact with low-grade metasedimentary country rock. The Big Bull Gabbro occurs as the most mafic member in a full spectrum of rocks, varying from mafic to felsic (Sheep Station Creek Complex). In contrast, the Days Creek Gabbro occurs as two larger plutons (~ 1 and ~ 2 km in length), partially surrounded by the Tobermory Monzogranite (Hillgrove Suite), except at the southern margin where it borders turbidites of the Girrakool beds. It is dominated by medium to coarse grained gabbro and contains rare pegmatite (grain size 5–20 mm; Fig. 2a). The southern pluton exhibits a doleritic (~ 1 mm) chilled margin against turbidites, which are contact-metamorphosed and exhibit rare occurrences of melting. Widespread but poorly exposed pieces of dolerite are found at various locations across both plutons, some in association with metasedimentary rocks and felsic veins containing gabbro breccia

(Fig. 2b). The encompassing Tobermory Monzogranite is usually coarse (average grain size a few millimetres) and lacks foliation. Although most contacts are not exposed, it is often finer grained (to ~ 1 mm) nearer the gabbro, indicative of quenching and late emplacement relative to most other members of the Hillgrove Suite (Landenberger et al., 1995). The Tobermory Monzogranite is cut on the western side by younger, unrelated mid–late Permian to Triassic I-type granite (Li et al., 2012).

3   Analytical methods

Selected 30 μm thin sections of samples were polished and carbon coated for X-ray analysis of mineral phases by scanning electron microprobe (SEM) at the University of Newcastle (UoN) using a Phillips XL30 SEM, with an Oxford ISIS energy dispersive spectrometer (EDS), 15 kV accelerating voltage, and 3 nA beam current. Bulk-rock samples were crushed by a tungsten carbide mill and losses on ignition were determined by weighing before and after heating in air at $\sim 1000\,°C$. These powders were then diluted in lithium borate flux at 1050 °C to produce a glass disc. Major element oxides (Na_2O, MgO, Al_2O_3, SiO_2, P_2O_5, K_2O, CaO, TiO_2, MnO, and FeO) and trace elements (Pb, P, Ti, V, Mn, Zn, and Cr reported here) were analysed by X-ray fluorescence (XRF) spectrometry at the UoN (Spectro X-Lab 2000 XRF system with EDS, Pb anode tube, polarised beam, multiple targets). All Fe is reported as FeO. Glass XRF discs from this study and from Jenkins et al. (2002) were sectioned and polished for further trace element analysis (Cs, Rb, Ba, Th, U, Nb, Ta, Sr, Zr, Hf, Ga, Y, Sc, Co, Ni, and the rare earth elements La, Ce, Pr, Nd, Sm, Eu, Gd, Dy, Er, and Yb) by laser ablation inductively coupled plasma mass spectrometry at the Research School of Earth Sciences, Australian National University (ANU), using a quadrupole Agilent 7500s coupled to a 193 nm argon fluoride Excimer laser (Eggins, 2003). Samples were analysed in parallel with NIST 612 (primary normalisation standard) and BCR-2g (secondary external standard) glasses, and either ^{43}Ca or ^{29}Si were used as internal standards, depending on bulk silica content (using CaO or SiO_2 from XRF). Data were reduced using an in-house spreadsheet. Further details are given in Supplement S5.

Magmatic zircon $^{238}U / ^{206}Pb$ ages of gabbroic and dioritic samples were determined at the ANU using sensitive high-resolution ion microprobes (SHRIMP). The gabbroic samples (Barney House and Days Creek gabbros) were analysed using SHRIMP-RG (reverse geometry) against the reference standard TEMORA, while dioritic samples (Bakers Creek Complex and Charon Creek Diorite) were analysed using SHRIMP-I against the AS3 reference material. Rejection of analyses was made on the basis of measurable common Pb, loss of Pb, or contribution to an unreasonably high mean square of weighted deviates (MSWD). In reviewing other U-Pb data for the NEO in the literature, it is noted that they were obtained against a range of reference materials over many years. The comparative study of Black et al. (2003) showed that some zircon ion-probe reference materials yielded small biases, with ages calculated against AS3 being $\sim 1\%$ too high and ages calculated against SL13 being variably (although on average $\sim 1\%$) too low. To account for this, we made corrections of -1% to our AS3 ages and $+1\%$ to SL13 ages assembled in our age compilation; other relevant AS3 ages in the literature were verified by other standards (Roberts et al., 2004, 2006). Although these corrections are significant in terms of precision, they have little influence on tectonic conclusions.

More importantly, some of the U-Pb ages for early NEO magmatism, the S-type Rockvale Granodiorite and Tia Granodiorite, as well as the I-type Halls Peak Volcanics, appear biased towards younger ages by rejection criteria. Cawood et al. (2011) presented ages for these and other igneous bodies, undertaken against reliable standards (CS3) that do not require corrections of the kind discussed above. However, they included an arbitrary criterion for recognition of zircon inheritance, namely that analyses older than 300 Ma should be excluded. Because individual zircon U-Pb determinations for these samples have approximately Gaussian distributions centred near 300 Ma, this has led to an excessive number of rejections and naturally to ages < 300 Ma (Rockvale Granodiorite: 292.6 ± 2.4 Ma, MSWD 1.5, 10 from 30 rejected; Tia Granodiorite: 295.7 ± 2.8 Ma, MSWD 0.37, 14 from 27 rejected; and Halls Peak Volcanics: 292.6 ± 2.0 Ma, MSWD 0.68; 11 from 26 rejected). An alternative criterion for recognition of zircon inheritance follows from the observation of Jeon et al. (2012) that the Th / U ratios of obviously inherited zircon in the Bundarra Supersuite are generally greater than ~ 0.3, while new magmatic zircon extends to as low as ~ 0.05. We have recalculated the ages of these samples with an emphasis on including zircon with low Th / U and maintaining Gaussian distributions.

4   Petrography

Fine-grained, doleritic Barney House, Big Bull (Sheep Station Creek Complex) and Days Creek gabbros exhibit granular and flow-foliated (Fig. 3a) to ophitic–subophitic textures (Fig. 3b) and occasionally contain phenocrystic or glomerocrystic plagioclase (Fig. 3c). Plagioclase is elongate, subrectangular, or lath-like, sharing irregular edges with or enclosing olivines, and it is typically normally zoned or unzoned, but sometimes contains distinct cores (An_{72-60} in groundmass; mostly $\sim An_{80}$ but up to An_{86} in cores). Phenocryst rims are sodic (to An_{50}) and texturally interlock with the fine gabbro groundmass. Olivine (Fo_{62-72}) is common and exhibits rounded, irregular, and embayed morphologies or is very rarely interstitial and mantled by pyroxene or hornblende. Pyroxene occurs as oikocrysts and interstitial crystals. High-Ca clinopyroxene (diopside–augite Mg# 78)

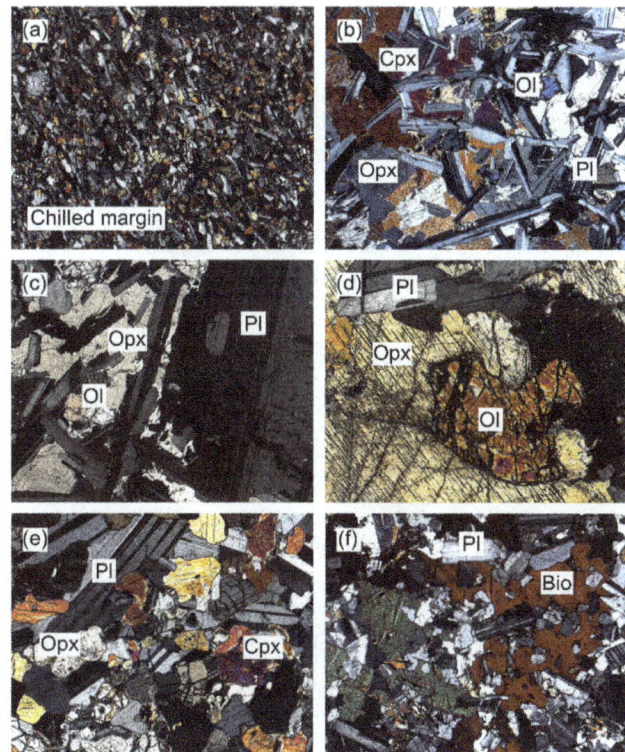

Figure 3. Cross-polarised light images of gabbroic rocks; all image field of views are 2 mm. **(a)** Chilled, flow-foliated margin of the Days Creek Gabbro DC65. **(b)** Ophitic micro-gabbro in Barney House Gabbro BH30. **(c)** Phenocrystic plagioclase in Barney House gabbro BH45. **(d)** Relict olivine in massive Days Creek Gabbro DC19. **(e)** Granular Days Creek Gabbro DC36. **(f)** Poikilitic biotite–diorite associated with Days Creek Gabbro DC98.

is more common than low-Ca orthopyroxene (Mg# 71) and contains exsolution lamellae of the latter. Pyroxene shares interstices with calcic magnesio-hastingsite hornblende and pargasite hornblende (Mg# 71) with high TiO_2 (~ 3.2 wt %) and Al_2O_3 (~ 11.0 wt %), as well as small titanium-rich phlogopite (~ 3.9 wt % TiO_2; average Mg# 75). Ilmenite and rare magnetite (sometimes intergrown) is usually associated with amphibole and phlogopite, often being mantled by them or included in their interstitial domains.

Coarsely crystalline gabbro (millimetre to centimetre crystals) is more typical of the Days Creek Gabbro, with orthocumulate or mesocumulate textures (Fig. 3d) comprising plagioclase, rare resorbed olivine, high-Ca clinopyroxene, very rare low-Ca orthopyroxene, ilmenite, and amphibole (latter often secondary). Massive coarse-grained gabbro also has rare granular texture (Fig. 3e). Plagioclase (An_{80-47}) ranges from isolated, equant euhedral crystals to subhedral crystals in an interlocking network, defining the ortho- or mesocumulate texture. They are commonly normally zoned and rarely exhibit oscillatory zoning or scissor deformation twins. Olivine is Fo_{65-59} and is anhedral or

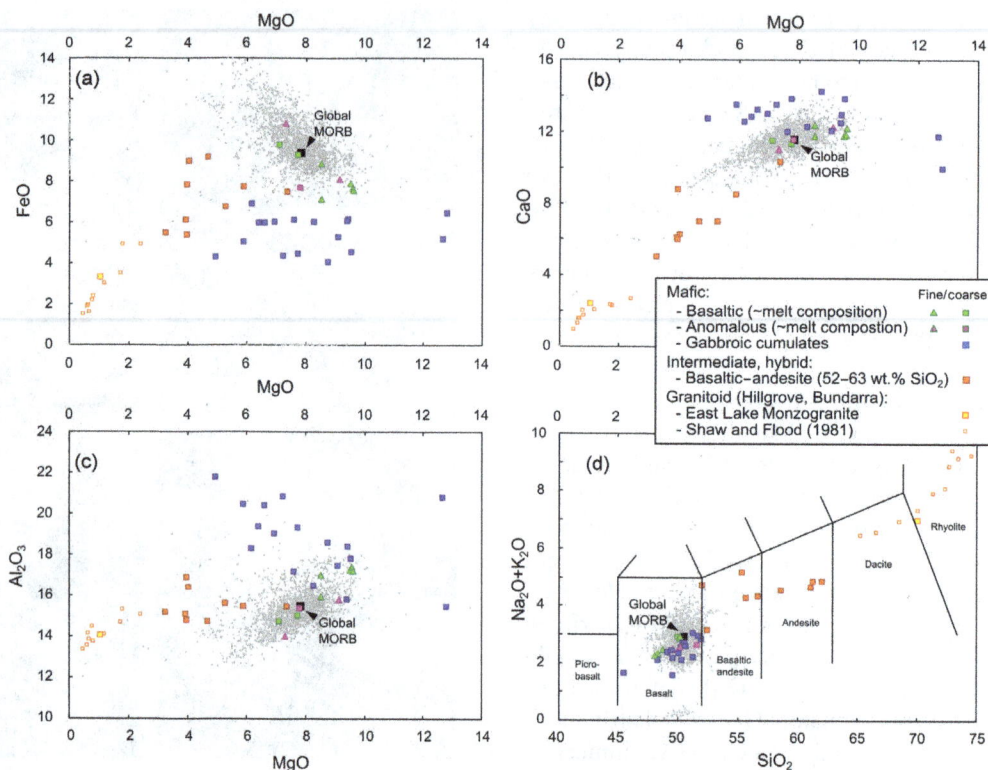

Figure 4. Major element chemistry of bulk samples. Finely crystalline, doleritic samples in triangles and coarsely crystalline samples in squares. Samples likely to represent melt compositions and anomalous samples (on the basis of unusual trace element contents; see Fig. 5) also likely to represent melt compositions are given in green and pink respectively. Cumulate gabbroic samples are in blue and mid-silica (compositionally basaltic andesite or andesite) rocks of probable hybrid origin (mantle–crust mixtures) are given in orange. Yellow squares indicate a single granitic sample of the East Lake Monzogranite (Hillgrove Suite). Other Hillgrove as well as Bundarra Suite samples from the classic study of Shaw and Flood (1981) are given as small orange squares. The global MORB identified by Arevalo and McDonough (2010) is presented as a large black square; MORB data from that study and from Jenner and O'Neill (2012) and the PetDB screening by Class and Lehnert (2012) are presented as small grey points. Fenner diagrams present MgO versus **(a)** FeO, with the trend defined by inferred melt compositions (in green field) indicating a tholeiitic association; **(b)** CaO; and **(c)** Al_2O_3. **(d)** Total alkalis vs. silica (TAS) diagram.

embayed. Secondary clinozoisite and serpentinite after plagioclase and olivine was not observed in fine-grained gabbros but is present in coarse-grained samples. Pyroxenes are subhedral or interstitial and are rarely optically continuous across multiple domains. In coarse gabbros, high-Ca clinopyroxene is diopside–augite (average Mg# 79), while low-Ca orthopyroxene is very rare, possibly because of uralitisation (Mg# 70 with exsolved clinopyroxene at Mg# 78). Very fine orthopyroxene exsolution is also present in clinopyroxenes. Amphibole is present in abundance approximately equal to that of pyroxenes and occurs as primary interstitial magnesio-hornblendes (pale brown and green; average Mg# 63) and secondary fibrous or radiating irregular actinolitic hornblende, magnesio-hornblende, or tschermakitic hornblende (green to green-blue varieties; average Mg# 60). Anhedral or interstitial ilmenite shares intercumulus spaces with pyroxenes and amphiboles.

Bakers Creek Suite rocks with higher silica contents (geochemically intermediate between gabbros and granitoids of

the Hillgrove Suite) display a wide range of textures and variation in mineralogy. Poikilitic, equigranular, and foliated textures are observed in rocks from parts of the Days Creek Gabbro, Camperdown Complex, and Woodburn Diorite. In poikilitic biotite–diorite associated with the Days Creek Gabbro, small rectangular or subhedral plagioclase and granular orthopyroxene are randomly enclosed within large oikocrysts of biotite, quartz, and orthopyroxene (sample DC98; Fig. 3f). Equigranular diorite from the Camperdown Complex (CC11) is dominated by subhedral plagioclase with green amphibole in large interstitial quartz domains, with possibly secondary green amphibole and calcite. More felsic varieties of the Bakers Creek Suite are closer in composition to and continuous with that of the Hillgrove Suite granitoids and have developed tectonic foliations or sub-gneissic textures, e.g. the Woodburn Diorite is composed of subhedral plagioclase and amphibole, folded or kinked biotites, and interstitial or ophitic quartz domains (WB32). Quartz and plagioclase are occasionally graphically intergrown. Preferentially aligned biotite is the main contributor to foliation.

5 Geochemistry

Samples of the Bakers Creek Suite cover a broad geochemical range (Fig. 4a and Supplement S1). Mafic samples with finely crystalline gabbroic or doleritic textures exhibit generally increasing FeO, Na_2O, P_2O_5, and TiO_2 as MgO, CaO, and Al_2O_3 decrease (Fig. 4a, b, and c); they define a trend that passes across mid-ocean ridge basalt (MORB) compositions for major elements (e.g. Jenner and O'Neill, 2012; Class and Lehnert, 2012; and the global MORB identified by Arevalo and McDonough, 2010). The Big Bull Gabbro (CC26A) is a close but resolvable outlier to this trend for some elements (e.g. FeO). From their petrography they likely represent liquid compositions; a range of FeO / MgO for similar SiO_2 is therefore indicative of tholeiitic style evolution (Arculus, 2003; Zimmer et al., 2010). In contrast, coarsely crystalline gabbros do not follow this trend for most oxides. With decreasing MgO, they have increasing Al_2O_3, P_2O_5, and TiO_2 but FeO, CaO, Na_2O, SiO_2, and K_2O are poorly related to MgO (e.g. Fig. 4a, b, and c). All gabbroic samples, whether coarse or finely crystalline, have low K_2O (< 0.3 wt %; low-K association of Gill, 1981) and are basaltic (Le Bas and Streckeisen, 1991). With our focus on the mafic end member, most of our samples are basaltic (i.e. doleritic) or gabbroic (45.5–52 wt % SiO_2), but some dioritic samples were also collected (basaltic–andesitic or andesitic compositions of up to 61.3 wt % SiO_2) and we also report one granitic sample from the East Lake Monzogranite, associated with the Woodburn Diorite ($\sim$ 70 % SiO_2). Dioritic samples span the range between Bakers Creek gabbroic and Hillgrove Suite granitic compositions through "basaltic andesite" to "dacite–rhyolite" (e.g. $\sim$ 52–62 wt % SiO_2 and 0.5–2.0 wt % K_2O; Fig. 4d). Some samples deviate from this trend to higher or lower levels of minor element contents (e.g. samples MP2, CCD, and GK5, FHB respectively). The granitic sample (EA31) falls within or near the main group of Hillgrove Suite samples for all major elements (e.g. Shaw and Flood, 1981).

There is a general correlation between major element and trace element geochemistry, with gabbroic samples having MORB-like trace and rare earth element (REE) abundances, while samples with intermediate and/or granitic geochemistry are enriched in incompatible trace and REEs (Fig. 5; multi-element plots normalised to global MORB of Arevalo and McDonough, 2010). In detail, most of the finely crystalline gabbros display flat and very MORB-like trace element patterns, especially for high-field-strength elements (HFSEs) Zr, Hf, Ti, Y as well as REE, but with positive Cs, Rb, Ba, Th, U, and Pb anomalies and depletions in Nb and Ta (e.g. Barney House and Big Bull gabbros). In combination with finely crystalline micro-gabbroic or doleritic tex-

tures, we interpret these samples to reflect melt compositions, rather than cumulate excesses of one or more minerals. Some coarsely crystalline gabbros (DC15, DC16) are geochemically similar to the fine-grained gabbros and may also reflect melt compositions via mostly in situ crystallisation. However, most gabbros are variably depleted in HFSE, REE, Th, Nb, Ta, and P, with positive anomalies for the same elements as in the finely crystalline gabbros (Cs, Rb, Ba, Th, U, and Pb) but also including Eu. Cr and Ni are variable, with some exhibiting clear enrichments (e.g. GK5 and FHB). Coarsely crystalline samples from the Days Creek Gabbro with middle and high-range Mg# display erratic, concave up patterns, with elevated lare ion lithophile elements (LILE) and Sr; low Nb, Ta, and HFSEs; and higher abundances of Cr and Ni. REE patterns are flat with considerable variation in absolute abundances, and as for finely crystalline samples, they are generally correlated with FeO. For samples with higher Mg#, REE abundances are lower and distinct Eu anomalies and light REE depletions are apparent. Some gabbros from the Days Creek Gabbro appear to be anomalous (D12, DC104, and DC65) with variable depletion in Cs, Th, U, Nb, Ta, K, P, Zr, Hf, and light REE and potential enrichment in Ba.

Trace and REE concentrations for higher silica, geochemically intermediate rocks of the Bakers Creek Suite, as well as the East Lake Monzogranite sample (Hillgrove Suite; EA31), are variably higher and patterns are inclined; peaks in Cs, Th, U, K, Pb, Zr, and Hf alternate with negative Nb, Ta, Sr, P, and Ti anomalies. The Hillgrove Suite sample is the most enriched in incompatible trace elements, and other geochemically intermediate samples also exhibit generally intermediate concentrations of such elements, i.e. they are correlated with SiO_2. Negative anomalies are common for Ba, Nb, and Ta, with some variation in HFSE where concentrations are similar to the granites. Cr and Ni likewise display a range intermediate to the gabbros and granites. Though higher silica samples of the Bakers Creek Suite have compositions intermediate to the gabbros and the monzogranite for most elements, there are important exceptions for Sr, P, Ti, Eu, and heavier REEs for certain samples (CCD, MP2, DC98). These characteristics are consistent with a hybrid origin via mixing of basaltic and granitic components.

6 Zircon chronology

We find zircon $^{206}Pb / ^{238}U$ ages of 303.9 $\pm$ 3.2 Ma (15 points with no rejections, MSWD 1.7) for the Barney House Gabbro and 305.1 $\pm$ 2.9 Ma (18 points with 1 rejection, MSWD 1.5) for the Days Creek Gabbro. These are the oldest ages for intrusive rocks in the Southern NEO, with the possible exceptions of the Rockvale and Tia granodiorites (see below). The Bakers Creek Complex has a similar age of 299.3 $\pm$ 3.1 Ma (corrected for AS3; uncorrected age is 302.3 $\pm$ 3.1 Ma, 18 points with 2 rejections, MSWD 1.3), while the Charon Creek Diorite has a younger age of 290.4 $\pm$ 3.2 Ma (corrected

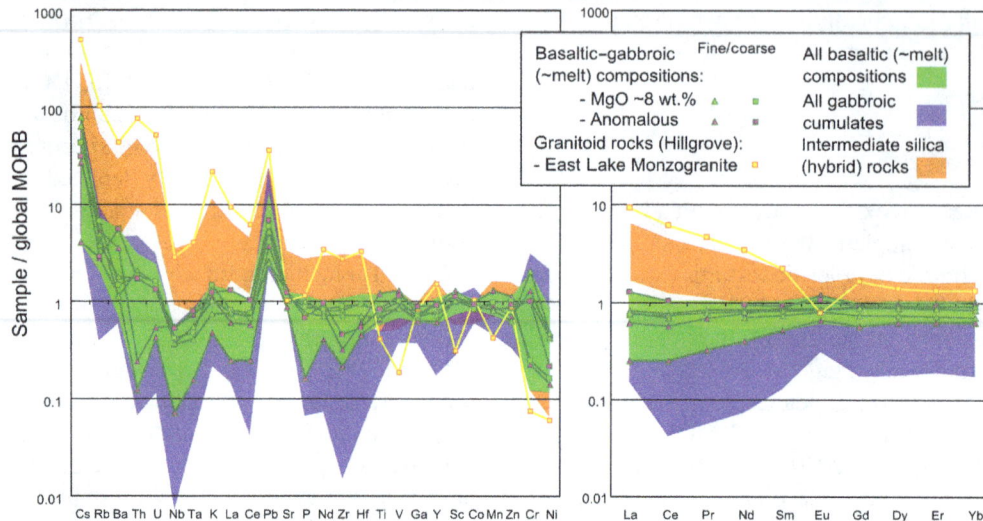

Figure 5. Trace element chemistry of bulk samples, normalised to global MORB (Arevalo and McDonough, 2010). Finely and coarsely crystalline gabbros (triangles and squares respectively) that approximate melt compositions with ∼8 wt% MgO are in green; anomalous melt compositions in pink; East Lake Monzogranite (Hillgrove Suite) in yellow. Range of compositions for basaltic melts, cumulates, and higher silica (basaltic andesite or andesitic) hybrid melts are given by green, blue, and orange fields respectively.

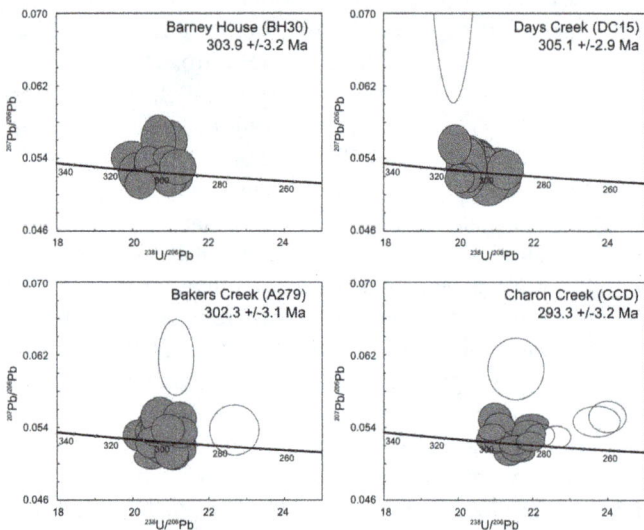

Figure 6. Tera–Wasserburg concordia plots of U-Pb data for Bakers Creek samples. Individual spot error ellipses are 68.3 % confidence limits. Unfilled ellipses were not included in weighted mean age calculations; age intervals are 95 % confidence and include error on standards.

for AS3; uncorrected age is 293.3 ± 3.2 Ma, 15 points and 5 rejections on the basis of Pb loss or measurable common Pb, MSWD 1.6). U-Pb data are given in Fig. 6 and Supplement S2.

With our Th / U criteria, the recalculated age of the Rockvale Granodiorite is 296.7 ± 2.3 Ma (MSWD 1.8), ∼4 Myr older than the age given by Cawood et al. (2011; Supplement S3). We revisited the original U-Pb age for the

Rockvale Granodiorite reported by Kent (1994), which at 303 ± 3 Ma, is older than other igneous rocks of the Southern NEO. His rejection criteria were fundamentally in accord with ours, although the age may suffer from variable bias from the SL13 standard, which would have depressed the age. If bias were, in this case, insignificant (SL13 behaviour is not consistent and sometimes does not bias ages at all; Black et al., 2003), then a discrepancy of 1.0 Myr would remain between Kent (1994) and the age recalculated from the data of Cawood et al. (2011). If, however, bias is present, then the age could be up to ∼306 Ma, with associated discrepancy of up to ∼4 Myr. Hence, U-Pb data for the Rockvale Granodiorite remain poorly understood.

Our recalculated age of the Tia Granodiorite is 299.7 ± 2.0 Ma (MSWD 0.92), again ∼4 Myr older than reported by Cawood et al. (2011). This age is consistent with previous U-Pb determinations (∼300 and ∼302 Ma from Dirks et al., 1993 and Kemp et al., 2009), which together with ages for the Rockvale Granodiorite, imply that the intrusion of some Hillgrove Supersuite plutons considerably predated the main S-type flux represented by the Bundarra Supersuite and most of the Hillgrove Supersuite. For the Halls Peak Volcanics, the recalculated age is 295.7 ± 2.2 Ma (MSWD 1.6), ∼3 Myr older than given by Cawood et al. (2011). Selected and rejected zircon analyses from Cawood et al. (2011) are given in Supplement S3.

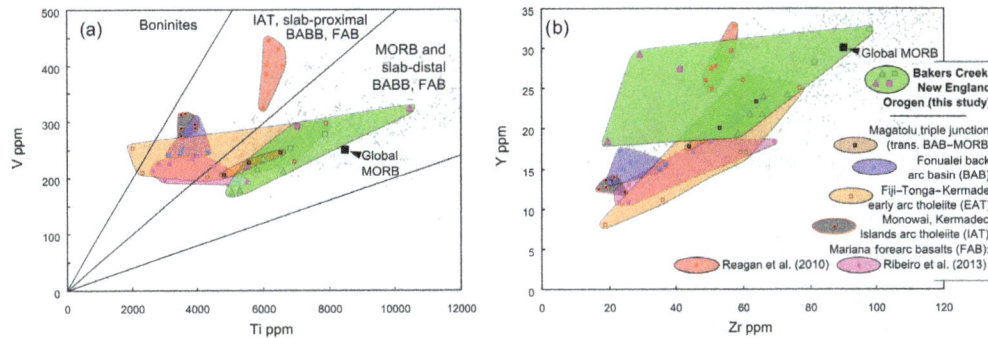

Figure 7. Ti-V and Zr-Y systematics of Bakers Creek Suite melt compositions and selected primitive melts from western Pacific arc systems (Keller et al., 2008; Reagan et al., 2010; Timm et al., 2011; Escrig et al., 2012; Ribeiro et al., 2013; Todd et al., 2012; Kemner et al., 2015). Ti-V fields from Shervais (1982) and Pearce (2014).

7   Discussion

7.1   Tectonic setting: BAB, EAT, FAB, or something else?

The wide range of compositions present in chilled margins, including anomalous (e.g. DC65) and main-group samples (e.g. BH30) of various MgO contents, indicates trapping of melt compositions after magmatic differentiation that occurred before or during emplacement of magmas during ascent through the mantle wedge and overlying crust. Some differentiation also seems to have occurred in situ, indicated by coarsely crystalline samples DC15 and DC16, which are geochemically similar to quenched or finely crystalline gabbros. We identify likely melt compositions by green and purple in Figs. 4, 5, 7, and 8.

Our melt compositions lie in the MORB and slab-distal BAB and FAB fields in terms of Ti and V (Shervais, 1982; Pearce, 2014; Fig. 7a), and they seem to have only been subtly affected by subduction zone influences on these elements (McCulloch and Gamble, 1991; Woodhead et al., 2001). Alternately, Y / Zr can be used to identify previously depleted mantle sources (e.g. Arculus et al., 2015). Despite the geochemical similarity of these elements to V / Ti under typical subduction zone redox conditions (trivalent and tetravalent respectively), there is clear distinction between main-group Bakers Creek Suite melt compositions and anomalous melts in Y / Zr space (Fig. 7b; samples DC65, DC104, and D12). These have much lower Zr (and Hf) for similar Y contents, which is a characteristic shared by FABs, e.g. Izu-Bonin FAB (Reagan et al., 2010; Ribeiro et al., 2013; Arculus et al., 2015).

Despite MORB-like Ti-V and Zr-Y systematics for most Bakers Creek Suite melt compositions, a subduction-derived component is clear for slab flux elements. In Nb / Yb and Th / Yb space (Pearce, 2014) the main group of samples are well clear of the MORB and ocean island basalt array and are high in the "oceanic arcs" region (Fig. 8a), while two anomalous samples again share similarities with FAB-type

basalts in having very low Th / Yb, consistent with Zr-Y (but not Ti-V) systematics. Multi-element plots in Fig. 5 suggest that this is due to a combination of (1) Nb depletion, either by retention in the mantle-wedge source or an under-contribution from the slab, and (2) addition of Th (and U) to main-group Bakers Creek Suite melts via addition of a sedimentary component (e.g. Woodhead et al., 2001). The latter might have been derived from subducted sediments in the undergoing slab or by simple contamination with Lachlan Orogen accretionary prism material as metasediments or S-type granitic melts (i.e. Hillgrove Suite). Anomalous Bakers Creek Suite melts DC65 and DC104 are in the extension of the MORB array to very low Th / Yb and low or very low Nb / Yb. Along with Th, the similarly incompatible indicators of sedimentary melting U and light REE (especially La and Ce) seem to have been under-contributed (no melting of zoisite or allanite; Spandler and Pirard, 2013), although the Th / U ratio of DC65 and DC104 is much lower than other samples. DC65 and D12 received unusually high Ba contributions, which may indicate a distinct fluid component (Woodhead et al., 2001). Anomalous samples are therefore associated with a peculiar elevated Ba / La (Fig. 8b). Additionally, DC104 has much lower Cs values (and K_2O) than the others, despite similar levels of Rb in main-group and anomalous Bakers Creek Suite samples. This strongly indicates decoupling of Rb from other trace alkalis Cs and K_2O, as well as from Ba, elements that are ordinarily associated in sub-arc settings, e.g. via phengite and paragonite melting (Spandler and Pirard, 2013).

A direct comparison is made of the multi-element plots for the melt compositions of the Bakers Creek Suite, with the FAB basalts of Reagan et al. (2010) and Ribeiro et al. (2013) in Fig. 9, for similar major element compositions (especially for MgO, in the range ∼ 6.6–8.6 wt %). They share some relative and absolute abundance trace element characteristics, especially those of anomalous composition (D12, DC104, and DC65). Low abundances of certain slab-flux elements, such as Th and U; the light to mid-REEs, especially La and Ce (and consequently low light REE to heavy REE ratio);

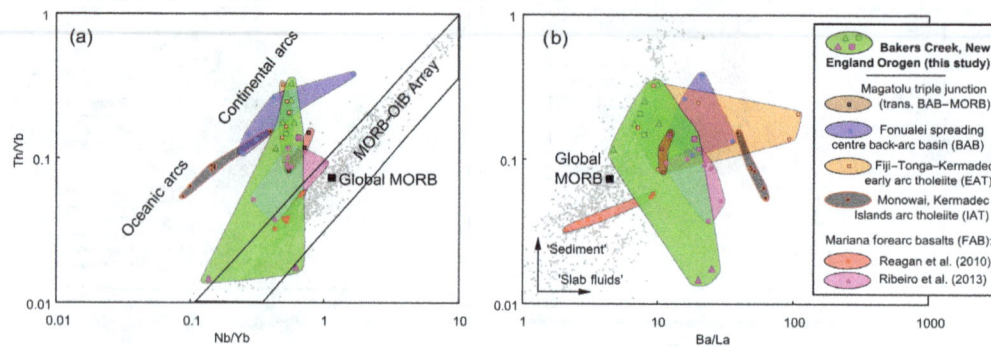

Figure 8. Indicators for subduction zone components, using Th / Yb as a proxy for sedimentary melt, versus **(a)** Nb / Yb identifying depletion of Nb, after classical subduction zone signatures (Pearce, 2014). **(b)** Ba / La as a proxy for a fluid-mobile component (Woodhead et al., 2001). Bakers Creek Suite compositions and selected primitive melts from western Pacific arc systems as for Fig. 7.

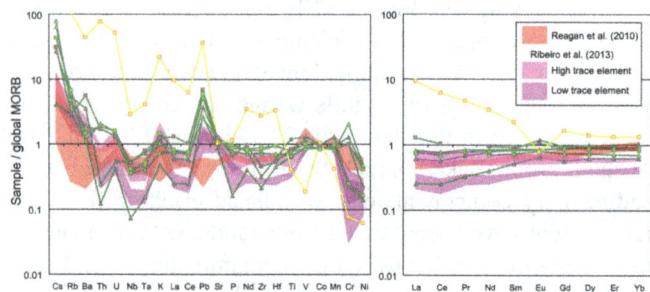

Figure 9. Bakers Creek basaltic melt compositions compared to forearc basalts from Reagan et al. (2010) and Ribeiro et al. (2013) for samples with ∼ 6.6–8.6 wt % MgO.

and the HFSEs Zr and Hf, all indicate involvement of a FAB component in otherwise arc- or back-arc-like basaltic compositions.

The trace element geochemistry of Bakers Creek Suite samples, and of forearc basalts in general, therefore indicates separation of some components, especially Ti and Zr (Fig. 7), in addition to those attributed to sedimentary or slab-fluid components (Th / Yb and Ba / La; Fig. 8) that are ordinarily associated, or correlated, with subduction zone associations. The unusual compositions found in the chilled margin of the Days Creek Gabbro may represent early forearc-style magmas, especially the more extreme characteristics of high Y / Zr and low Th / Yb. As chilled margins, these have been specifically sampled in the field and might be less often captured by random undersea sampling (e.g. Arculus et al., 2015). This particular type of magma seems to have been overwhelmed by later arc–back-arc style magmas (main group Bakers Creek) and the larger Days Creek Gabbro might be an example of a feeder pipe, capturing an early forearc component on its margin and a late back-arc component in its core.

7.2   Chronology of early NEO magmatism

The oldest dated Southern NEO intrusives clearly comprise the gabbroic plutons of the Bakers Creek Supersuite (Barney House and Days Creek gabbros); they are clearly resolved by U-Pb dating from other intrusive bodies. Other early samples that are not so clearly resolved include the largest compositionally intermediate pluton of the Bakers Creek Supersuite (Bakers Creek Complex), as well as isolated members of the S-type Hillgrove Supersuite (Tia and Rockvale granodiorites, with the age of the latter not well known but still older than 294 Ma; possibly also the Blue Knobby Monzogranite and Henry River Granite). The Tia Granodiorite constrains the age of the HTLP Tia Complex (Phillips et al., 2008) to greater than 299.7 ± 2.0 Ma (recalculated here from data of Cawood et al., 2011); metamorphic zircon in the HTLP Wongwibinda Complex records a similar, or slightly younger, U-Pb age of 296.8 ± 1.5 Ma (Craven et al., 2012).

As the magmatic pulse accelerated, diverse compositions continued with intrusion of the Jibbinbar Granite at ∼ 298 Ma (Cross et al., 2009), followed by the Rockisle Granite, Dorrigo Mountain Complex, and Mount You You Granite at ∼ 295 Ma (Rosenbaum et al., 2012). This diverse magmatism is also reflected at the same time in volcanic rocks, with the I-type Halls Peak Volcanics and various basaltic flows near the base of the newly opened Barnard Basin (Cawood et al., 2011) and the Alum Rock Volcanics (Roberts et al., 1996).

The major phase of pre-Hunter-Bowen magmatism in the Southern NEO occurred with the climactic emplacement of S-type granites and granodiorites of the Bundarra Supersuite at ∼ 292–285 Ma and many larger plutons of the Hillgrove Supersuite at ∼ 293–288 Ma (e.g. Hillgrove Monzogranite). Some diversity in magmatic compositions continued throughout this period, with emplacement of the ungrouped Kaloe Granodiorite (Cawood et al., 2011), Bullaganang Granite (Donchak et al., 2007), and Gandar Granodiorite (Rosenbaum et al., 2012), as well as our own AS3-

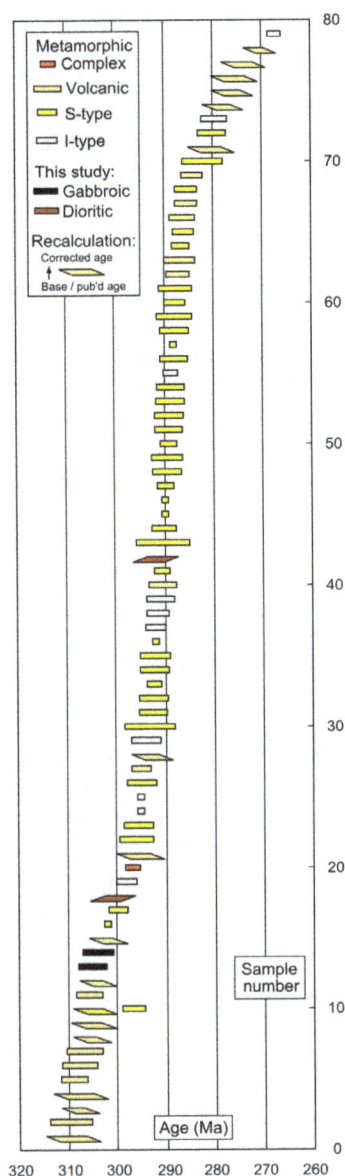

Figure 10. Summary of age determinations for latest Carboniferous to early Permian magmatic rocks of the Southern NEO, corresponding to the first, predominantly S-type granitic magmatism in the orogen. Sources of data in Supplement S4.

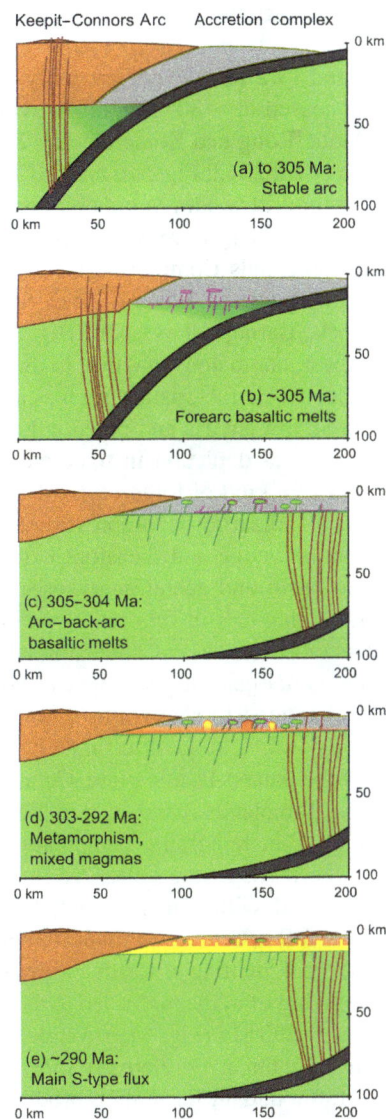

Figure 11. Chronology of early NEO magmatism: **(a)** Lachlan Orogen subduction zone (Keepit–Connors Arc) at ∼ 310–305 Ma; **(b)** early extension and production of FAB-type melts, emplaced as rare, anomalous chilled margins of the Bakers Creek Suite at ∼ 305 Ma; **(c)** continued extension at 305–304 Ma, inducing mantle melting and production of main Bakers Creek Suite gabbros with back-arc (BAB) to arc-like (EAT, IAT) affinities; **(d)** peak metamorphism in metamorphic complexes (Tia and Wongwibinda) and diverse magmatism, including mafic and felsic components (Bakers Creek and Hillgrove suites) at 303–292 Ma; and **(e)** main flux of Hillgrove and Bundarra S-type granites at ∼ 290 Ma (see Li et al., 2014, 2015).

corrected age for the Charon Creek Diorite (Bakers Creek Suite).

S-type magmatism persisted until ∼ 280 Ma for the Cheyenne Complex of the Hillgrove Supersuite, and possibly until ∼ 282 Ma for part of the Banalasta Monzogranite of the Bundarra Supersuite (Phillips et al., 2011). This last emplacement of S-type magma was contemporaneous with another burst of I-type magmatism in the form of the Alum Mountain Volcanics (∼ 274 Ma; Roberts et al., 1995; Li et al., 2014); the more conspicuous low-K, HREE-depleted Greymare Granodiorite (Donchak et al., 2007), similar to the

Clarence River Supersuite; and finally the ∼ 267 Ma Barrington Tops Granodiorite (Cawood et al., 2011). This chronology is illustrated in Fig. 10 (see also Supplement S4); thereafter, magmatism following the Hunter-Bowen Orogeny is reviewed by Li et al. (2012).

7.3 Tectonic implications

Magmatism related to a long lived, probably west-dipping subduction zone ceased at ∼ 305 Ma and provided the base of the NEO (Claoué-Long and Korsch, 2003; Roberts et al., 2004, 2006; Jeon et al., 2012; Figs. 10 and 11a); at about the same time, quenching of a new, anomalous or forearc-type mantle-derived magma (Days Creek Gabbro chilled margin) occurred in the Tablelands Complex. This was shortly followed by the main group of Bakers Creek Suite gabbroic melts (Days Creek, Barney House, and Big Bull) at 305–304 Ma. The earliest, anomalous magmas in the chilled margin have some unusual characteristics, such as a fluid-mobile high Ba component; low Zr, Th, and potentially lower Ti and higher V; and a general depletion in trace elements. These can be considered as a kind of forearc or ∼ FAB-type component (e.g. DC65), indicating decompression melting of the old, depleted mantle wedge and therefore extension of the overlying forearc basin and accretionary prism (Fig. 11b). Evolution to BAB magmatism reflects a combination of a more enriched mantle source and conventional EAT or island arc tholeiite component. It also suggests continued extension (Fig. 11c) related to slab rollback, or perhaps less likely by slab break-off, and could ultimately be driven by reorganisation of the palaeo-Pacific plate. Continued heating and melting of the Tablelands accretion complex during rifting generated high-T metamorphic complexes and the earliest S-type granitic melts of the Hillgrove Suite at ∼ 302–300 Ma. These, mixed with Bakers Creek Suite gabbroic melts, produced a full spectrum of mafic to felsic compositions (Fig. 11d). Peak melting of fertile greywackes, probably due to underplating of mafic melts, led to the main flux of S-type granites at ∼ 290 Ma (Fig. 11e). While modelling of such processes usually indicates relatively short timescales of ∼ 1 Myr or less for abundant felsic melt production (Annen and Sparks, 2002; Solano et al., 2012), the chronology constructed for the NEO implies that melt mobilisation takes significantly longer, perhaps due to rapid stratification of mafic and felsic melts (preventing later mafic melts from ascending), and importantly involving high melt fractions.

The events above raise the question as to whether the early Permian arc is preserved anywhere in continental Australia or elsewhere. Li et al. (2015) investigated detrital zircons in the Gympie terrane immediately overlying the basal Highbury Volcanics (clinopyroxene-rich basalts; Sivell and Waterhouse, 1988), finding an age cluster at 302 Ma. If these zircons are indeed derived from the Highbury Volcanics, which represents the newly established arc, then one can outline (within uncertainty limits of the U-Pb data) the following chronology: (1) the mafic Bakers Creek Suite plutons track the migration of magmatism, with early forearc-like magma (chilled margin of Days Creek Gabbro) crystallising at or just before ∼ 305 Ma; (2) these transition to back-arc-like magmas in this area (main Days Creek, Big Bull, and Barney House) at 304–305 Ma, while the arc magmas themselves may be less likely to be observed as they migrate; and (3) arc-like magmatism is finally established at ∼ 302 Ma (Highbury Volcanics), albeit so far only indirectly dated by overlying detrital zircons (Li et al., 2015). Obviously, a direct U-Pb age for the Highbury Volcanics would provide the best information on its relationship with the Bakers Creek and Hillgrove suites.

8 Conclusions

- The Bakers Creek Suite gabbros, associated with the Hillgrove Suite S-type granitoids, record an evolution from an early forearc-like component through normal arc–back-arc style gabbroic magmatism to hybrid melts, with a wide spectrum of intermediate compositions (continuous with the S-type Hillgrove Suite).

- Capture of a FAB-type component occurred in the early basaltic melts of the Bakers Creek Suite. This earliest intrusive magmatism in the NEO occurred at ∼ 305 Ma and was subsequently replaced by incipient arc and back-arc-type magma at ∼ 305–304 Ma.

- Rifting, extension of the overlying sedimentary complex, melting of the mantle wedge, and transport of the resulting melts were responsible for the high-temperature, low-pressure metamorphic complexes in the mid-crust and abundant S-type granitic magmas at depth (∼ 300 Ma and onwards), with peak migration and emplacement of the latter at ∼ 290 Ma.

- Ultimately, establishment of the New England Batholith and the Southern NEO, from forearc basin rocks outboard of the Lachlan Orogen, occurred by rifting and rapid evolution through a forearc to a back-arc environment, probably due to slab rollback.

Competing interests. The authors declare that they have no conflict of interest.

Acknowledgements. We thank Jenny Zobec for XRF analyses (UoN); Stephen Eggins for LA–ICP–MS analysis (ANU); Daniela Rubatto, Trevor Ireland, Peter Holden, and Peter Lanc for assistance with SHRIMP preparation and measurements; and Ian Williams, Richard Arculus, Joerg Hermann, Colleen Bryant, Pengfei Li, Gordon Lister, Bill Collins, and Robin Offler for lengthy discussion about zircon standards, magmatic geochemistry, the NEO, and tectonics. This research was partially supported by an Australian National University PhD Research Scholarship to Seann J. McKibbin, who is currently a postdoctoral fellow of the Research Foundation – Flanders (Fonds Wetenschapplijke Onderszoek; FWO). We thank Pengfei Li, Heejin Jeon, Kim Jessop, and the two anonymous reviewers for constructive comments on the paper, as well as Andrea Di Muro for editorial handling.

Edited by: A. Di Muro

References

Annen, C. and Sparks, R. S. J.: Effects of repetitive emplacement of basaltic intrusions on thermal evolution and melt generation in the crust, Earth Planet. Sc. Lett., 203, 937–955, 2002.

Arculus, R. J.: Use and abuse of the terms calcalkaline and calcalkalic, J. Petrol., 44, 929–935, 2003.

Arculus, R. J., Ishizuka, O., Bogus, K. A., Gurnis, M., Hickey-Vargas, R., Aljahdali, M. H., Bandini-Maeder, A. N., Barth, A. P., Brandl, P. A., Drab, L., do Monte Geurra, R., Hamada, M., Jiang, F., Kanayama, K., Kender, S., Kusano, Y., Li, H., Loudin, L. C., Maffione, M., Marsaglia, K. M., McCarthy, A., Meffre, S., Morris, A., Neuhaus, M., Savov, I. P., Sena, C., Tepley III, F. J., van der Land, C., Yogodzinski, G. M., and Zhang, Z.: A record of spontaneous subduction initiation in the Izu-Bonin-Mariana arc, Nat. Geosci., 8, 728–733, 2015.

Arevalo, R. and McDonough, W. F.: Chemical variations and regional diversity observed in MORB, Chem. Geol., 271, 70–85, 2010.

Black, L. P., Kamo, S. L., Williams, I. S., Mundil, R., Davis, D. W., Korsch, R. J., and Foudoulis, C.: The application of SHRIMP to Phanerozoic geochronology, a critical appraisal of four zircon standards, Chem. Geol., 200, 171–188, 2003.

Brown, M.: Hot orogens, tectonic switching, and creation of continental crust: Comment and Reply, Geology, 31, doi:10.1130/0091-7613-31.1.e9, 2003.

Caprarelli, G. and Leitch, E. C.: Magmatic changes during the stabilisation of a cordilleran fold belt: the Late Carboniferous-Triassic igneous history of eastern New South Wales, Australia, Lithos, 45, 413–430, 1998.

Cawood, P. A., Leitch, E. C., Merle, R. E., and Nemchin, A. A.: Orogenesis without collision: Stabilizing the Terra Australis accretionary orogen, eastern Australia, Geol. Soc. Am. Bull., 123, 2240-2255, 2011.

Chappell, B. W. and White, A. J. R.: Two contrasting granite types: 25 years later, Aust. J. Earth Sci., 48, 489–499, 2001.

Claoué-Long, J. C. and Korsch, R. J.: Numerical time measurement in the DM Tangorin DDH1 drillcore, in: Geology of the Cranky Corner Basin, edited by: Facer, R. A. and Foster, C. B., New South Wales Department of Mineral Resources, Coal and Petroleum Bulletin, 4, 179–205, 2003.

Class, C. and Lehnert., K.: PetDB expert MORB (mid-ocean ridge basalt) compilation, EarthChem Library, doi:10.1549/IEDA/100060, 2012.

Collins, W. J.: Hot orogens, tectonic switching, and creation of continental crust, Geology, 30, 535–538, 2002.

Collins, W. J. and Richards, S. W.: Geodynamic significance of S-type granites in circum-Pacific orogens, Geology, 36, 559–562, 2008.

Collins, W. J., Offler, R., Farrell, T. R., and Landenberger, B.: A revised Late Palaeozoic-Early Mesozoic tectonic history for the southern New England Fold Belt, NEO '93 Conference Proceedings, 69–84, 1993.

Craven, S. J., Daczko, N. R., and Halpin, J. A.: Thermal gradient and timing of high-T-low-P metamorphism in the Wongwibinda Metamorphic Complex, southern New England Orogen, Australia, J. Metamorph. Geol., 30, 3–20, 2012.

Cross, A. J., Purdy, D. J., Bultitude, R. J., Dhnaram, C. R., and von Gnielinski, F. E.: Joint GSQ-GA NGA geochronology project, New England Orogen and Drummond Basin, 2008, Queensland Geological Record 2009/03, 2009.

Dilek, Y. and Furnes, H.: Ophiolites and their origins, Elements, 10, 93–100, 2014.

Dirks, P. H. G. M., Lennox, P. G., and Shaw, S. E.: Tectonic implications of two Rb/Sr biotite dates for the Tia Granodiorite, southern New England Fold Belt, NSW, Australia, Aust. J. Earth Sci., 39, 111–114, 1992.

Dirks, P. H. G. M., Offler, R., and Collins, W. J.: Timing of emplacement and deformation of the Tia Granodiorite, southern New England Fold Belt, NSW: Implications for the metamorphic history, Aust. J. Earth Sci., 40, 103–108, 1993.

Donchak, P. J. T., Bultitude, R. J., Purdy, D. J., and Denaro, T. J.: Geology and mineralisation of the Texas Region, south-eastern Queensland, Queensland Geology, 11, 25–32, 2007.

Eggins, S. M.: Laser ablation-ICPMS analysis of geological materials prepared as Lithium-borate glasses, Geostandards Newsletter 27, 147–162, 2003.

Escrig, S., Bezos, A., Langmuir, C. H., Michael, P. J., and Arculus R.: Characterizing the effect of mantle source, subduction input and melting in the Fonualei Spreading Center, Lau Basin: Constraints on the origin of the boninitic signature of the back-arc lavas, Geochem. Geophys. Geosys., 13, Q10008, doi:10.1029/2012GC004130, 2012.

Farrell, T. R.: Structural geology and tectonic development of the Wongwibinda Metamorphic Complex, in: New England Orogen, Tectonics and Metallogenesis, edited by: Kleeman, J. D., Department of Geology and Geophysics, University of New England, Armidale, 117–124, 1988.

Flood, R. H. and Shaw, S. E.: Two "S-type" granite suites with low initial 87Sr / 86Sr ratios from the New England Batholith, Australia, Contrib. Mineral. Petr., 61, 163–173, 1977.

Glen, R. A. and Roberts, J.: Formation of oroclines in the New England Orogen, Eastern Australia, in: Stephen Johnston and Gideon Rosenbaum, edited by: Johnston, S. and Rosenbaum, G., J. Virtual Explor., 43, doi:10.3809/jvirtex.2012.00305, 2012.

Gill, J. B.: Geophysical setting of volcanism at convergent plate boundaries, in: Orogenic andesites and plate tectonics, Springer, Berlin, Heidelberg, 44–63, 1981.

Hand, M.: Structural analysis of a deformed subduction-accretion complex sequence, Nowendoc, NSW, in: New England Orogen, Tectonics and Metallogenesis, edited by: Kleeman, J. D., Department of Geology and Geophysics, University of New England, Armidale, 105–116, 1988.

Hensel, H.-D., McCulloch, M. T., and Chappell, B. W.: The New England Batholith: constraints on its derivation from Nd and Sr isotopic studies of granitoids and country rocks, Geochim. Cosmochim. Ac., 49, 369–384, 1985.

Jenkins, R. B., Landenberger, B., and Collins, W. J.: Late Palaeozoic retreating and advancing subduction boundary in the New England Fold Belt, New South Wales, Aust. J. Earth Sci., 49, 467–489, 2002.

Jenner, F. E. and O'Neill, H. Ş. C.: Analysis of 60 elements in 616 ocean floor basaltic glasses, Geochem. Geophys. Geosys., 13, Q02005, doi:10.1029/2011GC004009, 2012.

Jeon, H., Williams, I. S., and Chappell, B. W.: Magma to mud to magma: Rapid crustal recycling by Permian granite magmatism near the eastern Gondwana margin, Earth Planet. Sc. Lett., 319–320, 104–117, 2012.

Keller, N. S., Arculus, R. J., Hermann, J., and Richards, S.: Submarine back-arc lava with arc signature: Fonualei Spreading Center, northeast Lau Basin, Tonga, J. Geophys. Res., 113, B08S07, doi:10.1029/2007JB005451, 2008.

Kemner, F., Haase, K. M., Beier, C., Krumm, S., and Brandl, P. A.: Formation of andesite melts and Ca-rich plagioclase in the submarine Monowai volcanic system, Kermadec arc, Geochem. Geophys. Geosys., 16, 4130–4152, doi:10.1002/2015GC005884, 2015.

Kemp, A. I. S., Hawkesworth, C. J., Collins, W. J., Gray, C. M., Blevin, P. L., and EIMF: Isotopic evidence for rapid continental growth in an extensional accretionary orogen: The Tasmanides, eastern Australia, Earth Planet. Sc. Lett., 284, 455–466, 2009.

Kent, A. R. J.: Geochronology and geochemistry of Palaeozoic intrusive rocks in the Rockvale region, southern New England Orogen, New South Wales, Aust. J. Earth Sci., 41, 365–379, 1994.

Korsch, R. J.: A framework for the Palaeozoic geology of the southern part of the New England Geosyncline, J. Geol. Soc. Aust., 25, 339–355, 1977.

Landenberger, B., Farrell, T. R., Offler, R., Collins, W. J., and Whitford, D. J.: Tectonic implications of Rb-Sr biotite ages for the Hillgrove Plutonic Suite, New England Fold Belt, N.S.W., Australia, Precambrian Res., 71, 251–263, 1995.

Langmuir, C. H., Bézos, A., Escrig, S., and Parman, S. W.: Chemical systematics and hydrous melting of the mantle in back-arc basins. In: Back-arc spreading systems: Geological, biological, chemical, and physical interactions, Geophys. Monogr. Ser., 166, 87–146, 2006.

Le Bas, M. J. and Streckeisen, A. L.: The IUGS systematics of igneous rocks, J. Geol. Soc. London, 148, 825–833, 1991.

Leitch, E. C.: The Geological Development of the Southern Part of the New England Fold Belt, J. Geol. Soc. Aust., 21, 133–156, doi:10.1080/00167617408728840, 1974.

Leitch, E. C.: Plate tectonic interpretation of the Paleozoic history of the New England Fold Belt, Bull. Geol. Soc. Am., 86, 141–144, 1975.

Li, P.-F., Rosenbaum, G., and Rubatto, D.: Triassic asymmetric subduction rollback in the southern New England Orogen (eastern Australia): the end of the Hunter-Bowen Orogeny, Aust. J. Earth Sci., 59, 965–981, 2012.

Li, P.-F., Rosenbaum, G., and Vasconcelos, P.: Chronological constraints on the Permian geodynamic evolution of eastern Australia, Tectonophysics, 617, 20–30, 2014.

Li, P.-F., Rosenbaum. G., Yang, J.-H., and Hoy, D.: Australian-derived detrital zircons in the Permian-Triassic Gympie terrane (eastern Australia): Evidence for an autochthonous origin, Tectonics, 34, 858–874, doi:10.1002/2015TC003829, 2015.

McCulloch, M. T. and Gamble, J. A.: Geochemical and geodynamical constraints on subduction zone magmatism, Earth Planet. Sc. Lett., 102, 358–374, 1991.

Meffre, S., Falloon, T. J., Crawford, T. J., Hoernle, K., Hauff, F., Duncan, R. A., Bloomer, S. H., and Wright, D. J.: Basalts erupted along the Tongan fore arc during subduction initiation: Evidence from geochronology of dredged rocks from the Tonga fore arc and trench, Geochem. Geophys. Geosys., 13, 1–17, 2012.

Pearce, J.: Immobile element fingerprinting of ophiolites, Elements, 10, 101–108, 2014.

Pearce, J. A. and Stern, R. J.: Origin of back-arc basin magmas: Trace element and isotope perspectives, In: Back-arc spreading systems: Geological, biological, chemical, and physical interactions, Geophys. Monogr. Ser., 166, 63–86, 2006.

Phillips, G., Hand, M., and Offler, R.: P-T-t deformation framework of an accretionary prism, southern New England Orogen, eastern Australia: Implications for blueschist exhumation and metamorphic switching, Tectonics, 27, TC6017, doi:10.1029/2008TC002323, 2008.

Phillips, G., Landenberger, B., and Belousova, E. A.: Building the New England Batholith, eastern Australia – Linking granite petrogenesis with geodynamic setting using Hf isotopes in zircon, Lithos, 122, 1–12, 2011.

Reagan, M. K., Ishizuka, O., Stern, R. J., Kelley, K. A., Ohara, Y., Blichert-Toft, J., Bloomer, S. H., Cash, J., Fryer, P., Hanan, B. B., Hickey-Vargas, R., Ishii, T., Kimura, J.-I., Peate, D. W., Rowe, M. C., and Woods, M.: Fore-arc basalts and subduction initiation in the Izu-Bonin-Mariana system, Geochem. Geophys. Geosys., 11, 1–17, 2010.

Ribeiro, J. M., Stern, R. J., Kelley, K. A., Martinez, F., Ishizuka, O., Manton, W. I., and Ohara, Y.: Nature and distribution of slab-derived fluids and mantle sources beneath the Southeast Mariana forearc rift, Geochem. Geophys. Geosys., 14, 4584–4607, 2013.

Roberts, J., Claoué-Long, J. C., Jones, P. J., and Foster, C. B.: SHRIMP zircon age control of Gondwanan sequences in Late Carboniferous and Early Permian Australia, in: Dating and correlating biostratigraphically barren strata, edited by: Dunnay, R. E. and Hailwood, E. A., Geological Society of London Special Publication, 89, 145–174, 1995.

Roberts, J., Claoué-Long, J. C., and Foster, C. B.: SHRIMP zircon dating of the Permian System of eastern Australia, Aust. J. Earth Sci., 43, 401–421, 1996.

Roberts, J., Offler, R., and Fanning, M.: Upper Carboniferous to Lower Permian volcanic successions of the Carroll-Nandewar region, northern Tamworth Belt, southern New England Orogen, Australia, Aust. J. Earth Sci., 51, 205–232, 2004.

Roberts, J., Offler, R., and Fanning, M.: Carboniferous to Lower Permian stratigraphy of the southern Tamworth Belt, southern New England Orogen, Australia: boundary sequences of the Werrie and Rouchel blocks, Aust. J. Earth Sci., 53, 249–284, 2006.

Rosenbaum, G., Li, P.-F., and Rubatto, D.: The contorted New England Orogen (eastern Australia): New evidence from U-Pb geochronology of early Permian granitoids, Tectonics, 31, TC1006, doi:10.1029/2011TC002960, 2012.

Shaanan, U., Rosenbaum, G., and Wormald, R.: Provenance of the early Permian Nambucca block (eastern Australia) and implications for the role of trench retreat in accretionary orogens, Geol. Soc. Am. Bull., 127, 1052–1063, 2015.

Shaw, S. E. and Flood, R. H.: The New England Batholith, eastern Australia: Geochemical vartiations in time and space, J. Geophys. Res., 86, 10530–10544, 1981.

Shervais, J. W.: Ti-V plots and the petrogenesis of modern and ophiolitic lavas, Earth Planet. Sc. Lett., 59, 101–118, 1982.

Sivell, W. J. and Waterhouse, J. B.: Petrogenesis of Gympie Group volcanics: evidence for remnants of an early Permian volcanic arc in eastern Australia, Lithos, 21, 81–95, 1988.

Solano, J. M. S., Jackson, M. D., Sparks, R. S. J., Blundy, J. D., and

Annen, C.: Melt segregation in deep crustal hot zones: A mechanism for chemical differentiation, crustal assimilation and the formation of evolved magmas, J. Petrol., 53, 1999–2026, 2012.

Spandler, C. and Pirard, C.: Element recycling from subducting slabs to arc crust: A review, Lithos, 170–171, 202–223, 2013.

Timm, C., Graham, I. J., de Ronde, C. E. J., Leybourne, M. I., and Woodhead, J.: Geochemical evolution of Monowai volcanic center: New insights into the northern Kermadec arc subduction system, SW Pacific, Geochem. Geophys. Geosys., 12, Q0AF01, doi:10.1029/2011GC003654, 2011.

Todd, E., Gill, J. B., and Pearce, J. A.: A variably enriched mantle wedge and contrasting melt types during arc stages following subduction initiation in Fiji and Tonga, southwest Pacific, Earth Planet. Sc. Lett., 335–336, 180–194, 2012.

Woodhead, J. D., Hergt, J. M., Davidson, J. P., and Eggins, S. M.: Hafnium isotope evidence for "conservative" element mobility during subduction zone processes, Earth Planet. Sc. Lett., 192, 331–346, 2001.

Zimmer, M. M., Plank, T., Hauri, E. H., Yogodzinski, G. M., Stelling, P., Larsen, J., Singer, B., Jicha, B., Mandeville, C., and Nye, C. J.: The role of water in generating the calc-alkaline trend: New volatile data for Aleutian magmas and a new tholeiitic index, J. Petrol., 51, 2411–2444, 2010.

12

Synchrotron FTIR imaging of OH in quartz mylonites

Andreas K. Kronenberg[1], Hasnor F. B. Hasnan[1,a], Caleb W. Holyoke III[1,b], Richard D. Law[2], Zhenxian Liu[3], and Jay B. Thomas[4]

[1]Center for Tectonophysics, Department of Geology and Geophysics, MS 3115, Texas A&M University, College Station, TX 77843-3115, USA

[2]Department of Geosciences, MC 0420, Derring Hall RM 4044, Virginia Polytechnic Institute and State University, Blacksburg, VA 24061, USA

[3]Geophysical Laboratory, Carnegie Institution of Washington, 5251 Broad Branch Rd., NW Washington, D.C. 20015, USA

[4]Department of Earth Sciences, 204 Heroy Geology Laboratory, Syracuse University, Syracuse, NY 13244, USA

[a]now at: Department of Advanced Geophysics, PETRONAS, Carigali Sdn. Bhd., PETRONAS Twin Towers, Kuala Lumpur City Centre, 50088 Kuala Lumpur, Malaysia

[b]now at: Department of Geosciences, University of Akron, Akron, OH 44325-4101, USA

Correspondence to: Andreas K. Kronenberg (kronenberg@geo.tamu.edu)

Abstract. Previous measurements of water in deformed quartzites using conventional Fourier transform infrared spectroscopy (FTIR) instruments have shown that water contents of larger grains vary from one grain to another. However, the non-equilibrium variations in water content between neighboring grains and within quartz grains cannot be interrogated further without greater measurement resolution, nor can water contents be measured in finely recrystallized grains without including absorption bands due to fluid inclusions, films, and secondary minerals at grain boundaries.

Synchrotron infrared (IR) radiation coupled to a FTIR spectrometer has allowed us to distinguish and measure OH bands due to fluid inclusions, hydrogen point defects, and secondary hydrous mineral inclusions through an aperture of $10\,\mu m$ for specimens > $40\,\mu m$ thick. Doubly polished infrared (IR) plates can be prepared with thicknesses down to 4–$8\,\mu m$, but measurement of small OH bands is currently limited by strong interference fringes for samples < $25\,\mu m$ thick, precluding measurements of water within individual, finely recrystallized grains. By translating specimens under the $10\,\mu m$ IR beam by steps of 10 to $50\,\mu m$, using a software-controlled $x - y$ stage, spectra have been collected over specimen areas of nearly $4.5\,mm^2$. This technique allowed us to separate and quantify broad OH bands due to fluid inclusions in quartz and OH bands due to micas and map their distributions in quartzites from the Moine Thrust (Scotland) and Main Central Thrust (Himalayas).

Mylonitic quartzites deformed under greenschist facies conditions in the footwall to the Moine Thrust (MT) exhibit a large and variable $3400\,cm^{-1}$ OH absorption band due to molecular water, and maps of water content corresponding to fluid inclusions show that inclusion densities correlate with deformation and recrystallization microstructures. Quartz grains of mylonitic orthogneisses and paragneisses deformed under amphibolite conditions in the hanging wall to the Main Central Thrust (MCT) exhibit smaller broad OH bands, and spectra are dominated by sharp bands at 3595 to $3379\,cm^{-1}$ due to hydrogen point defects that appear to have uniform, equilibrium concentrations in the driest samples. The broad OH band at $3400\,cm^{-1}$ in these rocks is much less common. The variable water concentrations of MT quartzites and lack of detectable water in highly sheared MCT mylonites challenge our understanding of quartz rheology. However, where water absorption bands can be detected and compared with deformation microstructures, OH concentration maps provide information on the histories of deformation and recovery, evidence for the introduction and loss of fluid inclusions, and water weakening processes.

1 Introduction

Quartz mylonites sheared at middle to lower levels of the continental crust exhibit microstructural and textural evidence of dislocation creep, a process that is widely believed to require water weakening in framework silicates. The effects of water on dislocation creep of quartz, including the nucleation, glide, climb and recovery of dislocations, and recrystallization are well known from (1) experimental studies of natural crystals, in which water was introduced into grain interiors (e.g., Griggs, 1967; Blacic, 1975, 1981; FitzGerald et al., 1991), (2) studies of synthetic and natural quartz varieties with large initial water contents (e.g., Griggs and Blacic, 1965; Hobbs, 1968; Baeta and Ashby, 1970; Kekulawala et al., 1978; Kirby and McCormick, 1979; McLaren et al., 1983; Linker et al., 1984; Gerretsen et al., 1989; Muto et al., 2011; Holyoke and Kronenberg, 2013; Stünitz et al., 2017), and (3) quartzites and polycrystalline quartz aggregates with water added or removed before or during experiments (e.g., Jaoul et al., 1984; Kronenberg and Tullis, 1984; Tullis and Yund, 1989; Hirth and Tullis, 1992; Gleason and Tullis, 1995; Post et al., 1996; Chernak et al., 2009). IR spectroscopy has played a key role in experimental studies of water weakening, through the characterization and measurement of OH absorption bands due to different hydrogen defects and forms of molecular water within quartz interiors (e.g., Kats, 1962; Griggs and Blacic, 1965; Aines and Rossman, 1984; Aines et al., 1984; Stipp et al., 2006).

Water weakening in the continental crust is inferred because of the high laboratory strengths exhibited by quartz and feldspars in the absence of water (Griggs, 1967; Heard and Carter, 1968; Tullis and Yund, 1977, 1980; Tullis, 1983; Blacic and Christie, 1984) and the postulated effects of water on point defects and disruptions of fully linked Si–O bonds (Griggs, 1974; Hirsch, 1979; Hobbs, 1981; Paterson, 1989). With the advent of FTIR and IR microscopes, water and hydrogen defects in naturally deformed quartz have been reported with OH contents of 300 to > 10 000 ppm (molar $H/10^6Si$; Kronenberg and Wolf, 1990; Kronenberg et al., 1990; Nakashima et al., 1995; Gleason and DeSisto, 2008; Seaman et al., 2013; Finch et al., 2016; Kilian et al., 2016), comparable to those required for water weakening in laboratory experiments. In small granitic shear zones deformed at greenschist conditions, water contents of quartz grains appear to correlate with finite strain (Kronenberg et al., 1990; Gleason and DeSisto, 2008), and in the much larger Median Tectonic Line of Japan, intragranular water contents increase towards its center (Nakashima et al., 1995). However, OH concentrations of quartz of granitic rocks deformed at higher temperatures can be much lower (20–100 ppm; Han et al., 2013; Kilian et al., 2016), with IR spectra dominated by small sharp OH bands of hydrogen point defects (Kilian et al., 2016) that are not thought to weaken quartz. Moreover, quartz water contents have been reported that show a trend of decreasing OH content towards the center of a high-grade

Table 1. Structural distances of samples below the Moine Thrust at the Stack of Glencoul (Law et al., 1986, 2010) and above the Main Central Thrust on the NW and Eastern Sutlej transects (Law et al., 2013). Deformation temperatures estimated by Law et al. (2013) using the Kruhl (1998) quartz c-axis fabric opening angle thermometer, and temperatures and pressures of metamorphism estimated by Stahr (2013) using THERMOCALC multi-equilibria thermometry (Powell and Holland, 1994) are indicated.

Moine Thrust – Stack of Glencoul mylonitic quartzites.		
Sample	Distance below MT	
SG-7	2.5 m	
SG-8	2.9 m	
SG-10	4.6 m	
Main Central Thrust – NW Sutlej transect orthogneisses		
Sample	Distance above MCT	T
S09-30	∼ 750 m	∼ 600 °C[1]
S09-35	75 m	∼ 540 °C[1]
Main Central Thrust – Eastern Sutlej transect paragneisses			
Sample	Distance above MCT	T	P
S09-58	∼ 4500 m	735 °C[2]	900 MPa[2]
S09-63	∼ 1500 m	675 °C[2]	850 MPa[2]
S09-71B	25 m	610 °C[1]	

[1] Deformation temperature estimated from quartz c-axis fabric opening angle (Law et al., 2013). [2] Temperature and pressure of metamorphism estimated from THERMOCALC multi-equilibra thermobarometry (Stahr, 2013, p. 67; Law et al., 2013).

shear zone (Finch et al., 2016). Maps of OH content, constructed from FTIR spectra of deformed granitic rocks (Seaman et al., 2013), show compelling relationships between water content and microstructures generated during deformation, suggesting that water contents are reduced during recrystallization and partial melting.

Much as deformation microstructures and textures provide a link between our understanding of deformation mechanisms activated in deformation experiments at high laboratory strain rates and the mechanisms governing plasticity and creep of shear zones at low natural strain rates (Snoke et al., 1998; Heilbronner and Barrett, 2014), IR spectroscopy can provide a link between our understanding of water weakening in the lab and in nature. In this paper, we report on methods of FTIR to characterize OH absorption bands and image OH contents in quartz (and other nominally anhydrous minerals) at higher resolution than is possible using conventional instruments, coupling synchrotron IR radiation with FTIR. We apply these methods to mylonitic quartzites in the footwall to the Moine Thrust in NW Scotland and to mylonitic quartz-rich orthogneisses and paragneisses in the hanging wall to the Main Central Thrust in NW India.

2 Selected quartz mylonites

The quartz mylonites selected to test synchrotron FTIR and imaging of OH come from the footwall of the Moine Thrust (MT) at the Stack of Glencoul, NW Scotland (Christie, 1963; Law et al., 1986, 2010), and the hanging wall of the Main Central Thrust (MCT) exposed in the Sutlej Valley, NW India (Law et al., 2013; Stahr, 2013; Law, 2014). Both of these thrust faults are orogen-scale shear zones with penetrative deformation on the MT accommodating shortening at the foreland edge of the Caledonian orogeny (e.g., Peach et al., 1907; Elliot and Johnson, 1980; Law et al., 1986, 2010; Butler, 2010; Law, 2010; Law and Johnson, 2010; Dewey et al., 2015) and penetrative shear strains on the MCT accommodating southward-directed Oligocene–Miocene extrusion/exhumation of the overlying Greater Himalayan slab (e.g., Grujic et al., 1996; Grasemann et al., 1999; Godin et al., 2006; Law et al., 2013). Mylonitic grain shape foliations are well developed in rocks of both fault zones and mineral stretching lineations are parallel to the fault transport directions. Deformation microstructures and textures of quartz in these mylonites indicate that dislocation creep was the predominant deformation mechanism involving both basal and prism slip systems, internal recovery, and dynamic recrystallization.

Structural distances of samples below the MT and above the MCT, together with available information on deformation temperatures and temperatures/pressures of metamorphism, are summarized in Table 1. All samples from the footwall to the MT are mylonitic Cambrian quartzites. Samples from the hanging wall to the MCT on the NW Sutlej transect are penetratively deformed orthogneisses in which intensities of grain shape fabrics, traced downwards towards the thrust surface, increase and dynamically recrystallized grain sizes decrease (Law et al., 2013; their Fig. 6). Samples from the more hinterland-positioned Eastern Sutlej transect are less obviously foliated paragneisses with more granular textures due to extreme quartz grain boundary mobility.

Mylonitic Cambrian quartzites in the footwall to the MT at the Stack of Glencoul (Assynt region) display highly flattened relict grains aligned parallel to foliation, with grain shape aspect ratios up to 50 : 1 to 100 : 1 and smooth undulatory extinction between crossed polarizers that have been described as quartz ribbons (Bonney, 1883; Christie, 1960, 1963; Weathers et al., 1979; Law et al., 1986). At the margins of the larger quartz grains, more equant, finely recrystallized grains overprint these elongate high-strain grains, with the proportion of new to old grains increasing structurally upwards towards the MT plane (Christie, 1960; Weathers et al., 1979; Law et al., 1986). Rare grains of feldspar and quartz aligned in "mechanically strong" orientations are relatively equant, and appear as augen or globular quartz grains. Quartz c axes exhibit strong lattice preferred orientations in both deformed old grains and recrystallized grains, with symmetrical type 1 (Lister, 1977) cross-girdle fabrics at distances > 150 mm beneath the thrust plane and increasingly asymmetric cross-girdle to single girdle fabrics closer to the thrust plane (Law et al., 1986, 2010). These fabrics reflect general flattening strains accommodated by quartz basal and prism slip, with variations in estimated flow vorticities and partitioning of strain between original and recrystallized grains (Law et al., 2010; Law, 2010). Micas are highly aligned parallel to foliation, with coarse muscovite grains at quartz grain boundaries and fine, dispersed micas within quartz grain interiors. Optical microstructures show evidence of quartz recrystallization by bulge nucleation and subgrain rotation with relatively uniform mean recrystallized grain sizes of ∼ 15 μm (Christie et al., 1954; Weathers et al., 1979). Microstructures imaged by transmission electron microscopy (TEM) show dense arrays of curved free dislocations, subgrain walls, and fine fluid inclusions that decorate dislocations (Weathers et al., 1979; Ord and Christie, 1984). Earlier FTIR measurements of MT mylonites from a number of locations in the Assynt region revealed large OH absorption bands characteristic of milky quartz with OH contents of 1500 to 7500 ppm (Kronenberg and Wolf, 1990).

Quartz-rich Greater Himalayan Series orthogneisses and paragneisses in the hanging wall to the MCT exposed in NW and Eastern Sutlej transects, respectively, include quartz mylonites, quartz–mica schists, and quartz–garnet schists (Vannay and Grasemann, 1998; Grasemann et al., 1999; Law et al., 2013). Quartz grain shapes are not as highly elongate as observed in the MT mylonites, owing to their extensive recrystallization, with mean grain sizes on the NW transect that vary with structural level from 200 to 250 μm (grain boundary migration microstructures) at ∼ 1000 m above the MCT to 75–95 μm at 200–750 m above the thrust, and 35–60 μm (dominantly subgrain rotation microstructures) at ∼ 75 m above the thrust surface (Law et al., 2013). As noted above, quartz recrystallized grain sizes are much larger at a given structural height above the thrust plane on the hinterland-positioned eastern transect with grain sizes commonly ranging from 250 to > 1000 μm (0.25–1.0 mm) at ∼ 25 m above the thrust to > 1–2 mm at 200 m and higher above the thrust. Quartz grain interiors on the eastern transect display mild undulatory extinction with highly aligned subgrain boundaries that give the appearance of chessboard extinction and irregular, non-planar grain boundaries that envelope neighboring mica grains, suggestive of high grain boundary mobilities at the time of peak metamorphism and deformation. Quartz c axes show strong lattice preferred orientations with symmetrical and slightly asymmetrical cross-girdle fabrics on both transects, providing evidence of simultaneous basal and prism slip during plane strain deformation, with varying amounts of pure shear and simple shear (Law et al., 2013). Coarse-grained muscovite and biotite are highly aligned parallel to foliation, both at the boundaries of quartz grains and within quartz grain interiors.

While both the MT and the MCT mylonites were deformed by dislocation creep, their deformation temperatures were very different and the mechanisms of recrystallization, accommodating dislocation creep differ accordingly. Deformation temperatures for the MT footwall mylonites at the Stack of Glencoul are estimated at between 300 and 350 °C based on illite crystallinity (Johnson et al., 1985). Opening angles of c-axis fabrics measured separately on old and recrystallized grains (Law et al., 2010) indicate far higher apparent deformation temperatures using the Kruhl (1998) fabric opening angle thermometer (390–440 and 475–530 °C, respectively) than are compatible with their prehnite-pumpellyite to lower greenschist (chlorite) facies phyllosilicates (Law, 2014). Deformation temperatures for MCT hanging wall mylonites range between ∼ 535 and 610 °C on the NW Sutlej transect, using the Kruhl (1998) opening angle thermometer, and ∼ 610–> 735 °C based on fabric opening angles and petrologic constraints on the Eastern Sutlej transect (Table 1; Law et al., 2013; Stahr, 2013).

Dynamic quartz recrystallization microstructures in the MT mylonites are similar to microstructures developed under experimental Regime II creep conditions as defined by Hirth and Tullis (1992), with a combination of bulge nucleation recrystallization (BLG) at grain boundaries and subgrain rotation recrystallization (SGR) within quartz grain interiors, and they are consistent with deformation under greenschist facies conditions (Stipp et al., 2002, 2010; Law, 2014). Quartz microstructures in the MCT mylonites (particularly on the Eastern Sutlej transect and at large distances above the thrust on the NW transect) indicate more internal recovery and more extensive grain boundary migration (GBM) than is apparent in the Regime III creep experiments of Hirth and Tullis (1992). The microstructures are more similar to those observed by Stipp et al. (2002, 2010) in quartz veins deformed at natural strain rates and the equivalent of middle–upper amphibolite facies conditions (∼ 550–700 °C; see review by Law, 2014).

The small recrystallized quartz grain sizes (16–9 µm), subgrain sizes, and dislocation densities of the MT footwall mylonites at the Stack of Glencoul imply large differential stresses (40–250 MPa) during shearing (Weathers et al., 1979; Ord and Christie, 1984); differential stresses of ∼ 55–85 MPa are recalculated using these grain sizes and the recrystallized grain size piezometer of Stipp et al. (2006; their Fig. 8) modified by the stress correction of Holyoke and Kronenberg (2010). In contrast, quartz recrystallized grain sizes for MCT mylonites on the NW Sutlej transect of 60–35 µm at 75 m above the thrust to 95–75 µm at 750–200 m above the thrust indicate lower differential stresses of 19–30 and 13–16 MPa, respectively, extrapolating the Stipp et al. (2006) grain size piezometer, adjusted by the same stress correction (Francsis, 2012; Law et al., 2013). Flow stresses inferred for the Eastern Sutlej transect mylonites would presumably be even lower than for mylonites of the NW Sutlej transect, but their grain sizes are far greater than those encompassed by any experimental grain size piezometer.

3 Methods

IR spectroscopy has been an important tool for studying OH defects in nominally anhydrous minerals and, in coordination with deformation experiments, to study the effects of water and hydrogen defects on mechanical properties (e.g., Kats, 1962; Griggs and Blacic, 1965; Kekulawala et al., 1978, 1981; Aines and Rossman, 1984; Cordier and Doukhan, 1991; Mackwell and Kohlstedt, 1991; Bai and Kohlstedt, 1996; Kohlstedt et al., 1996). With the introduction of efficient FTIR spectrometers and IR microscopes, studies of intragranular water and hydrogen defects in naturally deformed rocks have been enabled using apertures of 50–100 µm (Kronenberg and Wolf, 1990; Nakashima et al., 1995; Gleason and DeSisto, 2008; Finch et al., 2016; Kilian et al., 2016) as well as FTIR mapping of OH contents (Seaman et al., 2013). Molecular water contents of naturally deformed quartz commonly have large grain-to-grain variations at this scale of observation, and it has not been possible to measure OH contents of fine, dynamically recrystallized grains without including OH bands associated with grain boundaries.

Synchrotron-generated IR radiation is much brighter (∼ 1000×) than conventional globar sources of commercial FTIR instruments, and high-quality IR spectra can readily be measured through small (10 µm) apertures (Lobo et al., 1999; Carr et al., 2008; Ma et al., 2013). By coupling synchrotron radiation to a FTIR instrument and IR microscope, OH absorption bands can be measured through a 10 µm aperture with higher signal-to-noise ratio than through a 100 µm aperture using a standard FTIR system. We have measured OH concentrations using both conventional and synchrotron FTIR and we have mapped OH distributions using synchrotron FTIR within quartz grains and across multiple grains of MT and MCT mylonites.

3.1 Preparation of IR plates

Doubly polished IR plates of uniform thickness were prepared perpendicular to foliation and parallel to lineation for MT and MCT mylonites. Images of large IR plates were collected using a high-resolution (4000 dpi) Nikon slide scanner (Coolscan 8000 ED), with and without polarizing filters on both sides of the sample (Fig. 1a, b). Images of smaller IR plates were recorded with a Zeiss Axioplan 2 petrographic microscope and AxioCam HRc imaging system (Fig. 1c–f). Image contrast was enhanced digitally (using Adobe Photoshop) for crossed-polarized light images of ultrathin samples with small optical retardations. Throughout the preparation of IR plates, impregnation of samples by epoxy or other insoluble resins was avoided to prevent the introduction of OH or CH absorption bands of mounting media, which might be difficult to distinguish from OH bands of samples. While central regions of the IR plates remain fragile, metal frames were mounted to IR plate extremities for mechanical support, to avoid catastrophic loss and facilitate handling.

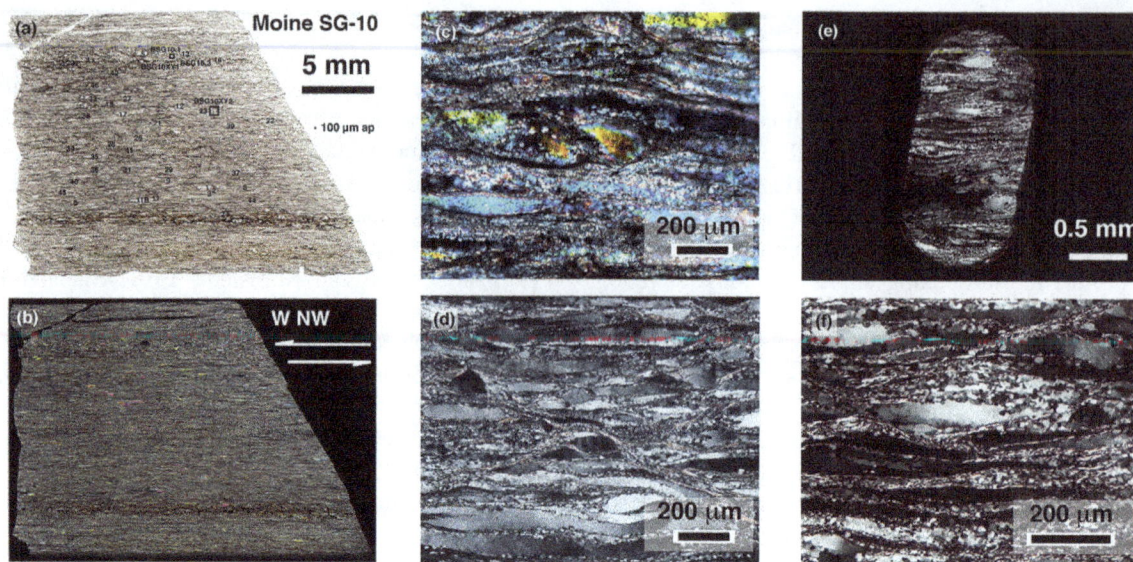

Figure 1. Doubly polished IR plates of Moine Thrust (MT) mylonites prepared perpendicular to foliation and parallel to lineation. **(a)** Low-magnification image of large IR plate prepared from Stack of Glencoul sample SG-10 (unpolarized light), with a mean thickness $t = 120\,\mu m$. **(b)** Same IR plate of SG-10 shown in **(a)** but with crossed-polarized light. The MT top to WNW shear sense is shown in the plane of the IR plate (top to the left). **(c)** Higher-magnification optical micrograph of large IR plate of sample SG-10, with crossed polarized light and local plate thickness ($117\,\mu m$) determined from IR interference fringes. **(d)** Optical micrograph of Stack of Glencoul sample SG-7 in crossed-polarized light (normal $30\,\mu m$ section thickness), illustrating deformation and recovery microstructures with higher resolution than in thick IR plates; note undulatory extinction in ribbon quartz grains, fine recrystallized grains, and aligned muscovite grains localized at quartz grain boundaries. **(e)** Low-magnification image of ultrathin IR plate of Stack of Glencoul sample SG-7-5 (crossed-polarized light) mounted on a copper TEM slot ring ($t = 4\text{–}8\,\mu m$, based on IR interference fringes). **(f)** Higher-magnification optical micrograph of the same IR plate SG-7-5 (crossed-polarized light) as shown in **(e)** with deformation and recrystallization microstructures shown more clearly than in normal $30\,\mu m$ thin section.

Oriented rock chips were first mounted on a glass thin section plate using CrystalBond 509 and a low-temperature hot plate. The top surfaces of samples were polished using a sequence of successively finer grits (400 and 600 mesh; 9.5, 3, and $0.3\,\mu m$). Polished specimens were removed from the glass plate and remounted, this time on their polished surfaces. Sample plates were cut parallel to the first surface, and the second (cut) surface was ground and polished to the desired thickness by the same method. A micrometer was used to measure the compound thickness of sample, glass plate, and mounting medium, checking thickness at sample extremities. In addition to micrometer measurements, sample thicknesses were tested during grinding, using interference colors for thicker samples (first-order colors for quartz plates 50–$90\,\mu m$ thick) and maximum detectable birefringence (grey-white scale) while polishing ultrathin ($<20\,\mu m$) samples.

Before removing doubly polished samples from the glass thin section plate, metal frames were mounted to sample extremities using a thin bead of epoxy resin to provide support. For larger ($>10\,mm$) samples with thicknesses of 25–$100\,\mu m$, metal frames made of Ni wire ($\sim 1\,mm$ diameter) were custom-fit to each sample. For thin samples ($<25\,\mu m$), several metal frames normally used to mount TEM samples (3 mm outer diameter copper rings or slot rings with $1 \times 2\,mm$ internal dimensions; supplied by Ted Pella, Inc., Fig. 1e) were mounted on each polished sample. Once the metal frames were attached and the epoxy cured, samples were removed from the glass thin section over a low-temperature hot plate, and the CrystalBond resin removed using acetone.

The fragile quartz IR plates, 4 to $150\,\mu m$ thick, were soaked in static acetone baths to dissolve the CrystalBond resin, exchanging the acetone and repeating this procedure three times. Even without ultrasonic agitation, some samples were damaged by air bubbles caught beneath them, leading to specimen warping and disintegration.

3.2   FTIR

IR absorption spectra were collected for quartz mylonites of varying plate thickness using a Nicolet Magna 560 FTIR instrument with Omnic software, a conventional globar source and NicPlan IR microscope (at Texas A&M University), and using a Bruker Hyperion FTIR instrument, OPUS software, and IR microscope at the U2A synchrotron beamline of the National Synchrotron Light Source (NSLS) I (at Brookhaven National Laboratory, Upton, NY). Both the Nicolet and Bruker instruments make use of liquid nitrogen-cooled MCT detectors. IR spectra were collected at a wavenumber reso-

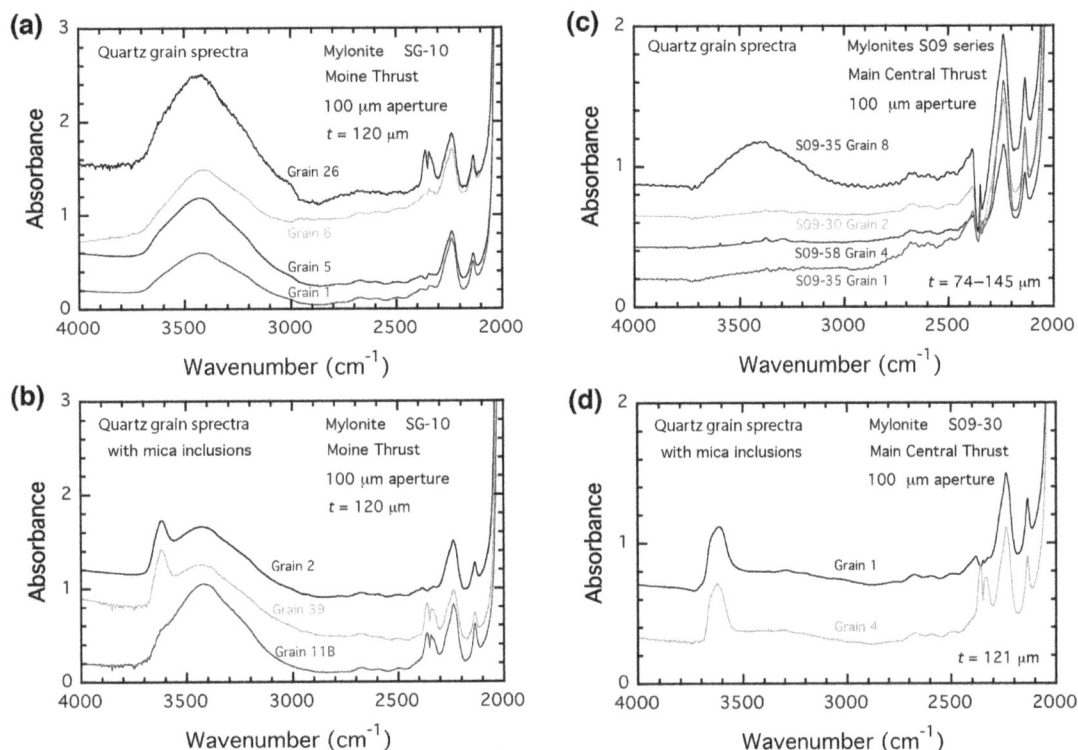

Figure 2. IR spectra of quartz grains in mylonites from the MT (Stack of Glencoul) and the Main Central Thrust (MCT, Sutlej Valley) using conventional FTIR with globar IR source through a 100 μm aperture. **(a)** Absorption spectra of quartz grains of MT IR plate SG-10 (shown in Fig. 1a–b) show a broad OH band at 3400 cm⁻¹ due to fluid inclusions. **(b)** Absorption spectra of MT quartz grains of the same IR plate (SG-10) showing the same broad OH band and an OH absorption band (or shoulder) at 3600 cm⁻¹ due to mica inclusions. **(c)** Absorption spectra of quartz grains of MCT IR plates S09-30 ($t = 121$ μm), S09-35 ($t = 74$ μm), and S09-58 ($t = 145$ μm), with many grains showing small sharp OH absorption bands due to hydrogen point defects, and less common grains with a large broad absorption band at 3400 cm⁻¹. **(d)** Absorption spectra of quartz grains of MCT sample S09-30 ($t = 121$ μm) with a prominent OH band at 3600 cm⁻¹ due to mica inclusions and smaller OH bands at lower wavenumbers. Absorbance values plotted vertically are normalized to represent their values for a uniform sample thickness of 1 mm.

lution of 4 cm⁻¹, compiling 512 scans for each spectrum (or more if needed using the Nicolet FTIR), and spectra were stored over wavenumbers of 4000 to 2000 cm⁻¹. For purposes of comparison, all absorbances shown in spectra for quartz grains of varying thickness have been normalized to a common thickness of 1 mm.

Representative IR spectra of individual quartz grains from MT and MCT samples measured with the Nicolet FTIR (100 μm aperture) exhibit OH absorption bands of variable size and character (Fig. 2). MT quartz spectra show a broad OH stretching band at 3400 cm⁻¹ of large but variable magnitude (Fig. 2a), characteristic of molecular water in fluid inclusions of milky quartz (Kekulawala et al., 1978; Aines and Rossman, 1984; Stünitz et al., 2017). In addition to this absorption band, some quartz grains from MT samples show an additional absorption band (or subtle shoulder) at ∼ 3600 cm⁻¹ (Fig. 2b) due to fine-scale micas (sometimes visible optically and sometimes too fine to resolve) dispersed within quartz grains. MCT quartz spectra tend to have smaller OH absorption bands (Fig. 2c), with some grains

showing small sharp bands at 3595, 3482, 3431, 3408, and 3379 cm⁻¹ due to hydrogen interstitial defects (Kats, 1962; Aines and Rossman, 1984) and less common grains with a larger broad band at 3400 cm⁻¹. Finely dispersed micas are less common in these coarser-grained mylonites, but some quartz grains also exhibit a 3600 cm⁻¹ OH band due to micas (Fig. 2d).

Aside from differences in the Nicolet and Bruker FTIR instruments and software, the most significant difference between these facilities is the IR source, so that OH absorption measurements with the Nicolet FTIR and its conventional IR source could not practically be made with apertures < 50–100 μm, while OH absorption bands could routinely be made with the Bruker FTIR and synchrotron IR source through a 10 μm aperture. In both cases, the IR sources are unpolarized, leading to differences in those OH absorption bands that are anisotropic in quartz and mica grains of varying orientation. Small sharp OH bands due to hydrogen interstitials of quartz (between 3595 and 3379 cm⁻¹) and intrinsic OH bands of micas (at ∼ 3600 cm⁻¹) are strongly anisotropic (Kats, 1962;

Beran, 2002). However, the primary OH band of interest for water weakening of milky quartz (at $3400\,\text{cm}^{-1}$) is broad and isotropic (Kekulawala et al., 1978; Aines et al., 1984), associated with dispersed fluid inclusions. Variations in absorbance of this band for neighboring quartz grains represent real variations in water content and not variations in quartz grain orientation.

Direct comparisons of intragranular water of a given quartz grain using the two FTIR facilities and a common aperture size could not be made because of (1) the poor signal-to-noise ratio of spectral measurements through a $10\,\mu\text{m}$ aperture with the broad globar IR radiation, and (2) the inability to make spectral measurements for a $100\,\mu\text{m}$-apertured area with the narrow synchrotron IR beam (which is not much broader at the microscope stage than the $10\,\mu\text{m}$ aperture). Samples with uniform OH contents might serve as standards to compare OH absorption bands for different spectrometers, irrespective of aperture size, but our observations indicate that OH contents of the quartz grains in the mylonites we measured are highly variable. Measurements through finer apertures led to larger variances in OH absorption bands, within grains as well as between neighboring grains.

We also observed larger-amplitude interference fringes in spectra measured with the synchrotron IR source than those observed in spectra measured with the globar IR source for a given sample. These fringes are caused by internal reflection within doubly polished, parallel-sided IR plates, and they have larger magnitudes for the highly collimated IR synchrotron beam than for the broad, confocal globar IR radiation.

3.3 Interference fringes

The appearance of interference fringes in spectra can be useful to determine the optical path length in thin samples, as long as their amplitude is small compared with vibrational absorption bands of interest, or they are distinguishable from absorption bands by their wavenumber spacing. Interference fringes are present in many spectra we collected for quartz mylonite plates (Fig. 2c, grains 1 and 8), which are of manageable size for our thicker samples and problematic for thinner samples.

Interference fringes were routinely observed for all samples measured with synchrotron IR radiation and the Bruker FTIR instrument. We therefore made use of these fringes to determine local sample plate thicknesses. For sample plates with large interference fringes, we attempted to reduce their size by tilting samples by $45°$ in the IR beam or using a Cassegrain objective lens with large numerical apertures (0.6). However, neither of these methods was effective in reducing amplitudes of some of the very large interference fringes sufficiently to detect small OH absorption bands. We had greater success reducing interference fringes by fitting them where the baseline was free of absorption bands, using

DatLab software (similar to fringe modeling of Clark and Moffatt, 1978; Pistorius and DeGrip, 2004; Konevskikh et al., 2015), and subtracting the model fringes from the spectral data. This improved the quality of spectra when interference fringes had modest amplitudes, but interference fringes for thin samples were very large and resulting backgrounds were sufficiently irregular that we could not resolve small OH absorption bands.

3.4 IR plate thickness and OH absorbance determinations

Mean IR plate thicknesses were determined by focusing on imperfections in the top and bottom specimen surfaces, recording the numerical graduations on the focusing knob of the IR microscope stage for each surface, and converting to vertical displacement.

Local IR plate thicknesses were also determined from interference fringes measured in spectra collected with the synchrotron–Bruker FTIR microscope system, where thickness t is given by

$$t = 1/(2n\delta v), \tag{1}$$

where n is the mean refractive index of quartz ($n = 1.55$) and δv is the measured peak-to-peak fringe spacing (Stuart et al., 1996). The two measures of thickness were in agreement within resolution ($\sim 5\,\%$) for a given IR plate and location within the specimen, with thick IR specimens ($\sim 100\,\mu\text{m}$) showing real variations in local thickness of $\pm 10\,\mu\text{m}$ and thin samples varying in local thickness from 4 to $8\,\mu\text{m}$.

IR spectra were collected and integrated absorbances of OH bands were measured above an assumed straight-line background, where backgrounds were fit and integration limits were chosen at the same wavenumber values, from ~ 3705 to $2880\,\text{cm}^{-1}$ for OH bands of quartz grains and ~ 3702 to $3544\,\text{cm}^{-1}$ for OH bands of micas (Fig. 3a). We are satisfied by our ability to separate, to first order, the OH absorption bands due to fluid inclusions in quartz and due to micas in MT samples, because IR spectra of coarse muscovite grains in MT mylonites consist of a single OH absorption band at $3620\,\text{cm}^{-1}$ (Fig. 3b). This strong OH stretching band is well known from multiple spectroscopic studies of muscovite (Beran, 2002; Tokiwai and Nakashima, 2010a, b; Heller-Kallai and Lapides, 2015).

Our ability to distinguish OH absorbance due to fluid inclusions and micas in MCT samples, by contrast, is poor. In addition to the primary OH absorption bands of muscovite (Fig. 3c) and biotite (Fig. 3d) grains at 3638 and $3614\,\text{cm}^{-1}$, respectively, MCT muscovite spectra show smaller OH bands at 3311, 3146, and $3035\,\text{cm}^{-1}$ and biotite spectra show shoulders at both sides of the primary OH absorption band (at ~ 3679 and $\sim 3561\,\text{cm}^{-1}$) and significant OH bands at 3258, 3043, and $2829\,\text{cm}^{-1}$. All of these OH bands are anisotropic, and complexly so. The primary OH band of MCT muscovite at $3638\,\text{cm}^{-1}$, measured in polarized IR ra-

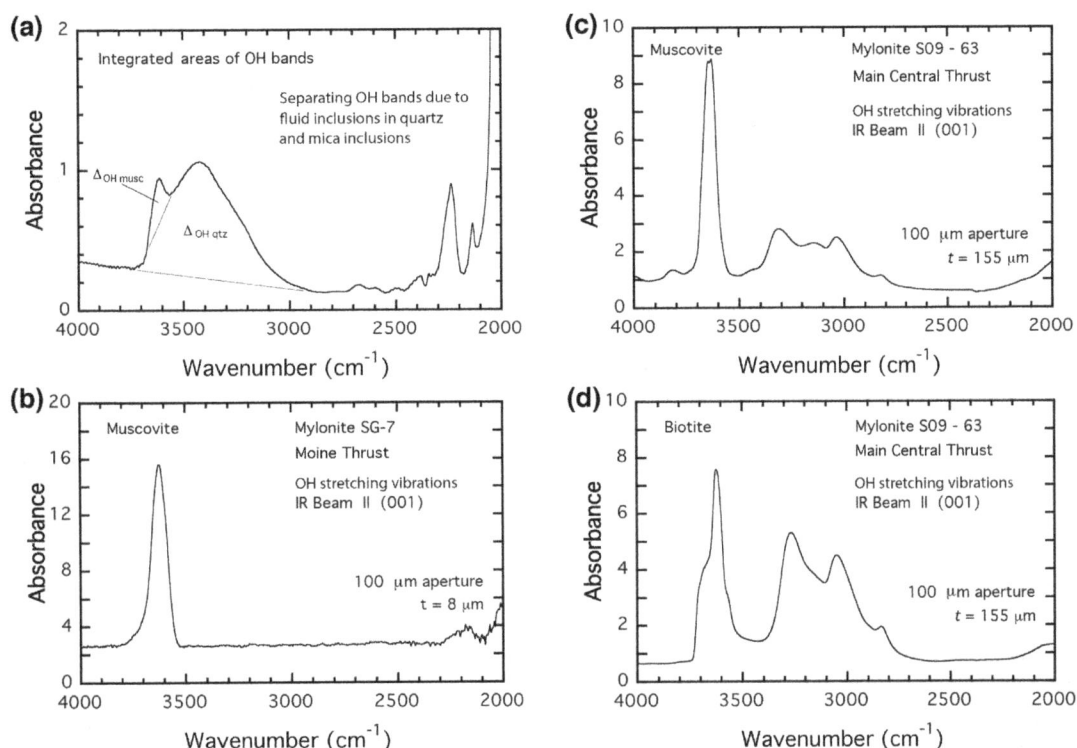

Figure 3. OH absorption spectra of quartz and mica grains. **(a)** IR spectrum of MT quartz grain (SG-10, Grain 2) with a broad 3400 cm^{-1} band due to fluid inclusions and a sharper OH band at 3600 cm^{-1} due to fine, intragranular mica inclusions. The integrated OH absorbances due to molecular water (and lesser hydrogen defects) of quartz ($\Delta_{OH\,qtz}$ of the broad 3400 cm^{-1} band, in cm^{-2}) and due to mica inclusions ($\Delta_{OH\,mica}$ of the 3600 cm^{-1} band) can be distinguished using approximate straight-line fits. **(b)** IR spectrum of a large muscovite grain of MT sample SG-7. Muscovite grains in MT samples exhibit a simple spectrum with a prominent OH absorption band at 3620 cm^{-1} ($t = 8\,\mu$m, absorbance normalized to $t = 0.1$ mm). **(c)** The prominent OH band of a large muscovite grain in MCT sample S09-63 ($t = 155$um) appears at 3638 cm^{-1} with smaller OH bands at 3311, 3146, and 3035 cm^{-1}. **(d)** The primary OH band of a large biotite grain in the same MCT sample appears at 3614 cm^{-1} (with shoulders at 3674 and 3561 cm^{-1}) and secondary OH bands at 3258, 3043, and 2829 cm^{-1}. The complex nature of secondary OH bands of MCT micas presents significant difficulties in separating OH absorptions of quartz grains due to molecular water and hydrogen defects from those of mica inclusions dispersed within quartz grains. All absorption spectra shown here were measured using unpolarized IR radiation with a conventional FTIR spectrometer and a 100 µm aperture. Absorbances for all spectra but **(b)** are normalized to a uniform thickness t of 1 mm.

diation, is strongest when the vibration direction E is parallel to the basal plane (001), consistent with idealized hydrogen positions in dioctahedral micas and the polarization of OH bands normally reported for muscovite (Beran, 2002). However, our polarized IR measurements of muscovite OH bands at 3311, 3146, and 3035 cm^{-1} indicate that they are nearly isotropic. The primary OH band of MCT biotite grains at 3614 cm^{-1} is strongest when E is perpendicular to (001), consistent with hydrogen positions of trioctahedral micas and OH band polarizations observed for phlogopite (Beran, 2002). However, the OH bands at 3258, 3043, and 2829 cm^{-1} are only weakly polarized and in the opposite sense of the primary 3614 cm^{-1} band.

As a result, we only feel confident in our determinations of OH absorbances of quartz grains in MCT mylonites (Fig. 2c), when mica inclusions are absent (the 3600 cm^{-1} band is undetectable). When the 3600 cm^{-1} band is present (Fig. 2d),

we cannot readily interpret absorption bands of molecular water or hydrogen defects in quartz; at best, quartz OH contents are overestimated.

Integrated absorbances of OH bands of quartz were used to determine OH contents based on the Beer–Lambert relation (Stuart et al., 1996),

$$A = kct, \qquad (2)$$

where integrated absorbance Δ ($= A/t$, determined in cm^{-2}) is related to the concentration c of OH, assuming that k for the broad isotropic OH band at 3400 cm^{-1} due to fluid inclusions in milky quartz is the same as for the broad isotropic OH absorption band of molecular water in wet synthetic quartz (Aines et al., 1984):

$$c \text{ (in molar ppm, OH}/10^6\text{Si)} = 1.05\Delta \text{ (in cm}^{-2}). \qquad (3)$$

To the extent that spectra include sharp OH absorption bands due to hydrogen point defects, this calibration will overestimate OH concentrations due to H interstitials, given that the value of k for these bands is larger than for molecular water (Kats, 1962; Thomas et al., 2009) for quartz grains oriented for maximum OH absorbance (vibrational directions perpendicular to the c axis). However, apparent OH concentrations due to H interstitials will appear smaller for quartz grains oriented for minimum OH absorbance in unpolarized IR radiation. These errors are not serious for quartz grains with large molecular water contents that dominate over hydrogen point defect concentrations. However, we expect that our determinations of OH concentrations for dry quartz grains (< 100 ppm) with spectra dominated by sharp anisotropic OH bands are not as accurate as for wet quartz grains with spectra dominated by isotropic broad OH bands. We also acknowledge that the calibration for molecular water absorption of synthetic quartz used here differs from the calibration of the $3400 \, cm^{-1}$ OH band absorbance reported by Stipp et al. (2006) for milky quartz grains with fluid inclusions. If this alternative calibration is correct, OH contents reported here are smaller by a factor of 0.56. The calibration of Stipp et al. (2006) is based on FTIR measurements of individual grains of a pure quartzite and independent Karl Fischer titration of disaggregated quartz particles, and we do not understand the source of discrepancies in absorbance calibrations.

Integrated absorbances Δ were also measured for the $3600 \, cm^{-1}$ OH band of micas, but no attempt was made to convert these to OH (or mica inclusion) contents. Integrated OH absorbances due to unseen (below optical resolution) mica inclusions will depend on the path length through the micas, the mica orientations, and whether they consist of muscovite, biotite, or both within the volume measured.

3.5   Mapping OH absorption bands

Integrated areas of OH were measured as a function of spatial x and y dimensions within IR plates measured on the Bruker FTIR system, distinguishing absorbances of OH bands due to fluid and mica inclusions to the extent that this was possible. Samples were translated under the $10 \, \mu m$ IR beam using a motorized $x - y$ stage, controlled by OPUS software, and spectra were collected for each translation step, usually moving the stage by 10, or 30 or $50 \, \mu m$ in order to map larger regions (up to $\sim 4.5 \, mm^2$). Integrated absorbance measurements were made over 3705–$2880 \, cm^{-1}$ (to include the $3400 \, cm^{-1}$ broad band of milky quartz) and 3702–$3544 \, cm^{-1}$ (to determine the $3600 \, cm^{-1}$ mica band) for each scanned area (compiling absorbance measurements for 660 to 4950 spectra). The integrated absorbances of the $3600 \, cm^{-1}$ mica band were subtracted from the first of the two integrated absorbances to determine a representative measure of the OH bands of molecular water and hydrogen defects of quartz.

Given that the same IR background was used to reference all spectral measurements made during $x - y$ mapping of OH bands, long step-scan mapping projects were susceptible to

changes in aperture and condenser lens centering, IR beam drift, and beam outages while the synchrotron ring was refilled. As a result, we modified our IR absorbance measurement methods for step-scan mapping to obtain reasonably accurate integrated absorbances of broad OH bands without requiring high-resolution or precision measurements of small, sharp OH bands, with the goal of limiting total measurement times.

IR spectra used to map OH bands were made at a wavenumber resolution of $8 \, cm^{-1}$, reducing the interferometer mirror translation times with no detectable degradation in integrated absorbance measurements. More significantly, we reduced the number of scans to measure each IR spectrum from 512 to just 16, resulting in measurement times for each spectrum of only $\sim 10 \, s$. Tests of spectral quality with reduced numbers of scans indicated that integrated areas determined after 16 scans were within 8 % of values determined after 512 scans (for $\Delta \sim 3000 \, cm^{-2}$ and $t \sim 120 \, \mu m$). OH maps were constructed for sample areas of $0.066 \, mm^2$ ($22 \times 30 = 660$ spectra) to nearly $4.5 \, mm^2$ ($50 \times 99 = 4950$ spectra) requiring measurement times of 110 min to nearly 14 h, respectively. The larger step-scan maps include some poor spectral data corresponding to beam drift or outages; these spectra were identified and removed from the data set.

Maps of integrated OH absorbance were contoured (using SigmaPlot) to form images of water contents and micas, and superposed on optical micrographs of the measured regions to look for correlations between OH content and microstructure. OH contents of quartz, due primarily to the broad $3400 \, cm^{-1}$ OH band of fluid inclusions and secondarily to sharp hydrogen defect bands, were mapped as molar ppm ($OH/10^6 Si$), while the OH absorption band at $\sim 3600 \, cm^{-1}$ due to micas was mapped in units of integrated absorbance (cm^{-2}). Logarithmic contour intervals of $\log_{10}(OH/10^6 Si)$ were chosen to image water contents in quartz over a wide range of concentrations, and to provide visual images of water distributions at low (and high) values. OH contents were contoured in color with blue (and other cool colors) corresponding to large water contents, and red (and other warm colors) corresponding to low water contents. Logarithmic contours were also chosen to image distributions of micas, plotting $\log_{10}(\Delta)$ for the integrated absorbance of the $\sim 3600 \, cm^{-1}$ OH band, using a similar key for the contour interval, where cool colors correspond qualitatively to high mica contents and warm colors correspond to low mica contents.

4   Results

4.1   Synchrotron FTIR measurements

IR spectra of quartz in MT and MCT samples measured with the synchrotron–FTIR system through a $10 \, \mu m$ aperture (for samples $\sim 100 \, \mu m$ thick) exhibit OH absorption bands of similar character as measured using a conventional FTIR system with a 100 μm aperture. OH bands of MT quartz grains generally include a large broad absorption band at 3400 cm^{-1} due to fluid inclusions, with some grains showing a secondary absorption band at 3600 cm^{-1} due to micas (Fig. 4a). Given the intense synchrotron IR source, we did not encounter any losses in spectral quality due to signal-to-noise ratios for the smaller sample volumes measured, though interference fringes are more apparent. Synchrotron FTIR measurements of MCT samples through a 10 μm aperture show the same sharp OH bands due to hydrogen point defects as measured for larger sampling volumes. With the exception of some unusual grains, broad OH bands at 3400 cm^{-1} in MCT samples are small to negligible.

Integrated absorbances of the broad 3400 cm^{-1} OH band measured with a 10 μm aperture are highly variable spatially, yielding water contents for individual spots within quartz grains for a given MT sample (SG-10) of 280 to 9000 ppm, with a comparable mean (2430 ppm) but more variance among individual measurements than measured with a 100 μm aperture (from 1130 to 8590 ppm for 45 measurements). For dry MCT samples (S09-71B), OH contents of quartz measured using a small (10 μm) aperture varied from 50 to 300 ppm, while OH contents measured with the larger (100 μm) aperture varied from 85 to 240 ppm. For unusually wet MCT samples (S09-35), OH contents of quartz measured using a small (10 μm) aperture varied from 150 to 7500 ppm, while OH contents measured with the larger (100 μm) aperture varied from 160 to 4620 ppm. The variations in size of sharp OH bands associated with hydrogen point defects, by comparison, are much smaller. The sizes of these absorption bands may be explained by differing quartz grain orientations without calling upon any variations in hydrogen point defect concentrations (Thomas et al., 2009).

4.2 Limits of IR plate thickness

IR spectra of quartz were measured through a 10 μm aperture, varying sample plate thickness from ∼ 100 to 4–8 μm, with the hope that we might be able to measure the OH bands of individual recrystallized grains in ultrathin IR plates. Owing to the coarse recrystallized grain sizes of MCT mylonites, we were able to measure IR spectra of most quartz grains in these samples for plate thicknesses of > 50 μm. However, our IR measurements of individual grains in MT samples (with recrystallized grains of ∼ 10–15 μm) are limited to larger porphyroclastic and deformed ribbon quartz grains. While the signal-to-noise ratio of our measurements continues to be acceptable to measure the small OH absorption bands in very thin IR plates, interference fringes increase in size as the IR plate thickness t is decreased. Interference fringes become very large at $t < 25$ μm and OH bands cannot be detected at $t = 13$ and 6.5 μm (Fig. 4b, c).

The magnitude of interference fringes can be modified by tilting the sample within the IR beam and increasing the nu-

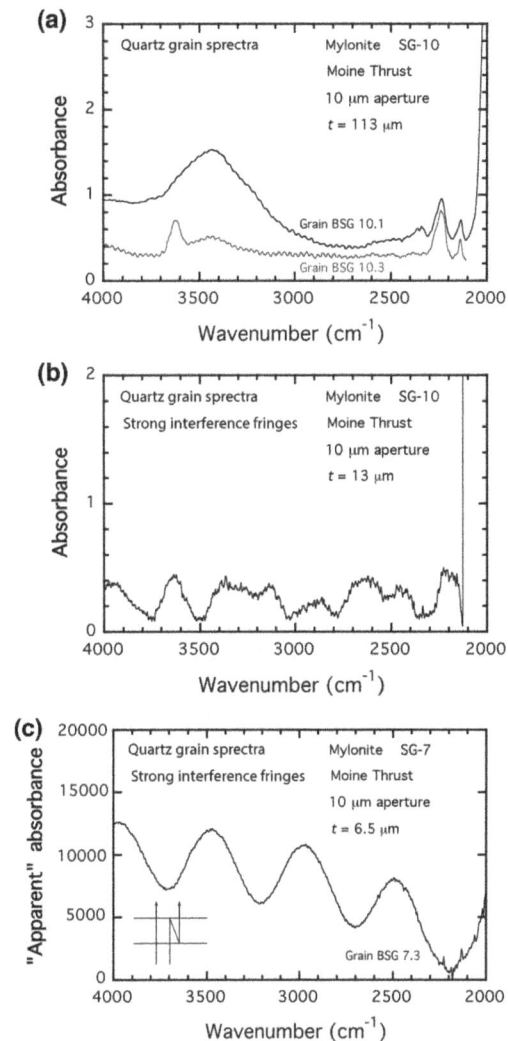

Figure 4. IR spectra of quartz grains in MT mylonite samples (Stack of Glencoul), measured with synchrotron–FTIR system using a 10 μm aperture with varying IR plate thicknesses **(a)** $t = 113$ μm (BSG 10, local plate thickness determined from interference fringes), **(b)** $t = 13$ μm (SG-10.2t), and **(c)** $t = 6.5$ μm (BSG 7.3, sample plate SG-7-1). **(a)** IR spectra of MT sample (SG-10) show OH absorption bands of similar character at the same wavenumbers for a 10 μm aperture as OH bands measured through a larger (100 μm) aperture, including a large broad absorption band at 3400 cm^{-1} due to dispersed fluid inclusions (both BSG 10.1 and BSG 10.3) and a sharper band at 3600 cm^{-1} due to mica inclusions (shown by BSG 10.3). Interference fringes in samples ∼ 100 μm thick are apparent, allowing determination of local IR plate thickness, but they do not obscure the OH absorption bands. **(b)** Interference fringes for samples < 25 μm thick are large, and make detection of small OH absorption bands difficult. The only detectable absorbance bands in sample SG-10.2t ($t = 13$ μm) are due to strong primary SiO vibrations (at $\upsilon < 2200$ cm^{-1}). **(c)** Interference fringes are very large for thin IR plates ($t = 6.5$ μm; SG-7-1); neither SiO nor OH absorption bands are observed, even after attempts to model them and remove fringes numerically. All absorbance values (and apparent absorbance values of interference fringes exhibited by SG-7) are normalized to a uniform sample thickness of 1 mm.

merical aperture of the Cassegrain objective on the IR microscope. However, with peak-to-trough fringe magnitudes 10^4 times greater than the OH bands in our mylonite samples, the modest reductions in fringe amplitude realized by these methods are not significant for the measurement of OH absorptions. Our efforts to model interference fringes were largely successful, reducing their magnitudes by a factor of ~ 100, but the resulting backgrounds were not flat enough at the scale of the known OH absorption bands of the MT (or MCT) samples to allow absorption band measurements for our thinnest sample plates. Interference fringes vary in magnitude according to IR plate thickness, surface polish, and scattering by internal flaws, but OH absorption bands could only be measured for samples of $\sim 40\,\mu m$ thickness or greater.

4.3 Optical microstructures and plate thickness

Although the painstaking efforts to prepare ultrathin IR plates were not rewarded by spectral measurements of OH content within individual, finely recrystallized grains, optical imaging of deformation and recrystallization microstructures in ultrathin samples was improved over that using conventional thin sections. Optical microstructures of $100\,\mu m$ thick IR plates are poorly resolved by comparison with those imaged in $30\,\mu m$ thick sections (Fig. 1c, d), with interference colors that reflect greater optical retardation, grain boundaries that are not as clearly defined, and greater numbers of overlapping grains. Remarkably, ultrathin IR plates of quartz mylonites, just 4–$8\,\mu m$ thick (Fig. 1f), continue to exhibit contrast between grains and within grains, with first-order black to grey birefringence that can be enhanced by increasing image contrast.

High-magnification optical micrographs of ultrathin MT mylonites reveal microstructures (Fig. 5) that correspond better to TEM observations (Weathers et al., 1979; Ord and Christie, 1984) than to optical microstructures of conventional thin sections. While highly flattened ribbon quartz grains observed in $30\,\mu m$ sections show smoothly varying undulatory extinction, ultrathin sections exhibit well developed subgrains within grain interiors, with sharply defined changes in extinction marking the locations of distinct subgrain walls (Fig. 5a). Prior TEM of MT mylonites revealed significant densities of free dislocations, which may be associated with smooth changes in extinction. However, TEM observations also show dense, sharply defined, low-angle subgrain walls (Weathers et al., 1979; Ord and Christie, 1984) that are often not resolved as optical microstructures in normal thin sections. These observations suggest that ultrathin sections may be useful to distinguish smooth changes in extinction (due to free dislocations and strains internal to subgrains) from discrete changes in extinction of overlapping and neighboring subgrains.

Recrystallized grains are apparent at some sutured grain boundaries in MT samples (Fig. 5a) with grain sizes that

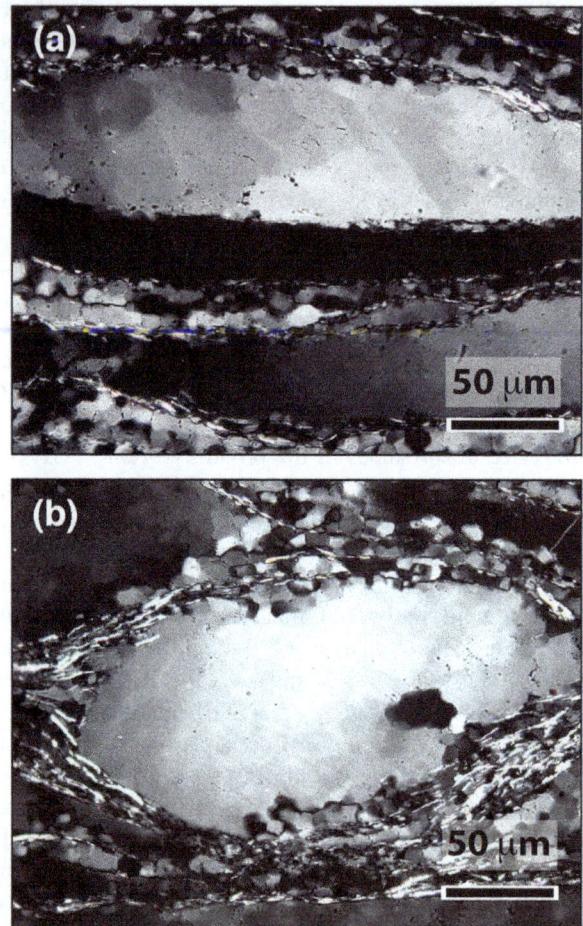

Figure 5. Ultrathin IR plates ($t = 4$–$8\,\mu m$) exhibit deformation microstructures in crossed-polarized light with greater clarity than those of conventional thin sections ($t = 30\,\mu m$). **(a)** Deformed ribbon quartz grains of MT sample SG-7-6 ($t = 6$–$8\,\mu m$) show distinct subgrains with sharply defined subgrain walls, while extinction in conventional thin sections of ribbon quartz is smoothly varying and subgrain boundaries are difficult to detect. Finely recrystallized quartz grains at sutured grain boundaries are smaller than subgrains within large deformed quartz grains. **(b)** Less deformed quartz porphyroclast of MT sample SG-7-5 ($t = 5$–$6\,\mu m$) with internal subgrains, and surrounding recrystallized quartz grains of similar dimensions to those of internal subgrains.

are significantly smaller than the dimensions of subgrains within the ribbon quartz grains. Microstructures of these new grains suggest that they form by grain boundary bulge mechanisms. Newly recrystallized grains surrounding other grains (Fig. 5b) have sizes in common with nearby subgrains, and the associated microstructures suggest that these recrystallized grains develop by subgrain rotation. While prior optical examination of MT mylonites using $30\,\mu m$ thin sections has led to the conclusion that both BLG and SGR recrystallization were important during deformation (Law, 2014), quantitative evaluations of these processes and their contributions to dislocation creep could be improved by the higher microstructural resolution offered by ultrathin sections.

4.4 High-resolution imaging of OH

By translating IR plates under the $10\,\mu m$ apertured IR beam and measuring spectra over many steps (spaced by 10 to $50\,\mu m$) we were able to compile integrated absorbances over spatial areas of up to $4.455\,mm^2$ ($50 \times 99 = 4950$ spectra, step size $= 30\,\mu m$). Integrated absorbances of OH bands were determined for the collection of spectra, choosing limits of 3705 to $2880\,cm^{-1}$ to include the broad OH band at $3400\,cm^{-1}$ and sharp OH bands between 3595 and $3379\,cm^{-1}$ due to fluid inclusions and hydrogen defects, respectively, within quartz grains. The integrated absorbance of the $3600\,cm^{-1}$ band of mica grains was determined (between 3702 and $3544\,cm^{-1}$) and subtracted from the first integrated absorbance of OH bands (as illustrated in Fig. 3a) to determine $\Delta_{OH\,qtz}$ of water and hydrogen defects of quartz grains.

These values were plotted spatially for MT and MCT samples, and contoured on common logarithm scales, to form high-resolution images (Figs. 6–9) of OH absorbance of quartz (converting to molar ppm, $OH/10^6Si$) and OH absorbance of micas ($\Delta_{OH\,musc}$ in cm^{-2}). In all cases, we superposed the contoured OH maps on optical micrographs to make sense of OH distributions in terms of fluid inclusions, defects, and micas, and their relationships to deformation and recrystallization microstructures. Images of the $3600\,cm^{-1}$ mica OH absorbance were placed over plane light micrographs that highlight scattering (and sometimes color) due to micas, grain boundaries, and fluid inclusions, while images of OH absorbance due to fluid inclusions (principally the broad $3400\,cm^{-1}$ band at large integrated absorbances) and lesser hydrogen defects of quartz were placed over cross-polarized light micrographs that emphasize undulatory extinction, deformation microstructures and recrystallization of quartz grains.

FTIR maps of OH of quartz grains show that water contents of deformed mylonites (Figs. 6–8) are extremely heterogeneous when absorbance is dominated by the $3400\,cm^{-1}$ band, with water contents that vary from 300 to $> 10\,000\,ppm$ between neighboring grains and within grain interiors. This result helps explain the wide variations in water contents measured by conventional FTIR through a $100\,\mu m$ aperture. OH contents of quartz grains of relatively dry mylonites (Fig. 9) are more nearly constant (50–150 ppm) when IR spectra are dominated by sharp OH bands of hydrogen point defects. FTIR maps of mylonite samples, constructed for the $3600\,cm^{-1}$ OH absorbance, show that micas are heterogeneously distributed, with concentrations apparent at quartz grain boundaries in optical micrographs (Figs. 6–9) and finely dispersed micas within quartz grain interiors (Figs. 7, 8), some of which can be difficult to detect optically. We show these relationships in several contoured OH maps, fo-

cusing on different types of quartz grains and deformation microstructures.

While most quartz grains in MT mylonites are highly deformed, some quartz grains remained nearly equant (globular grains or quartz augen with c axes aligned perpendicular to foliation; Law et al., 1986), despite their high original OH contents in the form of fluid inclusions (Fig. 6). The lack of $3600\,cm^{-1}$ OH absorbances within globular quartz grains indicates that finely dispersed micas in original relict quartz grains are absent, while large $3600\,cm^{-1}$ OH absorbances at quartz grain extremities indicate that micas are localized at quartz grain boundaries (Fig. 6a, b). Broad-band OH contents of quartz of $> 1000\,ppm$, thought to be sufficient for water weakening, are present in undeformed and deformed ribbon quartz grains (Fig. 6c, d) with very large water contents ($> 10\,000\,ppm$) marking a healed crack in the quartz augen, made up of a planar array of fluid inclusions. OH contents due to fluid inclusions are also very large at the globular augen quartz boundaries, coincident with high mica concentrations, and in some ribbon and recrystallized quartz grains. With a sample plate thickness of $56\,\mu m$, these maps represent OH absorbances within the interiors of larger quartz grains, while OH absorbances of fine micas and recrystallized quartz represent composite spectra of polycrystalline fault rock. However, we are confident that the quartz OH contents reflect fluid inclusions, even in these fine-grained regions, because of the simple spectral quality of muscovite (Fig. 3) in the MT samples and our ability to distinguish between the 3400 and $3600\,cm^{-1}$ OH absorbances.

High-strain ribbon quartz grains in the MT samples have large OH contents ($> 1000\,ppm$) comparable to those of water-weakened synthetic and milky quartz (samples SG-10 and SG-8, Figs. 7 and 8, respectively), with some reductions in OH at recrystallized margins of original grains. Micas, as imaged by the $3600\,cm^{-1}$ OH absorbance (Fig. 7a, b), continue to be highly localized at the grain boundaries of deformed quartz grains, with a mixture of fine-grained mica and quartz grains providing evidence for some redistribution during recrystallization (Fig. 8a, b). The broad $3400\,cm^{-1}$ OH absorbance in marginal recrystallized regions surrounding ribbon quartz grains are locally smaller than those of the original deformed quartz grains in some regions (Fig. 7c, d), while broad OH absorbances of recrystallized quartz continue to be large where mica contents (as evidenced by the $3600\,cm^{-1}$ band absorbance) are large (Fig. 8c, d).

FTIR maps of coarse-grained MCT mylonites yield spectral measurements of individual grains of quartz, muscovite, and biotite, even for relatively thick IR plates and larger step sizes (Fig. 9). MCT mica grains are readily detected by optical microscopy, and they are apparent as large OH contents based on the $3600\,cm^{-1}$ absorption band (Fig. 9a, b). Small, sharp absorption bands of quartz grains yield OH contents of $\sim 100\,ppm$ (Fig. 9c, d), with only local regions of quartz with larger OH contents near contacts with coarse-grained muscovite and biotite grains. Quartz grain interiors generally

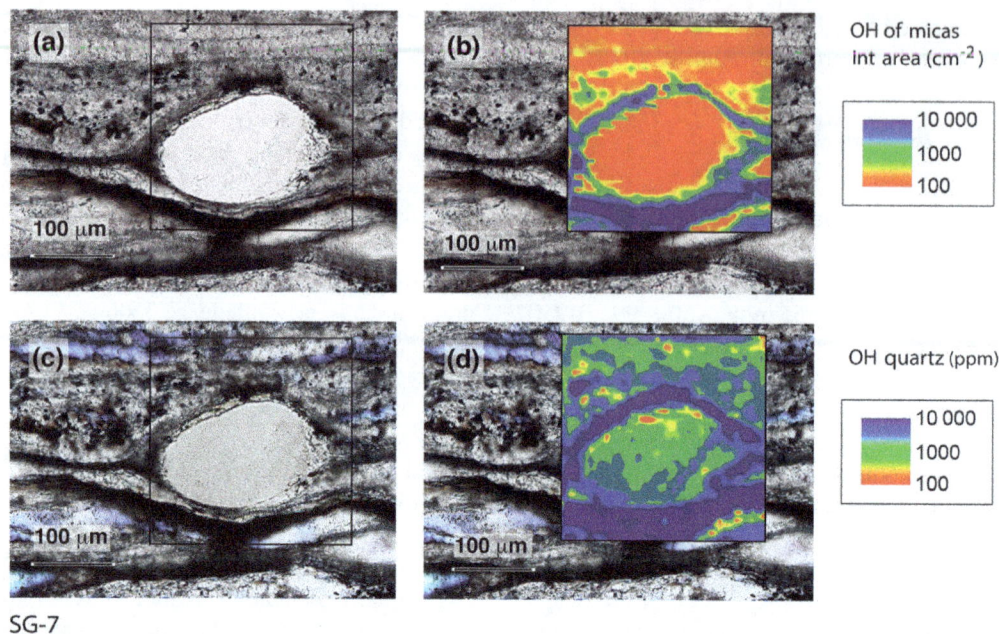

Figure 6. OH absorbance maps of MT sample SG-7 constructed from 900 IR spectra (SG-7t-1map, 30×30 steps, $10\,\mu m\,step^{-1}$) for a doubly polished plate prepared ($56\,\mu m$ thick) perpendicular to foliation and parallel to mineral lineation/transport direction (lineation horizontal in all panels). **(a)** Plane-polarized micrograph of IR plate SG-7t showing nearly equant globular augen quartz grain surrounded by highly deformed ribbon and recrystallized quartz grains. Light scattering is mostly due to micas and fluid inclusions. Outlined box is the region imaged by integrated IR absorbances. **(b)** The same plane-polarized light micrograph as in **(a)** with superposed map of OH absorbance of the $3600\,cm^{-1}$ band due to micas (OH of micas in integrated area, cm^{-2}). Contours are given in $\log_{10}(\Delta$ in $cm^{-2})$ for the integrated absorbance of the $\sim 3600\,cm^{-1}$ OH band. **(c)** Cross-polarized light micrograph of the same region of IR plate SG-7t as shown in **(a)**, with subtle undulatory extinction of ribbon quartz shown in white and first-order interference colors. **(d)** The same cross-polarized light micrograph as in **(c)** with superposed map of OH absorbance of the broad $3400\,cm^{-1}$ band due to molecular water in fluid inclusions (and smaller OH bands due to hydrogen defects) of quartz grains (OH of quartz in molar ppm, $OH/10^6 Si$). Contours given in $\log_{10}$ (ppm) for integrated absorbance of the broad $3400\,cm^{-1}$ OH band (and lesser sharp OH bands) of quartz. OH contents of quartz and those associated with micas are contoured in color with blue (and cool colors) corresponding to large water (OH) contents, and red (and warm colors) corresponding to low water (OH) contents.

lack the absorbance band at $3600\,cm^{-1}$; thus, there is no evidence for finely dispersed micas within these coarse-grained deformed (and recrystallized) quartz grains. Contours of integrated OH absorptions at $3400\,cm^{-1}$ are considerably larger for coarse-grained micas and near their contacts with quartz grains. However, these bands cannot be attributed to fluid inclusions where they coincide with large $3600\,cm^{-1}$ mica bands, given that muscovite and biotite grains in the MCT samples exhibit complex secondary OH bands between 3311 and $2829\,cm^{-1}$.

5   Discussion

Our IR spectra collected from quartz-rich mylonites in the footwall to the MT (Scottish Caledonides) and the hanging wall to the MCT (Himalaya of NW India) using synchrotron IR radiation through a $10\,\mu m$ aperture are comparable to IR spectra we collected for the same samples using a conventional FTIR-microscope system through a $100\,\mu m$ aperture. The broad OH band and large water contents of the MT my-

lonites deformed at 300–350 °C are in line with previous FTIR studies of OH in quartz deformed under greenschist facies conditions (Kronenberg et al., 1990; Gleason and De-Sisto, 2008). The sharp OH bands and low water contents of the MCT mylonites deformed at 535–735 °C are consistent with FTIR studies of quartz in other shear zones deformed at amphibolite conditions (Han et al., 2013; Kilian et al., 2016).

5.1   Limits on measurement volume

While synchrotron IR radiation has enabled spectral measurements through small apertured areas, we have not succeeded in measuring OH absorption bands for ultrathin samples when their interference fringes are larger than the absorption bands. Modeling interference fringes helped but was not sufficient to measure OH spectra of individual, recrystallized quartz grains of the MT. The physical means by which we attempted to reduce interference fringes (rotating the sample within the IR beam, and changing numerical aperture of the IR objective) were also unsuccessful. However, interference fringes might be reduced in future stud-

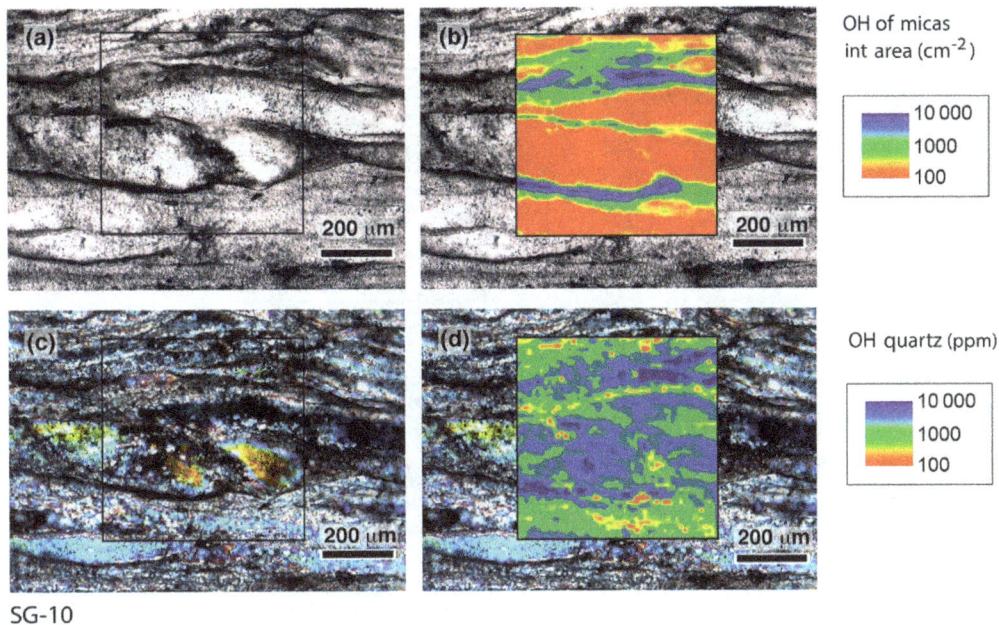

Figure 7. OH absorbance maps of MT sample SG-10 constructed from 3600 IR spectra (SG-10-XY2, 60×60 steps, $10 \, \mu m \, step^{-1}$) for a doubly polished plate prepared ($117 \, \mu m$ thick) perpendicular to foliation and parallel to mineral lineation/transport direction (lineation horizontal in all panels). **(a)** Plane-polarized micrograph of IR plate SG-10-2 showing deformed ribbon quartz grains surrounded by recrystallized quartz grains. **(b)** The same plane-polarized light micrograph as in **(a)** with superposed map of OH absorbance of the $3600 \, cm^{-1}$ band due to micas. **(c)** Cross-polarized light micrograph of the same region of IR plate SG-10-2 as shown in **(a)**, with undulatory extinction of deformed ribbon quartz grains, incipient recrystallized grains at quartz ribbon margins, and recrystallized matrix grains shown by first-order interference colors. **(d)** The same cross-polarized light micrograph as in **(c)** with superposed map of OH absorbance of the broad $3400 \, cm^{-1}$ band due to molecular water in fluid inclusions of quartz grains (OH of quartz in molar ppm, $OH/10^6 Si$). OH contents of quartz and micas are contoured in color using the same convention as in Fig. 6.

ies by a number of other methods. Interference fringes could be eliminated if ultrathin samples are mounted on a substrate with a matching refractive index n (S. Marti, personal communication, 2014). Such a substrate would need to be IR-transparent, facilitate bonding between the sample and substrate (by low-temperature melting or casting, to eliminate any residual air gaps between the sample and substrate), and lack OH bonds of its own. Alternatively, internal reflections could be reduced if thin samples are immersed in polychlorotrifluoroethylene oil ($n = 1.41$), which exhibits strong absorption bands at $< 2500 \, cm^{-1}$ but has only small bands (between 3600 and $3200 \, cm^{-1}$) that might interfere with OH absorptions of the sample (J. Mosenfelder and G. Rossman, personal communication, 2017). Internal reflections could also be reduced if only one surface of the ultrathin IR plate is polished, leaving the other surface precision-ground for a given plate thickness t (as employed by Woodhead et al., 1991). Spectra measured with just one polished surface will suffer some signal loss and spectra will need to be corrected for background. Another method of reducing interference fringes that obscure OH absorption bands might be developed by focused ion beam (FIB) milling of one side of a doubly polished sample to eliminate reflections at that surface over the IR wavenumbers of interest (R. Christoffersen, personal communication, 2014).

5.2 Wide variations in OH content

All previous FTIR studies of quartz mylonites have revealed large variations in water content for different grains within the same fault rock, and FTIR mapping of OH (Seaman et al., 2013, and this study) has revealed significant variations within grains. Our FTIR measurements of quartz OH contents using a $10 \, \mu m$ aperture show that these variations depend on sampling volume, as observed by Kilian et al. (2016), who showed that broad OH band absorptions scale with size of the measurement area due to the inhomogeneous distributions of fluid inclusions.

The large and variable water contents of quartz mylonites are far above equilibrium solubilities (e.g., Paterson, 1986; Kronenberg et al., 1986; Cordier and Doukhan, 1989), and the variations in non-equilibrium OH content probably reflect some part of the history of water migration during deformation. Images of contoured OH absorbances of the broad $3400 \, cm^{-1}$ band for quartz and the $3600 \, cm^{-1}$ bands of micas constructed for MT and MCT samples show relationships with optical deformation and recovery microstructures that suggest mechanisms by which water is incorporated in quartz

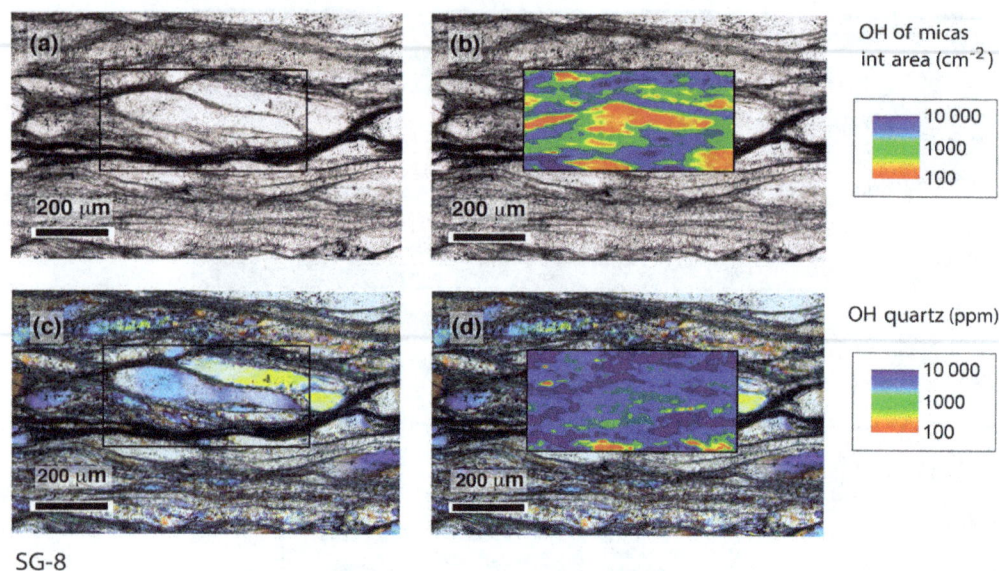

Figure 8. OH absorbance maps of MT sample SG-8 constructed from 1800 IR spectra (SG-8t-map1, 30×60 steps, $10 \, \mu m \, step^{-1}$) for a doubly polished plate prepared (73 µm thick) perpendicular to foliation and parallel to mineral lineation/transport direction (lineation horizontal in all panels). **(a)** Plane-polarized micrograph of IR plate SG-8t showing deformed ribbon quartz grains and regions of finely dispersed mica and recrystallized quartz grains. **(b)** The same plane-polarized light micrograph as in **(a)** with superposed map of OH absorbance of the $3600 \, cm^{-1}$ band due to micas. **(c)** Cross-polarized light micrograph of the same region of IR plate SG-8t as shown in **(a)**, with undulatory extinction of deformed ribbon quartz grains and regions of recrystallized grains shown by first-order interference colors. **(d)** The same cross-polarized light micrograph as in **(c)** with superposed map of OH absorbance of the broad $3400 \, cm^{-1}$ band due to molecular water in fluid inclusions of quartz grains (OH of quartz in molar ppm, $OH/10^6$ Si). OH contents of quartz and micas are contoured in color using the same convention as in Fig. 6.

grain interiors, how water becomes redistributed during deformation and recovery, and how water is lost from quartz grain interiors.

OH contours within relatively undeformed quartz augen of the MT reveal planar zones of high water content that correspond to secondary fluid inclusions at healed microcracks (Fig. 6). We know little of this early brittle deformation, but these fluid inclusion arrays have microstructures similar to those generated during hydrothermal diffusional healing of cracks (Smith and Evans, 1984; Beeler and Hickman, 2015). As a result, early brittle deformation, infiltration of water along open cracks, and crack healing appear to be important to the early introduction of water to quartz grain interiors (Kronenberg et al., 1986, 1990; FitzGerald et al., 1991; Diamond et al., 2010; Tarantola et al., 2010, 2012; Stünitz et al., 2017).

OH contours in plastically deformed regions of MT samples are complex and water contents appear to vary with strain patterns and recrystallization (Figs. 6–8). Clearly defined planar arrays of course fluid inclusions are absent from these regions, although the measured high water contents indicate redistribution of water within quartz grains, rather than the loss of water. Processes of fluid inclusion decrepitation under deviatoric stresses have been studied experimentally (Diamond et al., 2010; Tarantola et al., 2010, 2012, Stünitz et al., 2017), and they include simultaneous shrink-

age of coarse (> 10 µm, optical-scale) inclusions, generation of dislocations at fluid inclusion walls, and formation of a new population of very fine (< 100 nm) fluid inclusions (visible only by transmission electron microscopy), which reside at dislocations and resemble water clusters and inclusions of deformed and heat-treated synthetic quartz (McLaren and Hobbs, 1972; White, 1973; Kirby and McCormick, 1979; Christie and Ord, 1980). The loss of coarse fluid inclusions and growth of fine inclusions require diffusive transport, which may occur along interconnected and mobile dislocations (McLaren et al., 1983, 1989; Cordier et al., 1988, 1994; Bakker and Jansen, 1990, 1994; Hollister, 1990; Kronenberg et al., 1990; Tarantola et al., 2010, 2012; Stünitz et al., 2017). Once formed, very fine fluid inclusions may also coarsen by pipe diffusion and processes documented by McLaren et al. (1983) and Cordier et al. (1988), leading to continuous changes in inclusion densities and size distributions.

Some regions of recrystallized quartz grains in MT samples appear to have somewhat lower OH contents than original ribbon quartz grains (Fig. 7), while highly recovered and recrystallized quartz grains in the MCT samples have little or no detectable molecular water (Fig. 9). Reductions in intragranular water during dynamic recrystallization have been attributed to sweeping of fluid inclusions by mobile grain boundaries and losses of water from the fault rock by rapid

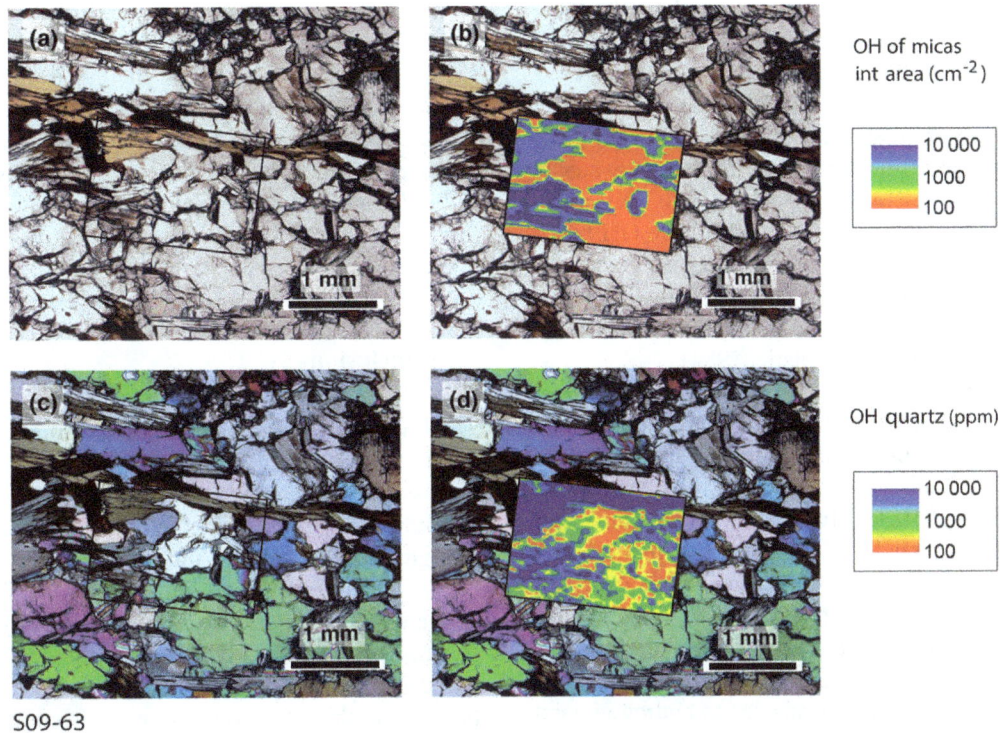

Figure 9. OH absorbance maps of MCT sample S09-63 constructed from 1200 IR spectra (S09-63-map1, 30×40 steps, $50\,\mu m$ step^{-1}) for a doubly polished plate prepared ($155\,\mu m$ thick) perpendicular to foliation and parallel to mineral lineation/transport direction (lineation horizontal in all panels). **(a)** Plane-polarized micrograph of IR plate S-09-63 showing coarse, clear quartz grains, and coarse muscovite and biotite grains with readily distinguishable color and pleochroism. Scattering of light is primarily due to grain boundaries, with little evidence for dense fluid inclusions or finely dispersed micas within quartz grains. **(b)** The same plane-polarized light micrograph as in **(a)** with superposed map of OH absorbance of the $3600\,cm^{-1}$ band due to micas. **(c)** Cross-polarized light micrograph of the same region of IR plate SG-09-63 as shown in **(a)**, with high-temperature deformation and recovery microstructures (in higher-order interference colors) that are characterized by subtle (to absent) undulatory extinction of quartz, subgrain walls, and coarse recrystallized grains. **(d)** The same cross-polarized light micrograph as in **(c)** with superposed map of OH absorbance over 3705 and $2880\,cm^{-1}$ to capture the broad and sharp bands of quartz, deducting the large $3600\,cm^{-1}$ band of micas, but including smaller OH bands of micas between 3311 and $2920\,cm^{-1}$. Contours in this absorbance can only be attributed unambiguously to fluid inclusions and hydrogen defects in quartz where micas (and their $3600\,cm^{-1}$ absorbances) are absent. OH contents of quartz and micas are contoured in color using the same convention as in Fig. 6.

grain boundary diffusion (Faleiros et al., 2010; Seaman et al., 2013; Finch et al., 2016; Kilian et al., 2016).

5.3 Water weakening in nature?

This study adds to an emerging impression that quartz tectonites deformed at greenschist facies temperatures and natural strain rates are wet. Water contents of quartz mylonites from other locations within the MT zone deformed at greenschist facies conditions vary from 1400 to 7500 ppm (Kronenberg and Wolf, 1990). Water contents of deformed quartz across the Median Tectonic Line (Japan) vary from 300 to 2500 ppm (Nakashima et al., 1995), depending on metamorphic grade and shear displacement. Water contents of quartz deformed in granitic shear zones and mylonites at greenschist conditions reach values of 1100 and > 10 000 ppm (Gleason and DeSisto, 2008; Kronenberg et al., 1990).

The water contents of MT samples and other quartz mylonites deformed at greenschist conditions are comparable to (and even larger than) those of wet varieties of synthetic and natural milky quartz (350–4000 ppm) that exhibit water weakening in laboratory studies (e.g., Griggs and Blacic, 1965; Kekulawala et al., 1978; Stünitz et al., 2017). Highly deformed ribbon quartz grains, less deformed quartz augen, and recrystallized quartz grains of the MT exhibit OH bands of similar character to those of quartzites deformed in laboratory experiments, a result that validates applications of wet quartzite rheologies to evaluate rates of dislocation creep in middle to upper crustal shear zones (e.g., Hirth et al., 2001; Behr and Platt, 2011; Law, 2014).

In contrast, water contents of highly sheared and recrystallized quartz in the hanging wall of MCT samples and other mylonites deformed at amphibolite conditions are much smaller than measured for wet varieties of quartz deformed under greenschist facies conditions. Large shear strains accumulated over $\sim$ 150–250 km of displacement along the MCT (e.g., Srivastava and Mitra, 1994; Hodges, 2000; Mitra et al., 2010; Tobgay et al., 2012; Law et al., 2013) and yet quartz OH contents (50–150 ppm) in our MCT samples are far lower than required for water weakening, thereby challenging our understanding of dislocation creep and the role of water in deformation deep in the continental crust. With only a few exceptions, IR spectra of our MCT samples have OH bands of the same character and size as dry natural quartz crystals, which are strong and have not been deformed by dislocation processes in laboratory experiments (e.g., Heard and Carter, 1968; Blacic, 1975; Blacic and Christie, 1984).

The low water contents of MCT quartz grains are consistent with the results of other IR studies of quartz deformed at amphibolite facies conditions (Nakashima et al., 1995; Han et al., 2013; Kilian et al., 2016). Han et al. (2013) reported water contents of just 10 to 110 ppm for quartz grains of granitic mylonites deformed at 400–500 °C within the Longmenshan tectonic zone (Sichuan, China). Kilian et al. (2016) measured OH contents of just 20 to 100 ppm for quartz grains of granitic mylonites deformed at upper amphibolite conditions (Truzzo meta-granite, Central Alps, Italy), with IR spectra dominated by sharp OH bands due to hydrogen point defects and no detectable broad band due to molecular water. Nakashima et al. (1995) found that quartz water contents of Sambagawa metamorphic rocks (Shikoku, Japan) depend on metamorphic grade, with the lowest water contents measured for rocks subjected to the highest temperatures.

Significant water contents (1400–4400 ppm) have been observed for quartz deformed at amphibolite conditions in mylonites of the El Pichao shear zone (NW Argentina), though reductions in water content are evident with progressive deformation (Finch et al., 2016). Fluid inclusions may be lost and intragranular water contents reduced during high temperature deformation by a variety of processes, ranging from pipe diffusion (Bakker and Jansen, 1990, 1994; Hollister, 1990; Cordier et al., 1994; Mavrogenes and Bodnar, 1994) to recrystallization and grain boundary sweeping (Faleiros et al., 2010; Seaman et al., 2013; Finch et al., 2016; Kilian et al., 2016), and partial melting (Seaman et al., 2013).

Given sufficiently high temperatures, it is possible that quartz may deform at tectonic strain rates without critical hydrogen defects at dislocations and water weakening (Kilian et al., 2016). However, this implies that we have not measured flow laws for appropriately dry quartzites that we can apply to amphibolite conditions and natural strain rates. Alternatively, water may have been lost from quartz interiors following deformation. It is also possible that hydrogen defects that enhance dislocation motion at high temperatures and natural strain rates may be sourced from grain boundaries or micas, diffusing over longer distances than are possible at greenschist conditions or laboratory strain rates.

Spatial variations in OH content of quartz in natural shear zones, as mapped in this study, may provide key insights into the role of water weakening and changes in wa-

ter content during deformation and recrystallization. High-resolution FTIR imaging of OH in MT and MCT samples shows that water contents are increased, fluid inclusions are redistributed, and water contents are decreased during brittle deformation, plastic creep, recovery, and recrystallization. Changes in OH contents of quartz mylonites and the history of fluid migration during deformation may lead to changes in governing flow laws, non-steady rates of creep, and shifting zones of localized shear.

6 Conclusions

The brightness of synchrotron IR radiation enables measurement of IR spectra for much smaller sampling volumes than is possible using conventional globar IR sources of FTIR instruments. In this study, we have used this improvement in signal to characterize and measure small OH absorption bands in quartz mylonites with an aperture size of 10 μm and to map water contents spatially. The ability to measure IR spectra for small, individual recrystallized grains by methods described in this study is limited by samples that must be > 40 μm thick to avoid internal reflections that lead to interference fringes that are larger than OH absorption bands. High-resolution images of OH in quartz mylonites, based on spectra collected through a 10 μm aperture as samples are translated under the beam, reveal large variations in OH content that correspond to the distributions of fluid inclusions and layer silicates, and to deformation and recrystallization microstructures. The OH contents of quartz in MT mylonites deformed at greenschist conditions are comparable to wet quartzites deformed in the laboratory by processes of water weakening. By comparison, OH contents of quartz in MCT mylonites deformed at amphibolite conditions are small, and molecular water, as required to deform quartz at experimental strain rates, is absent. What role water plays in deformation at these conditions is unclear, calling for further studies of water weakening in natural shear zones. High-resolution FTIR mapping of OH offers a new method of tracking changes in water content during deformation, recovery, and recrystallization.

Competing interests. The authors declare that they have no conflict of interest.

Special issue statement. This article is part of the special issue "Analysis of deformation microstructures and mechanisms on all scales". It is a result of the EGU General Assembly 2016, Vienna, Austria, 17–22 April 2016.

Acknowledgements. This study benefitted from helpful and enjoyable discussions with Kyle Ashley, Renee Heilbronner, Rüdiger Kilian, Stephen Kirby, Sina Marti, Michael Stipp,

Holger Stünitz, and Robert Tracy. The manuscript was greatly improved by the thoughtful review of Jed Mosenfelder, comments of a second anonymous reviewer, and the editorial oversight by Renee Heilbronner and Fabrizio Storti. We thank Nicholas Davis for his outstanding work preparing beautifully thin, doubly polished IR plates; his skill, care, and patience were invaluable to our work. We are indebted to Randy Smith of Brookhaven National Laboratory for sharing his expertise with OPUS imaging subroutines and teaching us how to compile and analyze integrated absorbances for multiple IR spectra collected over large scanned areas. Many thanks go to the leadership and staff of Brookhaven National Laboratory for operating the National Synchrotron Light Source (NSLS) I and awarding access to the U2A Beamline and Bruker Hyperion-2000 FTIR microscope. We thank the Consortium for Materials Properties Research in Earth Sciences COMPRES for their coordination of Earth science pursuits with other sciences done at NSLS. The National Science Foundation funded this work through a collaborative research grant awarded to the PIs at Virginia Tech (NSF EAR 1220345), Texas A&M University (NSF EAR 1220138), and Renssellaer Polytechnic Institute (transferred to Syracuse University, NSF EAR 1543627); their support is gratefully acknowledged.

Edited by: Renée Heilbronner

References

Aines, R. D. and Rossman, G. R.: Water in minerals? a peak in the infrared, J. Geophys. Res., 89, 4059–4071, 1984.

Aines, R. D., Kirby, S. H., and Rossman, G. R.: Hydrogen speciation in synthetic quartz, Phys. Chem. Miner., 11, 204–212, 1984.

Baeta, R. D. and Ashby, K. H. G.: Mechanical deformation of quartz I. constant strain-rate compression experiments, Phil. Mag., 22, 601–623, 1970.

Bai, Q. and Kohlstedt, D. L.: Effects of chemical environment on the solubility and incorporation mechanism for hydrogen in olivine, Phys. Chem. Miner., 19, 460–471, 1996.

Bakker, R. J. and Jansen, J. B. H.: Preferential water leakage from fluid inclusions by means of mobile dislocations, Nature, 345, 58–60, 1990.

Bakker, R. J. and Jansen, J. B. H.: A mechanism for preferential H_2O leakage from fluid inclusions in quartz, based on TEM observations, Contrib. Mineral. Petrol., 116, 7–20, 1994.

Beeler, N. M. and Hickman, S. H.: Direct measurement of asperity contact growth in quartz at hydrothermal conditions, J. Geophys. Res., 120, 3599–3616, https://doi.org/10.1002/2014JB011816, 2015.

Behr, W. M. and Platt, J. P.: A naturally constrained stress profile through the middle crust in an extensional terrane, Earth Planet. Sc. Lett., 303, 181–192, https://doi.org/10.1016/j.epsl.2010.11.044, 2011.

Beran, A.: Infrared spectroscopy of micas, in: Micas: Crystal Chemistry and Metamorphic Petrology, Reviews in Mineralogy and Geochemistry, v. 46, edited by: Mottana, A., Sassi, F. P., Thompson Jr., J. B., Guggenheim, S., Mineral. Soc. Amer., P. H. Ribbe Series Editor, Mineral. Soc. America, Washington, DC., chap. 7., 351–369, 2002.

Blacic, J. D.: Plastic deformation mechanisms in quartz: the effect of water, Tectonophysics, 27, 271–294, 1975.

Blacic, J. D.: Water diffusion in quartz at high pressure: tectonic implications, Geophys. Res. Lett., 8, 721–723, 1981.

Blacic, J. D. and Christie, J. M.: Plasticity and hydrolytic weakening of quartz single crystals, J. Geophys. Res., 89, 4223–4239, 1984.

Bonney, T.: Notes on some rocks collected by C. Callaway, Quarterly Journal of the Geological Society of London, 39, 414–422, 1883.

Butler, R. W. H.: The role of thrust tectonic models in understanding structural evolution in NW Scotland, in: Continental Tectonics and Mountain Building: The Legacy of Peach and Horne, edited by: Law, R. D., Butler, R. W. H., Holdsworth, R. E., Krabbendam, M., and Strachan R. A., Geological Society, London, Special Publications, 335, 293–320, https://doi.org/10.1144/SP335.14, 2010.

Carr, G. L., Smith, R. J., Mihaly, L., Zhang, H., Reitze, D. H., and Tanner, D. B.: High-resolution far-infrared spectroscopy at NSLS beamline U121R, Infrared Physics and Technology, 51, 404–406, https://doi.org/10.1016/j.infrared.2007.12.034, 2008.

Chernak, L. J., Hirth, G., Selverstone, J., and Tullis, J.: Effects of aqueous and carbonic fluids on the dislocation creep strength of quartz, J. Geophys. Res., 114, B04201, https://doi.org/10.1029/2008JB005884, 2009.

Christie, J. M.: Mylonitic rocks of the Moine Thrust-zone in the Assynt region, north-west Scotland, Trans. Geol. Soc. Edinburgh, 18, 79–93, 1960.

Christie, J. M.: The Moine thrust zone in the Assynt region, northwest Scotland, University of California Publication Geological Science, 40, 345–440, 1963.

Christie, J. M. and Ord, A.: Flow stress from microstructures of mylonites: example and current assessment, J. Geophys. Res., 85, 6253–6262, 1980.

Christie, J. M., McIntyre, D. B., and Weiss, L. E.: Appendix to McIntyre, D. B.: The Moine thrust – its discovery, age and tectonic significance, Proceedings of the Geologists' Association, 65, 219–220, 1954.

Clark, F. R. S. and Moffatt, D. J.: The elimination of interference fringes from infrared spectra, Appl. Spectrosc., 32, 547–549, 1978.

Cordier, P. and Doukhan, J.-C.: Water solubility in quartz and its influence on ductility, Eur. J. Mineral., 1, 221–237, 1989.

Cordier, P. and Doukhan, J.-C.: Water speciation in quartz: a near infrared study, Am. Mineral., 76, 361–369, 1991.

Cordier, P., Boulogne, B., and Doukhan, J.-C.: Water precipitation and diffusion in wet quartz and wet berlinite, $AlPO_4$, B. Mineral., 111, 113–137, 1988.

Cordier, P., Weil, J. A., Howarth, D. F., and Doukhan, J.-C.: Influence of the $(4H)_{Si}$ defect on dislocation motion in crystalline quartz, Eur. J. Mineral., 6, 17–22, 1994.

Dewey, J. F., Dalziel, I. W. D., Reavy, R. J., and Strachan, R. A.: The neoproterozoic to mid-Devonian evolution of Scotland: a review and unresolved issues, Scottish J. Geol., 51, 5–30, https://doi.org/10.1144/sjg2014-007, 2015.

Diamond, L. W., Tarantola, A., and Stünitz, H.: Modification of fluid inclusions in quartz by deviatoric stress, II: experimentally induced changes in inclusion volume and composition, Contrib. Mineral. Petr., 160, 845–864, https://doi.org/10.1007/s00410-010-0510-6, 2010.

Elliot, D. and Johnson, M. R. W.: The structural evolution of the northern part of the Moine thrust zone, T. Roy. Soc. Edin.-Earth, 71, 69–96, 1980.

Faleiros, F. M., Campanha, G. A. C., Bello, R. M. S., and Fuzikawa, K.: Quartz recrystallization regimes, c-axis texture transitions and fluid inclusion reequilibration in a prograde greenschist to amphibolite facies mylonite zone (Ribeira Shear Zone, SE Brazil), Tectonophysics, 485, 193–214, https://doi.org/10.1016/j.tecto.2009.12.014, 2010.

Finch, M. A., Weinberg, R. F., and Hunter, N. J. R.: Water loss and the origin of thick ultramylonites, Geology, 44, 599–602, https://doi.org/10.1130/G37972.1, 2016.

FitzGerald, J. D., Boland, J. N., McLaren, A. C., Ord, A., and Hobbs, B. E.: Microstructures in water-weakened single crystals of quartz, J. Geophys. Res., 96, 2139–2155, 1991.

Francsis, M. K.: Piezometry and Strain Rate Estimates along Mid-crustal Shear Zones, MS thesis, Virginia Tech, USA, available at: http://scholar.lib.vt.edu/theses/available/etd-05032012-162325/ (last access: 26 June 2017), 2012.

Gerretsen, J., Paterson, M. S., and McLaren, A. C.: The uptake and solubility of water in quartz at elevated pressure and temperature, Phys. Chem. Min., 16, 334–342, 1989.

Gleason, G. C. and DeSisto, S.: A natural example of crystal-plastic deformation enhancing the incorporation of water into quartz, Tectonophysics, 446, 16–30, https://doi.org/10.1016/j.tecto.2007.09.006, 2008.

Gleason, G. C. and Tullis, J.: A flow law for dislocation creep of quartz aggregates determined with the molten-salt cell, Tectonophysics, 247, 1–23, 1995.

Godin, L., Grujic, D., Law, R. D., and Searle, M. P.: Channel flow, ductile extrusion and exhumation in continental collision zones: an introduction, in: Channel Flow, Ductile Extrusion and Exhumation in Continental Collision Zones, edited by: Law, R. D., Searle, M. P., and Godin, L., Geological Society, London, Special Publications, 268, 1–23, https://doi.org/10.1144/GSL.SP.2006.268.01.01, 2006.

Grasemann, B., Fritz, H., and Vannay J. C.: Quantitative kinematic flow analysis from the Main Central Thrust Zone (NW-Himalaya, India): implications for a decelerating strain path and the extrusion of orogenic wedges, J. Struct. Geol., 21, 837–853, https://doi.org/10.1016/S0191-8141(99)00077-2, 1999.

Griggs, D. T.: Hydrolytic weakening of quartz and other silicates, Geophys. J. Roy. Astr. S., 14, 19–31, 1967.

Griggs, D. T.: A model of hydrolytic weakening in quartz, J. Geophys. Res., 79, 1653–1661, 1974.

Griggs, D. T. and Blacic, J. D.: Quartz: anomalous weakness of synthetic crystals, Science, 147, 292–295, 1965.

Grujic, D., Casey, M., Davidson, C., Hollister, L. S., Kundig, K. Pavlis, T., and Schmid, S.: Ductile extrusion of the Higher Himalayan crystalline in Bhutan: evidence from quartz microfabrics, Tectonophysics, 260, 21–43, https://doi.org/10.1016/0040-1951(96)00074-1, 1996.

Han, L., Zhou, Y. S., and He, C. R.: Water-enhanced plastic deformation in felsic rocks, Science China-Earth Sciences, 56, 203–216, https://doi.org/10.1007/s11430-012-4367-6, 2013.

Heard, H. C. and Carter, N. L.: Experimentally induced "natural" intragranular flow in quartz and quartzite, Am. J. Sci., 266, 1–42, 1968.

Heilbronner, R, and Barrett, S.: Image Analysis in Earth Sciences,

Microstructures and Textures of Earth Materials, Springer Verlag, Heidelberg, 520 pp., 2014.

Heller-Kallai, L. and Lapides, I.: Dehydroxylation of muscovite: study of quenched samples, Phys. Chem. Miner., 42, 835–845, 2015.

Hirsch, P. B.: A mechanism for the effect of doping on dislocation mobility, J. Phys. Colloque ,C6 40, C6-117–C6-121, 1979.

Hirth, G. and Tullis, J.: Dislocation creep regimes in quartz aggregates, J. Struct. Geol., 14, 145–159, 1992.

Hirth, G., Teyssier, C., and Dunlap, W. J.: An evaluation of quartzite flow laws based on comparisons between experimentally and naturally deformed rocks, Int. J. Earth Sci., 90, 77–87, 2001.

Hobbs, B. E.: Recrystallization of single crystals of quartz, Tectonophysics, 6, 353–401, 1968.

Hobbs, B. E.: The influence of metamorphic environment upon the deformation of minerals, Tectonophysics, 78, 335–383, 1981.

Hodges, K. V.: Tectonics of the Himalaya and southern Tibet from two perspectives, Geol. Soc. Am. Bull., 112, 324–350, https://doi.org/10.1130/0016-7606(2000)112<0324:TOTHAS>2.3.CO;2, 2000.

Hollister, L. S.: Enrichment of CO_2 in fluid inclusions in quartz by removal of H_2O during crystal-plastic deformation, J. Struct. Geol., 12, 895–901, 1990.

Holyoke III, C. W., and Kronenberg, A. K.: Accurate differential stress measurement using the molten salt cell and solid salt assemblies in the Griggs apparatus with application to strength, piezometers and rheology, Tectonophysics, 494, 17–31, https://doi.org/10.1016/j.tecto.2010.08.001, 2010.

Holyoke III, C. W. and Kronenberg, A. K.: Reversible water weakening of quartz, Earth Planet. Sc. Lett., 374, 185–190, https://doi.org/10.1016/j.epsl.2013.05.039, 2013.

Jaoul, O., Tullis, J., and Kronenberg, A. K.: The effect of varying water contents on the creep behavior of Heavitree quartzite, J. Geophys. Res., 89, 4298–4312, 1984.

Johnson, M. R. W., Kelly, S. P., Oliver, G. J. H., and Winter, D. A.: Thermal effects and timing of thrusting n the Moine thrust zone, J. Geol. Soc. London, 142, 863–874, 1985.

Kats, A.: Hydrogen in alpha quartz, Philips Research Reports, 17, 1–31, 133–195, 201–279, 1962.

Kekulawala, K. R. S. S., Paterson, M. S., and Boland, J. N.: Hydrolytic weakening in quartz, Tectonophysics, 46, T1–T6, 1978.

Kekulawala, K. R. S. S., Paterson, M. S., and Boland, J. N.: An experimental study of the role of water in quartz deformation, in: Mechanical Behavior of Crustal Rocks (The Handin Volume), edited by: Carter, N. L., Friedman, M., Logan, J. M., and Stearns, D. W., Geophys. Monograph 24, Amer. Geophys. Union, Washington, DC, 49–60, 1981.

Kilian, R., Heilbronner, R., Holyoke III, C. W., Kronenberg, A. K., and Stünitz, H.: Dislocation creep of dry quartz, J. Geophys. Res., 121, 3278–3299, https://doi.org/10.1002/2015JB012771, 2016.

Kirby, S. H. and McCormick, J. W.: Creep of hydrolytically weakened synthetic quartz crystals oriented to promote 2110 <0001> slip: a brief summary of work to date, B. Mineral., 102, 124–137, 1979.

Kohlstedt, D. L., Keppler, H., and Rubie, D. C.: Solubility of water in the α, β and γ phases of $(Mg,Fe)_2SiO_4$, Contrib. Mineral. Petrol., 123, 345–357, 1996.

Konevskikh, T., Ponossov, A., Blümel, R., Lukacs, R., and Kohler,

A.: Fringes in FTIR spectroscopy revisted: understanding and modeling fringes in infrared spectroscopy of thin films, Analyst, 140, 3969–3980, https://doi.org/10.1039/c4an02343a, 2015.

Kronenberg, A. K. and Tullis, J.: Flow strengths of quartz aggregates: grain size and pressure effects due to hydrolytic weakening, J. Geophys. Res., 89, 4281–4297, 1984.

Kronenberg, A. K. and Wolf, G. H.: FTIR determinations of intragranular water content in quartz-bearing rocks: implications for hydrolytic weakening in the laboratory and within the Earth, Tectonophysics, 172, 255–271, 1990.

Kronenberg, A. K., Kirby, S. H., Aines, R. D., and Rossman, G. R.: Solubility and diffusional uptake of hydrogen in quartz at high water pressures: implications for hydrolytic weakening, J. Geophys. Res., 91, 12723–12744, 1986.

Kronenberg, A. K., Segall, P., and Wolf, G. H.: Hydrolytic weakening and penetrative deformation within a natural shear zone, in: The Brittle-Ductile Transition in Rocks (The Heard Volume), edited by: Duba, A. G., Durham, W. B., Handin, J. W., and Wang, H. F., Geophys. Monograph 56, Amer. Geophys. Union, Washington, DC, 21–36, 1990.

Kruhl, J. H.: Reply: Prism- and basal-plane parallel subgrain boundaries in quartz: a microstructural geothermobarometer, J. Metamor. Geol., 16, 142–146, 1998.

Law, R. D.: Moine Thrust zone mylonites at the Stack of Glencoul: II – results of vorticity analyses and their tectonic significance, in: Continental Tectonics and Mountain Building: The Legacy of Peach and Horne, edited by: Law, R. D., Butler, R. W. H., Holdsworth, R. E., Krabbendam, M., and Strachan R. A., Geological Society, London, Special Publications, 335, 579–602, https://doi.org/10.1144/SP335.23, 2010.

Law, R. D.: Deformation thermometry based on quartz c-axis fabrics and recrystallization microstructures: A review, J. Struct. Geol., 66, 129–161, https://doi.org/10.1016/j.jsg.2014.05.023, 2014.

Law, R. D. and Johnson, M. R. W.: Microstructures and crystal fabrics of the Moine Thrust zone and Moine Nappe: history of research and changing tectonic interpretations, in: Continental Tectonics and Mountain Building: The Legacy of Peach and Horne, edited by: Law, R. D., Butler, R. W. H., Holdsworth, R. E., Krabbendam, M., and Strachan R. A., Geological Society, London, Special Publications, 335, 443–503, https://doi.org/10.1144/SP335.21, 2010.

Law, R. D., Casey, M., and Knipe, R. J.: Kinematic and tectonic significance of microstructures and crystallographic fabrics within quartz mylonites from the Assynt and Eriboll regions of the Moine thrust zone, NW Scotland, T. Roy. Soc. Edin.-Earth, 77, 99–125, 1986.

Law, R. D., Mainprice, D., Casey, M., Lloyd, G. E., Knipe, R. J., Cook, B., and Thigpen, J. R.: Moine Thrust zone mylonites at the Stack of Glencoul: 1 – microstructures, strain and influence of recrystallization on quartz crystal fabric development, in: Continental Tectonics and Mountain Building: The Legacy of Peach and Horne, edited by: Law, R. D., Butler, R. W. H., Holdsworth, R. E., Krabbendam, M., and Strachan R. A., Geological Society, London, Special Publications, 335, 543–577, https://doi.org/10.1144/SP335.23, 2010.

Law, R. D., Stahr III, D. W., Francsis, M. K., Ashley, K. T., Grasemann, B., and Ahmad, T.: Deformation temperatures and flow vorticities near the base of the Greater Himalayan Series, Sutlej Valley and Shimla Klippe, NW India, J. Struct. Geol., 54, 21–53, 2013.

Linker, M. F., Kirby, S. H., Ord, A., and Christie, J. M.: Effects of compression direction on the plasticity and rheology of hydrolytically weakened synthetic quartz crystals at atmospheric pressure, J. Geophys. Res., 89, 4241–4255, 1984.

Lister, G. S.: Crossed-girdle c-axis fabrics in quartzites plastically deformed by plane strain and progressive simple shear, Tectonophysics, 39, 51–54, https://doi.org/10.1016/0040-1951(77)90087-7, 1977.

Lobo, R. P. S. M., LaVeigne, J. D., Reitze, D. H., Tanner, D. B., and Carr, R. L.: Performance of new infrared beamline U121R at the National Synchrotron Light Source, Rev. Sci. Instrum., 70, 2899–2904, https://doi.org/10.1063/1.1149846, 1999.

Ma, M., Liu, W., Chen, Z., Liu, Z., and Li B.: Compression and structure of brucite to 31 GPa from synchrotron X-ray diffraction and infrared spectroscopy studies, Am. Mineral., 98, 33–40, https://doi.org/10.2138/am.2013.4117, 2013.

Mackwell, S. J. and Kohlstedt, D. L.: Diffusion of hydrogen in olivine – implications for water in the mantle, J. Geophys. Res., 95, 5079–5088, 1991.

Mavrogenes, J. A. and Bodnar, R. J.: Hydrogen movement into and out of fluid inclusions in quartz: experimental evidence and geological implications, Geochim. Cosmochim. Ac., 58, 141–148, 1994.

McLaren, A. D. and Hobbs, B. E.: Transmission electron microscope investigations of some naturally deformed quartzites, in: Flow and Fracture of Rocks, edited by: Heard, H. C., Borg, I. Y., Carter, N. L., and Raleigh, C. B., Geophys. Monograph, 16, 55–66, Amer. Geophys. Union, Washington, D.C., 1972.

McLaren, A. C., Cook, R. F., Hyde, S. T., and Tobin, R. C.: The mechanisms of the formation and growth of water bubbles and associated dislocation loops in synthetic quartz, Phys. Chem. Miner., 9, 79–94, 1983.

McLaren, A. C., FitzGerald, J. D., and Gerretsen, J.: Dislocation nucleation and multiplication in synthetic quartz: relevance to water weakening, Phys. Chem. Miner., 16, 465–482, 1989.

Mitra, G., Bhattacharyya, K., and Mukul, M.: The lesser Himalayan duplex in Sikkim: implications for variations in Himalayan shortening, J. Geol. Soc. India, 75, 289–301, 2010.

Muto, J., Hirth, G., Heilbronner, R., and Tullis, J.: Plastic anisotropy and fabric evolution in sheared and recrystallized quartz single crystals, J. Geophys. Res., 116, 1–18, 2011.

Nakashima, S., Matayoshi, H., Yuko, T., Michibayashi, K., Masuda T., Kuroki, N., Yamagishi, H., Ito, Y., and Nakamura, A.: Infrared microspectroscopy analysis of water distribution in deformed and metamorphosed rocks, Tectonophysics, 245, 263–276, https://doi.org/10.1016/0040-1951(94)00239-6, 1995.

Ord, A. and Christie, J. M.: Flow stresses from microstructures in mylonitic quartzites of the Moine Thrust zone, Assynt area, Scotland, J. Struct. Geol., 6, 639–654, 1984.

Paterson, M. S.: The thermodynamics of water in quartz, Phys. Chem. Miner., 13, 245–255, 1986.

Paterson, M. S.: The interaction of water with quartz and its influence in dislocation flow – an overview, in: Rheology of Solids and of the Earth, edited by: Karato, S.-I. and Toriumi, M., Oxford University Press, Oxford, 107–142, 1989.

Peach, B. N., Horne, J., Gunn, W., Clough, C. T., and Hinxman, L. W.: The Geological Structure of the Northwest Highlands of Scotland, Memoir Geological Survey Great Britain, HMSO, Glasgow, 1907.

Pistorius, A. M. A. and DeGrip, W. J.: Deconvolution as a tool to remove fringes from an FT-IR spectrum, Vibrational Spectroscopy, 36, 89–95, https://doi.org/10.1016/j.vibspec.2004.04.001, 2004.

Post, A. D., Tullis, J., and Yund, R. A.: Effects of chemical environment on dislocation creep of quartzite, J. Geophys. Res., 101, 22143–22155, 1996.

Powell, R. and Holland, T.: Optimal geothermometry and geobarometry, Am. Mineral., 79, 120–133, 1994.

Seaman, S. J., Williams, M. L., Jercinovic, M. J., Koteas, G. C., and Brown, L. B.: Water in nominally anhydrous minerals: implications for partial melting and strain localization in the lower crust, Geology, 41, 1051–1054, https://doi.org/10.1130/G34435.1, 2013.

Smith, D. L. and Evans, B.: Diffusional crack healing in quartz, J. Geophys. Res., 89, 4125–4135, 1984.

Snoke, A. W., Tullis, J., and Todd, V. R.: Fault-related Rocks, A Photographic Atlas, Princeton University Press, Princeton, 617 pp., 1998.

Srivastava, P. and Mitra, G.: Thrust geometries and deep structure of the outer and lesser Himalaya, Kumaon and Garhwal (India): implications for evolution of the Himalayan fold-and-thrust belt, Tectonics, 13, 89–109, 1994.

Stahr, D. W.: Kinematic Evolution, Metamorphism and Exhumation of the Himalayan Series, Sutlej River and Zanskar Regions of NW India, PhD thesis, Virginia Tech, USA, available at: https://vtechworks.lib.vt.edu/handle/10919/23081 (last access: 26 June 2017), 2013.

Stipp, M., Stünitz, H., Heilbronner, R., and Schmid, S. M.: The eastern Tonale fault zone: a "natural laboratory" for crystal plastic deformation of quartz over a temperature range from 250 to 700 degrees C, J. Struct. Geol., 24, 1861–1884, https://doi.org/10.1016/S0191-8141(02)00035-4, 2002.

Stipp, M., Tullis, J., and Behrens, A.: Effect of water on the dislocation creep microstructure and flow stress of quartz and implications for the recrystallized grain size piezometer, J. Geophys. Res., 111, 201–220, 2006.

Stipp, M., Tullis, J., Scherwarth, M., and Behrmann, J. H.: A new perspective on paleopiezometry: Dynamically recrystallized grain size distributions indicate mechanism changes, Geology, 38, 759–762, https://doi.org/10.1130/G31162.1, 2010.

Stuart, B. H., George, B., and McIntyre, P.: Modern Infrared Spectroscopy, J. Wiley and Sons, 200 pp., 1996.

Stünitz, H., Thust, A., Heilbronner, R., Behrens, H., Kilian, R., Tarantola, A., and FitzGerald, J. D.: Water redistribution in experimentally deformed natural milky quartz single crystals – implications for H_2O-weakening processes, J. Geophys. Res., 122, 866–894, https://doi.org/10.1002/2016JB013533, 2017.

Tarantola, A., Diamond, L. W., and Stünitz, H.: Modification of fluid inclusions in quartz by deviatoric stress I: experimentally induced changes in inclusion shapes and microstructures, Contrib. Mineral. Petr., 160, 825–843, https://doi.org/10.1007/s00410-010-0509-z, 2010.

Tarantola, A., Diamond, L. W., Stünitz, H., Thust, A., and Pec, M.: Modification of fluid inclusions in quartz by deviatoric stress, III: influence of principal stresses on inclusion density and orientation, Contrib. Mineral. Petr., 164, 537–550, https://doi.org/10.1007/s00410-012-0749-1, 2012.

Thomas, S.-M., Koch-Müller, M., Reichert, P., Rhede, D., Thomas, R., Wirth, R., and Matsyuk, S.: IR calibrations for water determination in olivine, r-GeO_2, and SiO_2 polymorphs, Phys. Chem. Miner., 36, 489–509, https://doi.org/10.1007/s00269-009-0295-1, 2009.

Tobgay, T., McQuarrie, N., Long, S., Kohn, M. J., and Corrie, S. L.: The age and rate of displacement along the Main Central Thrust in the western Bhutan Himalaya, Earth Planet. Sc. Lett., 319–320, 146–158, https://doi.org/10.1016/j.epsl.2011.12.005, 2012.

Tokiwai, K. and Nakashima, S.: Integral molar absorptivities of OH in muscovite at 20 to 650°C by in-situ high-temperature IR microspectroscopy, Am. Mineral. 95, 1052–1059, 2010a.

Tokiwai, K. and Nakashima, S.: Dehydration kinetics of muscovite by in situ infrared microspectroscopy, Phys. Chem. Miner., 37, 91–101, 2010b.

Tullis, J.: Deformation of feldspars, Chap. 13, in: Feldspar Mineralogy, Second Edition, edited by P.H. Ribbe, Reviews in Mineralogy, v.2, Min. Soc. Amer., Washington DC, 297–324, 1983.

Tullis, J. and Yund, R. A.: Experimental deformation of dry Westerly granite, J. Geophys. Res., 82, 5705–5718, 1977.

Tullis, J. and Yund, R. A.: Hydrolytic weakening of experimentally deformed Westerly granite and Hale albite rock, J. Struct. Geol., 2, 439–451, 1980.

Tullis, J. and Yund, R. A.: Hydrolytic weakening of quartz aggregates: the effects of water and pressure on recovery, Geophys. Res. Lett., 16, 1343–1346, 1989.

Vannay, J. C. and Grasemann, B.: Inverted metamorphism in the High Himalaya of Himachal Pradesh (NW India): phase equilibria versus thermobarometry, Schweiz. Miner. Petrog., 78, 107–132, 1998

Weathers, M. S., Bird, J. M., Cooper, R. F., and Kohlstedt, D. L.: Differential stress determined from deformation-induced microstructures of the Moine thrust zone, J. Geophys. Res., 84, 7495–7509, 1979.

White, S.: Dislocations and bubbles in vein quartz, Nature Phys. Sci., 243, 11–14, 1973.

Woodhead, J. A., Rossman, G. R., and Thomas, A. P.: Hydrous species in zircon, Am. Mineral., 76, 1533–1546, 1991.

13

Development of a composite soil degradation assessment index for cocoa agroecosystems in Southwestern Nigeria

Sunday Adenrele Adeniyi[1,2], **Willem Petrus de Clercq**[3], **and Adriaan van Niekerk**[1,4]

[1]Department of Geography and Environmental Studies, Stellenbosch University, Private Bag XI, Matieland 7602, Stellenbosch, South Africa
[2]Department of Geography, Osun State University, P.M.B 4494, Osogbo, Nigeria
[3]Department of Soil Science, Stellenbosch University, Private Bag XI, Matieland 7602, Stellenbosch, South Africa
[4]School of Plant Biology, University of Western Australia, Crawley WA 6009, Australia

Correspondence to: Sunday Adenrele Adeniyi (releadegeography@yahoo.com)

Abstract. Cocoa agroecosystems are a major land-use type in the tropical rainforest belt of West Africa, reportedly associated with several ecological changes, including soil degradation. This study aims to develop a composite soil degradation assessment index (CSDI) for determining the degradation level of cocoa soils under smallholder agroecosystems of southwestern Nigeria. Plots where natural forests have been converted to cocoa agroecosystems of ages 1–10, 11–40, and 41–80 years, respectively representing young cocoa plantations (YCPs), mature cocoa plantations (MCPs), and senescent cocoa plantations (SCPs), were identified to represent the biological cycle of the cocoa tree. Soil samples were collected at a depth of 0 to 20 cm in each plot and analysed in terms of their physical, chemical, and biological properties. Factor analysis of soil data revealed four major interacting soil degradation processes: decline in soil nutrients, loss of soil organic matter, increase in soil acidity, and the breakdown of soil textural characteristics over time. These processes were represented by eight soil properties (extractable zinc, silt, soil organic matter (SOM), cation exchange capacity (CEC), available phosphorus, total porosity, pH, and clay content). These soil properties were subjected to forward stepwise discriminant analysis (STEPDA), and the result showed that four soil properties (extractable zinc, cation exchange capacity, SOM, and clay content) are the most useful in separating the studied soils into YCP, MCP, and SCP. In this way, we have sufficiently eliminated redundancy in the final selection of soil degradation indicators. Based on these four soil parameters, a CSDI was developed and used to clas-

sify selected cocoa soils into three different classes of degradation. The results revealed that 65 % of the selected cocoa farms are moderately degraded, while 18 % have a high degradation status. The numerical value of the CSDI as an objective index of soil degradation under cocoa agroecosystems was statistically validated. The results of this study reveal that soil management should promote activities that help to increase organic matter and reduce Zn deficiency over the cocoa growth cycle. Finally, the newly developed CSDI can provide an early warning of soil degradation processes and help farmers and extension officers to implement rehabilitation practices on degraded cocoa soils.

1 Introduction

Healthy soil is vital to successful agriculture and global food security (Virto et al., 2014; Lal, 2015). Soil performs several ecosystem functions such as carbon sequestration and regulation (Novara et al., 2011; Brevik et al., 2015; Muñoz-rojas et al., 2017), buffering and filtering of pollutants (Keesstra et al., 2012), climate control through the regulation of C and N fluxes (Brevik et al., 2015; Zornoza et al., 2015; Al-Kaisi et al., 2017), and supporting biodiversity (Schulte et al., 2015). Nonetheless, misuse of soils, arising from intensive agricultural production and unsustainable land use practices has resulted in soil degradation, particularly in developing countries with poor infrastructure and financial capacity to manage natural resources (Tesfahunegn, 2016). Studies have reported that 500 Mha of land in the tropics (Lal, 2015), and more than 3500 Mha of global land area (Karlen and Rice, 2015), is currently affected by soil degradation, with serious implications for food security and the likelihood of malnutrition, ethnic conflict, and civil unrest (Lal, 2009). In response to these problems, an increasing interest in soil degradation has been observed among researchers and policymakers (Scherr, 1999; Lal, 2001; Bindraban et al., 2012; Baumhardt et al., 2015; Lal, 2015; Krasilnikov et al., 2016; Nezomba et al., 2017).

Soil degradation is a measurable loss or reduction of the current or potential capability of soils to produce plant materials of desired quantity and quality (Chen et al., 2002). Many scientists viewed soil degradation as a decline in soil quality (SQ; Lal, 2001; Adesodun et al., 2008; Beniston et al., 2016), and, in turn, SQ as the capacity of a soil to function within ecosystem and land use boundaries (Doran and Parkin, 1994; Doran and Zeiss, 2000; Doran, 2002). Unfortunately, when soil degradation reaches an advanced stage, soil quality restoration is difficult (Lal and Cummings, 1979). Therefore, good knowledge of SQ is important for developing appropriate conservation measures (Tesfahunegn et al., 2011). Since soil degradation and soil quality are interlinked through many processes (Lal, 2015), scholars have suggested that soil degradation can be assessed using soil quality assessment strategies (Tesfahunegn, 2014; Pulido et al., 2017). However, an essential step when assessing soil degradation based on soil quality assessment strategies is the careful selection of appropriate indicators relevant to degradation processes under investigation.

Degradation of soils is complex, often the consequence of many interacting processes (Prager et al., 2011). However, major processes include accelerated erosion (Lal, 2001; Cerda et al., 2009; Bindraban et al., 2012; Rodrigo Comino et al., 2016a, b; Xu et al., 2016), deforestation (De la paix et al., 2013), poor pasture management (De Souza Braz et al., 2013), decline in soil structure (Cerda, 2000), salinization associated with inadequate irrigation management (Prager et al., 2011; Ganjegunte et al., 2014), alkalinization and sodification (Condom et al., 1999), depletion of soil organic matter (SOM; Jordán et al., 2010), reduction in the activity of soil microorganisms (Lal, 2009), soil compaction (Pulido et al., 2017), and unsustainable agricultural practices (Krasilnikov et al., 2016). For sustainable soil management in agricultural regions, it is essential for farmers and scientists to identify major dominant degradation processes and their indicators.

Cocoa (*Theobroma cacao* L.) agroecosystems are a major agricultural land use type in the tropical rainforest belt of West Africa (Tondoh et al., 2015), covering an estimated total area of about 6 million ha in Côte d'Ivoire, Ghana, Nigeria, and Cameroon (Sonwa et al., 2014). Unfortunately, cocoa landscapes are often associated with a range of ecological changes including deforestation, biodiversity loss, destruction of soil flora and fauna from pesticide usage, and accelerated soil degradation (Critchley and Bruijnzeel, 1996;

Salami, 1998, 2001; Rice and Greenberg, 2000; Asare, 2005; Ntiamoah and Afrane, 2008; Mbile et al., 2009; Adeoye and Ayeni, 2011; Jagoret et al., 2012; Akinyemi, 2013; Schoneveld, 2014; Sonwa et al., 2014; Tondoh et al., 2015). Until present, soil degradation assessments on a plot scale in regions undergoing farmland conversion to cocoa agroecosystems have been limited.

Worldwide, agricultural practices have been regarded as one of the major causes of soil degradation (Rahmanipour et al., 2014; Karlen and Rice, 2015; Zornoza et al., 2008). It is widely acknowledged that agricultural practices or land use changes in agricultural regions alter key soil properties such as SOM, total nitrogen (TN), cation exchange capacity (CEC), exchangeable cations, water-holding capacity (WHC), bulk density (BD), and total porosity (TP; Lemenih et al., 2005; Awiti et al., 2008; Trabaquini et al., 2015; Dawoe et al., 2010, 2014; Ameyan and Ogidiolu, 1989; Hadgu et al., 2009; Thomaz and Luiz, 2012; Zhao et al., 2014; Tesfahunegn, 2014). Although many of these soil properties are regularly used as indicators of soil degradation (Trabaquini et al., 2015), the use of single soil characteristics often provides an incomplete representation of soil degradation (De la Rosa, 2005; Puglisi et al., 2005, 2006; Sione et al., 2017). To overcome this shortcoming, an integration of soil properties into numeric indices has been proposed (Doran and Parkin, 1994; Leirós et al., 1999; Bastida et al., 2006; Gómez et al., 2009; Puglisi et al., 2005, 2006; Sharma et al., 2008; Xu et al., 2016; Pulido et al., 2017).

Multivariate statistical techniques such as principal component analysis (PCA), canonical discriminant analysis (CDA), cluster analysis (CA), partial least squares (PLS), principal component regression (PCR), ordinary least squares regression (OLS), and multiple linear regression analysis (MLRA) have been applied to assess soil quality (Parras-Alcántara and Lozano-García, 2014; Xu et al., 2016; Sione et al., 2017; Biswas et al., 2017; Renzi et al., 2017; Khaledian et al., 2017). These statistical techniques can assist researchers in selecting important soil quality indicators that are useful for developing an overall soil quality or degradation index for effective land resource management and planning (Khaledian et al., 2017). Regardless of the techniques used, the selection of a minimum data set (MDS) of soil quality and degradation parameters has been widely supported in the literature (Biswas et al., 2017). For instance, Sione et al. (2017) used a soil quality index (SQI) to evaluate the impact of rice production systems that use irrigation with groundwater on soil degradation on the field scale in Argentina. They selected six soil quality indicators including aggregate stability, water percolation, SOM, exchangeable sodium content (ESC), pH, and electrical conductivity in saturated paste extract. Their results showed that the use of soil quality indicators can provide an early assessment of soil degradation processes and help land managers to implement soil conservation practices (Sione et al., 2017). In South Asia, Biswas et al. (2017) combined PCA and multiple regression analysis to create MDSs of physical, chemical, and biological indicators, which were integrated to develop a unified SQI for rice–rice cropping systems. Thus, Sánchez-Navarro et al. (2015) developed an overall SQI suitable for monitoring soil degradation in semi-arid Mediterranean ecosystems. Pulido et al. (2017) developed a soil degradation index for rangelands of Extremadura southwestern Spain based on six indicators, namely CEC, available potassium, SOM, water content at field capacity, soil depth, and the thickness of the Ah horizon. Another example is Gómez et al. (2009), who developed three soil degradation indexes (obtained through a PCA) of soils under organic olive farms in southern Spain. One of the indices used only three soil properties, namely organic C, water stable macroaggregates, and extractable P. According to these authors, this index had the highest potential to be used as a relatively easy and inexpensive screening test of soil degradation. Very little attention has been given to the development of numeric indices for monitoring soil degradation under crop-specific land use management systems in tropical countries. Such indices can serve as the basis for integrating and interpreting several soil measurements, thereby indicating whether a particular land use management system (e.g agroecosystems) is sustainable or not.

Therefore, the aim of the present study is to develop a CSDI for shaded cocoa agroecosystems under tropical conditions in southwestern Nigeria. This area is currently suffering from soil degradation arising from low input cocoa agroecosystems. Soil conditions under age-sequenced peasant cocoa agroecosystems are investigated. The cocoa agroecosystem ages of 1–10, 11–40, and 41–80 years – hereafter referred to as young cocoa plantation (YCP), mature cocoa plantation (MCP), and senescent cocoa plantation (SCP), respectively – were targeted as this is in line with the biological cycle of the cocoa tree (Isaac et al., 2005; Jagoret et al., 2011, 2012; Saj et al., 2013). Our goals are to (i) identify the most important soil degradation processes, (ii) select a MDS of soil degradation indicators using multivariate statistical techniques, (iii) integrate the MDS into a CSDI, and (iv) statistically validate the CSDI and evaluate to what extent the CSDI can be used as a tool by researchers, farmers, agricultural extension officers, and government agencies involved in rehabilitating degraded cocoa soils in southwestern Nigeria (and similar environments).

2   Materials and methods

2.1   Study area

This study was carried out in the Ife region of southwestern Nigeria between 6°50′27″–7°38′33″ N and 4°21′33″–4°45′55″ E (Fig. 1), where most soils have been under cocoa plantations for more than 80 years (Abiodun, 1971; Berry, 1974). The climate is humid tropical with a mean daily minimum temperature of 25 °C and a mean maximum tempera-

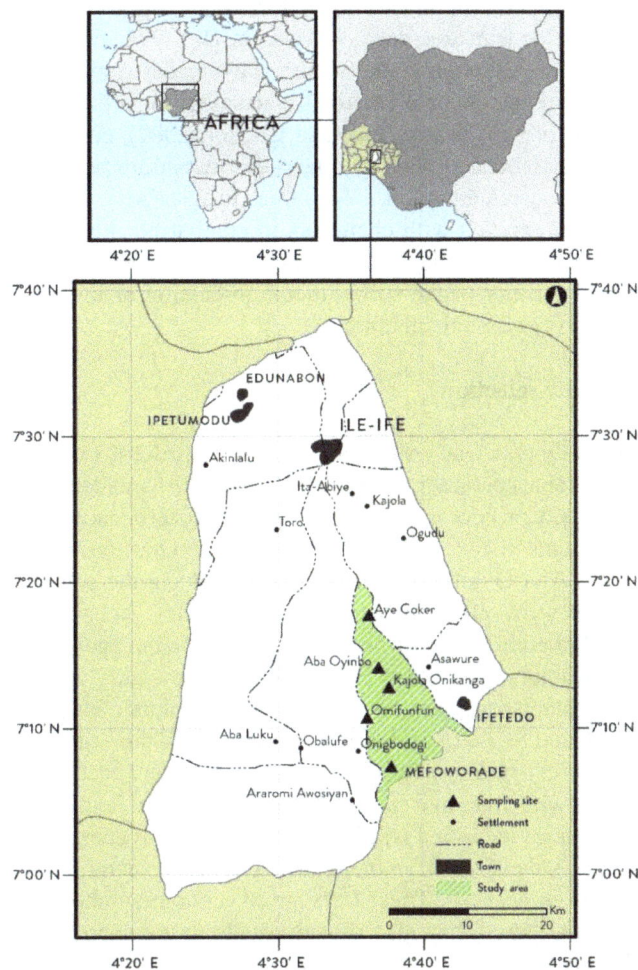

Figure 1. Location map of the study area.

ture of 33 °C. The mean annual rainfall ranges between 1400 and 1600 mm, with a long wet season lasting from April to October and a relatively short dry season that lasts from November to March. The natural vegetation is dominated by humid tropical rainforests of the moist evergreen type, characterized by multiple canopies and lianas. The area is underlain by rocks from the basement complex, which are exposed as outcrops in several areas, of the Precambrian age. The soils are mainly Alfisols, classified as Kanhaplic Rhodustalf (Soil Survey Staff, 2014) or Luvisols (IUSS Working Group WRB, 2015) and locally known as the Egbeda association (Smyth and Montgomery, 1962). The area of study lies within the Egbeda soil series, characterized by sandy loam soils, with increasing clay content in the lower horizons. The soils are slightly acidic to neutral in reaction (pH 6.5). With the exception of the areas set aside as forest reserves, the natural vegetation has been replaced with perennial and annual crops. Cocoa farmers in the region traditionally established their cocoa farms by planting cocoa trees where primary or secondary forests are selectively cleared. Cocoa trees are then planted along with understory food crops and a range of forest or fruit tree species (Isaac et al., 2005; Jagoret et al., 2017). Although some farmers have recently shifted towards full-sun cocoa plantations, particularly in areas where natural forest is scarce (Oke and Chokor, 2009), ecological changes associated with such land use transitions are yet to attract research attention. Cocoa trees in agroecosystems are regularly sprayed with chemicals to combat black pod disease (*Phytophthora sp.*), but farmers depend entirely on the natural fertility of the soil without application of inorganic fertilizers or organic manure.

2.2   Site selection

The study area was visited in March and April 2013 to identify suitable cocoa agroecosystems and locate candidate sample sites. Considering soil variability and heterogeneity, five settlements of cocoa farmers (Mefoworade, Omifunfun, Aye Coker, Aba Oyinbo, and Kajola-Onikanga) in the southern Ife area were randomly selected as study sites. At each site, a total of eight cocoa agroecosystems of different ages (since site clearance) were randomly selected and assigned to three cocoa plantation age categories: YCP (10 plots), MCP (15 plots), and SCP (15 plots). For the purpose of this study, cocoa agroecosystems are conceived as areas where cocoa trees coexist with other tree species on the same plot of land. Some tree species identified within selected cocoa agroecosystems include kola (*Cola acuminata* and *Cola nitida*) and oil palm (*Elaeis guineensis)*. These trees are of economic importance to the farmers. They also provide shade to the cocoa trees. The selected cocoa agroecosystems are between 2 and 3 ha in size, with a tree spacing of 3×3 m as recommended by good agricultural practices for sustainable cocoa production in the West African subregion. All sampled plots were restricted to upper slope positions of a catena where the slope angle did not exceed $2°$ to ensure that catenary variation in soil properties between the farms studied was minimal. Local farmers served as the main source of information on the age distribution of the cocoa plantations and their permission was also sought to use their farms as research plots. Each research plot was visited at least once before soil sampling. During the field visits no evidence of substantial soil erosion was observed on any of the plots, as the floors of the selected cocoa agroecosystems are covered with leaves and plant litter.

2.3   Soil sample collection for laboratory analysis

Soil sampling was conducted in May 2013. A quadrat measuring $1000 \, m^2$ was demarcated at the centre of each cocoa agroecosystem. Each quadrat was subdivided into 10 sub-quadrats of $100 \, m^2$ and serially labelled. Soil samples were drawn at the centre of the even-numbered sub-quadrats, resulting in a total of five soil samples per plot. Measurements were deliberately restricted to a depth of 0 to 20 cm for the

following reasons: (i) most significant changes in soil characteristics in any vegetation (especially in a tropical environment) are confined to the topmost layer of the soil profile (Aweto, 1981; Aweto and Iyanda, 2003; Tondoh et al., 2015); (ii) these depths cover the main distribution of roots and soil nutrient stocks of cocoa plantations (Hartemink, 2005) and is therefore usually used in soil surveys for fertilizer recommendations in West African cocoa-based agroecosystems (Snoeck et al., 2010); (iii) several studies (e.g. Isaac et al., 2007) demonstrated that cacao trees tend to have shallow root activity within the topsoil (0–20 cm); (iv) biological processes, such as earthworm activities, are restricted to 0–10 cm layer of tropical soils; (v) measurements were restricted to facilitate future replication of the methodology as routine soil samples are usually taken from the topsoil layer (plough layer); and (vi) the soil degradation index developed in this study is expected to be used by farmers and extension officers for rehabilitating degraded cocoa plantations in the study area and similar environments, and by confining the samples to the topsoil, the likelihood of adoption by the end users is greater.

Two categories of soil samples were taken at each sampling point to promote a detailed investigation of soil-property differences. The first was an undisturbed sample using a BD ring measuring 5×5 cm (diameter and height), whereas the other sample was taken using a soil auger. The first sample was used to determine BD, WHC, and saturated hydraulic conductivity (SHC), and the second sample was used to determine the other studied soil properties. The soil samples were stored in labelled polythene bags and taken to the laboratory for analysis. The composite soil samples aggregated from the five samples collected in each plot were air-dried for 2 weeks, hand ground in a ceramic mortar, passed through a 2 mm sieve and analysed for chemical properties and particle-size distribution. For analysis 22 soil properties were selected. The analytical methods are summarized in Table 1, and average values (in range) of all the soil degradation parameters considered are provided in Table S1 (Supplement).

2.4   Statistical analyses and index development

Based on an extensive review of literature on soil quality and degradation assessment indexing, the CSDI was developed using a range of statistical techniques and procedures. The methodology consisted of eight steps as outlined below:

Step (1) involved selection of relevant indicators of soil degradation. Here, we selected 22 analytical soil properties widely acknowledged as soil quality and degradation indicators.

In Step (2) a factor analysis was performed to group all the soil data into statistical factors with PCA as the method of factor extraction (Tesfahunegn et al., 2011). Factors were subjected to varimax rotation with Kaiser normalization in order to generate factor patterns that load highly significant

Table 1. Methods and field analysis of soil data.

Soil properties	Method of determination and reference
*Particle size distribution (Sand, silt, and clay content (%))	Pipette method (Gee and Or, 2002)
Bulk density (g cm^3)	Core method (Grossman and Reinsch, 2002)
Total porosity (%)	Computed from value of bulk density (Vomocil, 1965)
Water-holding capacity (%)	Oven-dry method
Saturated hydraulic conductivity (cm h^{-1})	Determined in the laboratory using a constant head permeameter (Reynolds and Elrick, 2002)
pH (KCl)	Potentiometrically in 0.1 M CaCl$_2$ solution (Peech, 1965)
Organic matter (%)	Walkley and Black (1934)
Available phosphorus (mg kg^{-1})	Olsen and Sommer (1982)
Total nitrogen (%)	Kjeldahl method (Bremner, 1996)
Exchangeable Ca and Mg (mg kg^{-1})	Atomic absorption spectrophotometer
Exchangeable Na and K (mg kg^{-1})	Flame photometer
Cation exchange capacity (cmol$_c$ kg^{-1})	Summation method (Juo et al., 1976)
Base saturation (%)	Calculated as the percentage of the CEC occupied by basic cations
Extractable Zn, Mn, Mg, and Cu (mg kg^{-1})	Atomic absorption spectrophotometer
Earthworm population (per m^2)	Anderson and Ingram (1993)

Ca: calcium; Mg: magnesium; Na: sodium; K: potassium; Zn: zinc; Mn: manganese; Cu: copper. * For determining the particle size distribution, samples were treated with H$_2$O$_2$ (6 %) to remove organic matter (OM) as described by Parras-Alcántara et al. (2015).

Table 2. Rotated factor loadings for the first five factors including proportion of variance, eigenvalues, and communalities of measured soil properties.

Eigenvalue	8.545	3.964	2.088	1.265	1.113	
Total variance (%)	23.702	16.382	14.642	9.131	13.300	
Cumulative variance	23.702	40.083	54.725	63.856	77.155	

	Principal component, PC					
Soil degradation indicators	PC 1	PC 2	PC 3	PC 4	PC 5	Communalities
Sand (%)	−0.510	−0.282	−0.093	−0.094	−0.688	0.830
Silt (%)	**0.838**	−0.060	−0.154	0.217	−0.014	0.777
Clay content (%)	−0.097	0.378	0.235	−0.070	**0.812**	0.871
Bulk density (g cm^{-3})	−0.393	−0.051	−0.143	−0.633	0.055	0.582
Total porosity (%)	0.128	−0.016	**0.801**	−0.087	0.233	0.719
Base saturation (%)	0.397	0.104	0.355	0.272	0.661	0.806
pH (KCl)	0.104	0.008	−0.029	**0.791**	0.143	0.658
Cation exchange capacity (cmol$_c$ kg^{-1})	−0.081	**0.884**	−0.124	−0.094	−0.067	0.816
Water-holding capacity (%)	0.721	−0.147	0.358	0.367	0.278	0.882
Saturated hydraulic conductivity (cm h^{-1})	0.060	−0.442	0.603	0.480	0.204	0.835
Total nitrogen (%)	0.667	0.196	0.583	0.187	0.225	0.908
Available phosphorus (mg kg^{-1})	0.016	0.144	**0.810**	0.063	0.075	0.686
Exchangeable potassium (mg kg^{-1})	0.219	−0.249	0.099	0.094	0.624	0.518
Exchangeable calcium (mg kg^{-1})	0.022	**0.871**	−0.007	0.028	0.084	0.767
Exchangeable magnesium (mg kg^{-1})	0.295	0.481	0.260	0.079	0.508	0.650
Extractable zinc (mg kg^{-1})	**0.875**	0.315	0.037	0.062	0.162	0.896
Extractable manganese (mg kg^{-1})	**0.857**	0.114	0.152	−0.007	0.313	0.868
Extractable copper (mg kg^{-1})	−0.632	0.247	−0.382	−0.463	−0.168	0.849
Extractable magnesium (mg kg^{-1})	0.679	−0.232	0.518	0.210	0.078	0.834
Exchangeable sodium (mg kg^{-1})	−0.001	0.601	0.032	0.289	0.393	0.600
Organic matter (%)	0.472	**0.711**	0.142	−0.209	0.231	0.846
Earthworm population (per m^2)	0.459	−0.401	0.552	0.144	0.282	0.776

Rotation method: varimax with Kaiser normalization. Boldface factor loadings are considered highly weighted; extraction method: principal component analysis.

variables into one factor, thereby producing a matrix with a simple structure that is easy to interpret (Ameyan and Ogidiolu, 1989; de Lima et al., 2008; Momtaz et al., 2009). Factors with eigenvalues of less than 1 were ignored. The order in which the factors were interpreted was determined by the magnitude of their eigenvalues. Under each factor, soil properties regarded as highly important were retained. These were defined as those that had a loading value within 10 % of the highest loading within an individual factor (Andrews et al., 2002). Soil properties that are widely acknowledged as good indicators of soil quality, but with factor loading scores ≤ 0.70, were also retained.

Soil physical, chemical, and biological properties that have been suggested as important soil quality indicators include soil organic carbon, available nutrients and particle size, BD, pH, soil aggregate stability, CEC, and available water content (Doran and Parkin, 1994; Larson and Pierce, 1994; Karlen et al., 1997; Zornoza et al., 2007, 2015; García-Ruiz et al., 2008; Qi et al., 2009; Marzaioli et al., 2010; Fernandes et al., 2011; Lima et al., 2013; Merrill et al., 2013; Rousseau et al., 2012, 2013; Singh et al., 2014). In cases in which more than one soil property was found to be of high importance under a single PC, Pearson's correlation coefficients were used to determine if any of these variables are redundant (Qi et al., 2009). When two highly important variables were found to be strongly correlated ($r^2 > \pm 0.70$; $p < 0.05$), the one with the highest factor loading (absolute value) was retained (Andrews and Carroll, 2001; Andrews et al., 2002, 2004; Montecchia et al., 2011).

In Step (3) of the CSDI development, the highly important soil properties under each factor were subjected to stepwise discriminant analysis (STEPDA) to select key soil properties (variables). In principle, stepwise discriminant analysis generates two or more linear combinations of the discriminating variables, often referred to as discriminant functions (Tesfahunegn et al., 2011). Conversely, the discriminant functions can be represented as

$$D_i = d_{i1}Z_1 + d_{i2}Z_2 + d_{iP}Z_P, \tag{1}$$

in which D_i is the score on discriminant function i, d's are weighting coefficients, and Z's are the standardized values of the p discriminating variables used in the analysis (Awiti et al., 2008). In this study, STEPDA was used to select variables with the highest power to discriminate between the treatments. The validity of the result was evaluated using the Wilks' lambda value. This value is an index of the discriminating power ranging between 0 and 1 (the lower the value, the higher the discriminating power). At each step of STEPDA, the variable that minimizes the overall Wilks' lambda was selected. One of the advantages of STEPDA is that the final model contains the variables that are considered useful. The result of this process was an MDS consisting of the most important variables for quantifying soil degradation in the selected plantations.

Step (4) involved the normalization of the MDS variables to numerical scores between 0 and 1 using a linear scoring function (Masto et al., 2008; Ngo-mbogba et al., 2015). The "more is better" scoring curve was used to determine the linear score of soil variables:

$$S_L = \left(\frac{X - l}{h - l}\right), \tag{2}$$

in which S_L is the linear score (between 0 and 1) of a soil variable, x is the soil variable value, l is the minimum value, and h is the maximum value of the soil variable.

During Step (5), the normalized MDS values were transformed into degradation scores (D) as described by Gómez et al. (2009) and obtained from

$$D = 1 - S_L, \tag{3}$$

in which D is the degradation score and S_L is the normalized MDS value. Here, a score of 1 signifies the highest possible soil degradation score and 0 represents complete absence of degradation for a particular soil property.

In Step (6) the degradation scores (D) were integrated into an index using the weighted additive method:

$$CSDI = \sum_{i=1}^{n} (W_i D_i), \tag{4}$$

in which CSDI represents the composite soil degradation index, W_i is the weight of variable i, D_i represents the degradation scores of the parameters in the MDS for each of the cocoa farms, and n is the number of indicators in the MDS. W_i in Eq. (4) was derived by the percentage of the total variance explained by the factor in which the soil property had the highest load divided by the total variance explained by all the factors with eigenvalues ≥ 1 (Masto et al., 2008; Armenise et al., 2013).

In Step (7) CSDI values were categorized into number of desired (3) classes of degradation using their z score value as obtained by

$$z = \frac{x - \mu}{\sigma}, \tag{5}$$

in which Z is the z score, x is the CSDI value of each plot, μ is the mean value, and σ is the standard deviation. In principle, z scores explain the standard deviations of input values from the mean (Hinton, 1999). For this purpose, Z values between -1 and 1 were regarded as having a moderate degradation status, while values of more than 1 were regarded as high and less than -1 as low (see the results section for further explanation on this categorization).

In Step (8) the CSDI classification was statistically validated using a CDA. CDA is a multivariate statistical technique whose objective is to discriminate among pre-specified groups of sampling entities. The technique involves deriving linear combinations of two or more discriminating variables (canonical variates) that will best discriminate among the a priori defined groups. In this study, we used the "leave-one-out" cross-validation procedure of CDA. Using this procedure, a given observation is deleted (excluded) and the remaining observations are used to compute a canonical discriminant function that is used to assign the observation into a degradation class with the highest probability. For instance, a sample with a probability of 0.003, 0.993, or 0.004 belonging to the low, moderate, or high degradation classes, respectively, was assigned to medium (see Table S2 for detail). This procedure is repeated for all observations and the result is a "hit ratio" or confusion matrix, which indicates the proportions of observations that are correctly classified. Additionally, CDA was used to confirm the significance of the explanatory variables that discriminate between the three soil degradation classes. In this study, the threshold (T) for the selection of variables correlating significantly with the canonical discriminant functions was taken as $T = 0.2/\sqrt{}$ (eigenvalue) as suggested by Hadgu et al. (2009). Scoring and indexing were performed using Microsoft Excel 2013. All statistical analyses were performed using XLSTAT version 2016 (Addinsoft New York, USA).

3   Results and discussion

3.1   Identification of soil degradation processes using factor analysis

Table 2 shows the results of the factor analysis and reveals that the first five PCs had eigenvalues > 1. Each PC explained 5 % or more of the variation in the data set. The first five PCs jointly accounted for more than 77 % of the total variance in the data set. In addition, they explained 68 % of the variance in available phosphorus, 84 % in SOM, 76 % in calcium, 65 % in pH, 87 % in clay content, 90 % in TN, 77 % in silt, 83 % in magnesium, 83 % in sand, and 58 % in BD. The high communalities among the soil properties suggest that variability in selected soil properties is well accounted for by the extracted factors (Tesfahunegn et al., 2011).

Extractable zinc, extractable manganese, and silt had high positive loadings on PC1 (0.875, 0.857, and 0.838, respectively). Because a significant correlation exists between extractable zinc and extractable manganese ($r = 0.834$, $p < 0.001$; Table 3), the latter variable was excluded. For ease of association, PC1 was labelled soil micronutrient degradation factor. PC2 was loaded highly by CEC (0.884) and exchangeable calcium (0.871), but given that the correlation analysis showed a strong relationship ($r = 0.870$, $p < 0.001$; Table 3) between CEC and exchangeable calcium, the latter was also excluded. SOM, with a relatively high factor loading (0.711), was retained owing to its relevance in monitoring soil quality degradation (Brejda et al., 2000; Sharma et al., 2009; Masto et al., 2008, 2009; Zornoza et al., 2015). Because the correlation coefficient between SOM and CEC was relatively low ($r = 0.578$; $p < 0.001$;

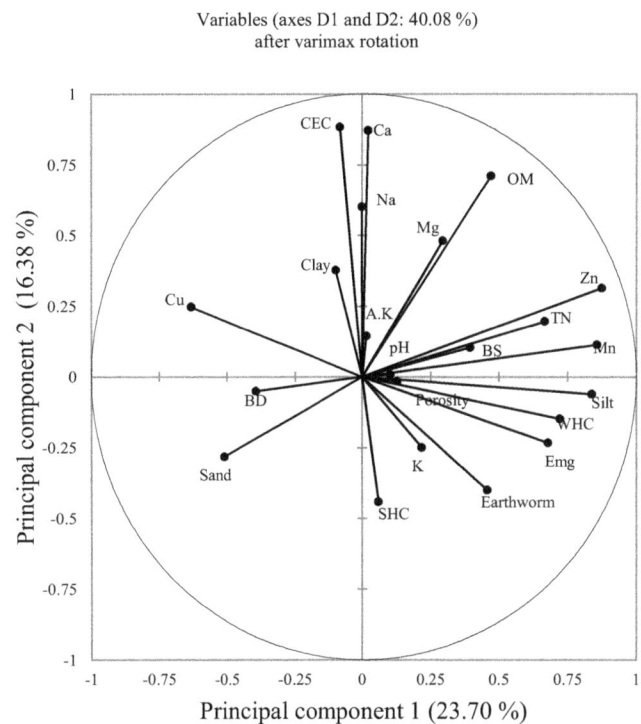

Figure 2. Principal component distribution of the investigated soil properties in age-sequenced peasant cocoa plantations. BD – bulk density; Clay – clay content; WHC – water-holding capacity; SHC – saturated hydraulic conductivity; OM – organic matter; AP – available phosphorus; TN – total nitrogen; Ca – exchangeable calcium; Mg – exchangeable magnesium; K – exchangeable potassium; Na – exchangeable sodium; CEC – cation exchange capacity; BS – base saturation; Cu – extractable copper; Zn – extractable zinc; Mn – extractable manganese; EMg – extractable magnesium; earthworm population.

Table 3), both were retained as highly important variables. Given that SOM was significantly correlated with several of the eliminated soil properties in the group, the second component factor was labelled the soil organic matter degradation factor.

The third component factor (PC3) was highly loaded on available phosphorus (0.810) and TP (0.801). Because the correlation coefficient between the two variables is relatively low ($r = 0.578$; $p < 0.001$; Table 3), both properties were retained. The group of variables associated with the third factor was termed the available phosphorus degradation factor. The fourth factor was labelled as the soil acidity degradation factor because it was highly loaded on pH (0.791) only. Similarly, the fifth factor was labelled as the soil textural degradation factor because it was dominated by clay content (0.812).

So far, the PCA result suggests that soil degradation in the study region is mainly linked to four degradation processes, namely (1) decline in soil nutrients, (2) loss of soil organic matter, (3) increase in soil acidity, and (4) the breakdown of soil textural characteristics arising from differences

Table 3. Correlation coefficient between highly weighted variables under PCs with high factor loading.

PC1 variables	Extractable zinc	Extractable manganese	Silt
Extractable zinc	1.000	0.834**	0.653*
Extractable manganese	0.834**	1.000	0.612*
Silt	0.653*	0.612*	1.000
PC2 variables	**Cation exchange capacity**	**Exchangeable calcium**	**Organic matter**
Cation exchange capacity	1.000	0.870**	0.523*
Exchangeable calcium	0.870**	1.000	0.619*
Organic matter	0.523*	0.619*	1.000
PC3 variables	**Available phosphorus**	**Total porosity**	
Available phosphorus	1.000	0.578*	
Total porosity	0.578*	1.000	
PC4 variable	**pH**		
pH	1.000		
PC5 variable	**Clay content**		
Clay content	1.000		

* Significant difference at $P = 0.05$. ** Significant difference at $P = 0.01$.

in eluviation of clay content. Figure 2 summarizes the results of the interrelationship among the 22 soil properties as a correlation circle. The figure shows that the first two PCA axes jointly accounted for 40.08 % of the total variance, with the first axis (eigenvalue = 8.545) representing mainly micronutrients with extractable manganese, zinc, silt, and TN in contrast to bulk density, copper, and sand. The second axis (eigenvalue = 3.96) is represented by CEC and exchangeable calcium as opposed to the pH content of the soils. Figure 3 represents the percentage contributions of the investigated soil properties in selected cocoa plantation chronosequence (CPC).

3.2 Selecting a MDS of soil degradation indicators

The PCA results presented thus far suggest that eight indicators (extractable zinc, silt, SOM, CEC, available phosphorus, TP, pH, and clay content) can be used to assess soil degradation in the study area. However, the collection and analysis of such a large number of indicators is not viable for monitoring programmes covering extensive areas and the identification of key soil degradation indicators will be very useful. The eight soil properties were consequently subjected to forward STEPDA to determine which of them are most important for soil degradation monitoring in the study area. Figure 4 and Table 4 show that STEPDA separated CPC into three groups (YCP, MCP, and SCP), based on the explanatory variables (eight soil parameters) included in the model. The first discriminant function separates the MCP from YCP and SCP, while the second discriminant function separates YCP from MCP and SCP. The overall Wilks' lambda test ($\lambda = 0.047$,

Table 4. Result of stepwise discriminant analysis (STEPDA) separating YCP, MCP, and SCP.

	Discriminant function	
	1	2
Significance	0.000	0.000
Eigenvalue	6.826	1.696
% of variance	80.101	19.899
Cumulative % variance	80.101	100.000
Canonical correlation coefficient	0.934	0.793

Variables	Canonical coefficient correlation	
Silt	0.353	−0.520
Clay content	0.373**	−0.139
Porosity	0.158	−0.309
pH	0.029	−0.211
Cation exchange capacity	0.611*	0.622
Available phosphorus	0.186	−0.035
Extractable zinc	0.806*	−0.527
Organic matter	0.952*	0.096

* Significant at $p < 0.05$. ** Significant at $p < 0.001$.

$p < 0.001$) confirms that the means of the CPC were significantly different for the two discriminant functions.

Table 4 shows that the first discriminant function, which accounts for more than 80 % of the variance in soil properties, is positively correlated with organic matter (0.952, $p < 0.001$), extractable zinc (0.806, $p < 0.001$), and CEC (0.611, $p < 0.001$); thus, it is labelled as the soil organic mat-

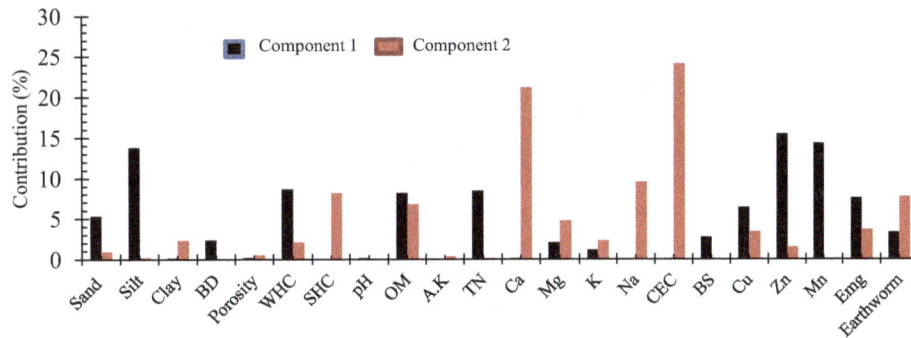

Figure 3. Percentage contributions of the investigated soil properties in age-sequenced peasant cocoa plantations. Clay – clay content; BD – bulk density; WHC – water-holding capacity; SHC – saturated hydraulic conductivity; OM – organic matter; AP – available phosphorus; TN – total nitrogen; Ca – exchangeable calcium; Mg – exchangeable magnesium; K – exchangeable potassium; Na – exchangeable sodium; CEC – cation exchange capacity; BS – base saturation; Cu – extractable copper; Zn – extractable zinc; Mn – extractable manganese; EMg – extractable magnesium; earthworm population.

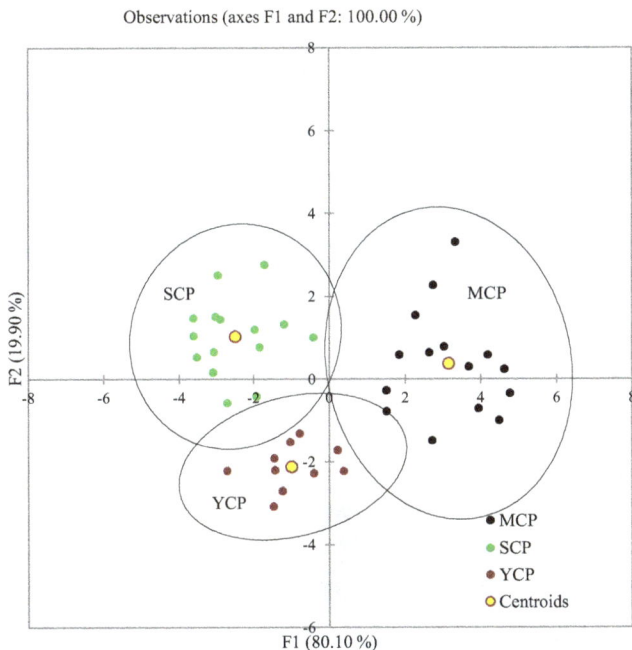

Figure 4. First and second discriminant function separating different cocoa plantations in southwestern Nigeria.

ter and macronutrients dimension. This result suggests that the plots in MCP have higher concentrations of soil nutrients than YCP and SCP. Similarly, the second discriminant function, which accounts for more than 19 % of the variance in soil properties, is positively correlated with CEC (0.622, $p < 0.001$) and SOM (0.096), but negatively correlated with silt (0.520), clay content (0.139), porosity (0.309), zinc (0.527), and available phosphorus (0.035). This suggests that the YCP cases have poor physical soil properties compared to MCP and SCP. This function is labelled as the soil physical and micronutrient dimension.

The result of STEPDA confirmed that only four soil properties are significant in discriminating between the CPC. These soil properties and their partial regression (R^2) are SOM ($R^2 = 0.797$, $p < 0.001$; Wilks' Lambda $= 0.203$), extractable zinc ($R^2 = 0.548$, $p < 0.001$; Wilks' Lambda $= 0.259$), CEC ($R^2 = 0.379$, $p < 0.001$; Wilks' Lambda $= 0.432$), and clay content ($R^2 = 0.169$, $p < 0.05$; Wilks' Lambda $= 0.866$). The relative importance of these variables, as indicated by the length of their eigenvectors, is (in decreasing order) SOM, extractable zinc, CEC, and clay content. Consequently, these four soil properties constitute a MDS of soil degradation indicators in our study area.

3.3 MDS normalization, transformation, and integration into CSDI

The four selected indicators of the MDS were normalized and transformed into degradation scores (D) as described in Sect. 2.4. Weights were assigned to each degradation score using the result of the factor analysis (Table 2). As an example, the procedure to calculate the weighting factor for extractable zinc was as follows: the individual percentage variance for PC1 (23.70) was divided by 77.15 %, the cumulative percentage of variation explained by all the retained PCs (Table 3), to yield the weight of 0.31. After assigning different weights to each parameter, they were integrated into a CSDI. This index is the sum of the normalized and weighted values of each parameter. CSDI was computed for each cocoa agroecosystems as

$$CSDI = 0.21(DSOM) + 0.31(DZn) + 0.21(DCEC) + 0.17(DClay). \tag{6}$$

Ordering the variables included in the equation as a function of the loading of the coefficient gave

$$CSDI = 0.31(DZn) + 0.21(DSOM) + 0.21(DCEC)$$

Table 5. Classification of soils into degradation levels and their interpretations modified after Gómez et al. (2009).

Range	Classes of degradation degradation	Interpretation
<0.195	Low	Farms with little or no form of degradation and their nutrient deficiencies can be restored with moderate effort
0.195–0.383	Moderate	Farms with moderate soil quality degradation, where some action should be taken to improve soil conditions
>0.383	High	Farms are currently degraded and their soil quality restoration will require sustained management efforts

Table 6. Standardized and unstandardized coefficient functions of canonical discriminant analysis.

	Constant	Zn	OM	CEC	Clay content
Function 1^{Ψ}	−11.863	0.599*	1.225*	0.226*	0.054^{ns}
Function 2^{Ψ}	−5.248	−0.326*	0.092^{ns}	0.214^{ns}	0.365*
Classes of degradation					
Low	−145.980	6.851	10.885	6.634	3.977
Moderate	−104.651	5.889	7.806	5.776	3.459
High	−74.970	3.359	3.489	5.202	3.564

OM: organic matter (%); CEC: cation exchange capacity ($cmol_c\ kg^{-1}$); Zn: extractable zinc ($mg\ kg^{-1}$); clay content (%). $^{\Psi}$ Wilks' lambda test of functions ($F_{observed} = 22.576$ and $F_{critical} = 2.499$) shows that the discriminant model was significant at probability $P = 0.000$ for the two functions, indicating that these functions contributed more to the model. $^{\Psi}$ Eigenvalue for F1 = 3.506 and F2 = 0.426; threshold for F1 is $0.2/\sqrt{3.506} = 0.106$; F2 is $0.2/\sqrt{0.426} = 0.30$. * Significant. ^{ns} Not significant.

$$+ 0.17(\text{DClay}), \qquad (7)$$

in which CSDI is the composite soil degradation index and DZn, DSOM, DCEC, and DClay are the degradation scores of extractable zinc, organic matter, CEC, and clay content, respectively.

One significant result from this study is that Zn was identified as the most important degradation indicator and it plays a key role in maintaining soil quality in the study area. Zn deficiency has been widely reported in agricultural soils in Africa (Vanlauwe et al., 2015), and cocoa is highly sensitive to Zn deficiency (Ogeh and Ipinmoroti, 2013; Van Vliet and Giller, 2017). Our results suggest that there is a Zn deficiency in the study area with a potential effect on the growth and yield of cocoa over time.

3.4 Classification into degradation classes

Table 5 shows the soil degradation classification of CSDI scores by solving Eq. (5). In our case, μ and σ were calculated as 0.289 and 0.094, respectively, resulting in CSDI values of 0.195 when $Z = -1$ and 0.383 when $Z = 1$. Consequently, the CSDI classes are low (<0.0195) and high (>0.383). CSDI values between 0.195 and 0.383 were regarded as moderate. The interpretations of these classes is shown in Table 5 (modified from Gómez et al., 2009). Most of the selected cocoa agroecosystems (65 %) are moderately degraded, while 18 % have a high degradation status.

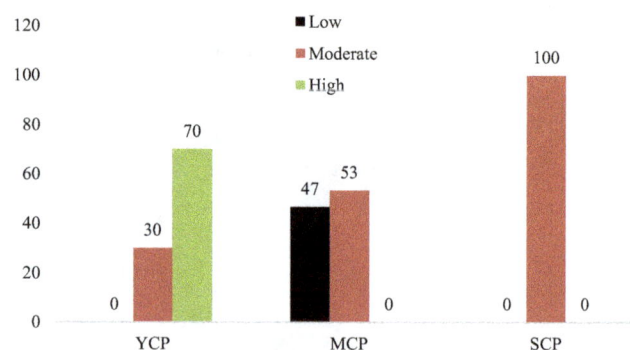

Figure 5. Percentages of degraded farms across cocoa chronosequence plantations (YCP, MCP, and SCP).

A significant difference was observed in the degradation status of YCP, MCP, and SCP (ANOVA test, $F_{2,39} = 57.59$; $P < 0.001$; Table not shown). Figure 5 shows that 30 % of YCP, 53.33 % of MCP, and 100 % of SCP are moderately degraded. However, 70 % of YCP is highly degraded and 47 % of MCP shows no sign of degradation. This implies that MCP plots are less degraded compared to YCP and SCP. This result is consistent with other studies in West Africa. For instance, Dawoe et al. (2014) reported that, in humid lowland Ghana, soil properties and quality parameters of a Ferric Lixisol improved under cocoa plantations that have been operating for 15–30 years and were better than that of a YCP with a

Table 7. Cross-validation results by canonical discriminant analysis.

Case	Actual group	Discriminant analysis of classification of predicted group membership				
Original group	From/to	Low	Moderate	High	Total	% correct
	Low	**6**	1	0	7	85.71 %
	Moderate	2	**23**	1	26	88.46 %
	High	0	0	7	7	100.00 %
	Total	8	24	8	40	90.00 %
Cross-validated	From/to	Low	Moderate	High	Total	% correct
	Low	**6**	1	0	7	85.71 %
	Moderate	2	**22**	2	26	84.62 %
	High	0	0	7	7	100.00 %
	Total	8	23	9	40	87.50 %

Percentage of grouped cases correctly classified is 87.50 %. Bold font in each group is the number of cases correctly classified by canonical discriminant analysis.

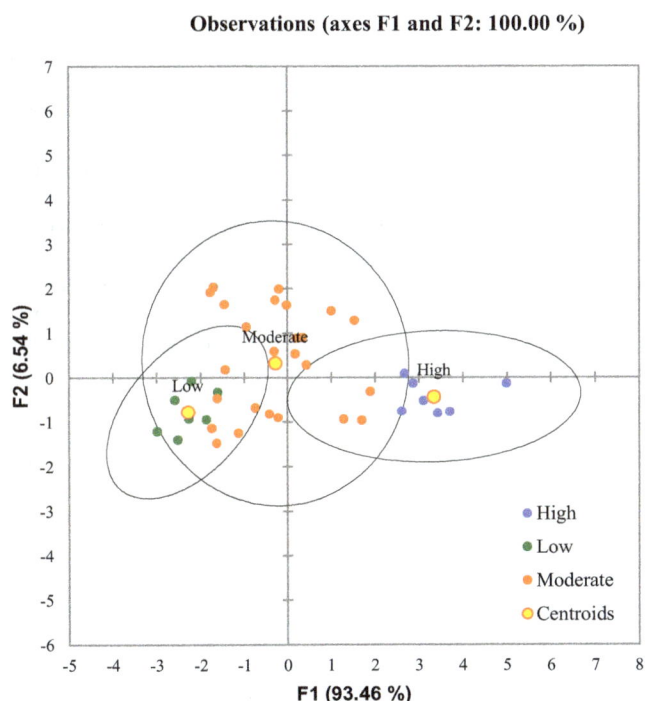

Figure 6. First and second canonical function of canonical discriminant analysis separating studied soils into three degradation classes (low, moderate, and high).

3-year production age. Similar results were obtained by Tondoh et al. (2015), who reported that, in Côte d'Ivoire, there was a steady degradation of soil quality over time in full-sun cocoa stands planted on Ferralsols for 10 years, but the degradation value was less pronounced in 20-year-old plantations. Comparing our results with those of Dawoe et al. (2014) and Tondoh et al. (2015) highlights the effects of poor and unsustainable land management practices on soil degradation

in peasant cocoa agroecosystems in West Africa. Traditionally, cocoa plots are cultivated with food crops in the first 3 to 5 years of development until the canopies have formed. Given that smallholder cacao farmers in the study area do not use chemical fertilizers to improve soil quality, degradation of the physical, chemical, and biological properties of cocoa soils are imminent during this phase of plantation establishment.

3.5 Statistical validation of CSDI

A CDA was used to validate the CSDI classification. The values of the four soil properties (organic matter, extractable zinc, CEC, and clay content) were used as data input. Figure 6 and Table 6 show that the three soil degradation classes (low, moderate, and high) were significantly separated on the first and second canonical functions (Wilks' lambda = 0.156, $F_{6,68} = 13.04$, $p < 0.0001$). Of the total variance, 93.46 % was accounted for by the first canonical function, which was significant at $p < 0.001$. The second canonical function accounted for 6.54 % of the total variance and was significant at $P < 0.005$. Extractable zinc, organic matter, and CEC significantly contributed to the distinction among soil degradation classes and were positively associated with the first canonical function (Table 6). Clay content also contributed significantly to the distinction among soil degradation classes, but was positively associated with the second canonical function (Table 6).

CDA classification results in Table 7 reveal that the CSDI model performs reasonable well, showing a low level of misclassification. The table shows that for the original grouped cases, the CDA correctly classified 6 of the 7 (85.7 %) low, 23 of 26 (88.4 %) moderate, and all of the high cases. The implication of the CDA accuracy assessment is that the proposed classes of soil degradation (low, moderate, and high) were significantly separated by the four canonical variables included in the model and that the model can consequently be used with a high degree of confidence. Results from this study indicate that the CSDI can effectively be used to monitor and evaluate the degree of soil (Alfisols) degradation under cocoa plantations in the study area (and similar environments). The results of this study also confirm that composite indicators, which are intended as tools for assessing the state and evolution of complex and multifaceted environmental phenomena (OECD, 2008), are generally easier to interpret than an array of individual indicators (Renzi et al., 2017). Therefore, the CSDI developed in this study represents a promising methodology for assessing soil degradation in cocoa agroecosystems. More work is needed to apply and evaluate the index on different soil types from different cocoa-producing regions and countries.

4 Conclusions

In this study, we developed a composite soil degradation index to cost-effectively assess the status of soil degradation under cocoa agroecosystems. Of the initial 22 soil properties evaluated, multivariate statistical analyses revealed that four soil properties (extractable zinc, SOM, CEC, and clay content) were the main indicators of soil degradation. This MDS of soil degradation indicators was used to produce a CSDI, which was classified into three classes of degradation. According to this classification, 65 % of the selected cocoa farms are moderately degraded, 17.5 % have a high degradation status, and 17.5 % show no sign of degradation. This classification corresponded well with a CDA classification performed on the same data set.

The findings suggest that the selection of a small set of relevant indicators will be more cost-efficient and less time consuming than using a large number of soil properties that may be irrelevant to the processes of degradation. They also suggest that soil degradation under cocoa agroecosystems (in this region at least) is mainly attributed to a decline in soil nutrients, loss of soil organic matter, increase in soil acidity, and the breakdown of soil textural characteristics over time. This study shows that both physical and chemical soil properties are degraded under long-term cocoa agroecosystems. The implications are serious for sustainability of cocoa agroecosystems on acidic Alfisols. While degradation of physical components of these soils poses serious risks to crop yields, degradation of chemical soil properties coupled with non-application of fertilizers will likely exacerbate soil degradation processes. To prevent smallholder cocoa production from becoming unsustainable in the long-term, it is critical to advise farmers of the need for the application of artificial (organic) fertilizers, particularly under YCP. Obviously, application of organic fertilizers will substantially improve the soil structure and nutrient conditions of cocoa soils (Van Vliet and Giller, 2017) but the poor transportation system in rural areas and prohibitive costs associated with artificial fertilizer application in cocoa groves remains a challenge to both farmers and governments. Therefore, alternative fertilizers in terms of organic residues, with the potential of increasing organic matter have been proposed in recent times (Van Vliet and Giller, 2017). Studies have reported that the addition of organic plant residues to crop soils helps to improve soil structure (Jordán et al., 2010). In addition, animal manure can be added to cocoa soils, but the potential effect on cocoa yield is yet to be reported in the literature. Although this study sets a basis for soil quality monitoring, more work is needed to improve our knowledge of changes in soil quality and health under cocoa agroecosystems of different ages. Hopefully this will lead to much-needed evidence-based recommendations for rehabilitation of degraded cocoa soils in West Africa.

Competing interests. The authors declare that they have no conflict of interest.

Acknowledgements. Financial support provided by the TETFund, administrated by the Osun State University Research Committee, is gratefully acknowledged. A special word of gratitude is owed to Kayode Are, soil physicist at the Institute of Agricultural Training, Obafemi Awolowo University, for his assistance during fieldwork. We are also grateful to www.linguafix.net for the language checking and editing services provided. The efforts of the technical and laboratory staff of Soil and Land Resource Management, Obafemi Awolowo University, Ile-Ife, Nigeria, are sincerely acknowledged. We are also grateful to the chiefs of the various villages for their support during the interviews and the 40 cocoa farmers for their permission to carry out this study on their farms.

Edited by: Antonio Jordán

References

Abiodun, J.: Service centres and consumer behaviour within the Nigerian Cocoa Area, Geografiska Annaler series B, Human Geography, 53, 78–93, 1971.

Adeoye, N. O. and Ayeni, B.: Assessment of deforestation, biodiversity loss and the associated factors: case study of Ijesa-Ekiti region of Southwestern Nigeria, GeoJournal, 76, 229–243, https://doi.org/10.1007/s10708-009-9336-z, 2011.

Adesodun, J. K., Davidson, D. A., and Mbagwu, J. S. C.: Soil quality assessment of an oil-contaminated tropical Alfisol amended with organic wastes using image analysis of pore space, Geoderma, 146, 166–74, https://doi.org/10.1016/j.geoderma.2008.05.013, 2008.

Akinyemi, F. O.: An assessment of landuse change in the cocoa belt of south-west Nigeria, Int. J. Remote Sens., 34, 2858–2875, 2013.

Al-Kaisi, M. M., Lal, R., Olson, K. R., and Lowery, B.: Fundamentals and functions of soil environment in Soil health and intensification of agroecosytems, edited by: Al-Kaisi, M. M, and Lowery, B., Academic press, 1–23, 2017.

Ameyan, O. and Ogidiolu, O.: Agricultural landuse and soil degradation in a part of Kwara State, Nigeria, Environmentalist, 9, 285–290, 1989.

Anderson, J. M. and Ingram, J. S. I. (Eds.): Tropical soil biology and fertility: a handbook of methods, CAB international, Wallingford, UK, 1993.

Andrews, S. S. and Carroll, C. R.: Designing a soil quality assessment tool for sustainable agro-ecosystem management, Ecol. Appl., 11, 1573–1585, 2001.

Andrews, S. S., Karlen, D. L., and Mitchell, J. P.: A comparison of soil quality indexing methods for vegetable production systems in Northern California, Agr. Ecosyst. Environ., 90, 25–45, https://doi.org/10.1016/S0167-8809(01)00174-8, 2002.

Andrews, S. S., Karlen, D. L., and Cambardella, C. A.: The soil management assessment framework: a quantitative soil quality evaluation method, Soil Sci. Soc. Am. J., 68, 1945–1962, https://doi.org/10.2136/sssaj2004.1945, 2004.

Armenise, E., Redmile-Gordon, M. A., Stellacci, A. M., Ciccarese, A., and Rubino, P.: Developing a soil quality index to compare soil fitness for agricultural use under different managements in the Mediterranean environment, Soil Till. Res., 130, 91–98, https://doi.org/10.1016/j.still.2013.02.013, 2013.

Asare, R.: Cocoa agroforests in West Africa: a look at activities on preferred trees in the farming systems. Forestry and Landscape Working Paper, Arboretum Working Paper, No. 6. Forest and Landscape Denmark, 2005.

Aweto, A. O.: Organic matter in fallow soil in a part of Nigeria and its effects on soil properties, J. Biogeogr., 8, 67–74, 1981.

Aweto, A. O. and Iyanda, A. O.: Effects of *Newbouldia Laevis* on soil subjected to shifting cultivation in the Ibadan Area, Southwestern Nigeria, Land Degrad. Dev., 56, 51–56, 2003.

Awiti, A. O., Walsh, M. G., Shepherd, K. D., and Kinyamario, J.: Soil condition classification using infrared spectroscopy: A proposition for assessment of soil condition along a tropical forest-cropland chronosequence, Geoderma, 143, 73–84, 2008.

Bastida, F, Luis M. J., and García, C.: Microbiological degradation index of soils in a semiarid climate, Soil Biol. Biochem., 38, 3463–3473, https://doi.org/10.1016/j.soilbio.2006.06.001, 2006.

Baumhardt, R. L., Stewart, B. A., and Sainju, U. M.: North American soil degradation: processes, practices, and mitigating strategies, Sustainability, 7, 2936–2960, 2015.

Beniston, J. W., Lal, R., and Mercer, K. L.: Assessing and managing soil quality for urban agriculture in a degraded vacant lot soil, Land Degrad. Dev., 27, 996–1006, https://doi.org/10.1002/ldr.2342, 2015.

Berry, S.: The concept of innovation and the history of cocoa farming in western Nigeria, J. Afr. Hist., 15, 83–95, 1974.

Bindraban, P. S., Velde, V. D. M., Ye, L., Berg, V. D. M., Materechera, S., Kiba, I. D., Tamene, L., Ragnarsdottir Vala Kristin Jongschaap, R., Hoogmoed, M., Hoogmed, W., Beek, C. V., and Lynden, G. V.: Assessing the impact of soil degradation on food production, Current Opinion in Environmental Sustainability, 4, 476–488, 2012.

Biswas, S., Hazra, G. C., Purakayastha, T. J., Saha, N., Mitran, T., Roy, S. S., Basak, N., and Mandal, B.: Establishment of critical limits of indicators and indices of soil quality in rice-rice cropping systems under different soil orders, Geoderma, 292, 34–48, 2017.

Brejda, J. J., Karlen, D. L., Smith, J. L., and Allan, D. L.: Identification of regional soil quality factors and indicators: II. Northern Mississippi Loess Hills and Palouse Prairie, Soil Sci. Soc. Am. J., 64, 2125–2135, 2000.

Bremner, J. M.: Total nitrogen, in: Methods of Soil Analysis: Chemical Methods, edited by: Sparks, D. L., Soil Science Society of America, Madison, WI, 1085–1086, 1996.

Brevik, E. C., Cerdà, A., Mataix-Solera, J., Pereg, L., Quinton, J. N., Six, J., and Van Oost, K.: The interdisciplinary nature of SOIL, SOIL, 1, 117–129, https://doi.org/10.5194/soil-1-117-2015, 2015.

Condom, N., Kuper, M., Marlet, S., Valles, V., and Kijne, J.: Salinization, alkalinization and sodification in punjab (pakistan): characterization of the geochemical and physical processes of degradation, Land Degrad. Dev., 10, 123–140, 1999.

Cerdà, A.: Aggregate stability against water forces under different climates on agriculture land and scrubland in southern Bolivia, Soil Till. Res., 57, 159–166, 2000.

Cerdà, A., Morera, A. G., and Bodi, M. B.: Soil and water losses from new citrus orchards growing on sloped soils in the western, Earth Surf. Processes, 34, 1822–1830, 2009.

Chen, J., Chen, J., Tan, M., and Gong, Z.: Soil degradation?: a global problem endangering sustainable development, J. Geogr. Sci., 12, 243–252, 2002.

Critchley, W. and Bruijnzeel, L. A.: Environmental impacts of converting moist tropical forest to agriculture and plantations, UNESCO International Hydrological Programme, available at: http://unesdoc.unesco.org/images/0010/001096/109608eo.pdf (last access: 22 November 2016), 1996.

Dawoe, E. K., Isaac, M. E., and Quashie-Sam, J.: Litterfall and litter nutrient dynamics under cocoa ecosystems in lowland humid Ghana, Plant Soil, 330, 55–64, 2010.

Dawoe, E. K., Quashie-Sam, J. S., and Oppong, S. K.: Effect of landuse conversion from forest to cocoa agroforest on soil characteristics and quality of a Ferric Lixisol in lowland humid Ghana, Agroforest. Syst., 88, 87–99, https://doi.org/10.1007/s10457-013-9658-1, 2014.

De la paix, M. J., Lanhai, L., Xi, C., Ahmed, S., and Varenyam, A.: Soil degradation and altered flood risk as a consequence of deforestation, Land Degrad. Dev., 24, 478–485, 2013.

De la Rosa, D.: Soil quality evaluation and monitoring based on land evaluation, Land Degrad. Dev., 16, 551–559, 2005.

de Lima, A. C. R., Hoogmoed W., and Brussaard, L.: Soil quality assessment in rice production systems: establishing a minimum data set, J. Environ. Qual., 37, 623–630, https://doi.org/10.2134/jeq2006.0280, 2008.

De Souza Braz, A. M., Fernandes, A. R., and Alleoni, L. R. F.: Soil attributes after the conversion from forest to pasture in Amazon, Land Degrad. Dev., 24, 33–38, 2013.

Doran, J. W.: Soil health and global sustainability?: translating science into practice, Agr. Ecosyst. Environ., 88, 119–127, 2002.

Doran, J. W. and Parkin, T. B.: Defining and Assessing Soil Quality, in: Defining soil quality for a sustainable environment, edited by: Doran, J. W., Coleman, D. F., Bezdicek, D. F., and Stewart, B. A., Soil Sci. Soc. Am., Special Publication 35, Madison, WI, 3–21, 1994.

Doran, J. W. and Zeiss, M. R.: Soil health and sustainability?: managing the biotic component of soil quality, Appl. Soil Ecol., 15, 3–11, 2000.

Fernandes, J. C., Gamero, C. A., Rodrigues, J. G. L., and Mirás-Avalos, J. M.: Determination of the quality index of a Paleudult under sunflower culture and different management systems, Soil Till. Res., 112, 167–174, 2011.

Ganjegunte, G. K., Sheng Z., and Clark, J. A.: Soil salinity and sodicity appraisal by electromagnetic induction in soils irrigated to grow cotton, Land Degrad. Dev., 25, 228–235, https://doi.org/10.1002/ldr.1162, 2014.

García-Ruiz, R., Ochoa, V., Hinojosa, M. B., and Carreira, J. A.: Suitability of enzyme activities for the monitoring of soil quality improvement in organic agricultural systems, Soil Biol. Biochem., 40, 2137–2145, 2008.

Gee, G. W. and Or, D.: Particle-size analysis, in: methods of soil analysis, Part 4. soil physical properties, agronomy monograph 5, edited by: Dane, J. H. and Topp, G. C., SSSA, Madison, WI, 225–275, 2002.

Gómez, J. A., Sonia, Á., and María-Auxiliadora, S.: Development of a soil degradation assessment tool for organic Olive groves in Southern Spain, Catena, 79, 9–17, 2009.

Grossman, R. B. and Reinsch, T. G.: Bulk density and linear extensibility: core method, in: Methods of soil analysis, Part 4, Physical methods, edited by: Dane, J. H. and Topp, G. C., Madison (WI), Soil Science Society of America, 208–228, 2002.

Hadgu, K. M., Rossing, W. A., Kooistra, L., and van Bruggen, A. H.: Spatial variation in biodiversity, soil degradation and productivity in agricultural landscapes in the highlands of Tigray, Northern Ethiopia, Food Security, 1, 83–97, https://doi.org/10.1007/s12571-008-0008-5, 2009.

Hartemink, A. E.: Nutrient stocks, nutrient cycling, and soil changes in cocoa ecosystems: A Review, Adv. Agron., 86, 227–253, 2005.

Hinton, P. R.: Statistics explained: A guide for social science students, NY: Routledge, 1999.

Isaac, M. E., Gordon, A. M., Thevathasan, N., Oppong, S. K., and Quashie-Sam, J.: Temporal changes in soil carbon and nitrogen in West African multistrata agroforestry systems: a chronosequence of pools and fluxes, Agroforest. Syst., 65, 23–31, 2005.

Isaac, M. E., Timmer, V. R., and Quashie-Sam, S. J.: Shade tree effects in an 8-year-old cocoa agroforestry system: Biomass and nutrient diagnosis of Theobroma cacao by vector analysis, Nutrient Cycling in Agro-ecosystems, 78, 155-165, 2007.

IUSS Working Group WRB: World Reference Base for Soil Resources 2014, update 2015, International soil classification system for naming soils and creating legends for soil maps, World Soil Resources Reports No. 106. FAO, Rome, 2015.

Jagoret, P., Michel-Dounias, I., and Malézieux, E.: Long-term dynamics of cocoa agroforests: a case study in central Cameroon, Agroforest. Syst., 81, 267–278, https://doi.org/10.1007/s10457-010-9368-x, 2011.

Jagoret, P., Michel-Dounias, I., Snoeck, D., Ngnogué, H. T., and Malézieux, E.: Afforestation of savannah with cocoa agroforestry systems: a small-farmer innovation in central Cameroon, Agroforest. Syst., 86, 493–504, https://doi.org/10.1007/s10457-012-9513-9, 2012.

Jagoret, P., Snoeck, D., Bouambi, E., Ngnogue, T. H., Nyasse, S., and Saj, S.: Rehabilitation practices that shape cocoa agroforestry systems in Central Cameroon?: key management strategies for long-term exploitation, Agroforest. Syst., 1–15, https://doi.org/10.1007/s10457-016-0055-4, 2017.

Jordán, A., Zavala, L. M., and Gil, J.: Effects of mulching on soil physical properties and runoff under semi-

arid conditions in southern Spain, Catena, 81, 77–85, https://doi.org/10.1016/j.catena.2010.01.007, 2010.

Juo, A. S. R., Ayanlaja, S. A., and Ogunwale, J. A.: An evaluation of cation exchange capacity measurements for soils in the tropics, Commun. Soil Sci. Plan., 7, 751–761, 1976.

Karlen, D. L. and Rice, C. W.: Soil degradation: Will humankind ever learn?, Sustainability, 7, 12490–12501, 2015.

Karlen, D. L., Mausbach, M. J., Doran, J. W., Cline, R. G., Harris, R. F., and Schuman, G. E.: Soil quality: a concept, definition, and framework for evaluation, Soil Sci. Soc. Am. J., 61, 4–10, 1997.

Keesstra, S. D., Geissen, V., Mosse, K., Piiranen, S., Scudiero, E., Leistra, M., and van Schaik, L.: Soil as a filter for groundwater quality, Current Opinions in Environmental Sustainability, 4, 507–516, 2012.

Khaledian, Y., Kiani, F., Ebrahimi, S., Brevik, E. C., and Aitkenhead-Peterson, J.: Assessment and monitoring of soil degradation during land use change using multivariate analysis, Land Degrad. Dev., 28, 128–141, https://doi.org/10.1002/ldr.2541, 2017.

Krasilnikov, P., Makarov, O., Alyabina, I., and Nachtergaele, F.: Assessing soil degradation in northern Eurasia, Geoderma Regional, 7, 1–10, 2016.

Lal, R.: Soil degradation by erosion, Land Degrad. Dev., 12, 519–39, 2001.

Lal, R.: Soil degradation as a reason for inadequate human nutrition, Food Security, 1, 45–57, 2009.

Lal, R.: Restoring soil quality to mitigate soil degradation, Sustainability, 7, 5875–5895, 2015.

Lal, R. and Cummings, D. J.: Clearing a tropical forest I. Effects on soil and micro-climate, Field Crop. Res., 2, 91–107, 1979.

Larson, W. E. and Pierce, F. J.: The dynamics of soil quality as a measure of sustainable mangement, in: Defining soil quality for a sustainable environment, edited by: Doran, J. W., Coleman, D. C., Bezdicek, D. F., and Stewart, B. A., SSSA-Special Publication 35, Soil Science Society of America, Madison, WI, 37–51, 1994.

Leirós, M. C., Trasar-Cepeda, C., García-Fernández, F., and Gil-Sotres, F.: Defining the validity of a biochemical index of soil quality, Biol. Fert. Soils, 30, 140–146, 1999.

Lemenih, M., Karltun, E., and Olsson, M.: Soil organic matter dynamics after deforestation along a farm field chronosequence in southern highlands of Ethiopia, Agr. Ecosyst. Environ., 109, 9–19, https://doi.org/10.1016/j.agee.2005.02.015, 2005.

Lima, A. C. R., Brussaard, L., Totola, M. R., Hoogmoed, W. B., and de Goede, R. G. M.: A functional evaluation of three indicator sets for assessing soil quality, Appl. Soil Ecol., 64, 194–200, 2013.

Marzaioli, R., D'Ascoli, R., De Pascale, R. A., and Rutigliano, F. A.: Soil quality in a Mediterranean area of Southern Italy as related to different land use types, Appl. Soil Ecol., 44, 205–212, 2010.

Masto, R. E., Chhonkar, P. K., Singh, D., and Patra, A. K.: Alternative soil quality indices for evaluating the effect of intensive cropping, fertilisation and manuring for 31 years in the semi-arid soils of India, Environ. Monit. Assess., 136, 419–435, 2008.

Masto, R. E., Chhonkar P. K., Singh, D., and Patra, A. K.: Changes in soil quality indicators under long-term sewage irrigation in a sub-tropical environment, Environ. Geol., 56, 1237–1243, https://doi.org/10.1007/s00254-008-1223-2, 2009.

Mbile, P., Ngaunkam, P., Besingi, M., Nfoumou, C., Degrande, A., Tsobeng, A., Sado, T., and Menimo, T.: Farmer management of cocoa agroforests in Cameroon: Impacts of decision scenarios on structure and biodiversity of indigenous tree species, Biodiversity, 10, 12–19, https://doi.org/10.1080/14888386.2009.9712857, 2009.

Merrill, S. D., Liebig, M. A., Tanaka, D. L., Krupinsky, J. M., and Hanson, J. D.: Comparison of soil quality and productivity at two sites differing in profile structure and topsoil properties, Agr. Ecosyst. Environ., 179, 53–61, 2013.

Momtaz, H. R., Jafarzadeh, A. A., Torabi, H., Oustan, S., Samadi, A., Davatgar, N., and Gilkes R. J.: An assessment of the variation in soil properties within and between landform in the Amol region, Iran, Geoderma, 149, 10–18, 2009.

Montecchia, M. S., Correa, O. S., Soria, M. A., Frey, S. D., García, A. F., and Garland, J. L.: Multivariate approach to characterizing soil microbial communities in pristine and agricultural sites in Northwest Argentina, Appl. Soil Ecol., 47, 176–183, https://doi.org/10.1016/j.apsoil.2010.12.008, 2011.

Muñoz-rojas, M., Abd-elmabod, S. K., Zavala, L. M., De la Rosa, D., and Jordán, A: Climate change impacts on soil organic carbon stocks of Mediterranean agricultural areas?: A case study in Northern Egypt, Agr. Ecosyst. Environ., 238, 142–152, 2017.

Ngo-mbogba, M., Yemefack, M., and Nyeck, B.: Assessing soil quality under different land cover types within shifting agriculture in South Cameroon, Soil Till. Res., 150, 124–131, 2015.

Nezomba, H., Mtambanengwe, F., Tittonell, P., and Mapfumo, P.: Practical assessment of soil degradation on smallholder farmers' fields in Zimbabwe: Integrating local knowledge and scientific diagnostic indicators, Catena, 156, 216–227, 2017.

Novara, A., Gristina, L., Bodì, M. B., and Cerdà, A. The impact of fire on redistribution of soil organic matter on a Mediterranean hillslope under maquia vegetation type, Land Degrad. Dev., 22, 530–536, https://doi.org/10.1002/ldr.1027, 2011.

Ntiamoah, A. and Afrane, G.: Environmental impacts of cocoa production and processing in Ghana: life cycle assessment approach, J. Clean. Prod., 16, 1735–1740, 2008.

OECD: Handbook on constructing composite indicators, ISPRA, Paris, 2008.

Oke, O. C. and Chokor, J. U.: Land snail populations in shade and full-sun cocoa plantations in South Western Nigeria, West Africa, African Scientist, 10, 19–29, 2009.

Ogeh, J. S. and Ipinmoroti, R. R.: Micronutrient assessment of cocoa, kola, cashew and coffee plantations for sustainable production at Uhonmora, Edo State, Nigeria, Journal of Tropical Soils, 18, 1–5, 2013.

Olsen, S. R. and Sommers, L. E.: Phosphorus, in: method of soil analysis: chemical and microbiological properties, edited by: Sparks, D. L., Page, A. L., Helmke, P. A., and Loeppert, R. H., Part 2, agronomy monograph 9, 403–430, Soil Science Society of America, Wisconsin, WI, 1982.

Parras-Alcántara, L. and Lozano-García, B.: Conventional tillage versus organic farming in relation to soil organic carbon stock in olive groves in Mediterranean rangelands (southern Spain), Solid Earth, 5, 299–311, https://doi.org/10.5194/se-5-299-2014, 2014.

Parras-Alcántara, L., Díaz-Jaimes, L., and Lozano-García, B.: Management effects on soil organic carbon stock in Mediterranean open rangelands – treeless grasslands, Land Degrad. Dev., 26, 22–34, 2015.

Peech, M.: Hydrogen-ion activity, in: methods of soil analysis, edited by: Black, C. A., American Society of Agronomy, Madison, 2, 914–926, 1965.

Prager, K., Schuler, J., Helming, K., Zander, P., Ratinger, T., and Hagedorn, K.: Soil degradation, farming practices, institutions and policy responses: an analytical framework, Land Degrad. Dev., 22, 32–46, 2011.

Puglisi, E., Nicelli, M., Capri, E., Trevisan, M., and Del Re, A. A. M.: A soil alteration index based on phospholipid fatty acids, Chemosphere, 61, 1548–1557, 2005.

Puglisi, E., Del Re, A. A. M., Rao, M. A., and Gianfreda, L.: Development and validation of numerical indexes integrating enzyme activities of soils, Soil Biol. Biochem., 38, 1673–1681, https://doi.org/10.1016/j.soilbio.2005.11.021, 2006.

Pulido, M., Schnabel, S., Contador, J. F. L., Lozano-Parra, J., and Gómez-Gutiérrez, Á.: Selecting indicators for assessing soil quality and degradation in rangelands of Extremadura (SW Spain), Ecol. Indic., 74, 49–61, 2017.

Qi, Y., Darilek, J. L., Huang, B., Zhao, Y., Sun, W., and Gu, Z.: Evaluating soil quality indices in an agricultural region of Jiangsu Province, China, Geoderma, 149, 325–334, https://doi.org/10.1016/j.geoderma.2008.12.015, 2009.

Rahmanipour, F., Marzaioli, R., Bahrami, H. A., Fereidouni, Z., and Bandarabadi, S. R.: Assessment of soil quality indices in agricultural lands of Qazvin Province, Iran, Ecol. Indic., 40, 19–26, 2014.

Renzi, G., Canfora, L., Salvati, L., and Benedetti, A.: Validation of the soil Biological Fertility Index (BFI) using a multidimensional statistical approach: A country-scale exercise, Catena, 149, 294–299, 2017.

Reynolds, W. D. and Elrick, D.: Constant head soil core (tank) method, in: Methods of soil analysis, edited by: Dane, J. H. and Topp, G. C., Part 4, Physical methods, Madison (WI): Soil Science Society of America, 804–808, 2002.

Rice, R. A. and Greenberg, R.: Cacao cultivation and the conservation of biological diversity, Ambio: A Journal of the Human Environment, 29, 20–25, 2000.

Rodrigo Comino, J., Quiquerez, A., Follain, S., Raclot, D., Le Bissonnais, Y., Casalí, J., Giménez, R., Cerdà, A., Keesstra, S. D., Brevik, E. C., Pereira, P., Senciales, J. M., Seeger, M., Ruiz Sinoga, J. D., and Ries, J. B.: Soil erosion in sloping vineyards assessed by using botanical indicators and sediment collectors in the Ruwer-Mosel valley, Agr. Ecosyst. Environ., 233, 158–170, https://doi.org/10.1016/j.agee.2016.09.009, 2016a.

Rodrigo Comino, J., Ruiz Sinoga, J. D., Senciales González, J. M., Guerra-Merchán, A., Seeger, M., and Ries, J. B.: High variability of soil erosion and hydrological processes in Mediterranean hillslope vineyards (Montes de Málaga, Spain), Catena, 145, 274–284, https://doi.org/10.1016/j.catena.2016.06.012, 2016b.

Rousseau, G. X., Deheuvels, O., Rodriguez, Arias I., and Somarriba, E.: Indicating soil quality in cacao-based agroforestry systems and old-growth forests: The potential of soil macrofaunal assemblage, Ecol. Indic., 23, 535–543, 2012.

Rousseau, L., Fonte, S. J., Téllez, O., van der Hoek, R., and Lavelle, P.: Soil macrofauna as indicators of soil quality and land use impacts in smallholder agro-ecosystems of western Nicaragua, Ecol. Indic., 27, 71–82, 2013.

Saj, S., Jagoret, P., and Ngogue, H. T.: Carbon storage and density dynamics of associated trees in three contrasting Theobroma cacao agroforests of Central Cameroon, Agroforest. Syst., 87, 1309–1320, https://doi.org/10.1007/s10457-013-9639-4, 2013.

Salami, A. T.: Vegetation modification and man-induced environmental change in rural southwestern Nigeria, Agric. Ecosyst. Environ., 70, 159–167, 1998.

Salami, A. T.: Agricultural colonisation and floristic degradation in Nigeria's rainforest ecosystem, Environmentalist, 21, 221–229, 2001.

Sánchez-Navarro, A., Gil-Vázquez, J. M., Delgado-Iniesta, M. J., Marín-Sanleandro, P., Blanco-Bernardeau, A., and Ortiz-Silla, R.: Establishing an index and identification of limiting parameters for characterizing soil quality in Mediterranean ecosystems, Catena, 131, 35–45, 2015.

Scherr, S. J.: Soil degradation: a threat to developing country food security by 2020? vision 2020: food, agriculture, and the environment discussion paper, International Food Policy Research Institute, 27, 14–25, 1999.

Schoneveld, G. C.: The politics of the forest frontier: Negotiating between conservation, development, and indigenous rights in Cross River State, Nigeria, Land Use Policy, 38, 147–162, https://doi.org/10.1016/j.landusepol.2013.11.003, 2014.

Schulte, R. P. O., Bampa, F., Bardy, M., Coyle, C., Creamer, R. E., Fealy, R., Gardi, C., Ghaley, B.B., Jordan, P., Laudon, H., O'Donoghue, C., Ó'hUallacháin, D., and O'Sullivan, L., Rutgers, M., Six, J., Toth, G. L., and Vrebos, D.: Making the most of our land: managing soil functions from local to continental scale, Front. Environ. Sci., 3, 1–14, 2015.

Sharma, K. L., Mandal, U. K., Srinivas, K., Vittal, K. P., Mandal, B., Grace, J. K., and Ramesh, V.: Long-term soil management effects on crop yields and soil quality in a dryland Alfisol, Soil Till. Res., 83, 246–259, 2005.

Sharma, K. L., Grace, J. K., Mandal, U. K., Gajbhiye, P. N., Srinivas, K., Korwar, G. R., Hima Bindu, V., Ramesh, V., Ramachandran, K., and Yadav, S. K.: Evaluation of long-term soil management practices using key indicators and soil quality indices in a semi-arid tropical Alfisol, Soil Res., 46, 368–37, 2008.

Sharma, K. L., Raju, K. R., Das, S. K., Rao, B. P., Kulkami, B. S., Srinivas, K., Grace, J. K., Madhavi, M., and Gajbhiye, P. N.: Soil fertility and quality assessment under tree-, crop-, and pasture-based landuse systems in a rainfed environment, Commun. Soil Sci. Plan., 40, 1436–1461, 2009.

Singh, A. K., Bordoloi, L. J., Kumar, M., Hazarika, S., and Parmar, B.: Land use impact on soil quality in eastern Himalayan region of India, Environ. Monit. Assess., 186, 2013–2024, 2014.

Sione, S. M. J., Wilson, M. G., Lado, M., and Gonzalez, A. P.: Evaluation of soil degradation produced by rice crop systems in a Vertisol, using a soil quality index, Catena, 150, 79–86, 2017.

Smyth, A. J. and Montgomery, R. F.: Soils and landuse in central western Nigeria, Government Printer, Ibadan, Nigeria, 1962.

Snoeck, D., Afrifa, A., Ofori-Frimpong, A. K., Boateng, E., and Abekoe, M. K.: Mapping Fertilizer Recommendations for Cocoa Production in Ghana Using Soil Diagnostic and GIS Tools West African, J. Appl. Ecol., 17, 97–107, 2010.

Soil Survey Staff.: Keys to soil taxonomy, 12th Edn., USDA-natural resources conservation service, Washington, DC, 2014.

Sonwa D. J., Weise, S. F., Schroth, G., Janssens, M. J. J., and Shapiro, H.: Plant diversity management in cocoa agro-

forestry systems in West and Central Africa—effects of markets and household needs, Agroforest. Syst., 88, 1021–1034, https://doi.org/10.1007/s10457-014-9714-5, 2014.

Tesfahunegn, G. B.: Soil quality assessment strategies for evaluating soil degradation in northern Ethiopia, Appl. Environ. Soil Sci., 2014, 1–14, 2014.

Tesfahunegn, G. B.: Soil quality indicators response to land use and soil management systems in northern Ethiopia's Catchment, Land Degrad. Dev., 27, 438–448, 2016.

Tesfahunegn, G. B., Tamene, L., and Vlek, P. L. G.: Evaluation of soil quality identified by local farmers in Mai-Negus catchment, northern Ethiopia, Geoderma, 163, 209–218, 2011.

Thomaz, E. L. and Luiz, J. C.: Soil loss, soil degradation and rehabilitation in a degraded land area in Guarapuava (BRAZIL), Land Degrad. Dev., 23, 72–81, 2012.

Tondoh, J. E., Kouamé, F. N., Guéi, A. M., Sey, B., Koné, A. W., and Gnessougou, N.: Ecological changes induced by full-sun cocoa farming in Côte d'Ivoire, Glob. Ecol. Conserv., 3, 575–595, https://doi.org/10.1016/j.gecco.2015.02.007, 2015.

Trabaquini, K., Formaggio, R. A., and Galvão, L. S.: Changes in physical properties of soils with land use time in the Brazilian savanna environment, Land Degrad. Dev., 26, 397–408, 2015.

Virto, I., Imaz, M., Fernández-Ugalde, O., Gartzia-Bengoetxea, N., Enrique, A., and Bescansa, P.: Soil degradation and soil quality in western Europe: Current situation and future perspectives, Sustainability, 7, 1, 313–365, 2014.

Van Vliet, J. A. and Giller, K, E.: Mineral nutrition of cocoa: A review, Adv. Agron., 141, 185–270, 2017.

Vanlauwe, B., Descheemaeker, K., Giller, K. E., Huising, J., Merckx, R., Nziguheba, G., Wendt, J., and Zingore, S.: Integrated soil fertility management in sub-Saharan Africa: unravelling local adaptation, SOIL, 1, 491–508, https://doi.org/10.5194/soil-1-491-2015, 2015.

Vocomil, J. A.: Porosity. In methods of soil analysis part 1, edited by: Black, C. A., American Society of Agronomy, Madison WI, 299–314, 1965.

Walkley, A. and Black I. A.: An examination of the Degtjareff method for determining soil organic matter and a proposed modification of the chromic acid titration method, Soil Sci., 37, 29–38, 1934.

Xu, M., Li, Q., and Wilson, G.: Degradation of soil physicochemical quality by ephemeral gully erosion on sloping cropland of the hilly Loess Plateau, China, Soil Till. Res., 155, 9–18, 2016.

Zhao, Q., Shiliang, L., Li, D., Shikui, D., and Wang, C.: Soil degradation associated with water-level fluctuations in the Manwan Reservoir, Lancang River Basin, Catena, 113, 226–235, 2014.

Zornoza, R., Mataix-Solera, J., Guerrero, C., Arcenegui, V., García-Orenes, F., Mataix-Beneyto, J., and Morugán, A.: Evaluation of soil quality using multiple lineal regression based on physical, chemical and biochemical properties, Sci. Total Environ., 378, 233–237, 2007.

Zornoza, R., Mataix-Solera, J., Guerrero, C., Arcenegui, V., Mataix-Beneyto, J., and Gómez, I.: Validating the effectiveness and sensitivity of two soil quality indices based on natural forest soils under Mediterranean conditions, Soil Biol. Biochem., 40, 2079–2087, 2008.

Permissions

List of Contributors

Stefano Tavani and Amerigo Corradetti
DiSTAR, Università degli Studi di Napoli "Federico II", Largo S. Marcellino 10, 80138 Naples, Italy

Pablo Granado, Pau Arbués and J. Anton Muñoz
Institut de Recerca Geomodels, Universitat de Barcelona, Martí i Franquès s/n, 08028 Barcelona, Spain
Departament de Dinàmica de la Terra i de l'Oceà, Universitat de Barcelona, Martí i Franquès s/n, 08028 Barcelona, Spain

Paul W. J. Glover
School of Earth and Environment, University of Leeds, Leeds, UK

Cheng Zeng
State Key Laboratory of Environmental Geochemistry, Institute of Geochemistry, Chinese Academy of Sciences, 99 Lincheng West Road, Guiyang 550081, Guizhou Province, PR China
School of Geographyical and Environmental Sciences, Guizhou Normal University, Guiyang 550001, China
Puding Karst Ecosystem Observation and Research Station, Chinese Academy of Sciences, Puding 562100, Guizhou Province, PR China

Shijie Wang, Xiaoyong Bai, Yichao Tian and Luhua Wu
State Key Laboratory of Environmental Geochemistry, Institute of Geochemistry, Chinese Academy of Sciences, 99 Lincheng West Road, Guiyang 550081, Guizhou Province, PR China
Puding Karst Ecosystem Observation and Research Station, Chinese Academy of Sciences, Puding 562100, Guizhou Province, PR China

Yangbing Li
School of Geographyical and Environmental Sciences, Guizhou Normal University, Guiyang 550001, China

Yue Li
Key Laboratory of State Forestry Administration on Soil and Water Conservation, Beijing Forestry University, Beijing 100083, China

Guangjie Luo
Puding Karst Ecosystem Observation and Research Station, Chinese Academy of Sciences, Puding 562100, Guizhou Province, PR China

Institute of Agricultural Ecology and Rural Development, Guizhou Normal College, Guiyang 550018, China

Alejandra Quintanilla-Terminel, Mark E. Zimmerman and David L. Kohlstedt
Department of Earth Sciences, University of Minnesota, Minneapolis, MN 55455, USA

Brian Evans
Earth, Atmospheric and Planetary Sciences, Massachusetts Institute of Technology, Cambridge, MA 02139, USA

Michael Rubey, Simon E. Williams and R. Dietmar Müller
Earthbyte Group, School of Geosciences, the University of Sydney, Sydney, Australia

Sascha Brune
Helmholtz Centre Potsdam, GFZ German Research Centre for Geosciences, Potsdam, Germany
Institute of Earth and Environmental Science, University of Potsdam, Potsdam, Germany

Christian Heine
Specialist Geosciences, Shell Projects and Technology, Rijswijk, the Netherlands

D. Rhodri Davies
Research School of Earth Sciences, Australian National University, Canberra, Australia

Yusong Deng, Chongfa Cai, Shuwen Ding, Jiazhou Chen and Tianwei Wang
Key Laboratory of Arable Land Conservation (Middle and Lower Reaches of Yangtze River) of the Ministry of Agriculture, College of Resources and Environment, Huazhong Agricultural University, Wuhan, 430070, People's Republic of China

Dong Xia
College of Hydraulic and Environmental Engineering, China Three Gorges University, Yichang 443002, People's Republic of China

Raphael Schneeberger, Daniel Egli, Alfons Berger and Marco Herwegh
Institute of Geological Sciences, University of Bern, Baltzerstrasse 1 + 3, 3012 Bern, Switzerland

Miguel de La Varga, Florian Wellmann
Graduate School AICES, RWTH Aachen University, Schinkelstrasse 2, 52062 Aachen, Germany

Florian Kober
Nagra, Hardstrasse 73, 5430 Wettingen, Switzerland

Elizaveta Kovaleva
Department of Geology, University of the Free State, Bloemfontein, 9300, 205 Nelson Mandela Drive, Free State, South Africa
Department of Lithospheric Research, Faculty of Earth Sciences, Geography and Astronomy, University of Vienna, Althanstrasse 14, Vienna, 1090, Austria

Håkon O. Austrheim
Section of Physics of Geological processes, Department of Geoscience, University of Oslo, Oslo, 0316, Norway

Urs S. Klötzli
Department of Lithospheric Research, Faculty of Earth Sciences, Geography and Astronomy, University of Vienna, Althanstrasse 14, Vienna, 1090, Austria

Thomas Chauve, Maurine Montagnat, Cedric Lachaud and David Georges
Université Grenoble Alpes, CNRS, IRD, G-INP, IGE, 38041 Grenoble, France

Pierre Vacher
Laboratoire SYMME, Université de Savoie Mont Blanc, BP 80439, 74944 Annecy le Vieux CEDEX, France

Dácil Unzué-Belmonte, Patrick Meire and Eric Struyf
EcosystemManagement Research Group, Department of Biology, University of Antwerp, Universiteitsplein 1C, 2610 Wilrijk, Belgium

Yolanda Ameijeiras-Mariño and Sophie Opfergelt
Earth and Life Institute, Environmental Sciences, Université catholique de Louvain, Croix du Sud 2 bte L7.05.10, 1348 Louvain-la-Neuve, Belgium

Jean-Thomas Cornelis
Department Biosystem Engineering (BIOSE), Gembloux Agro-Bio Tech (GxABT), University of Liège (ULg), Avenue Maréchal Juin, 27, 5030 Gembloux, Belgium

Lúcia Barão
ICAAM, Instituto de Ciências Agrárias e Ambientais Mediterrânicas, University of Évora, Apartado 94, 7002-554 Évora, Portugal

Jean Minella
Universidade Federal de Santa Maria (UFSM), Department of Soil Science, 1000 Avenue Roraima, Camobi, CEP 97105-900 Santa Maria, RS, Brazil

Seann J. McKibbin
School of Environmental and Life Sciences, University of Newcastle, University Drive, Callaghan, 2308, Australia
Research School of Earth Sciences, Australian National University, Bldg. 61, Mills Road, Canberra, 0200, Australia

Bill Landenberger
School of Environmental and Life Sciences, University of Newcastle, University Drive, Callaghan, 2308, Australia

C. Mark Fanning
Research School of Earth Sciences, Australian National University, Bldg. 61, Mills Road, Canberra, 0200, Australia

Andreas K. Kronenberg, Hasnor F. B. Hasnan and Caleb W. Holyoke III
Center for Tectonophysics, Department of Geology and Geophysics, MS 3115, Texas A&M University, College Station, TX 77843-3115, USA

Richard D. Law
Department of Geosciences, MC 0420, Derring Hall RM 4044, Virginia Polytechnic Institute and State University, Blacksburg, VA 24061, USA

Zhenxian Liu
Geophysical Laboratory, Carnegie Institution of Washington, 5251 Broad Branch Rd., NW Washington, D.C. 20015, USA

Jay B. Thomas
Department of Earth Sciences, 204 Heroy Geology Laboratory, Syracuse University, Syracuse, NY 13244, USA

Sunday Adenrele Adeniyi
Department of Geography and Environmental Studies, Stellenbosch University, Private Bag XI, Matieland 7602, Stellenbosch, South Africa
Department of Geography, Osun State University, P.M.B 4494, Osogbo, Nigeria

Willem Petrus de Clercq
Department of Soil Science, Stellenbosch University, Private Bag XI, Matieland 7602, Stellenbosch, South Africa

Adriaan van Niekerk
School of Plant Biology, University of Western Australia, Crawley WA 6009, Australia

Index

www.ingramcontent.com/pod-product-compliance
Lightning Source LLC
Chambersburg PA
CBHW080659200326
41458CB00013B/4911